Stefan Spange, Michael Mehring (Eds.)
Twin Polymerization

Also of Interest

Handbook of Colloid and Interface Science.
Set of Volumes 1–4
Tadros, 2017
ISBN 978-3-11-054050-5

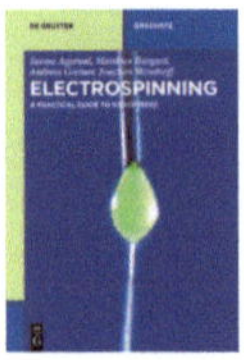

Electrospinning.
A Practical Guide to Nanofibers
Agarwal, Burgard, Greiner, Wendorff, 2016
ISBN 978-3-11-033180-6, e-ISBN 978-3-11-033351-0

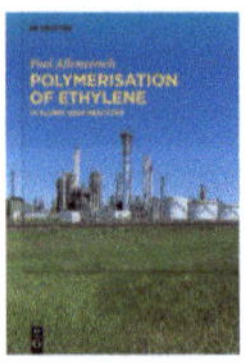

Polymerisation of Ethylene.
In Slurry Loop Reactors
Allemeersch, 2015
ISBN 978-3-11-029214-5, e-ISBN 978-3-11-029219-0

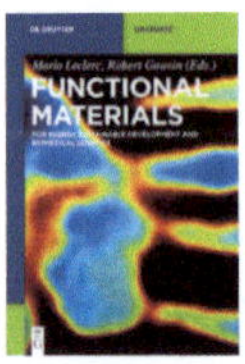

Functional Materials.
For Energy, Sustainable Development and Biomedical Sciences
Leclerc, Gauvin (Eds.), 2014
ISBN 978-3-11-030781-8, e-ISBN 978-3-11-030782-5

Twin Polymerization

New Strategy for Hybrid Materials Synthesis

Edited by
Stefan Spange, Michael Mehring

DE GRUYTER

Editors
Prof. Dr. Stefan Spange
Technische Universität Chemnitz
Institut für Chemie – Polymerchemie
Straße der Nationen 62
09111 Chemnitz
Germany

Prof. Dr. Michael Mehring
Technische Universität Chemnitz
Institut für Chemie – Koordinationschemie
Straße der Nationen 62
09111 Chemnitz
Germany

ISBN 978-3-11-050067-7
e-ISBN (PDF) 978-3-11-049936-0
e-ISBN (EPUB) 978-3-11-049801-1

Library of Congress Cataloging-in-Publication Data
Names: Spange, Stefan, 1950- editor. | Mehring, M., 1971- editor.
Title: Twin polymerization: new strategy for hybrid materials synthesis / edited by Stefan Spange, Michael Mehring.
Description: Berlin; Boston: De Gruyter, [2018] | Includes index.
Identifiers: LCCN 2018029226| ISBN 9783110500677 (print) | ISBN 9783110498011 (epub) | ISBN 9783110499360 (pdf)
Subjects: LCSH: Polymers. | Polymerization.
Classification: LCC QD381 .T945 2018 | DDC 668.9--dc23 LC record available at https://lccn.loc.gov/2018029226

Bibliographic information published by the Deutsche Nationalbibliothek
The Deutsche Nationalbibliothek lists this publication in the Deutsche Nationalbibliografie; detailed bibliographic data are available on the Internet at http://dnb.dnb.de.

Typesetting: Integra Software Services Pvt. Ltd.
Printing and binding: CPI books GmbH, Leck
Cover image: courtesy of Tina Mark, Mandy Göring, Lysann Kaßner and Kevin Nagel

www.degruyter.com

Preface

This book is the joint work of the scientists in the consortium of the DFG FOR 1497 (Research Unit of the Deutsche Forschungsgemeinschaft) with the topic: "Organic-Inorganic Nanocomposites through Twin Polymerization". The feature of twin polymerization is the targeted fabrication of inorganic/organic nanostructured hybrid materials by a simple procedure. The book is intended for material scientists and chemists with basic knowledge in organic chemistry and materials science. It describes the theoretical background, the physical chemistry and the combination of inorganic and organic chemistry of twin polymerization processes.

The book is structured into five main chapters: (1) Introduction, (2) Twin Polymerization – General Aspects, (3) Structure and Theory, (4) Materials, and (5) Applications.

The introduction generally explains polymerization reactions and various monomer structures. In this context, the classification of multifunctional monomers and complex polymerization cascades to hybrid materials are briefly discussed in comparison to common types of monomers and polymerization processes. Terms used for production of two polymers within one process, such as simultaneous polymerization and twin polymerization, are explained.

The variable synthetic concept of twin polymerization for the production of manifold molecular structure combinations in hybrid materials is introduced in Chapter 2. Also, various methodologies for the production of polymer organic/inorganic hybrid materials compared to twin polymerization are presented. The different types of twin polymerization and related processes are classified according to established IUPAC nomenclature. The differentiations and similarities of aspects of twin polymerization compared to simultaneous polymerization and other types of multifunctional monomer polymerizations are discussed in detail.

Twin polymerization is determined by the chemistry of a new class of monomers, so-called twin monomers, which are the topic of Chapter 3. The versatility of possible twin monomer molecule structures is addressed by structure investigations and quantum chemical calculations in subchapters of that chapter. Therefore, the structure–reactivity relationships of various twin monomers as functions of elemental compositions are demonstrated in detail. The general classes of precursors being suitable for twin polymerization and their specific structural features are presented in Chapter 3.1. The relationships between their structural features and their reactivity in the polymerization process are discussed. In addition, the influence of the chemical nature of the precursors with regard to the properties of the as-obtained polymerization products, e.g., organic/inorganic nanocomposites, is unveiled.

The fourth chapter describes the production of various nanostructured hybrid materials which are available by twin polymerization. Silicon-related and inorganic oxide hybrid materials, carbon materials and metal nanoparticles within hybrid materials are addressed in subchapters. Chapter 4.1 describes which types of twin

https://doi.org/10.1515/9783110499360-201

monomers are suitable to produce nanostructured silica and silicon nitride-related polymer hybrid materials. A special aspect is the characterization of the molecular composition of hybrid materials by solid state NMR spectroscopy and morphology investigations by various electron microscopy techniques. The application of the simultaneous twin polymerization by use of functional twin monomers for the fabrication of tailor-made ternary siliceous hybrid materials is particularly emphasized. Carbon materials derived from nanostructured hybrid materials of twin polymerization processes are the content of Chapter 4.2. In Chapter 4.3, the variety of porous inorganic oxides and mixtures of porous inorganic oxides that have been obtained using the concept of twin polymerization so far are presented. The different concepts for their synthesis exploiting distinguishable polymerization protocols are explained. The chapter closes with an overview of potential applications for these porous oxide materials. The twin polymerization of diverse main group and transition metal-containing twin monomers and metal carboxylates as well as tin alkoxides as a source for metal and metal oxide nanoparticles to produce porous carbon and silica hybrid materials, respectively, decorated with metal or metal oxide nanoparticles or tin alloys is discussed in Chapter 4.4. In addition, the encapsulation of metal and metal oxide nanoparticles within porous carbon shells by using twin polymerization is reported. Current trends in this field of chemistry are discussed as well.

In Chapter 5, applications of twin polymerization are compiled. Chapter 5.1 shows the application of carbon-containing materials as electrode materials for rechargeable lithium-ion batteries synthesized over the concept of twin polymerization. Electrochemical tests of anode materials based on germanium (Ge@C), tin ($Sn@SiO_2/C$), and tin alloys ($Sn_xM_y@SiO_2/C$; M = Co, Ni, Sb) as well as cathode materials based on sulfur-containing carbon hollow spheres are presented. A comparison with literature state of the art materials and values is given. Combination of various twin monomers with each other and/or with conventional monomers for polymerization is a suitable tool for the production of various hybrid material compositions on demand by a modular concept. The specific application of twin polymerization in combination with other polymer syntheses, such as polyepoxides and polyurethanes, is covered in Chapter 5.2.

Concluding remarks emphasize the perspectives of the twin polymerization process for the development of new hybrid materials as inspiring concept in materials science.

Contents

Preface — V

Acknowledgments — XI

List of contributors — XIII

Stefan Spange

1 **Introduction — 1**

1.1 Polymerization mechanisms — **4**

1.2 Characteristics of multifunctional monomers — **14**

1.3 Multifunctional monomers for hybrid materials synthesis — **18**

1.4 Conclusion — **21**

Andreas Seifert, Michael Mehring and Stefan Spange

2 **Twin Polymerization (TP) – General Aspects — 27**

2.1 Principles of twin polymerization — **27**

2.2 Comparison of twin polymerization with simultaneous polymerization of two monomers — **30**

2.3 Polymerization of heterobifunctional monomers — **33**

2.4 Inorganic network formation by twin polymerization, aqueous and non-aqueous sol-gel process — **35**

2.5 Types of twin polymerization processes — **39**

2.5.1 Condensative and ideal twin polymerization — **39**

2.5.2 Simultaneous twin polymerization — **40**

2.5.3 Apparent twin polymerization — **42**

2.6 Twin monomers – a modular molecular system — **43**

2.7 Transformation of hybrid materials to porous metal oxides or carbon — **44**

2.8 Twin polymerization in combination with additives — **45**

3 **Structure and Theory — 51**

Philipp Kitschke, Michael Mehring

3.1 Structure – reactivity relationships of the precursors for twin polymerization — **51**

3.1.1 Introduction — **51**

3.1.2 Precursor classes and their reactivity in TP — **56**

3.1.3 Summary — **114**

Alexander A. Auer, Giovanni Bistoni
3.2 Theory of the TP – computational studies on the mechanism of the proton-catalyzed and thermally induced TP — **116**
3.2.1 Introduction — **116**
3.2.2 Mechanistic studies on the proton-catalyzed TP — **118**
3.2.3 Mechanistic studies on the thermally induced TP — **128**
3.2.4 Final remarks — **135**

Janett Prehl, Benjamin Fiedler, Joachim Friedrich
3.3 Modeling and simulation of nanostructure formation of TP — **135**
3.3.1 Introduction — **135**
3.3.2 Basic concept of the proton-catalyzed reaction mechanism of TP — **136**
3.3.3 Reactive bond fluctuation model — **139**
3.3.4 rMD simulation — **144**
3.3.5 Extended quantum chemical modeling — **148**
3.3.6 Summary — **158**

4 Materials — 167
Stefan Spange, Mandy Göring, Lysann Kaßner, Kevin Nagel and Andreas Seifert
4.1 Silicon-containing hybrid materials — **167**
4.1.1 History — **167**
4.1.2 Silicon-containing twin monomers — **168**
4.1.3 Polymer/Silica hybrid materials by twin polymerization — **171**
4.1.4 Multiple polymerizations — **178**
4.1.5 Template-assisted twin polymerization — **179**
4.1.6 Ternary hybrid materials by simultaneous twin polymerization (STP) — **182**
4.1.7 Polymer/Silica hybrid foams by simultaneous twin polymerization — **187**
4.1.8 Functional twin monomers — **189**
4.1.9 Silicon monomers based on Si–N-containing moieties — **191**
4.1.10 Polyamide 6/Silica composite production — **196**
4.1.11 Summary — **205**

Thomas Ebert
4.2 Carbon materials — **206**
4.2.1 Synthesis of porous carbon materials — **206**
4.2.2 Carbon from hybrid materials synthesized by twin polymerization — **207**

4.2.3 Porous carbon composites by twin polymerization on surfaces — **210**
4.2.4 Structured porous carbon materials by twin polymerization on hard templates — **213**
4.2.5 Strategies for the synthesis of hierarchically porous carbon materials by consecutive twin polymerizations on hard templates — **216**

Philipp Kitschke, Michael Mehring
4.3 Inorganic oxides from twin polymerization — **217**
4.3.1 Synthesis of inorganic oxides and mixed inorganic oxides — **218**
4.3.2 Potential applications — **230**

Heinrich Lang
4.4 Twin polymerization – a unique and efficient methodology for decorating porous carbon and silica materials with nanoparticles — **231**
4.4.1 Introduction — **231**
4.4.2 Metal and metal oxide nanoparticles by thermal decomposition of *single-source* metal carboxylate precursors — **233**
4.4.3 Metal nanoparticle-decorated porous carbon and silicon dioxide materials by twin polymerization — **236**
4.4.4 Tin alloys embedded in carbon and silica materials by twin polymerization — **247**
4.4.5 Metal-loaded porous hollow carbon spheres by twin polymerization — **251**
4.4.6 Summary — **258**

5 Applications — 273
Felix Roschke and Michael Mehring
5.1 Porous materials from twin polymerization for energy storage — **273**
5.1.1 Introduction — **273**
5.1.2 Syntheses and electrochemical tests of porous materials — **275**
5.1.3 Summary — **293**

Katja Schreiter, Matthias Birkner, Mandy Göring, Susanne Höhne, Daniel Uhlig
5.2 Functionalized twin monomers and their application in materials synthesis — **295**
5.2.1 Characteristics of functionalized twin monomers — **296**
5.2.2 Amino-functionalized twin monomers — **298**
5.2.3 Interface design – functionalized TM as adhesion promoter — **299**

5.2.4 Post-functionalization – construction of layered systems — **305**
5.2.5 Summary — **325**

Michael Mehring and Stefan Spange
6 Concluding Remarks — 333

Index — 335

Acknowledgments

Twin polymerization has been funded by the Deutsche Forschungsgemeinschaft – DFG FOR 1497 – in the time period from 2011 to 2018. The consortium of project managers (principal investigators) of the DFG FOR 1497 in alphabetic order consisted of Alexander A. Auer (AU 206/5-1, AU 206/5-2), Joachim Friedrich (FR 2898/5-1), Michael Hietschold (HI 512/12-1), Karl Heinz Hofmann (HO 901/9-1), Heinrich Lang (LA 543/47-1, LA 543/47-2), Michael Mehring (ME 2284/4-1, ME 2284/4-2), Janett Prehl (PR 1507/1-1), Katja Schreiter (SCHR 1469/1-1), Andreas Seifert, Stefan Spange (SP 392/34-1, SP 392/34-2), and Dietrich R. T. Zahn. Funding is gratefully acknowledged by all members of the team.

Particular funding of special parts on twin polymerization was granted by the Federal Cluster of Excellence EXC 1075 "MERGE Technologies for Multifunctional Lightweight Structures," Fonds der Chemischen Industrie e.V., DFG SP 392/31-2 and SP 392/39-1, and the BASF SE Ludwigshafen as well as the SPP 1807, the MPG and the MPI CEC.

We would like to acknowledge the research groups of Lothar Kroll and Thomas Lampke (TU Chemnitz), and Petra Uhlmann (Leibniz-Institut für Polymerforschung Dresden e.V.) for fruitful cooperation and several helpful discussions. In this context, special thanks to Susann Anders and Isabelle Roth-Panke.

Helpful discussions and suggestions by Werner A. Goedel (TU Chemnitz) as well as the cooperative work of some colleagues from the BASF SE Ludwigshafen, which provided many valuable contributions to the topic of twin polymerization, are gratefully acknowledged.

We especially thank all PhD students, diploma students, master students, bachelor students, practical training students, technicians, and postdocs for their excellent work in the frame of the research unit DFG FOR 1497. Special thanks are given to Ute Stöß for assistance in preparing Chapter 4.4.

https://doi.org/10.1515/9783110499360-202

List of contributors

Professor Dr. Alexander Auer
Max-Planck-Institut für Kohlenforschung
D-45470 Mühlheim an der Ruhr
Technische Universität Chemnitz
D-09107 Chemnitz

Matthias Birkner
Technische Universität Chemnitz
D-09107 Chemnitz

Dr. Giovanni Bistoni
Max-Planck-Institut für Kohlenforschung
D-45470 Mühlheim an der Ruhr

Dr. Thomas Ebert
Technische Universität Chemnitz
D-09107 Chemnitz

Benjamin Fiedler
Technische Universität Chemnitz
D-09107 Chemnitz

Dr. Joachim Friedrich
Technische Universität Chemnitz
D-09107 Chemnitz

Mandy Göring
Technische Universität Chemnitz
D-09107 Chemnitz

Dr. Susanne Höhne
Leibniz-Institut für Polymerforschung
Dresden e.V.
D-01069 Dresden

Dr. Lysann Kaßner
Technische Universität Chemnitz
D-09107 Chemnitz

Dr. Philipp Kitschke
Technische Universität Chemnitz
D-09107 Chemnitz

Professor Dr. Heinrich Lang
Technische Universität Chemnitz
D-09107 Chemnitz

Professor Dr. Michael Mehring
Technische Universität Chemnitz
D-09107 Chemnitz

Kevin Nagel
Technische Universität Chemnitz
D-09107 Chemnitz

Dr. Janett Prehl
Technische Universität Chemnitz
D-09107 Chemnitz

Dr. Felix Roschke
Technische Universität Chemnitz
D-09107 Chemnitz

Dr. Katja Schreiter
Technische Universität Chemnitz
D-09107 Chemnitz

Dr. Andreas Seifert
Technische Universität Chemnitz
D-09107 Chemnitz

Professor Dr. Stefan Spange
Technische Universität Chemnitz
D-09107 Chemnitz

Daniel Uhlig
Technische Universität Chemnitz
D-09107 Chemnitz

https://doi.org/10.1515/9783110499360-203

Stefan Spange

1 Introduction

Polymerization reactions are of great importance for the fabrication of a variety of plastics, resins, composites, and rubberlike materials [1–7]. Hence, a variety of polymerization reactions and related processes have been developed for materials syntheses and related purposes over the last decades. Regardless of the mechanism of polymerization, the molecular structure of a polymer (and hence its properties) is also determined by the functionality and molecular composition of the monomers.

Furthermore, in polymerization reactions, organic monomers are often used in combination with diverse types of solid materials, such as metals, ceramics, and inorganic oxides, to fabricate hybrid and composite materials [8–12]. Hybrid materials are advantageous for many applications because they combine the properties of plastics, resins, and rubbers with those of solid inorganic materials in a synergistic manner. Presentations of concepts for hybrid material production are the main objective and motivation of this chapter. For this purpose, a novel polymerization principle is implemented.

Polymeric organic/inorganic hybrid materials consist of two or more solid components whose properties are determined by chemical composition, morphology, and topology of the interfaces. The ratio of the atoms at the surface corresponds approximately to that of the volume units of the components. Therefore, for hybrid materials, the domains of the individual components must be on a length scale of a few nanometers [8–10]. Thus, genuine hybrid materials show chemical and physical properties which are different from those of the individual components. That is the striking difference to traditional composite materials in which the characteristics are determined by the individual component but purposefully modified by the second component.

The key point in hybrid materials synthesis is that two different polymeric components must be simultaneously synthesized. For the production of nanostructured organic/inorganic hybrid materials, it is of importance that both polymers simultaneously develop within one overall polymerization process. That is a challenge when two polymerizable groups of two different monomer species are employed which are different in their kinetic behavior. The even arrangement of nanostructure only results when the originating polymer strands remain in vicinity to each other. In other words, phase separation of the two differently formed polymer components does not take place. This feature can be realized by the use of specifically constructed monomers or by polymerization techniques for the hybrid materials synthesis. The inorganic component of hybrid materials can be of a different nature, such as inorganic oxides, metal particles, or carbon allotropes. Organic polymers usually are polyolefins, polyesters, polyamides, and resins. Organic polymer syntheses by step-growth or chain polymerization are widely employed. Mostly, the sol-gel process is used for oxidic components, whereas metals and carbon materials are obtained by post-reactions of hybrid materials such as chemical reduction processes or thermal treatment. The

https://doi.org/10.1515/9783110499360-001

sol-gel process is a step-growth polymerization, which is not specifically addressed in this book, whereas other relating processes are considered.

The fundamentally different polymerization mechanisms leading to inorganic and organic polymers are discussed in Section 1.1.

Classification of hybrid materials is not trivial because various aspects play a role. Classification may include the composition, synthetic methodology, or kind of interaction of the targeted components. In literature, the type of interaction of the components has been accepted as prime criterion to classify a hybrid material [13–22].

Siliceous components play an important role in hybrid materials synthesis which is also an extensive part in this monography. For a better understanding, silica hybrid materials will be briefly used as instructive example to explain the nomenclature of hybrid materials.

To explain the conception of this book, we will demonstrate that the polymerization process of each monomer is decisively determined by its functionality, especially the number of functionalities. Functionality f means how many bonds the monomer can form to other partners or to itself (see Fig. 1.1). Molecules with $f=1$ are unsuitable for polymerization. For the case $f=2$, a linear chain or cycle results. Monomers with $f=3$ can deliver various polymer structures depending on reaction conditions and chemical composition of the monomer. Thus, trialkoxysilanes [R–Si(OR)$_3$] ($f=3$) can react in different ways under formation of polymeric silsesquioxanes or cubic silsesquioxanes [8, 11].

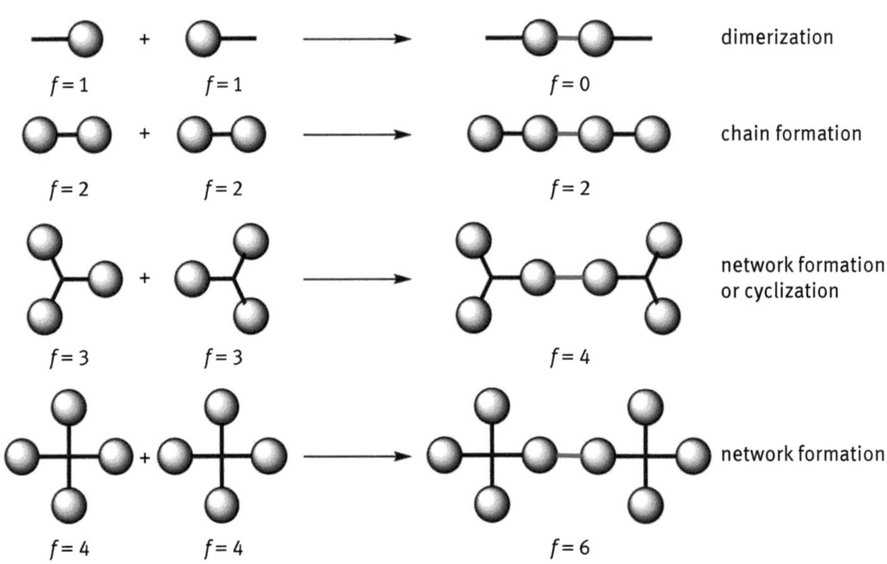

Fig. 1.1: Functionality of hypothetical reactant molecules and resulting functionality of the product. Each spherical part of each reactant is able to form one new bond to another sphere. The new bond formed is red marked in each product. The product of dimerization might result from addition or condensation reaction. Condensation by-products are not shown for clarity.

The case $f = 4$ is typical for silica network formation. The situation becomes more complex when monomer mixtures or multifunctional monomers are employed. But in principle, the concept of functionality remains valid.

In case that one generated component rapidly undergoes crosslinking, i.e., silica formation among the sol-gel process, simultaneously formed (linear) polymer strands can be physically entrapped and thus do not migrate. Those types of hybrid materials, containing one of the components physically bonded, are classified as type A (Mackenzie) or Class I (Sanchez) hybrid materials [13–17]. With regard to the synthesis, it is of importance for this type of hybrid materials to note that so far, two completely different polymerization processes, one for each component, were used within one procedure. For the production of this class of hybrid materials, the combination of two simultaneous polymerizations (SPs) of two common, but structurally different, monomers is suitable.

Mackenzie also introduced type B hybrid materials, which are characterized by forming strong hydrogen bonds between the components, but there is no covalent bond. If the two different polymeric components are covalently bonded within the hybrid material, they are classified as type C (Mackenzie) and Class II (Sanchez), respectively. The classification of hybrid materials with respect to the kind of interaction is illustrated in Figure 1.2. The blue points in the left drawing represent physically entrapped molecules. The yellow and orange sections are crosslinked inorganic or organic polymers.

For the production of Type C or Class II hybrid materials, the use of special so-called heterobifunctional monomers is required. They contain two structurally

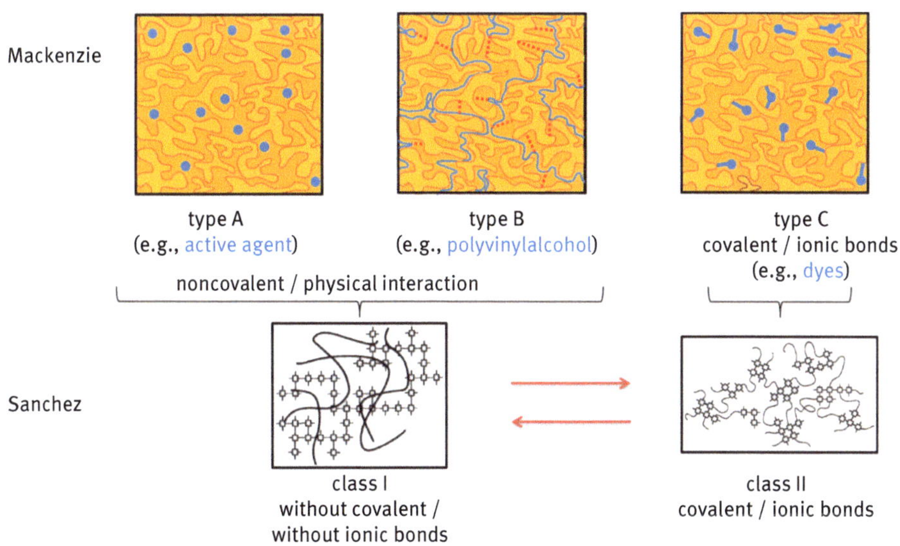

Fig. 1.2: Accepted nomenclature of hybrid materials with respect to the kind of interaction of the components according to Mackenzie and Sanchez [13–17].

different groups which can polymerize independently of each other. The different groups are connected to each other by strong covalent bonds which are not easily cleavable. The importance of both monomer design and functionality for hybrid material fabrication is set out in detail in the second part of the introduction.

To gain a better understanding of the types of polymerization reactions established in relation to the topic of this book, the classification of polymerization reactions will be briefly explained. For a more profound lesson on special aspects, the reader should revert to the references given in each chapter. In the second part of the introduction, the content is focused on different types of monomers and their functionalities because the main topic of this book deals with novel monomer architectures for the production of hybrid materials. In the third part, this introduction focuses on established and novel specifically designed monomers which are suitable to produce organic/inorganic hybrid materials within one integrated process. The two different types of complex monomers are explained with respect to functionality of the respective monomer fragment suitable for chain or step-growth polymerization. The special features of twin monomers are discussed.

1.1 Polymerization mechanisms

Polymerization reactions are very complex and several types of polymerization reactions are known. However, polymers are formed following two main mechanisms. These are step-growth polymerization and chain polymerization [1, 2, 7]. These modern classifications of polymerization types have been finally accepted, compared to the conventional perception originally introduced by *Carothers* [5, 6], because the kinetics of each of the reaction cascades are completely different [1, 2, 7]. Each polymerization reaction can be attributed to one of these basic mechanisms.

Step-reaction polymerizations (or step-growth polymerizations) are characterized by the random arrangement of monomers which can form dimers, trimers, tetramers, and longer molecular chains. In case of linear step-growth, the functionality f is at least two in both the growing polymer and the monomer. Each of the reactive monomers and intermediate products among the step-growth process contains functional groups that react with each other. Thus, dimeric chain-lengthening fragments and starting monomers can react spontaneously with each other to form macromolecules using about the same activation energy for each crucial growth step. The rate of a step-growth reaction can be increased significantly by applying an appropriate catalyst. In textbooks, these functional groups are mostly denoted as A and B [7].

Two types of monomers for step-growth polymerization are known: A–B-type monomers and a combination of A–A with B–B monomers. This established concept only considers the reaction of A with B. In practice, particular A–A-type monomers (or B–B) can also readily undergo step-growth polymerization. It depends on the chemical composition whether A–A or B–B monomers can polymerize spontaneously. Condensation polymerization of A–A monomers is important for producing inorganic polymers and networks because silanol groups, for example, actually react to siloxane moieties and water at ambient temperature. Likewise, methanediol (the hydrate of formaldehyde) spontaneously polymerizes to paraformaldehyde in water solution. Consequently, the polymer strands, instantaneously formed from such A–A monomers, undergo further condensation which makes their storage difficult. Usually, reactive A–A monomers are generated *in situ* by hydrolysis of the corresponding precursors, e.g., dialkoxysilane compounds, or by dissolution of formaldehyde in aqueous solution. However, dicarboxylic acids and diols also belong to the class of A–A or B–B monomers but are less reactive. In principle, dicarboxylic acids can react toward polycarboxylic anhydrides and diols to polyethers, but compared to the actual running polyester, these chemical reactions are not preferred.

Similar to readily polymerizable A–A monomers, shelf-life and storage conditions of A–B monomers are determined by the activation energy of the elementary step. Hence, for polyester synthesis, the A–B monomers are frequently generated *in situ* by ring-opening of lactones with traces of water. Otherwise, potential A–B monomers such as pure amino acids behave inertly due to their betaine structure. In this case, the ring-opening of lactams with a low portion of water is useful to generate the A–B monomer as intermediate during polymerization.

Often, step-growth polymerization of less reactive A–B monomers can only be induced by a catalyst. For instance, *p*-bromostyrene can be catalytically polymerized by step-growth to *p*-poly(phenylenevinylene), which is known as the Heck reaction. Furthermore, there are also particular step-growth polymerization processes, which essentially require both a catalyst and an additional reactant (oxidizing agent), i.e., the synthesis of poly(oxy-2,6-dimethyl-1,4-phenylene) from 2,6-dimethylphenol. In this category, production of several conjugated polymers, such as polyaniline and polythiophene, has to be considered as well [1].

Hypothetically, each organic molecule can serve as A–B monomer for step-growth polymerization. Hence, ethylene can theoretically polymerize to polyacetylene and dihydrogen. But this reaction does not take place. The reaction is neither thermodynamically favored nor does a suitable catalyst exist. Please remember that a thermodynamically favored reaction requires $\Delta_R G^\ominus$ to be negative at a given reaction temperature (Eq. 1).

$$\Delta_R G^\ominus = \Delta_R H^\ominus - T\Delta_R S^\ominus \tag{1}$$

Scenario $\Delta_R G^\ominus < 0$ is fulfilled, if $\Delta_R H^\ominus$ is negative and its contribution is stronger than that of the entropy term $T\Delta_R S^\ominus$. $\Delta_R S^\ominus$ for polymerization reactions are mostly negative excluding specific ring-opening polymerizations (ROPs) such as that of sulfur [11]. Further reading about thermodynamics of polymerizations is referred to textbooks for polymer chemistry [18, 19].

The combination of A–A with B–B monomers is the most used methodology for production of step-growth polymers because the reaction starts only when the monomer mixture (A–A + B–B) is prepared. Representative examples for step-growth polymerization reactions are shown in Scheme 1.1.

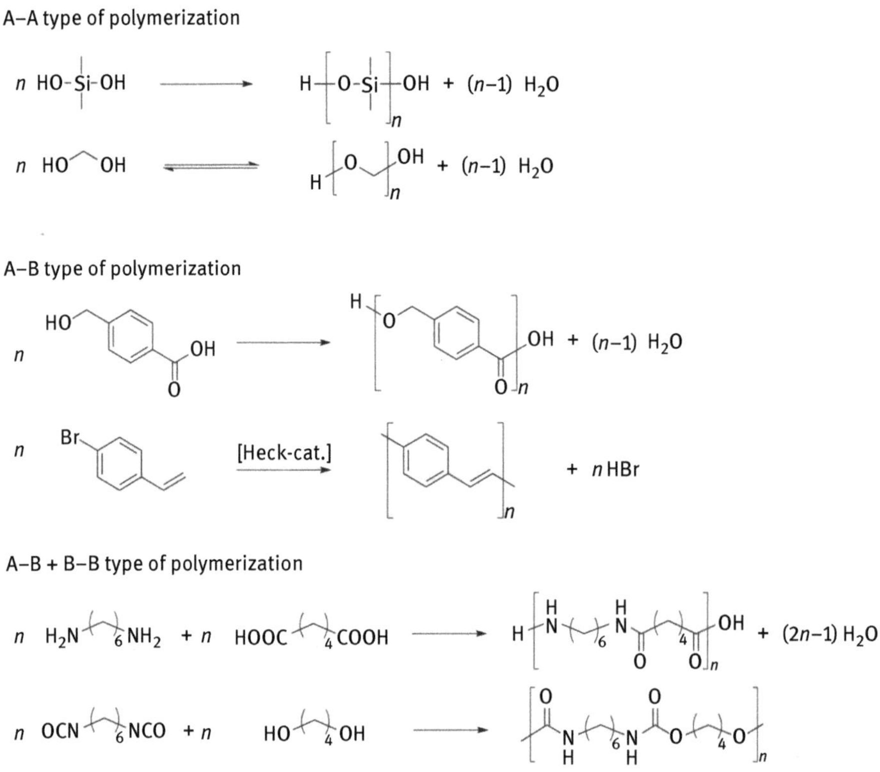

Scheme 1.1: Different types of step-growth polymerization processes to form linear polymers.

The dependence of the degree of polymerization DP_n as a function of the monomer conversion is an important relationship in polymer synthesis. The average degree of polymerization DP_n is defined as the number of monomer units in one macromolecule (Eq. 2):

$$DP_n = \frac{M_n}{M} \tag{2}$$

M_n is the number average molecular weight of the polymer and M the molecular weight of the starting monomer. The relationship between DP_n and overall conversion (p) of the functional monomer groups of step-growth processes can mathematically be expressed by the well-established *Carothers* equation (Eq. 3) [1, 2, 7].

$$DP_n = \frac{1}{1-p} \tag{3}$$

From Equation 3, it can be derived that nearly complete conversion is necessary to achieve a high DP_n. An important issue of step-growth polymerizations is that the polymerization rate steadily decreases with increasing conversion of functional groups due to their self-dilution. Importantly, Equation 2 is only fully valid for linear chain formations with $f = 2$. Cyclization reactions are not taken into consideration by Equation 2. The calculation of the proportion of cyclization reaction among linear polymer chain formation requires the use of a modified *Carothers* equation. This feature is explicated in special textbooks [4, 20].

For step-growth polymerization, monomers and monomer combinations with an average functionality higher than two as well as mixtures of monomers with different functionalities can be used. Then, the average functionality (f_{av}) must be taken into account. In case that f_{av} of monomers (A_xB_y, $x,y > 2$) is greater than two, crosslinking between growing chains might occur. Then, the extended *Carothers* equation (Eq. 4) allows the determination of conversion of the theoretical gel point at which an infinite polymer network ($DP_n \rightarrow \infty$) is formed ($p \approx 2/f_{av}$).

$$DP_n = \frac{2}{2 - f_{av} \cdot p} \tag{4}$$

$f_{av} = (xn + ym)/(n + m)$; x and y are the stoichiometric factors of functionality of the monomers with respect to A and B, n and m are the molar portions of the monomers A and B, respectively, in the polymerization mixture. For instance, if the stoichiometric factors f (or $x = f_A$ respectively $y = f_B$) for a common A–B monomer ($n = m = 1$) are both one, then it follows $f_{av} = 2$.

Network formation is already possible without complete conversion of functional groups if $f_{av} > 2$. Silica formation from tetraalkoxysilane [$Si(OR)_4$] in the sol-gel process is a typical network formation with $f = 4$ because the *ortho*-silicic acid [$Si(OH)_4$], which is considered as theoretical model, can be assumed as a monomer of an A_4-type. Then, ≥ 50 % conversion of Si–OR groups is theoretically sufficient for network formation to take place. Tetraethoxysilane, $Si(OC_2H_5)_4$, is the most frequently used precursor monomer for silica fabrication in sol-gel processing [21–23].

The active species directly suitable for step-growth polymerization, the Si–OH functional group, is only formed when $Si(OC_2H_5)_4$ reacts with water in an upstream reaction. This aspect must be considered for all types of M–O–alkyl species (M = Si, B, Ti, Sn, etc.) when used in hybrid materials synthesis.

Step-growth polymerization of AB_m monomers (f for A is 1, $m = 2, 3, 4, \ldots$) can yield branched polymers, yet only if both A and B cannot react with themselves (see Refs. [21, 24]).

Chain-reaction polymerizations do not start spontaneously at ambient temperature in contrast to common step-growth polymerizations. An initiator is necessary to induce the chain polymerization reaction. The most important monomer species for chain polymerization reactions are ethylene derivatives, mainly 1-olefins. Such ethylene-based monomers are often called vinyl monomers. Polymerization of olefins and several kinds of ROPs belong to the class of chain polymerization. Typical examples for chain polymerizations are demonstrated in Scheme 1.2.

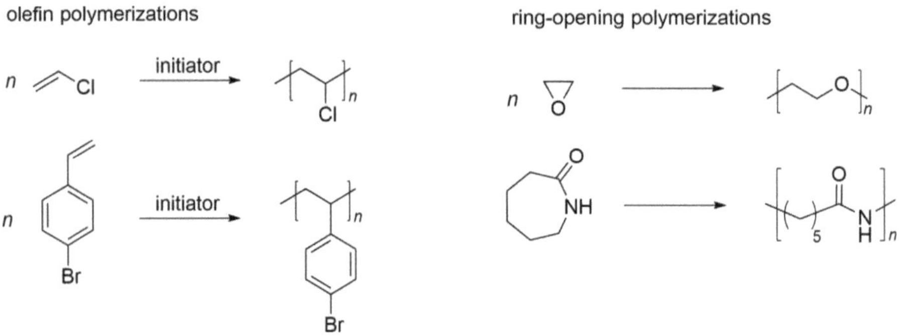

Scheme 1.2: Important types of chain-addition polymerizations. Each reaction equation illustrates only the gross propagation reaction. Compared with Scheme 1.1 (A–B type of polymerization), *p*-bromostyrene as monomer can form two different polymers, depending on whether a Heck catalyst or radical initiator is used.

The functionality of two ($f = 2$) of an olefin results from the double bond, which can form two σ bonds toward two other molecular fragments in a typical addition reaction as known from organic chemistry. In case olefins react with each other, a linear chain results.

For a better understanding in comparison to the step-growth polymerization, it might be assumed that a growing, active chain end with $DP_n{}^*$ has a functionality of one. After addition of one monomer with $f = 2$, the functionality of the $DP_{n+1}{}^*$ active growing chain end remains one (Scheme 1.3). Thus, f is one in an individual growing polymer chain but its value is two in the monomer. That is contrary to step-growth polymerization, where both are two. Therefore, *Carothers* equation cannot be applied to chain polymerization.

$$\sim\!\sim M_n{}^* \quad + \quad M \quad \longrightarrow \quad \sim\!\sim M_{n+1}{}^*$$

$$f_{CP}{}^* = 1 \qquad f_{CP} = 2 \qquad f_{CP}{}^* = 1$$

Scheme 1.3: Number of functionalities of a growing active polymer chain ($f_{CP}{}^*$) and corresponding monomer.

However, the situation for chain polymerization is more complex. A chain-reaction polymerization involves different individual chemical processes: initiation, propagation, termination, and transfer reactions. Each of these distinct reactions is different concerning the elementary chemical processes. Furthermore, there are four main initiation reactions of chain polymerizations to consider: radical, cationic, anionic, and transition metal complex coordination (i.e., Ziegler–Natta polymerization, metallocene polymerization, metathesis polymerization). The initiation, propagation, transfer, and termination reactions of each individual chain polymerization either radical, cationic, or anionic are completely different. Therefore, it is impossible to consider all known chain polymerization reactions in a universally accepted kinetic scheme.

There are several common aspects which can be discussed basically for chain-addition polymerization reactions. In principle, the average degree of polymerization DP_n of chain polymerization reactions is a function of both monomer $[M]$ and initiator concentration $[I]$ [1, 2, 4, 7]. The mathematical expression DP_n as a function of $[M]$ and $[I]$ is different for free radical polymerization (FRP), controlled radical polymerization, cationic polymerization, anionic polymerization, and transition metal-catalyzed polymerization. The reason for this fact is that DP_n as a function of monomer conversion ($d[M]/dt$) and initiator concentration depends on many factors which are determined by the detailed mechanism of polymerization. Often, the concentration of the active initiator is determined by complex equilibria of the reactants. Furthermore, DP_n depends on the extent to which transfer and termination reactions play a role. Notably, termination reactions of FRPs and ionic polymerizations are completely different in their nature, because radicals can annihilate each other, whereas equally charged chains do repel. The situation is still more entangled if ionic aggregates are involved.

Generally, DP_n (Eq. 5) is approximately the rate of propagation (v_p) divided by the sum of both rates of termination (v_t) and transfer reaction (v_{tr}).

$$DP_n \approx \frac{v_p}{v_t + v_{tr}} \tag{5}$$

In FRP, long chains are formed from the beginning of the reaction because propagation reaction (v_p) occurs much faster than initiation (v_i). Thus, the conversion of monomer has no great influence on average molecular weight.

Living chain polymerizations are characterized as those in which neither transfer nor termination reactions take place. That is $v_i \gg v_p$ and $v_t + v_{tr} = 0$. In living polymerization, average degree of polymerization increases linearly with conversion of the monomer (Fig. 1.3). In this case, DP_n at complete monomer conversion can be simply calculated by Equation 6. $[M]$ is the initial concentration of monomer and $[I]$ the real concentration of initiator.

$$DP_n = \frac{[M]}{[I]} \tag{6}$$

Equation 6 can only be applied when v_i is greater than v_p. That condition can be accomplished for several anionic and cationic polymerization reactions of olefins and cyclic compounds [25, 26].

Examples of Scheme 1.2 are selected according to the aspect that the propagation reaction delivers a structurally well-defined polymer with regioselectively linked

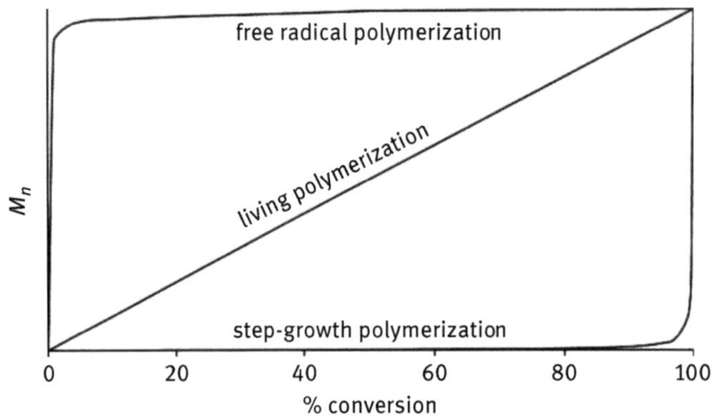

Fig. 1.3: Molecular weight M_n or degree of polymerization DP_n as a function of conversion of functional groups for free radical, living, and step-growth polymerization.

monomer units. That is not always the case. In other cases where an olefin monomer can particularly undergo isomerization reactions during polymerization, the resulting polymer constitution (structure) is often difficult to describe. The representative example for this occurrence is buta-1,3-diene polymerization, which can produce a mixture of different constitutional and configurational isomers along the same polymer backbone and between different polymer strands. That feature has to be considered independently of the mechanism of buta-1,3-diene and related to, i.e., isoprene polymerization. Altogether, 1,4-*cis*, 1,4-*trans*, 1,2-*isotactic*, 1,2-*syndiotactic*, and 1,2-*atactic* repeating units are theoretically possible. Thus, the sole term polybutadiene is not sufficient to describe the exact structure. Principally, "polybutadiene" isomers along the chain can also be classified as copolymer resulting from "1,2-butadiene" and "1,4-butadiene." But this description is not used. Commonly, the composition of each specific polymer sample is given as extra information.

ROPs show some peculiarities which are determined by the nature of the monomer and the mechanisms of polymerization. There are two types of ROP, that of cyclic compounds which contain breakable σ bonds or π bonds. The latter do polymerize by the ring-opening metathesis polymerization (ROMP) which will be discussed subsequently.

The functionality of two ($f = 2$) of a cyclic compound such as ethylene oxide or ε-caprolactam results from the breaking of a σ bond. The breaking of a σ bond in a cyclic compound provides two reactive sites ($f = 2$) which can form a chain (e.g., see Scheme 1.2). These basic considerations are independent of the polymerization mechanism.

The mechanism of addition-ring-opening polymerization of monomers is clearly a chain growth in the case that the active species is attached to the active polymer chain. For ROP, which occurs by activated monomer ring-opening mechanism, the situation is more complex [27]. Principally, nearly each polymer which can be produced by ROP can be alternatively synthesized by a condensation polymerization. Therefore, specific ROPs must be handled with care regarding their classification. But a distinction between chain- or step-growth mechanisms is always possible.

ROMP of cyclic olefins is also a chain polymerization of which the course is strongly dependent on the nature of catalyst used. Living ROMP has been established as a valuable tool for the production of tailor-made polymers [28]. The polymerization of acyclic olefins and di-olefins with a metathesis catalyst requires special attention because interference of chain-growth and step-growth reaction is possible as a function of monomer structure and resulting polymer [28–30]. The functionality of a monomer suitable for metathesis polymerization is two, as well, but with regard to both σ and π bond of the double bond, which can form only one double bond during polymerization.

Thus, in contrast to ROMP, the polymerization of nonconjugated, linear dienes (**a**cyclic **d**iene **met**athesis, ADMET) takes place by the use of identical catalyst and similar mechanism on the molecular level. But it requires a special consideration because in that case genuine step-growth polymerization takes place which is connected with ethylene (or ethylene derivative) elimination [30]. Thus, in olefin metathesis polymerization, the molecular structure of the monomer determines the polymerization process rather than the catalyst. This example is completely in contrast to the *p*-bromostyrene monomer example where the initiator or catalyst determines the course of polymerization and product formation (see Schemes 1.1 and 1.2). That comparison will show that any decision whether chain or step-growth polymerization takes place should be carefully reviewed.

Polymerization of monomer combinations

Molecular composition of polymers can be adjusted in a wide range by using the **copolymerization** techniques. Copolymerization is useful for both step-growth and chain polymerization. Copolymerization is a valuable tool for tailoring properties of industrial chain polymers. These aspects are the content of many established polymer chemistry textbooks [1–4]. There are different possible scenarios when two monomers are polymerized together within one procedure. Statistical copolymerization, alternating copolymerization, block copolymer formation, and SP can take place (Scheme 1.4).

Apart from the formation of copolymers, when two different monomers are chain polymerized within one composition, the formation of two homopolymers can also take place. This scenario is usually not specifically treated in polymer text books because it is undesirable in practice for chain polymerization. Another extreme scenario is the so-called selective polymerization. In this case, only one of the monomers (M_1) undergoes polymerization and monomer M_2 remains unaffected. That behavior is observed for anionic or cationic chain polymerizations of monomers which differ significantly in their reactivity. This procedure may be suitable for special application if M_2 can be used in a consecutive step to form a second polymer component.

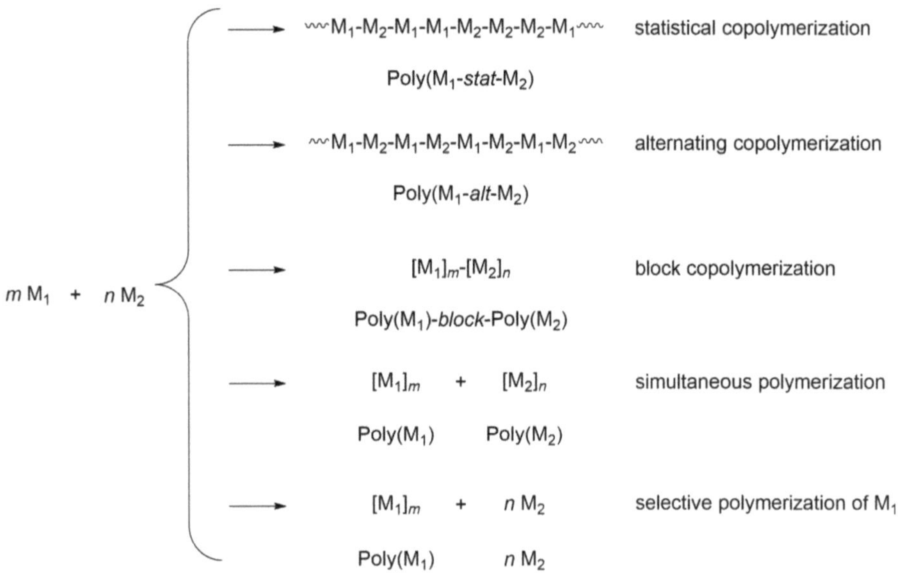

Scheme 1.4: Possible scenarios when two monomers are mixed together and then polymerization is started within one chemical process.

Let us now consider the scenario when two monomers do simultaneously polymerize but cannot copolymerize. SP of two monomers within one procedure is a suitable methodology to produce polymer blends or organic/inorganic hybrid materials. From the above arguments, the conclusion follows that DP_n as a function of conversion of monomer is completely different for FRP, step-growth, and living chain polymerization, which is illustrated in Figure 1.3.

Especially, DP_n versus conversion behavior of FRP is contrary to step-growth polymerization (see Fig. 1.3). High molecular weight polymers are obtained due to the very fast propagation speed of FRP already at low monomer conversion. The situation for step-growth polymerization is derivable from Equation 3 showing that a high degree of polymerization is only possible at high p.

To develop a material consisting of two different polymers requires a synthetic protocol where the two different polymer structures are generated simultaneously within the overall process. Therefore, combination of either two different chain or two different step-growth polymerizations is, as expected, advantageous in that both polymerizations are similarly fast and DP_n increases in a similar way. Therefore, combination of FRP with step-growth polymerization for the production of hybrid materials is expected to produce inhomogeneous hybrid materials due to the different time scales of the individual polymerization rates. This point is of great importance for discussing aspects of the SP.

That feature is hard to realize when only chain polymerizations of olefins are employed because often copolymers are formed or one monomer does not react and thus polymer/monomer mixtures result (see Scheme 1.4). Combination of chain polymerizations to produce hybrid materials is still a challenge. Only simultaneous chain polymerization of acrylate monomers together with epoxide monomers as a concrete beneficial example has been reported in literature. The acrylate monomer polymerizes by a radical mechanism and the epoxide by a cationic polymerization. Both active growing chain ends do not evidently interfere with each other. Thus, a polymer/polymer hybrid material of polyacrylate/polyepoxide is formed (Scheme 1.5) [31–33]. The advantage of this process is that two different chain polymerizations, olefin polymerization and ROP, can be beneficially combined.

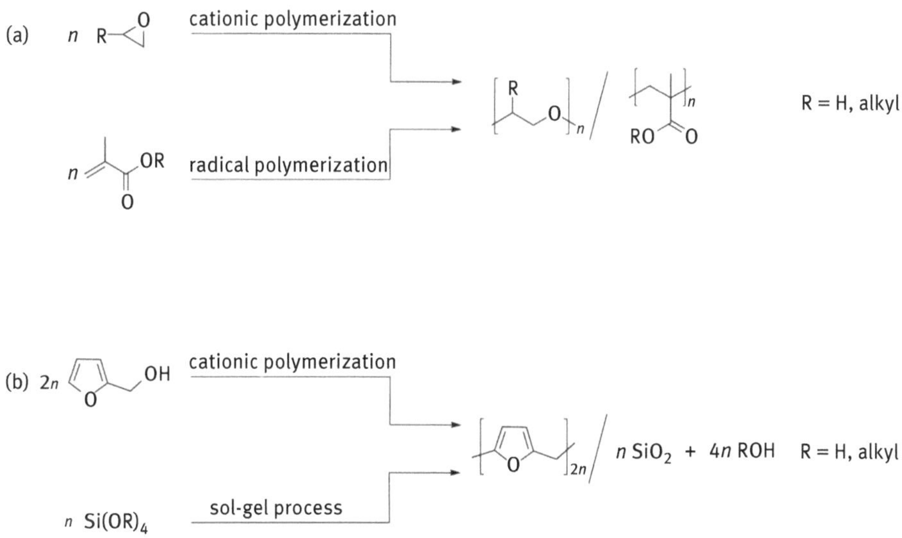

Scheme 1.5: Two established examples of simultaneous polymerizations for fabrication of hybrid materials. (a) Synthesis of interpenetrating polyepoxide/polyacrylate hybrid material by combination of two chain polymerizations. (b) Synthesis of a polyfurfuryl alcohol/silica hybrid material by combination of two step-growth polymerizations.

In contrast to SP of chain polymerizations, simultaneous step-growth polymerization processes are widely used for the production of organic/inorganic hybrid materials because the sol-gel process of silicon- and metal alkoxides belongs to the category of step-growth processes [21–23, 34]. For instance, combination of sol-gel chemistry and step-growth polymerization of furfuryl alcohol has often been used to fabricate polymer/silica hybrid materials [35, 36]. SP techniques will be discussed again as an alternative route compared to twin polymerization in Chapter 2.

The examples of SP to produce hybrid materials as demonstrated in Scheme 1.5 are particularly advantageous for the reason that two polymerization processes are combined which are even adapted in their kinetics.

The conception of polymeric hybrid materials synthesis requires the consideration of various polymers achievable from either step-growth or chain polymerization reactions. As mentioned before [1, 2, 7], step-growth polymerization, (free radical) chain polymerization, and "living" polymerization are totally different with regard to DP_n as a function of conversion of functional groups (see Fig. 1.3). Thus, their SPs to high molecular weight polymers are limited.

Principally, the drawback can be overcome if the functional group is both step-growth and chain polymerizable. There are only few chemically reactive groups which can undergo both chain and step-growth polymerization. Isocyanates and epoxides belong to this class of monomers. Formaldehyde is a special monomer molecule which can undergo chain polymerization and serve as component in step-growth polymerization due to its high carbonyl reactivity. Related carbonyl compounds (aldehydes and ketones) can also theoretically operate in this way, but their reactivity strongly decreases with increasing π-bonding energy of the carbonyl bond. Therefore, the combination of the chemical reaction of groups which are suitable for chain and step-growth polymerization may be advantageous to assemble both polymerization processes. That feature can be realized when two polymerizable groups are integrated within one molecule. These are multifunctional monomers. Their occurrence in the course of polymerization will be discussed in the next section.

1.2 Characteristics of multifunctional monomers

In polymer chemistry, multifunctional monomers are defined as molecule compounds that contain more than one polymerizable group with $f = 2$. There are two types of multifunctional monomers depending on whether the bond connecting the two polymerizable groups is cleavable or persists during the chemical processing. Cleavable monomers commonly react with water or other reagents at ambient temperature. This process results in two different monomers which can polymerize by different mechanisms (see Section 1.3).

The discussion starts with a class of compounds that have a not easily cleavable bond. Amide or ester groups are suitable for this purpose linking the polymerizing groups. The simplest case is a bifunctional monomer, e.g., bisacrylate, where two vinyl monomer units are covalently linked together *via* a spacer. Then, the overall functionality is four. In the case of the chemical structure of diverse polymerizable groups being identical in one molecule, the monomer should be classified as homo-*N*-functional. *N* denotes the number of identical functional groups. $N = 2$ corresponds to a homobifunctional monomer ($f = 4$), $N = 3$ to homotrifunctional monomer ($f = 6$), and so on. Thus, overall functionality of homo-*N*-functional monomers for chain polymerization is $f = 2N$. Typical homobifunctional monomers serve as crosslinking agents during chain polymerization (see Scheme 1.6) [2, 37, 38].

The feature $f > 2$ for homo-*N*-functional monomers is related to crosslinking and gel formation of step-growth polymerization as already explained earlier by

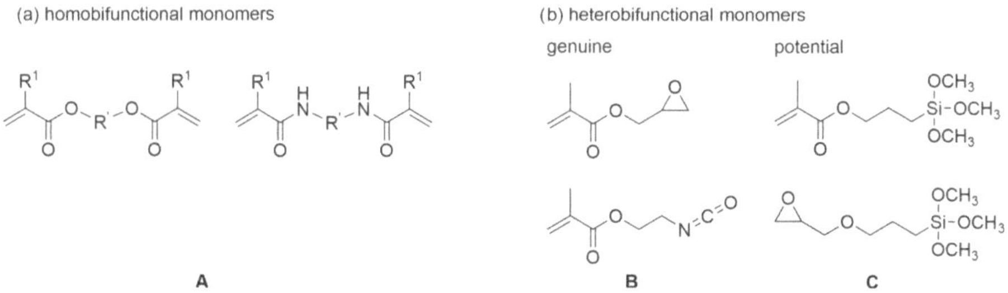

Scheme 1.6: Examples of homo- and heterobifunctional monomers. The homobifunctional monomers are bisacrylates and bisacrylamides ($R^1 = H, CH_3$; R^2 is usually an acyclic or cyclic alkyl group). Genuine heterobifunctional monomers are 2,3-epoxypropyl methacrylate and 2-isocyanoethyl methacrylate. The two potential heterobifunctional monomers are (3-glycidyloxypropyl)trimethoxysilane and 3-(trimethoxysilyl)propyl methacrylate.

Equation 2. In contrast, in the case of step-growth monomers, overall functionality of a homo-*N*-functional monomer is $f = N$.

Heterobifunctional monomers combine two polymerizable groups, which are completely different in their chemical structure, in one molecule. Scheme 1.6 shows some examples for homo- (**A**) and heterobifunctional monomers (**B** and **C**).

Heterobifunctional monomers of the type presented in Scheme 1.6 are potential candidates for the production of type C or class II hybrid materials (Fig. 1.2). In the broadest sense, class II hybrid materials can be classified as copolymers from two networks. The most important factor of heterobifunctional monomers is that the functionality of the non-polymerizing group increases during homopolymerization.

Polymerizable groups of genuine heterobifunctional monomers, such as acrylate, epoxide, or isocyanate, are immediately usable. The methoxysilyl group (or other M–O–R groups) is not directly usable because it must be transformed into Si–OH by reaction with water. Thus, for alkoxysilyl compounds, the actual functionality is also a function of conversion with water. Therefore, it must be distinguished between genuine and potential heterobifunctional monomers.

The nomenclature for heterobifunctional monomers in literature is not always according to rules of International Union of Pure and Applied Chemistry (IUPAC). Thus, 2,3-epoxypropyl methacrylate was named as "Jekyll and Hyde" monomer because it can enter into both vinyl and ROP [39]. Also, the name "tandem" monomer is sometimes used for this type of heterobifunctional monomer to demonstrate the concurrence of two differently polymerizable groups within one molecule [40, 41].

In case that a heterobifunctional monomer possesses one polymerizable group for chain polymerization and another one for step-growth polymerization, the determination of *f* of the single monomer depends on whether both groups can react with each other. If they do not react with each other, the functionality results from each chemically reactive group individually. For this purpose, two different abbreviations for the respective functionalities are useful; f_{SG} for the functionality of the step-polymerizable groups and f_{CP} for the chain-polymerizable groups.

In the following, the complex scenario for polymerization of 3-(trimethoxysilyl)-propyl methacrylate (TMMA) will be discussed as an example. The functionality *f* of TMMA with regard to potential step-growth polymerization of trimethoxysilyl group is three ($f_{SG} = 3$) and with regard to chain polymerization (acrylate double bond), it is two ($f_{CP} = 2$). The situation changes if either chain polymerization of TMMA by methacrylate groups or step-growth of trimethoxysilyl groups starts. The crucial aspect of heterobifunctional monomers is that the functionality of the group not involved in the polymerization increases linearly with the degree of polymerization. This scenario is shown for CH_2=CH–R–Si(OR′)$_3$ and CH_2=CH–R–SiR″(OH)$_2$ as hypothetical monomers in Scheme 1.7. The latter would result from CH_2=CH–R–SiR″(OR‴)$_2$ after hydrolysis.

The difference in functionality *f* of chain and step-growth polymerization ability as a function of water conversion of a TMMA-related model monomer (R′O)$_3$Si–R–CH=CH_2 is illustrated in Table 1.1 (see also Scheme 1.7). $x/1$ represents the water/monomer ratio corresponding to the resulting species if complete conversion of the –Si(OR′)$_3$ group takes place.

For instance, the overall functionality *f* of the potential step-growth moiety –Si(OR′)$_3$ increases with increasing degree of polymerization in the following manner. Assuming an average DP_n of 10 of the formed oligomer regarding the ethylenic moiety, then the overall functionality f_{SG} with respect to the fully hydrolyzed –Si(OH)$_3$ moiety is 30 according to Equation 2. Then, only about 6.7 % of the silanol group conversion to Si–O–Si linkages is sufficient for crosslinking of the formed polymer to take place. For a DP_n of 100, the conversion of silanol groups can be below 1 % to guarantee gel formation. Therefore, only a low quantity of generated silanol groups is sufficient for gel formation to take place when the vinyl group part begins to polymerize. The

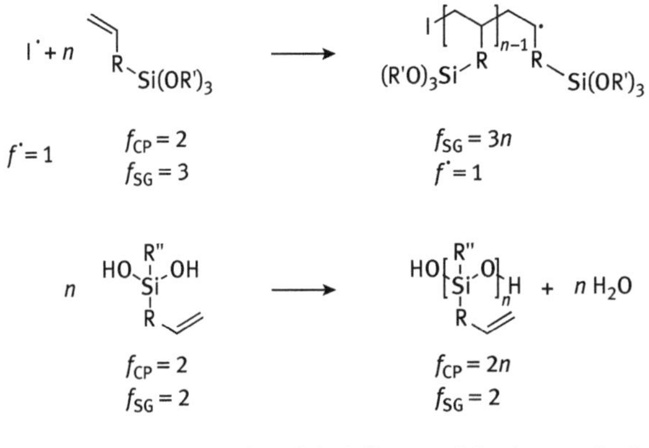

Scheme 1.7: Demonstration of the influence of the degree of polymerization *n* on the resulting functionality f_{SG} and f_{CP}, of the respective group not involved in radical chain polymerization of CH_2=CH–R–Si(OR′)$_3$ (f_{SG}) or step-growth polymerization of CH_2=CH–R–SiR″(OH)$_2$ (f_{CP}), after complete conversion.

Tab. 1.1: Assumed molecular species by stepwise reaction of a heterobifunctional monomer of the $(R'O)_3Si{-}R{-}CH{=}CH_2$ type with water illustrated for the trimethoxy group hydrolysis and impact of different degree of polymerization of the vinyl group ($DP_{n,CP}$) on resulting overall functionality f_{SG} of –Si–OH moieties at the formed polymer.

$x/1$	Assumed monomer species	f_{CP}	f_{SG}				
			$DP_{n,CP} = 1$	2	5	10	100
0/1	$(R'O)_3Si{-}R{-}CH{=}CH_2$	2	0	0	0	0	0
1/1	$HO(R'O)_2Si{-}R{-}CH{=}CH_2$	2	1	2	5	20	100
2/1	$(HO)_2(R'O)Si{-}R{-}CH{=}CH_2$	2	2	4	10	20	200
3/1	$(HO)_3Si{-}R{-}CH{=}CH_2$	2	3	6	15	30	300

same applies in the case of TMMA undergoing sol-gel reaction at first. Then, homo-*N*-functionality of the vinyl groups of the formed oligomer increases as a function of the degree of polymerization as well. Subsequently, N equals DP_n of the silicatic hybrid network. The numbers of functionalities ($f_{SG} + f_{CP}$) become completely complex if both groups are simultaneously polymerized by using initiator/catalyst mixture, etc. The fast increasing f of each polymerizing group would rapidly induce a crosslinking reaction already at the beginning of the reaction. The gel formation would hinder the further polymerization of the heterobifunctional monomer. This factor is the reason that hybrid materials syntheses which combine chain and step-growth polymerization by the sole use of heterobifunctional monomers are hard to control not only due to the different time scales at which each polymerization reaction takes place. Then, several polymerizable groups still remain intact but covalently bonded in the hybrid material. This is the typical occurrence which is observed in crosslinking polymerization and explained by the theory of gel point formation.

These considerations clearly demonstrate that the definition of individual functionalities of the heterobifunctional monomer is not helpful if the pure monomer is used in polymerization reactions.

In most cases, heterobifunctional monomers are only used in small amounts to covalently bond two polymer structures or to link polymers on surfaces. For this purpose, the use of heterobifunctional monomers has been established for the synthesis of a variety of organic/inorganic hybrid materials. Especially, TMMA and structurally related acrylic compounds are suitable as heterobifunctional monomers for producing a variety of poly(methyl methyacrylate)/silicatic hybrid materials.

Alkoxysilane compounds containing functional groups, suitable to react with polymers or surfaces, are widely available and find application in materials chemistry. (3-Mercaptopropyl)trimethoxysilane and (3-glycidoxypropyl)trimethoxysilane fall into this category.

(3-Glycidyloxypropyl)trimethoxysilane (Scheme 1.6) is a special monomer because the epoxy groups polymerize cationically and can react with silanol groups during the sol-gel reaction [41].

Mostly, trimethoxysilyl group-bearing heterobifunctional monomers such as (3-glycidyloxypropyl)trimethoxysilane and TMMA are used in combination with tetraalkoxysilanes, organic monomers or other multifunctional alkoxysilanes in step-growth copolymerization or as components in reactions with reactive polymers [42–47].

1.3 Multifunctional monomers for hybrid materials synthesis

Type A and B or class I hybrid materials (Fig. 1.2) are available by SP of two different monomers as already illustrated in Scheme 1.5. Due to the independently occurring polymerizations at SPs, phase separation of the polymers is mostly observed. In order to reduce the degree of separation, concepts have been developed to construct monomers in such a way that two polymers are coupled simultaneously to one another at the same time.

Two types of specially designed monomers are known which fulfill this requirement. The first group of monomers combines functional groups for sol-gel chemistry with those for chain polymerization in a way different from the monomers in Schemes 1.6 and 1.7. The difference is that a hydrolytically cleavable bond of the M–O–R type connects the two polymerizable moieties compared to the monomers in Scheme 1.6 [45–47]. Thus, cleavable multifunctional monomers (CMMs) contain M–O–R bonds which can readily react with water or related reagents to produce two separated monomers immediately suitable for chain and/or step-growth polymerization according to Scheme 1.8.

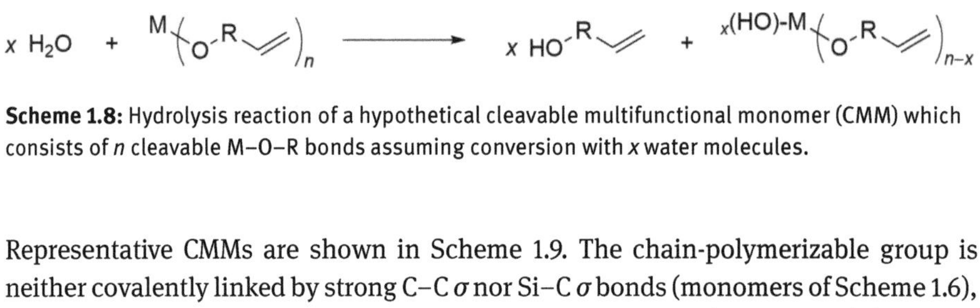

Scheme 1.8: Hydrolysis reaction of a hypothetical cleavable multifunctional monomer (CMM) which consists of *n* cleavable M–O–R bonds assuming conversion with *x* water molecules.

Representative CMMs are shown in Scheme 1.9. The chain-polymerizable group is neither covalently linked by strong C–C σ nor Si–C σ bonds (monomers of Scheme 1.6), respectively, but by a hydrolytically cleavable M–O–C bond (M = Ti, Si, B) (CMMs of Scheme 1.9). In the case of CMM, the formation of the inorganic component results from the hydrolytic instability of the M–O–C bond and its step-growth ability.

According to the definition of f, the monomers shown in Scheme 1.9 are homo-tetrafunctional with respect to chain polymerization ability ($N = 4$, $f_{CP} = 8$). Whether the monomers in Scheme 1.6 act as heterobifunctional monomers depends on the presence of a second component such as water (Scheme 1.8). This circumstance is comparable to the heterobifunctional monomer TMMA, but there is a decisive difference which will be explained in the following.

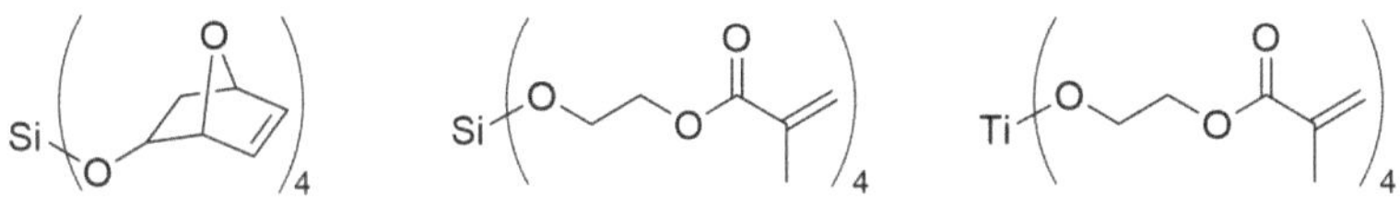

Scheme 1.9: Examples for specific heterobifunctional monomers which combine chemically reactive groups for chain and step-growth polymerization due to utilization for sol-gel process. These specific monomers are hydrolytically cleavable.

First, changes of both functionalities *f* for chain polymerization and step-growth ability of a common hypothetical M–(O–R–CH=CH_2)$_n$ monomer as a function of conversion with water *x* (Tab. 1.2) in comparison to a (R′O)$_3$Si–R–CH=CH_2 monomer (Tab. 1.1) have to be considered. Second, it is important to note that the *in situ* formed HO–R–CH=CH_2 monomer ($f = 2$) can also participate and copolymerize. In contrast to the potential heterobifunctional monomers of the type (R′O)$_3$Si–R–CH=CH_2, the functionality *f* for chain polymerization of M(O–R–CH=CH_2)$_n$ is dependent on the water/monomer ratio.

Tab. 1.2: Resulting molecular species by stepwise hydrolysis of composite monomers of the M–(O–R–CH=CH_2)$_n$ type illustrated for $n = 4$ and the resulting functionality of the appropriate polymerizing group. *x*/2 represents the water/monomer ratio corresponding to the resulting species if complete conversion takes place.

x/2	Assumed species	f_{CP}	f_{SG}
0/2	M(O–R–CH=CH_2)$_4$	8 (Crosslinking)	0 (No polymerization)
1/2	HO–M–(O–R–CH=CH_2)$_3$	6 (Crosslinking)	1 (Only dimerization)
2/2	(HO)$_2$–M–(O–R–CH=CH_2)$_2$	4 (Crosslinking)	2 (Linear chain formation)
3/2	(HO)$_3$–M–(O–R–CH=CH_2)	2 (No crosslinking)	3 (Crosslinking)
4/2	(HO)$_4$M	0	4 (Crosslinking)

This consideration is of importance to illustrate the different requirements for fabrication of nanostructured organic/inorganic hybrid materials when different types of heterobifunctional monomers are used.

The functionalities of the assumed species of Table 1.2 change abruptly when the resulting species (HO)$_x$–M–(O–R–CH=CH_2)$_{n-x}$ begin to polymerize. Then, the discussed scenarios of Scheme 1.7 and Table 1.1 additionally play a role. It is an important factor that mixtures of all assumed species in Table 1.2 can be simultaneously involved in the actual polymerization process. Therefore, whether a class I or class II hybrid material is formed from a CMM does also depend on the portion of water added and in which sequence chain and step-growth polymerization are externally triggered.

An advantage of this type of CMM type is the potential high functionality of each kind of the two polymerizing groups. Therefore, both formed polymers crosslink immediately which prevents the phase separation. But in contrast to the non-cleavable heterobifunctional monomers, the functionalities of f_{CP} and f_{SG} are coupled to each other in another way. With increasing conversion of M–O–R, they are opposite (Tab. 1.2).

With regard to monomers in Schemes 1.8 and 1.9, the chain and step-growth polymerization takes place individually for each polymerizing group. Both mechanisms, chain and step-growth, are not coupled with each other on the molecular level. Thus, an initiator (for the chain-addition polymerization) as well as a catalyst and water as reactant (for the step-growth reaction) are required for the synthesis. Thus, the consecutive or SP of this type of specifically designed monomers can produce a mixture of Class I and Class II hybrid materials, depending on the stage of polymerization. It is also possible that hydrolysis takes place as post-reaction altering the Class II hybrid material to Class I.

These considerations show the restrictions of the concept of those heterobifunctional monomers. However, the manifold combination of heterobifunctional monomers with different types of common co-monomers, either suitable for chain or step-growth polymerization, opens a wide field for synthesizing hybrid materials with desired composition. As mentioned, this subject has been discussed in many review articles and textbooks [8–12, 48, 49].

To produce fine-tuned nanostructured hybrid materials, both the inorganic and organic polymer component within the hybrid material must be generated at the same time. As shown, established heterobifunctional monomers (Scheme 1.6) and hydrolytically cleavable monomers (Scheme 1.9) combine two polymerizable moieties, one for step-growth and one for chain polymerization. Thus, the functionality of each polymerizable fragment during polymerization of CMM is dependent on conversion of the covalently attached polymerizable group at the formed polymer backbone. That is explained by data from Tables 1.1 and 1.2.

Common monomers for genuine step-growth polymerization have a fixed functionality. Therefore, complex monomers are required which contain two step-growth moieties combined in such a way that both polymerizations take place mechanistically coupled and both polymers are formed simultaneously at the same time scale. Conceptually, this feature can be realized by so-called twin monomers [50–58]. Twin monomers belong to the class of heterobifunctional monomers which contain two polymerizable groups suitable for step-growth polymerization. There are two step-growth polymerizable molecule fragments covalently linked in the twin monomer. The functionalities of the two polymerizable units are fixed because no second component, such as water, is required.

As representative example, the formation of phenolic resin/silica hybrid material from 2,2′-spirobi[4*H*-1,3,2-benzodioxasiline] twin monomer (Si-Spiro) is briefly mentioned in this introduction (Scheme 1.10).

According to the principles of polymerization processes, twin polymerization of 2,2′-spirobi[4*H*-1,3,2-benzodioxasiline] can also be considered as rearrangement condensation reaction finally providing phenolic resin and silica within one hybrid

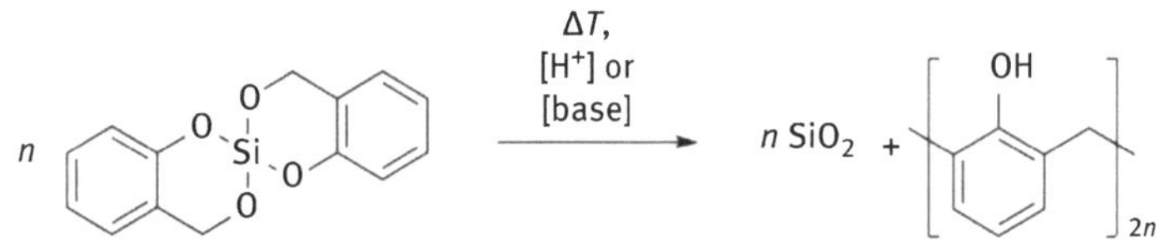

Scheme 1.10: The twin polymerization of 2,2′-spirobi[4*H*-1,3,2-benzodioxasiline]. For this twin polymerization, the functionalities of the silicon moiety with $f = 4$ and for the phenolic resin with $f = 3$ are fixed. Aspects of functionality of twin monomers are treated in Chapter 2.1. Please notice, various linkages (*o,o'*-, *o,m*-, *o,p*-) are possible.

material. A twin monomer is a specifically constructed multifunctional monomer that polymerizes in a step-growth reaction in one process to give two polymers. The formation of both polymers takes place mechanistically coupled with each other. Therefore, twin polymerization is examined as novel polymerization principle.

For this purpose, the availability of polymerizable twin monomers is a prerequisite. Twin monomers can be widely designed by using a variety of organic fragments suitable for step-growth polymerization and inorganic elements such as the metals and semimetals Ti, Sn, B, Si, etc. Furthermore, functional groups suitable for chain and step-growth polymerization can be chemically introduced and substitutions of heteroatoms (O) by other elements such as sulfur or nitrogen are possible. Thus, twin polymerization is a modular concept. The versatile concept, the theory, and the application of twin polymerization as a novel polymerization principle for the fabrication of various hybrid materials are addressed in this textbook. The general concept for twin polymerization is discussed in Chapter 2.

1.4 Conclusion

The introduction describes principle polymerization processes and aspects of monomers suitable for this purpose. Special attention is given to organic/inorganic hybrid materials synthesis and the monomers for their fabrication. To recapitulate, the definitions of the types of monomers suitable for the synthesis of linear polymers and hybrid materials production are summarized below.

(a) A–A, A–B, and B–B monomers are suitable for (linear or cyclic) step-growth polymerization. Each monomer possesses the functionality of two. Importantly, the molecular structure of the A- or B-chemical functional group decides whether A (or B) does only react with B or also with itself at chosen reaction conditions.
(b) An olefin with one double bond or an appropriate cyclic compound such as an epoxide is a monomer with the functionality of two for linear chain polymerization.
(c) A homobifunctional monomer contains two groups, according to (b), suitable for chain polymerization. Then, the overall functionality is four. Thus, a related homotrifunctional monomer possesses a functionality of six, and so on. Avoid

the confusion of homobifunctional monomer with A–A or B–B-type monomers because homo-*N*-functional monomers for step-growth polymerization are A_n or B_n monomers with $n = 3, 4, 5, \ldots$. An odd number of functionality is not possible for homo-*N*-functional monomers which are suitable for chain polymerization and ROP.

(d) Principally, a heterobifunctional monomer contains one or more groups suitable for chain polymerization and/or step-growth polymerization which are linked covalently together *via* a spacer or by ionic character. The number of functionality of each moiety is determined by the same criteria mentioned under (a), (b), and (c). However, functionalities are not only dependent on the initial monomer but are also dynamically changed as soon as the polymerization starts. Each individual polymerization, step-growth and chain polymerization, of the heterobifunctional monomer takes place mechanistically independent of each other.

(e) There are two types of heterobifunctional monomers to distinguish depending especially on whether the chemical bond is cleavable or not by water or other reagents during polymerization.

(f) Hydrolytically cleavable heterobifunctional monomers and twin monomers are specifically designed monomers which can form two polymers in the course of the cross-polymerization process.

(g) A twin monomer is a specially designed monomer which contains different inorganic and organic fragments suitable for two or more different step-growth polymerization processes. Both polymerization processes take place mechanistically coupled with each other by use of solely one catalyst or other stimuli like heat or light.

References

[1] Koltzenburg S, Maskos M, Nuyken O. Polymere: Synthese, Eigenschaften und Anwendungen. Berlin Heidelberg: Springer Spektrum; 2014.

[2] Odian GG. Principles of polymerization. 4th ed. Hoboken, N.J: Wiley-Interscience; 2004.

[3] Ulbricht J. Grundlagen der Synthese von Polymeren. 2., überarb. Aufl. Basel: Hüthig & Wepf; 1992.

[4] Kricheldorf H. Polycondensation: history and new results. Berlin: Springer; 2014.

[5] Carothers WH. Polymerization. Chem Rev 1931;8:353–426. doi:10.1021/cr60031a001.

[6] Carothers WH. Polymers and polyfunctionality. Transactions of the Faraday Society 1936;32:39. doi:10.1039/tf9363200039.

[7] Stevens MP. Polymer chemistry: an introduction. 3rd ed. New York: Oxford University Press; 1999.

[8] Kickelbick G. Hybrid materials synthesis, characterization, and applications. Weinheim: WILEY-VCH; 2007.

[9] Ozin GA, Arsenault AC, Cademartiri L. Nanochemistry: a chemical approach to nanomaterials. 2. ed. Cambridge: Royal Society of Chemistry; 2009.

[10] Gerasin VA, Antipov EM, Karbushev VV, Kulichikhin VG, Karpacheva GP, Talroze RV, et al. New approaches to the development of hybrid nanocomposites: from structural materials to high-tech applications. Russ Chem Res 2013;82:303–32. doi:10.1070/RC2013v082n04ABEH004322.
[11] Zou H, Wu S, Shen J. Polymer/Silica Nanocomposites: Preparation, Characterization, Properties, and Applications. Chem Rev 2008;108:3893–957. doi:10.1021/cr068035q.
[12] Lofgreen JE, Ozin GA. Controlling morphology and porosity to improve performance of molecularly imprinted sol–gel silica. Chem Soc Rev 2014;43:911–33. doi:10.1039/C3CS60276A.
[13] Sanchez C, Julián B, Belleville P, Popall M. Applications of hybrid organic–inorganic nanocomposites. J Mater Chem 2005;15:3559. doi:10.1039/b509097k.
[14] Gómez-Romero P, Sanchez C, editors. Functional hybrid materials. Weinheim: Wiley-VCH; 2004.
[15] Judeinstein P, Sanchez C. Hybrid organic–inorganic materials: a land of multidisciplinarity. J Mater Chem 1996;6:511–25. doi:10.1039/JM9960600511.
[16] Drisko GL, Sanchez C. Hybridization in Materials Science – Evolution, Current State, and Future Aspirations. Eur J Inorg Chem 2012; 5097–105. doi:10.1002/ejic.201201216.
[17] Mackenzie JD. Structures and properties of Ormosils: Code: B1. Journal of Sol-Gel Science and Technology 1994;2:81–6. doi:10.1007/BF00486217.
[18] Sawada H. Thermodynamics of polymerization. vol. 23. New York: Marcel Dekker, Inc.; 1976.
[19] Ivin KJ. Zur Thermodynamik von Additionspolymerisationsprozessen. Angew Chem 1973;85:533–41. doi:10.1002/ange.19730851203.
[20] Kricheldorf HR, Rabenstein M, Langanke D, Schwarz G, Schmidt M, Maskos M, et al. Ring-Closing Polycondensations. High Performance Polymers 2001;13:S123–36. doi:10.1088/0954-0083/13/2/312.
[21] Brinker CJ, Scherer GW. Sol-gel science: the physics and chemistry of sol-gel processing. Boston: Academic Press; 1990.
[22] Schmidt HK. Das Sol-Gel-Verfahren: Anorganische Synthesemethoden. Chemie in unserer Zeit 2001;35:176–84. doi:10.1002/1521-3781(200106)35:3<176::AID-CIUZ176>3.0.CO;2-B.
[23] Schubert U, Hüsing N. Synthesis of inorganic materials. 2., revised and updated ed. Weinheim: Wiley VCH; 2005.
[24] Mishra MK, Kobayashi S, editors. Star and hyperbranched polymers. New York: Marcel Dekker; 1999.
[25] Müller AHE, Matyjaszewski K, editors. Controlled and living polymerizations: methods and materials. Weinheim: Wiley-VCH; 2009.
[26] Szwarc M, Van Beylen M. Ionic Polymerization and Living Polymers. Springer Verlag; 2013.
[27] Dubois P, Coulembier O, Raquez J-M, editors. Handbook of ring-opening polymerization. Weinheim: Wiley-VCH Verlag GmbH & Co. KGaA; 2009.
[28] Bielawski CW, Grubbs RH. Living ring-opening metathesis polymerization. Prog Polym Sci 2007;32:1–29. doi:10.1016/j.progpolymsci.2006.08.006.
[29] Parker KA, Sampson NS. Precision Synthesis of Alternating Copolymers via Ring-Opening Polymerization of 1-Substituted Cyclobutenes. Acc Chem Res 2016;49:408–17. doi:10.1021/acs.accounts.5b00490.
[30] Baughman TW, Wagener KB. Recent Advances in ADMET Polymerization. In: Buchmeiser MR, editor. Metathesis Polymerization, Springer Berlin Heidelberg; 2005, p. 1–42. doi:10.1007/b101318.
[31] Decker C, Nguyen Thi Viet T, Decker D, Weber-Koehl E. UV-radiation curing of acrylate/epoxide systems. Polymer 2001;42:5531–41. doi:10.1016/S0032-3861(01)00065-9.
[32] Oxman JD, Jacobs DW, Trom MC, Sipani V, Ficek B, Scranton AB. Evaluation of initiator systems for controlled and sequentially curable free-radical/cationic hybrid photopolymerizations. J Polym Sci Part A: Polym Chem 2005;43:1747–56. doi:10.1002/pola.20641.

[33] Shin H, Nguyen C, Kim B, Han M, Kim JS, Kim J. Synthesis and characteristics of acrylol borate as new acrylic gelator for lithium secondary battery. Macromol Res 2008;16:134–8. doi:10.1007/BF03218842.

[34] Pinna N, Niederberger M. Surfactant-Free Nonaqueous Synthesis of Metal Oxide Nanostructures. Angew Chem Int Ed 2008;47:5292–304. doi:10.1002/anie.200704541.

[35] Müller H, Rehak P, Jäger C, Hartmann J, Meyer N, Spange S. A Concept for the Fabrication of Penetrating Carbon/Silica Hybrid Materials. Adv Mater 2000;12:1671–5. doi:10.1002/1521-4095(200011)12:22<1671::AID-ADMA1671>3.0.CO;2-M.

[36] Spange S, Müller H, Pleul D, Simon F. Structure formation of poly(furfuryl alcohol)/silica hybrids, vol. 132, 2001. doi:10.1016/S0167-2991(01)82092-9.

[37] Mane S, Ponrathnam S, Chavan N. Effect of Chemical Crosslinking on Properties of Polymer Microbeads: A Review. Can Chem Trans 2016:473–85. doi:10.13179/canchemtrans.2015.03.04.0245.

[38] Stansbury JW. Dimethacrylate network formation and polymer property evolution as determined by the selection of monomers and curing conditions. Dent Mater 2012;28:13–22. doi:10.1016/j.dental.2011.09.005.

[39] Chagneux N, Camerlynck S, Hamilton E, Vilela FML, Sherrington DC. Synthesis of Laterally Linked Poly(tetrahydrofuran)-Poly(methyl methacrylate) Block Copolymers via Use of a "Jekyll and Hyde" Comonomer. Macromolecules 2007;40:3183–9. doi:10.1021/ma062720h.

[40] Chemtob A, Belon C, Croutxé-Barghorn C, Rigolet S, Vidal L, Brendlé J, et al. Tandem cationic and sol-gel photopolymerizations of a vinyl ether alkoxysilane. Polymer Eng Sci 2011;51:1466–75. doi:10.1002/pen.21951.

[41] Ni L, Moreau N, Chemtob A, Croutxé-Barghorn C. Organic-inorganic tandem route to polymer nanocomposites: kinetic products versus thermodynamic products. J Solgel Sci Technol 2012;64:500–9. doi:10.1007/s10971-012-2882-8.

[42] Chan C-K, Peng S-L, Chu I-M, Ni S-C. Effects of heat treatment on the properties of poly(methyl methacrylate)/silica hybrid materials prepared by sol-gel process. Polymer 2001;42:4189–96. doi:10.1016/S0032-3861(00)00817-X.

[43] Moszner N, Klapdohr S. Nanotechnology for dental composites. Int J Nanotechnol 2004;1:130. doi:10.1504/IJNT.2004.003723.

[44] Ma J-Z, Hu J, Zhang Z-J. Polyacrylate/silica nanocomposite materials prepared by sol-gel process. Eur Polym J 2007;43:4169–77. doi:10.1016/j.eurpolymj.2007.06.051.

[45] Rodríguez Hernández JC, Salmerón Sánchez M, Gómez Ribelles JL, Monleón Pradas M. Polymer-silica nanocomposites prepared by sol-gel technique: Nanoindentation and tapping mode AFM studies. Eur Polym J 2007;43:2775–83. doi:10.1016/j.eurpolymj.2007.04.021.

[46] Connell LS, Romer F, Suárez M, Valliant EM, Zhang Z, Lee PD, et al. Chemical characterisation and fabrication of chitosan-silica hybrid scaffolds with 3-glycidoxypropyl trimethoxysilane. J Mater Chem B 2014;2:668–80. doi:10.1039/C3TB21507E.

[47] Innocenzi P, Brusatin G, Guglielmi M, Bertani R. New Synthetic Route to (3-Glycidoxypropyl) trimethoxysilane-Based Hybrid Organic-Inorganic Materials. Chem Mater 1999;11:1672–9. doi:10.1021/cm980734z.

[48] Buschow KHJ, editor. Encyclopedia of materials: science and technology. Amsterdam ; New York: Elsevier; 2001.

[49] Wan Y, Zhao. On the Controllable Soft-Templating Approach to Mesoporous Silicates. Chem Rev 2007;107:2821–60. doi:10.1021/cr068020s.

[50] Seifert A, Spange S. Aus eins mach zwei: die Zwillingspolymerisation. Nachrichten aus der Chemie 2013;61:1028–33. doi:10.1002/nadc.201390319.

[51] Ebert T, Seifert A, Spange S. Twin Polymerization – a New Principle for Hybrid Material Synthesis. Macromol Rapid Commun 2015;36:1623–39. doi:10.1002/marc.201500182.

[52] Auer AA, Richter A, Berezkin AV, Guseva DV, Spange S. Theoretical Study of Twin Polymerization – From Chemical Reactivity to Structure Formation. Macromol Theory Simul 2012;21:615–28. doi:10.1002/mats.201200036.

[53] Leonhardt C, Brumm S, Seifert A, Cox G, Lange A, Rüffer T, et al. Tin Oxide Nanoparticles and SnO_2/SiO_2 Hybrid Materials by Twin Polymerization Using Tin(IV) Alkoxides. ChemPlusChem 2013;78:1400–12. doi:10.1002/cplu.201200242.

[54] Spange S, Kempe P, Seifert A, Auer AA, Ecorchard P, Lang H, et al. Nanocomposites with Structure Domains of 0.5 to 3 nm by Polymerization of Silicon Spiro Compounds. Angew Chem Int Ed 2009;48:8254–8. doi:10.1002/anie.200901113.

[55] Schliebe C, Gemming T, Mertens L, Mehring M, Lang H. Twin Polymerization: A Unique and Efficient Tool for Supporting Silver Nanoparticles on Highly Porous Carbon and Silica. Eur J Inorg Chem 2014; 3161–3. doi:10.1002/ejic.201402299.

[56] Tchernook I, Prehl J, Friedrich J. Quantum chemical investigation of the counter anion in the acid catalyzed initiation of 2,2′-spirobi[4*H*-1,3,2-benzodioxasiline] polymerization. Polymer 2015;60:241–51. doi:10.1016/j.polymer.2015.01.042.

[57] Göring M, Seifert A, Schreiter K, Müller P, Spange S. A non-aqueous procedure to synthesize amino group bearing nanostructured organic–inorganic hybrid materials. Chem Commun 2014;50:9753–6. doi:10.1039/C4CC03640A.

[58] Kitschke P, Auer AA, Löschner T, Seifert A, Spange S, Rüffer T, et al. Microporous Carbon and Mesoporous Silica by Use of Twin Polymerization: An Integrated Experimental and Theoretical Approach to Precursor Reactivity. ChemPlusChem 2014;79:1009–23. doi:10.1002/cplu.201402029.

Andreas Seifert, Michael Mehring and Stefan Spange

2 Twin Polymerization (TP) – General Aspects

2.1 Principles of twin polymerization

Twin polymerization is a novel concept for the synthesis of organic/inorganic hybrid materials following both new as well as established general aspects of polymer materials synthesis. The following chapter is focused on general principles and several questions will be addressed: What is twin polymerization? What are its special features? What differentiates twin polymerization from other well-established polymerization techniques? Which types of twin polymerization processes exist? Which kinds of materials are accessible by twin polymerization? Do the products of twin polymerization show special properties? Are there any potential applications?

By definition, twin polymerization is the formation of two polymers starting from one monomer in one mechanistically coupled process. The monomer that can undergo twin polymerization is called twin monomer. It is noteworthy that the alternative denotation "mother monomer" was skipped after critical discussion, but it nicely reflects that two polymers are "born" from the same type of monomer within the same time period during polymerization [1]. The twin polymerization might be regarded as the opposite of copolymerization, in which two types of monomers are incorporated into one copolymer. The basic principle is depicted in Figure 2.1.

There are three important points to be considered: Firstly, the twin monomer must have two different moieties that are able to undergo step-growth polymerization reactions resulting in two different polymers. Thus, the twin monomer is heterobifunctional. Secondly, these polymerizable moieties are covalently bonded in the twin monomer but separated after the twin polymerization process. Thus, covalent bonds have to be cleaved in the course of the polymerization reaction. Thirdly, both functional groups have to be polymerizable and their polymerization must be started by the same external stimulus, e.g., thermally or with a cationic or anionic starter. Thus, twin polymerization is classified as polymerization of heterobifunctional monomers of special molecular structure, which is different from other well-established heterobifunctional monomers as mentioned in the previous chapter.

Usually, the twin monomer consists of an inorganic (A in Fig. 2.1) and an organic (C in Fig. 2.1) part resulting in the corresponding polymers. All examples of twin monomers in the book belong to this type. Twin polymerization is a suitable tool for the preparation of organic/inorganic hybrid materials and is of special interest as a tool to hinder phase separation of the two polymers. In established procedures, two polymers do not spontaneously mix with each other. They tend to phase separate for thermodynamical reasons resulting in a minimized boundary surface (interphase).

https://doi.org/10.1515/9783110499360-002

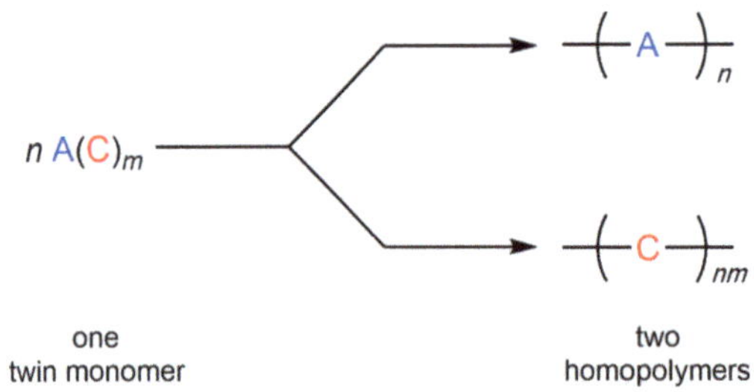

Fig. 2.1: Basic principle of twin polymerization [2]. Copyright Wiley-VCH Verlag GmbH & Co. KGaA. Reproduced with permission.

To overcome this problem, the tendency for diffusion of the two polymer strands has to be reduced and their close proximity must somehow be retained during the polymerization process to preserve the structure in the final material. There are different approaches to tackle this problem, but it has to be kept in mind that the two polymers must be in solid state after the polymerization. To illustrate this by some examples, the twin monomers synthesized from furfuryl alcohol (FA) and tetraethoxysilane, triethoxymethylsilane, diethoxydimethylsilane, or ethoxytrimethylsilane are discussed. These twin monomers serve as illustrative examples to demonstrate the principle of twin polymerization. Tetrafurfuryloxysilane (TFOS), trifurfuryloxymethylsilane, and difurfuryloxydimethylsilane (DFOS) undergo twin polymerization under acidic conditions with formation of polyfurfuryl alcohol (PFA) and silica or polymono- or polydimethylsiloxane (PDMS) (Scheme 2.1) [3]. Furfuryloxytrimethylsilane does polymerize to PFA and hexamethyldisiloxane (HMDS), but the process is not a twin polymerization because HMDS is a molecular siloxane, not a polymer.

The PFA resulting from an acid-catalyzed process is a dark colored, infusible solid showing high hardness, flexural strength, and chemical resistance to most solvents, acids, and alkaline solutions. The green/brown color of these materials is attributed to conjugated sequences that are formed by isomerization reactions during cationic polymerization and crosslinking of the furfuryloxy moiety [4]. Crosslinked PFA resins are available when strong acids are used as catalysts. PFA finds practical application in brake pads or as binder in sand casting [5, 6].

By contrast, polydimethylsiloxane (PDMS, silicon oil) is a colorless, temperature-stable liquid and it is widely used as hydraulic fluid, fluid for heat exchange, or as lubricant. In twin polymerization of difurfuryloxydimethylsilane, both polymers, PFA and PDMS, are formed within one process. The liquid PDMS easily flows out from the solid crosslinked PFA resin. After washing the solid, a pure PFA resin material results. The electron microscopy images show a porous structure of the PFA resin at the micrometer scale (Fig. 2.2) as the outflowing PDMS leads to the observed structuring.

The chemical structure of the twin monomer can easily be modified by substituting two methyl groups of DFOS with two additional furfuryloxy groups. As DFOS,

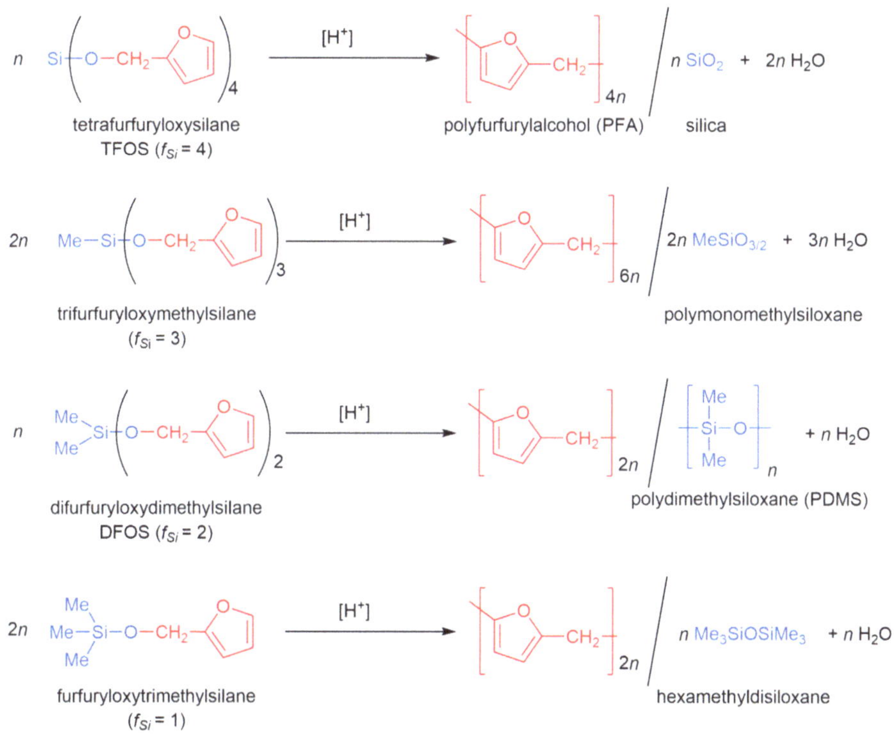

Scheme 2.1: Twin polymerization of furfuryloxysilane monomers with varying functionality at the silicon atom (f_{Si}), which alters the structure of the corresponding poly(siloxane) ($f_{Si}=4$, 3: networks; $f_{Si}=2$: chains, cycles; $f_{Si}=1$: dimer) [2]. Copyright Wiley-VCH Verlag GmbH & Co. KGaA. Reproduced with permission.

Fig. 2.2: The scanning electron microscopy (SEM) image of the washed solid material obtained from twin polymerization of difurfuryloxydimethylsilane which forms solid polyfurfuryl alcohol and liquid polydimethylsiloxane [3]. Copyright John Wiley & Sons, Inc. Reproduced with permission.

the resulting tetrafurfuryloxysilane (TFOS) is a twin monomer suitable for polymerization under acidic condition (Scheme 2.1) [2]. The difference here is the formation of silica (SiO_2) instead of PDMS. As the properties of PDMS and silica are different, the functionalities of the corresponding monomers are different. The silicon atom in DFOS has a functionality (*f*) of two, allowing it to form two new covalent Si–O–Si bonds with silicon atoms from other DFOS monomers during polymerization. Thus, only linear chains or cycles of different sizes are formed, and the buildup of networks is not possible. In this case, polymers with high molecular weight and a large degree of polymerization are obtained if the conversion of functional groups is nearly quantitative, because it is a step-growth polymerization. This phenomenon is described by the *Carothers* equation (see Chapter 1). The functionality of silicon in TFOS is increased to four. From a theoretical point of view, an average conversion of 50% of all functional groups is now sufficient to interconnect all monomers to one single macromolecule, and there are still lots of functional groups available for further crosslinking reactions. Twin polymerization of TFOS results in two polymers, PFA and silica, and both of these reach high molecular weights at moderate conversion of functional groups due to the functionality of both fragments being larger than 2. Thus, the viscosity increases rapidly in the course of the reaction, the mobility and diffusion of the polymer strains is reduced, and phase separation is minimized. As a result, both polymers are placed in close contact in an organic/inorganic hybrid material.

2.2 Comparison of twin polymerization with simultaneous polymerization of two monomers

Is twin polymerization really necessary to obtain two polymers in one step? Two polymers can in principle also be generated by simply mixing two different monomers and polymerize them in a simultaneous way. In the case of vinyl monomers, which bear a polymerizable C=C double bond, this strategy almost always fails and leads to the formation of copolymers rather than of polymer blends. Radical polymerization is often the method of choice for industrial production of polymers based on such vinyl monomers because it is applicable for lots of monomers, and is easy to handle in terms of the required reaction conditions (moderate temperatures, water-free conditions are not required). The growing polymer chain has an active radical that can add further monomer molecules until termination by recombination of two radicals or disproportionation occurs. There are different scenarios possible for a mixture of two different monomers under radical polymerization that depend mainly on the reactivity of each monomer, or to be more concrete, on the stabilization of the radical, the polarity, and the steric accessibility of the vinyl group [7]. It is more likely that both types of monomers react to give a copolymer rather than to form two pure homopolymers. If the reactivity of the monomers differs too much, only the polymerization of one type of monomer without the polymerization of the second one is observed. However, copolymerization is very important for many applications. For example, it is used to adjust the properties of the final materials such as pure poly(styrene), which is very brittle as homopolymer but gets shock-resistant by copolymerization with small amounts of butadiene. In general, simultaneous polymerization of a mixture of two different monomers is not particularly suitable for producing two polymers in one step based on the same polymerization mechanism and this is applicable for all chain polymerizations, regardless whether they are initiated radically, anionically, or cationically.

It is worth taking a closer look at the chemical structure of tetrafurfuryloxysilane (TFOS). It was the very first twin monomer discovered; in principle, it might be regarded as a combination of furfuryl alcohol and a tetraalkoxysilane, like tetraethoxysilane (or tetraethyl orthosilicate, TEOS), in one molecule [3]. TFOS can be synthesized by a transesterification reaction between TEOS and furfuryl alcohol (FA) under the release of ethanol (Scheme 2.2).

Scheme 2.2: Synthesis of the twin monomer tetrafurfuryloxysilane (TFOS) by transesterification of tetraethoxysilane (TEOS) with furfuryl alcohol under release of ethanol.

The polymerization of furfuryl alcohol is a step-growth polymerization with elimination of equimolar amounts of water per individual reaction step, and therefore called polycondensation reaction (Scheme 2.3a). Tetraalkoxysilanes are esters of silicic acid that get hydrolyzed by water to form alcohol and free silicic acid. The latter, $Si(OH)_4$, is only stable in very low concentrations and has a high tendency towards the formation of Si–O–Si bonds in a condensation reaction [8]. Finally, this results in amorphous SiO_2, silicon dioxide, *via* the so-called sol-gel process with water formed as by-product (Scheme 2.3b) [9]. The sol-gel process is widely used for several applications because it allows the formation of inorganic (metal) oxides under convenient conditions and reaction temperatures far below their melting points. Thus, the combination of inorganic chemistry of metal oxides with the "temperature sensitive" organic chemistry of organic or bioorganic molecules is possible and results in highly attractive hybrid materials [10]. For example, even enzymes were successfully incorporated in and mechanically stabilized by an inorganic SiO_2-matrix, and then the enzymes were still active in the final material. It is worth noting that enzymes do not survive at temperatures above 1000 °C, which are required for classical glass processing in molten state. The sol-gel process is catalyzed by bases, fluoride ions, and, similar to furfuryl alcohol polymerization by acids. Therefore, the polymerization of furfuryl alcohol and the sol-gel process of TEOS can be performed simultaneously on a mixture of both of the monomers by adding a protonic acid as a catalyst.

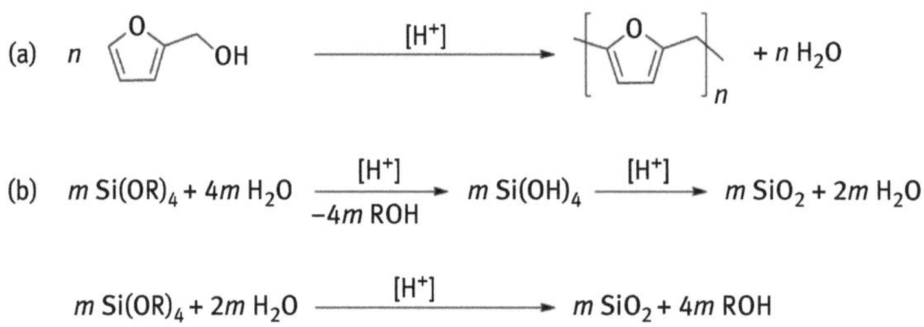

Scheme 2.3: (a) Polycondensation reaction of furfuryl alcohol delivers equimolar amounts of water. (b) Sol-gel process of tetraalkoxysilanes occurs *via* hydrolysis and condensation reactions. The net reaction consumes water.

Solid PFA/silica composites were readily obtained from equimolar mixtures of furfuryl alcohol and TEOS after addition of water and aqueous HCl [11]. TEOS is immiscible with water and the hydrolysis reaction is very slow when catalysts are absent. Fortunately, furfuryl alcohol mediates the compatibility of the system and the phase separation can be avoided. The miscibility of monomers and the phase separation processes can always be critical points when using the concept of "simultaneous polymerization of two monomers" for (nano)composite materials production. Generally, the two different rates of polymerization are independent of each other; but, they must be in the same order of magnitude to achieve the formation of interpenetrating networks. Phase separation is highly probable if, for example, one monomer is consumed in minutes and the polymerization of the other one lasts for days. In case of furfuryl alcohol/TEOS/water/acid from the literature, the assumption of two independent processes is valid in presence of excess amount of water. But, what happens if the simultaneous polymerization of furfuryl alcohol and TEOS is started in the absence of water? In this case, the condensation of furfuryl alcohol is necessary to produce water which then hydrolyses TEOS in the first step of the sol-gel process. The stoichiometric ratio of furfuryl alcohol and TEOS has to be at least 2:1 to achieve full conversion of TEOS (Scheme 2.3). Ideally, all water formed by furfuryl alcohol polymerization is consumed by TEOS. The ratio FA:Si (silicon content in TEOS) in the twin monomer TFOS is set intrinsically to 4:1 by its molecular structure (Scheme 2.2). Miscibility problems are completely eradicated by twin polymerization as only one twin monomer is necessary to produce two polymers and the polymerization mechanisms are mechanistically coupled. The silica formation can be considered as a condensation by-product of furfuryl alcohol polymerization, or, vice versa, the PFA is the by-product of the sol-gel process. As it is one process only, there is no need to carefully choose the experimental parameters in order to adjust two different reaction rates to end up with the desired interpenetrating networks. Thus, the formation of nanostructured hybrid materials by twin polymerization is a very robust process that works out under diverse experimental conditions.

A comprehensive study of materials obtained by twin polymerization in comparison to those obtained by simultaneous polymerization of two monomers was performed on titanium-based twin monomers or compounds, respectively [12]. The chemical structure of twin monomers can be altered in different ways. The functionality of the inorganic center atom can be varied as already shown in the example of furfuryloxysilanes (Scheme 2.1). It is also possible to modify the organic ligand sphere, or to use elements other than silicon as inorganic center atom. The design of twin monomers is discussed in detail in Chapter 3.1. The chemical structure of titanium-based twin monomers is much more complex than that of the silicon ones. The transesterification of tetraisopropyl orthotitanate, $Ti(O\textit{i}Pr)_4$, with furfuryl alcohol results in a tetranuclear titanium oxido cluster built up by four titanium atoms and 14 furfuryl alcoholate ligands [13]. Replacing furfuryl alcohol with thenyl alcohol (2-thiophenemethanol) results in the thenyl derivative with a similar structure [14]. The cationic twin polymerization of both titanium clusters produces hybrid materials of the corresponding organic polymer and TiO_2. The simultaneous polymerization of a combination of furfuryl alcohol/$Ti(O\textit{i}Pr)_4$ leads to the formation of hybrid materials very similar to those obtained by twin polymerization with regard to the composition obtained by elemental analysis and molecular structure as determined by solid state NMR spectroscopy. However, electron microscopy reveals that the TiO_2 phase domain sizes in the hybrid materials are very different. In the case of twin polymerization, these are 2–4 nm in size, whereas the alternative concept of simultaneous polymerization gives significantly coarser phase domains of 20–500 nm [15]. Hybrid materials composed of poly(2-thiophenemethanol) and TiO_2 are accessible by twin polymerization of the thenyl alcohol-based titanium cluster. In this example, the simultaneous polymerization of a mixture of thenyl alcohol and $Ti(O\textit{i}Pr)_4$ completely fails and hybrid materials are not obtained.

2.3 Polymerization of heterobifunctional monomers

Heterobifunctional monomers are generally monomers that bear two different polymerizable groups in one molecule. General aspects and conceptual background of that monomer class have been explained in Chapter 1. Here, we will discuss about different aspects of experimental criteria and practical implementation of this type of monomers with regard to simultaneous polymerization and twin polymerization. Miscibility problems, which may be an issue in simultaneous polymerization of two different monomers, are efficiently excluded. An intriguing example is discussed here: 3-(trimethoxysilyl)propyl methacrylate is a methacrylate monomer functionalized with a trialkoxysilane group. Both the functional groups can be polymerized independently of each other, but the covalent Si–C linkage is not cleaved during the polymerization procedures. Finally, the two polymers, poly(methacrylate) and poly(alkylsiloxane), form a copolymer (Scheme 2.4).

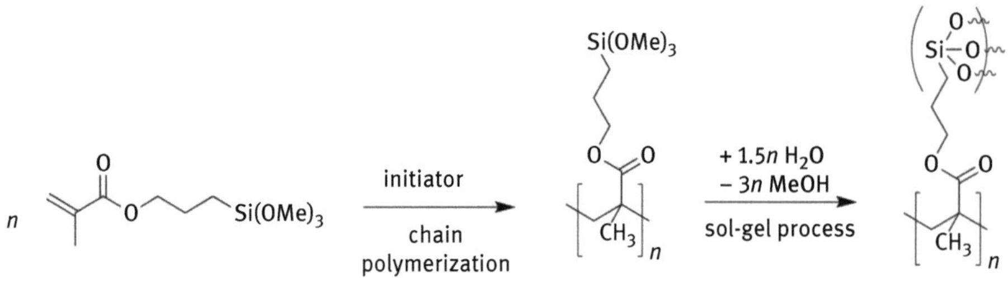

Scheme 2.4: The heterobifunctional monomer 3-(trimethoxysilyl)propyl methacrylate can undergo chain polymerization of the C=C double bond and sol-gel process of the trialkoxysilane group.

The molecular weight as well as the morphology of the methacrylate block can be precisely controlled by using anionic living polymerization or controlled radical polymerization techniques with the corresponding appropriate initiators [15, 16]. The subsequently catalyzed sol-gel process crosslinks the organic polymers to produce hybrid materials. The alkoxysilane groups of the heterobifunctional monomer can also be used to modify the surface of inorganic particles like silica that find application as fillers in dental composite materials [17]. The methacrylate groups of the modified particle surface are copolymerized with other olefinic monomers during the photo polymerization (hardening) of the material. The surface modification significantly improves the mechanical properties of the final dental composite materials because the filler and the organic matrix are connected by covalent bonds.

Novak and coworkers introduced a new class of heterobifunctional monomers based on tetraalkenyl orthosilicates in which both functional groups are detached in the course of the polymerization reaction, resulting in two homopolymers (Scheme 2.5) [18–20]. For this purpose, two polymerization processes are carried out simultaneously: sol-gel process and free radical polymerization or ROMP (ring-opening metathesis polymerization). The basic principle is as follows: Addition of water and fluoride ions as catalyst start the hydrolysis of the silicic acid ester to form silanol groups and a free polymerizable alcohol derivative. The silicic acid condenses to an inorganic network (sol-gel process) and simultaneously the alcohol derivative is polymerized. Shrinking, which is a major concern of the traditional sol-gel process caused by removing the alcohol, is drastically minimized. Both reaction rates have to be tuned to form both networks on the same time scale because the functionalities of both moieties are also a function of conversion with water. If the process is

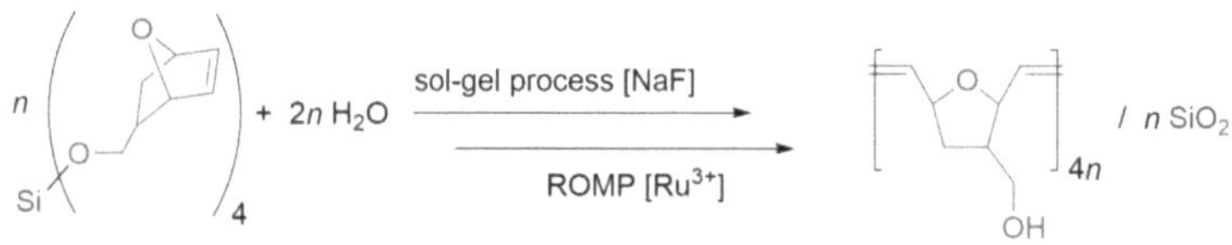

Scheme 2.5: The polymerization of Novak's heterobifunctional composite monomers consists of two independent processes [19].

successful, then transparent and non-shrinking organic/inorganic composites are obtained [18–20].

The main advantage of twin polymerization over the polymerization of heterobifunctional monomers of Novak's type is the mechanistically coupled polymerization process: There is no need for fine-tuning of different catalyst concentrations. Addition of a single type of starting agent is sufficient to end up with two polymers, making the twin polymerization process very robust against a wide range of experimental conditions. As already mentioned before, twin monomers are heterobifunctional monomers in general, but with special structural features. The key point is that both of the monomer fragments have to be polymerizable by the same initiator or catalyst. The characteristics of different polymerization methods are compared with twin polymerization in the following Table 2.1.

2.4 Inorganic network formation by twin polymerization, aqueous and non-aqueous sol-gel process

A major aspect of twin polymerization is the formation of inorganic networks, such as SiO_2, starting from a twin monomer. In previous subchapters, we already discussed that similar products are also accessible by the aqueous sol-gel process, which can be used in combination with the polymerization of an organic monomer to produce organic/inorganic composite materials. Novak's heterobifunctional monomers are in principle also tetraalkoxysilanes following the route of aqueous sol-gel process [18–20]. The source of oxygen needed to build up the inorganic oxide network (M–O–M) is delivered by additional water starting hydrolysis of the precursor molecule (see Scheme 2.6a) [9]. Water is therefore essential for the aqueous sol-gel process. In case of twin polymerization, the oxygen needed to build up the inorganic polymer is already present in the twin monomer. Twin polymerization can therefore be processed in organic solvents, which might be immiscible with water. On the contrary, water can disturb twin polymerization because it can hydrolyze the monomer, inducing a "normal" aqueous sol-gel process of the inorganic part of the monomer. Thus, twin polymerization is very similar to the so-called non-aqueous sol-gel process, a sophisticated synthesis methodology for metal oxides under water-free

Tab. 2.1: Comparison of production aspects of polymerization techniques to achieve the formation of two polymers within one procedure.

criterion	twin polymerization	simultaneous polymerization of two monomers	polymerization of heterobifunctional monomers
product	two polymers	two polymers	two polymers or copolymer
number of monomers	one twin monomer	two monomers	one monomer
number of polymerization processes	one mechanistically coupled process	two processes	two processes
number of catalysts	one initiator/catalyst	one or two catalysts or initiators	two catalysts or initiators
molar ratio of the two polymers formed	molar ratio of the two polymers formed is set in the molecular structure of the twin monomer	variable by setting the ratio of the two monomers	molar ratio of the two polymers formed is set in the molecular structure of the monomer
synthetic effort	synthesis of twin monomer is necessary	synthesis of special monomer is not necessary	synthesis of special monomer is necessary
miscibility problems	no, only one monomer	yes, miscibility of the two monomers must be considered	no, only one monomer
possibility of formation of hybrid materials	yes, very robust process	yes, but optimization of the reaction rates is necessary	yes, but optimization of the reaction rates is necessary

conditions [21–24]. The reaction of $TiCl_4$ with benzyl alcohol, which is also used as solvent, gives nanoparticles of TiO_2 in anatase modification and benzyl chloride [25]. The oxygen atoms for the metal–oxygen–metal bond origins from the solvent (see Scheme 2.6b).

Other metal oxide precursors besides metal halides are metal alkoxides, acetylacetonates, or acetates [24]. The most common routes of metal–oxygen–metal bond formation are alkyl halide, ether, and ester elimination. The condensation of two metal alkoxides under ether elimination is an example being very similar to twin polymerization: Under these circumstances, the oxygen needed to build up the metal oxide originates directly from the precursor, an aspect which is identical with twin polymerization [25]. In this context, twin polymerization might be classified as a non-aqueous sol-gel process. The unique feature of twin polymerization among other non-aqueous sol-gel processes is the formation of an organic polymer. This is achieved by using

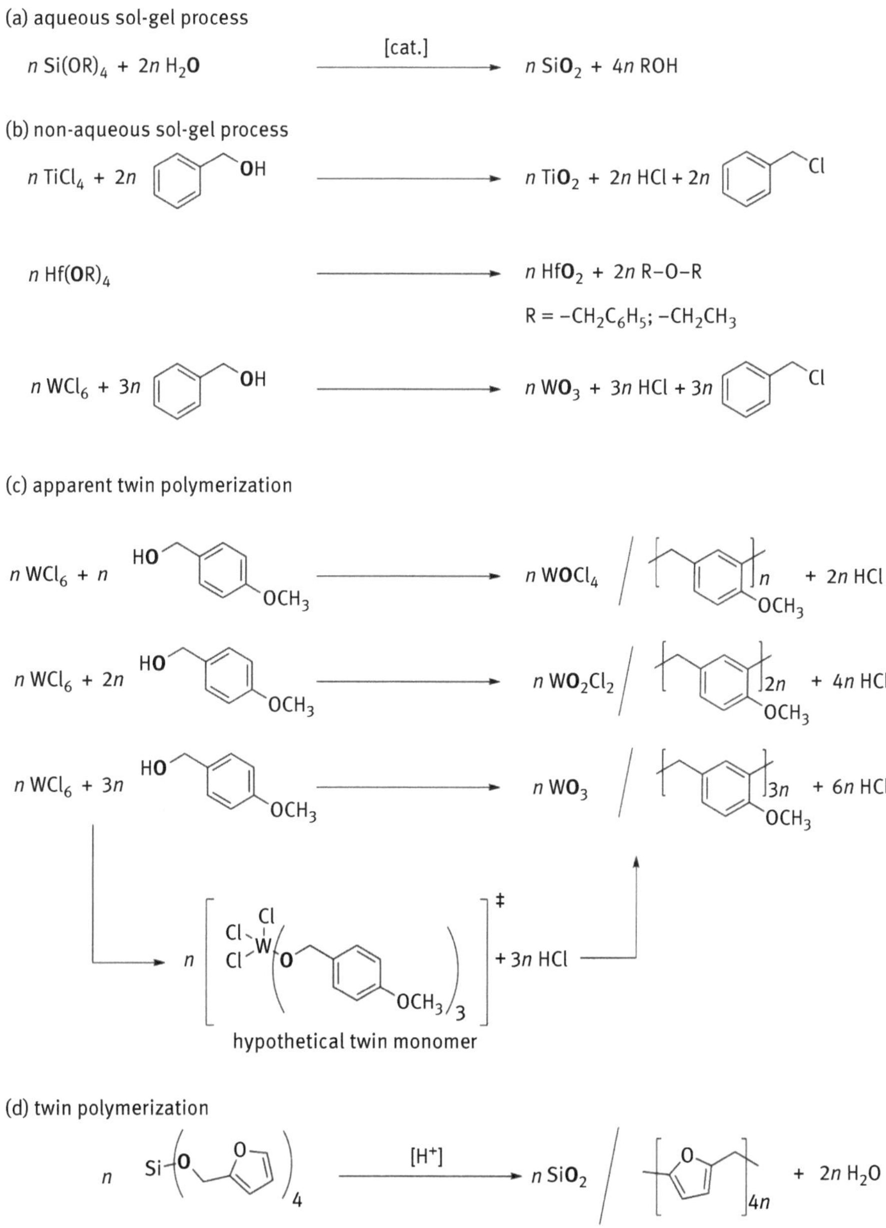

Scheme 2.6: Comparison of (a) aqueous and (b) non-aqueous sol-gel processes, and (c) apparent twin polymerization and (d) condensative twin polymerization.

polymerizable alcohols for the preparation of twin monomers. Thus, twin polymerization aims for organic/inorganic hybrid materials with interpenetrating networks, rather than for well-defined and shaped metal oxides. It must be pointed out that several examples of organic/inorganic hybrid nanostructures obtained by non-aqueous sol-gel processes are already mentioned in the literature [26–33]. In nearly all cases, benzyl alcohol or its derivatives were used as solvents, which can be easily oxidized to the aldehyde or acid in the course of the reaction with a metal oxide present. The aldehyde or acid forms complexes with the metal oxide, limits their growth, and prevents the inorganic particles from sintering, resulting in defined, mainly lamellar structures, after precipitation from solution. These materials differ significantly from the products obtained by twin polymerization because there is no simultaneous formation of two polymers, but rather an *in situ* formation of surface active compounds.

The crossover from the "classical" non-aqueous sol-gel process to twin polymerization happens smoothly. An elucidatory example is tungsten oxide as target compound. Bulk tungsten oxide exhibits manifold application potential, including catalytic activity, electrochromism, semiconductivity, and sensing properties. Using WCl_6 as precursor and benzyl alcohol as solvent results in tungsten oxide nanoparticles, having a platelet-like shape with sizes ranging from 30 to 100 nm [33]. It was found that each of these nanoparticles behaves like a single crystal, but they are mesocrystals composed of smaller crystallites, which are just a few nanometers in size. The particle morphology can be drastically influenced by adding a siderophore, namely deferoxamine mesylate, to the reaction mixture of WCl_6 and benzyl alcohol. Bundles of tungsten oxide nanowires were obtained, which are held together by benzaldehyde. They represent an organic/inorganic hybrid nanostructure [33]. But what happens, if a more electron-rich derivative of benzyl alcohol is used as solvent for the non-aqueous sol-gel process of WCl_6? An additional methoxy group in *ortho-* or *para*-position to the methylol-group stabilizes benzylic cations, which are formed in the course of the reaction. The benzylic cations then start a Friedel-Crafts polymerization, resulting in poly(methoxybenzyl alcohol) [34]. Finally, organic/inorganic hybrid materials of tungsten oxide and poly(methoxybenzyl alcohol) are obtained. It was found out that the process also works for other electron-rich aromatic compounds such as thenyl alcohol and furfuryl acetate in combination with WCl_6. Electron microscopic investigations with EDX analysis revealed a homogeneous distribution of tungsten and the other elements within the material. Thus, the phase domain sizes must be very small. The procedure to produce hybrid materials by combining WCl_6 with polymerizable alcohols differs from the classical non-aqueous sol-process of WCl_6 by the benzyl alcohol route. When these results were presented in 2009, the process was called "*in situ* twin polymerization," because the products are similar to those of twin polymerization with characteristic formation of two polymers [34]. The second aspect was that some experimental results pointed towards the formation of some kind of twin monomer as intermediate, a hypothetical twin monomer, in the course of the reaction of the metal halide with the alcohol. All attempts to isolate a defined twin monomer failed. In all cases, the polymerization of the aromatic alcohol could not be prevented. However, WCl_6 does not simply act as initiator for the Friedel-Crafts polymerization: Even if the amount of *para*-methoxybenzyl alcohol to a given amount of the tungsten halide was drastically increased from 1:1 to 15:1, the yield of hybrid material that precipitates from dichloromethane solution was nearly constant for all experiments. More importantly, the carbon content of the products was about 22 wt% in all cases. This correlates well with an equimolar stoichiometry of poly(*para*-methoxybenzyl alcohol) and tungsten oxychloride $[WOCl_4]_n$ in the material. The *in-situ* formation of species similar to twin monomers is assumed. These reactive intermediates are classified as hypothetical twin monomers and the process is called apparent twin polymerization [35]. The apparent twin polymerization combines elements of the non-aqueous sol-gel process and "real" twin polymerization with twin monomers. The apparent twin polymerization processes can be also used for other synthetic protocols due to the structural diversity of possible reactants.

2.5 Types of twin polymerization processes

There are two ways to classify various types of twin polymerizations: classification with respect to the stoichiometry of the overall process or with respect to the monomer structure. According to the stoichiometry of the gross process, the following basic twin polymerization processes are treated in this section: condensative twin polymerization, ideal twin polymerization, simultaneous twin polymerization, and apparent twin polymerization. Reference to following chapters of this book is made based on the application of the particular special form of twin polymerization for particular classes of materials.

2.5.1 Condensative and ideal twin polymerization

All of the twin monomers discussed so far in this chapter have one feature in common: In every case, low molecular weight by-products like water or HCl are formed during the polymerization process. Therefore, this process is called "condensative twin polymerization" [35]. The organic moiety of the corresponding twin monomers undergoes polycondensation reactions.

The silicon compound 2,2'-spirobi[4*H*-1,3,2-benzodioxasiline] (Si-Spiro) is an example of a twin monomer which polymerizes to give two polymers without the formation of any by-products. The reason for this feature is a ring-opening polymerization of the organic moiety. Si-Spiro is synthesized from two molecules of salicylic alcohol (2-hydroxybenzyl alcohol) and a silicon precursor like tetramethoxysilane or $SiCl_4$. The salicylic alcohol forms a 6-membered ring including the silicon atom and resulting in a spiro compound. The “ideal twin polymerization” of Si-Spiro to phenolic resin and silica starts at temperatures above 200 °C or it is catalyzed by acids or bases (Scheme 2.7) [36–38].

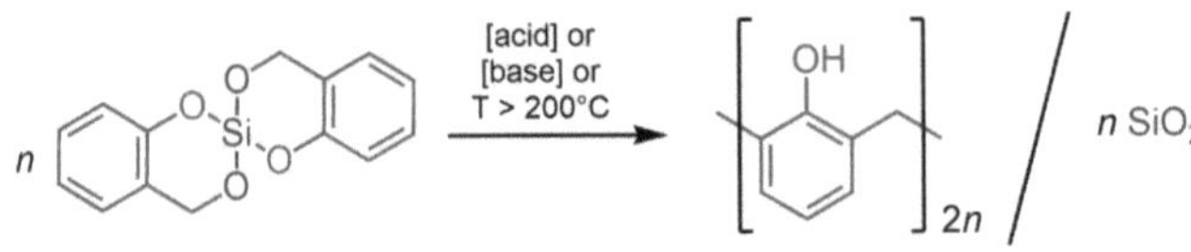

Scheme 2.7: Ideal twin polymerization of 2,2’-spirobi[4*H*-1,3,2-benzodioxasiline] (Si-Spiro) forms hybrid materials of phenolic resin and silica. Please notice, various linkages *(o,o’-, o,m-, o,p-)* are possible.

This ideal twin polymerization process, with regard to the molecular structure of the phenolic resin, is not always completely fulfilled. The formation of various molecular fragments within phenolic resin species depends on the reaction conditions and kind of catalysis process used, because minor molecular structural variations of the phenolic products can take place at high temperatures which are attributed to the established chemistry of the phenolic resin components [37]. Thus, in addition to the *ortho*-linkage, also the *para*-linkages and substitution reactions of alkyl groups are possible [37].

2.5.2 Simultaneous twin polymerization

Simultaneous twin polymerizations (STPs) are those processes which involve the use of two or more twin monomers within one procedure. Theoretically, *n* or *m* portions of the twin monomers A–C and B–D can produce four different polymer structures within one process. The schematic illustration of the product formations from A–C and B–D twin monomers is shown in Scheme 2.8a [39, 40]. At first sight, the overall process seems to be very complicated because of the many products formed. However, such a synthesis route has enormous potential for controlled hybrid materials synthesis. Consider that there are only two twin monomers which have the same polymerizable units to reduce the complexity. Thus, the STP of *n* A–C and *m* B–C can theoretically produce three polymers, or one homopolymer, and one copolymer (Scheme 2.8b). The particular aspect of this scenario is that a homopolymer is formed from the fragment C, although two different monomers are used. According to the IUPAC definition, a copolymer is a polymer that is derived from more than one species of monomer [41]. As a result, the STP of *n* A–C and *m* B–C to a homopolymer, $-(C)_{n+m}-$ seems to be a paradoxon with regard to established concepts of polymerizations. Therefore, the STP of twin monomers must be considered as a special process.

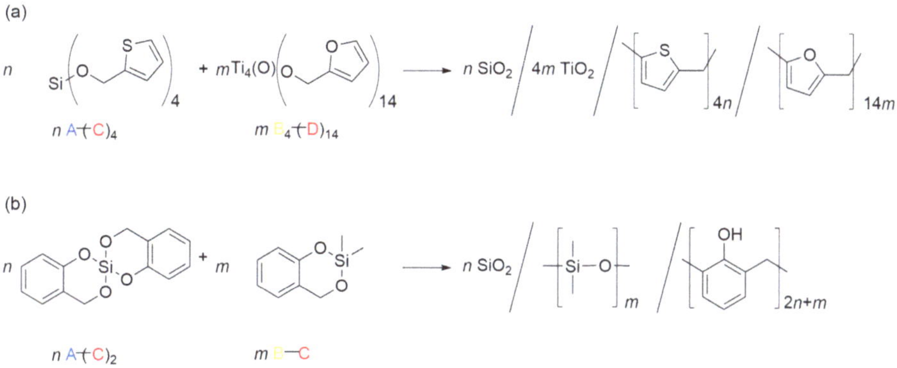

Scheme 2.8: General chemical reaction equations for STP of combinations A–C and B–D (a) as well as A–C and B–C twin monomers (b) within one process [40]. Copyright John Wiley & Sons, Inc. Reproduced with permission.

The formation of the copolymer $-(A_nB_m)-$ during the STP of n A–C and m B–C is usually the desired case for targeted hybrid materials synthesis. With this synthesis method, materials can be produced which are normally not easily accessible. An interesting situation occurs in the case of the formation of the AB copolymer. It is either thermodynamically or kinetically not preferred, but the formation of polymer C is largely favorable. Then, the formation of polymer C might mediate the formation of the AB copolymer. However, it is also possible that polymers A and B are encapsulated within the polymer matrix of C, but post reaction brings A and B together as suggested in Figure 2.3.

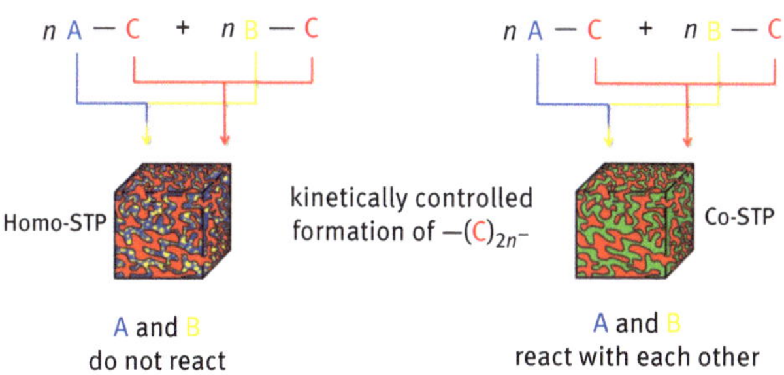

Fig. 2.3: Cartoon description of nanostructured hybrid material formation, where monomers A–C and B–C form polymer $-(C)_{2n}-$ and components A and B do not react (left) or they react with each other forming a copolymer (right) [35]. Copyright John Wiley & Sons, Inc. Reproduced with permission.

The STP is an important synthesis concept in addition to the simple twin polymerization as shown in some following chapters of this book. The STP is an excellent tool to manufacture ternary siliceous hybrid materials (Chapter 4.1), mixed inorganic oxides (Chapter 4.3), metal nanoparticles within polymer matrices (Chapter 4.4), and also to produce amino-functionalized and polyurethane-containing hybrid materials (Chapter 5.2).

2.5.3 Apparent twin polymerization

Apparent twin polymerization processes are characterized by the feature that the actual twin monomer species is generated *in situ* by reaction between two reactants. Therefore, the apparent twin polymerization is not a genuine twin polymerization, because a second component is required to fulfil the stoichiometry of the polymerization process of one monomer to two polymers. This is theoretically illustrated for the cationic polymerization of tetraphenoxysilane with 1,3,5-trioxane or hypothetically for the polymerization of 1,1',1'',1'''-silanetetrayltetrakis(azepan-2-one) with water in Scheme 2.9 and already mentioned for production of WO_3/polymer hybrid materials in Scheme 2.6.

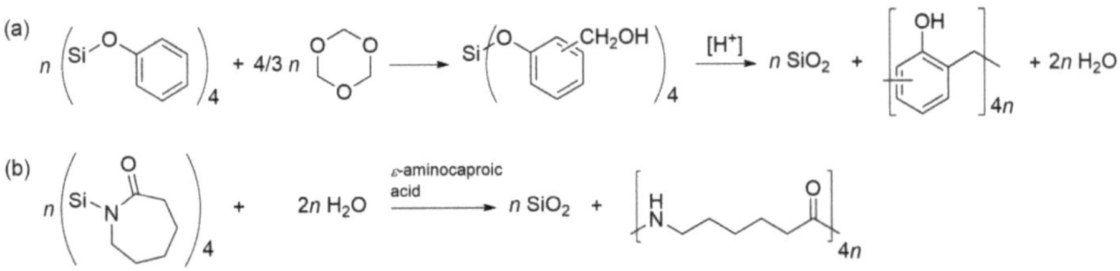

Scheme 2.9: (a) Schematic description and polymerization of tetraphenoxysilane with trioxane as an example of an apparent twin polymerization of a heterobifunctional compound which can react to an implicit twin monomer. (b) Hypothetical polymerization of 1,1',1'',1'''-silanetetrayltetrakis(azepan-2-one) with water in the presence of ε-aminocaproic acid to silica and polyamide 6.

Altogether, the processes can be classified as polymerization of a so-called deficient twin monomer [35]. However, there are distinctive differences between the individual examples.

Step-growth polymerization of tetraphenoxysilane with 1,3,5-trioxane proceeds *via* an implicit twin monomer. The implicit twin monomer is an intermediate which cannot be isolated due to its high reactivity, even reacting with itself. It is similar to the hypothetical monomers of Scheme 2.6. This process works well in practice and has been subject of patent literature for producing phenolic resin/metal oxide hybrid materials. In addition to silicon, other metals or semimetals can be used as part of phenoxy derivatives.

In contrast, the reaction of 1,1',1'',1'''-silanetetrayltetrakis(azepan-2-one) with water does not deliver a polyamide 6/silica hybrid material as given in Scheme 2.9. 1,1',1'',1'''-Silanetetrayltetrakis(azepan-2-one) spontaneously hydrolyzes to ε-caprolactam and $Si(OH)_4$, followed by condensation to SiO_2 and water, independent of the temperature or other reaction conditions. However, the combination of 1,1',1'',1'''-silanetetrayltetrakis(azepan-2-one) with ε-aminocaproic acid is suitable to produce polyamide 6/silica hybrid materials within one process. That process, demonstrating the combination of polyamide 6 production with silica formation, is presented in detail in Chapter 4.1.

2.6 Twin monomers – a modular molecular system

The composition of the twin monomer determines the molecular structure and the morphology of the resulting hybrid material. According to the ideal and condensative twin monomers Si-Spiro and TFOS, respectively, the inorganic element or the polymerizable organic fragment can be replaced by other appropriate bricks (Scheme 2.10). This concept works out for many combinations of metal and semimetal elements when combined with salicylic alcohol or benzylic compound, but not in every case. Twin monomers based on B, Ti, Ge, Sn, Hf, Zr, In, or Fe are presented in Chapters 4.3 and 4.4. They are useful for the production of inorganic oxides and mixed oxides as well as for the incorporation of noble metal fragments.

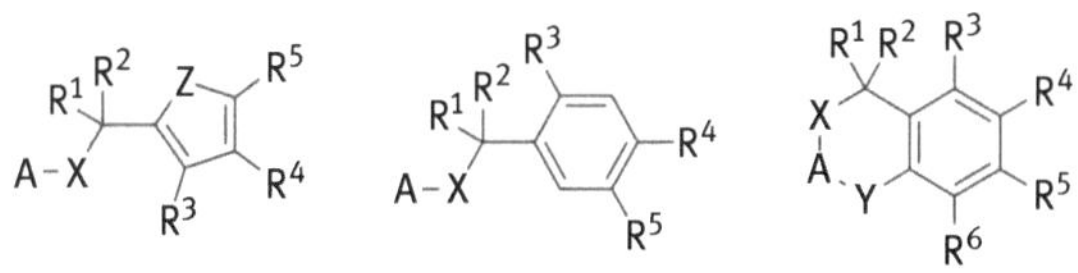

X, Y = NH, O, S
Z = O, S
A = central atom (e.g., Si, Sn, Ge, B, Ti)
R^1–R^6 = H, alkyl, functional groups

Scheme 2.10: Twin monomers with their general structural motifs of the organic part based on furfuryl-, benzyl-, and salicylic alcohol or their nitrogen and sulfur analogs [35]. Copyright John Wiley & Sons, Inc. Reproduced with permission.

In several cases, it is proved to be very difficult or impossible to produce structurally well-defined twin monomers. For example, aluminum trifurfurylate, which should be easily accessible from triethyl aluminum and furfuryl alcohol, is a pale yellow solid which is completely insoluble in organic solvents. A polymeric structure is assumed based on a solid state ^{27}Al NMR spectrum that indicates five- and six-fold coordinated aluminum atoms. Other types of well-defined aluminum twin monomers are still not available. The resulting solids are difficult to characterize and handle. We assume that they polymerize by acid catalysis to organic polymer/aluminum oxide hybrid materials based on preliminary results. Novel coordination compounds of germanium were also synthesized with a molecular structure related to twin monomers. In addition, the germanium sulfur analog of Si-Spiro was successfully prepared, but it does not undergo twin polymerization. The diversity of twin monomer structures and their reactivity in a twin polymerization process is discussed in detail in Chapter 3.1.

Another important aspect of twin monomer design is the incorporation of functional groups in the periphery of the molecule. Scheme 2.11 shows some representative examples of functionalized twin monomers. The functional group can be bonded either at the organic fragment or at the central metal or semimetal atom. Previously, only Si-containing functional twin monomers have been investigated because of the easy availability of suitable precursors for the synthesis. In comparison, incorporation of a functional group at the periphery of the organic fragment requires much more synthetic effort and is therefore not beneficial to produce hybrid materials in larger quantities.

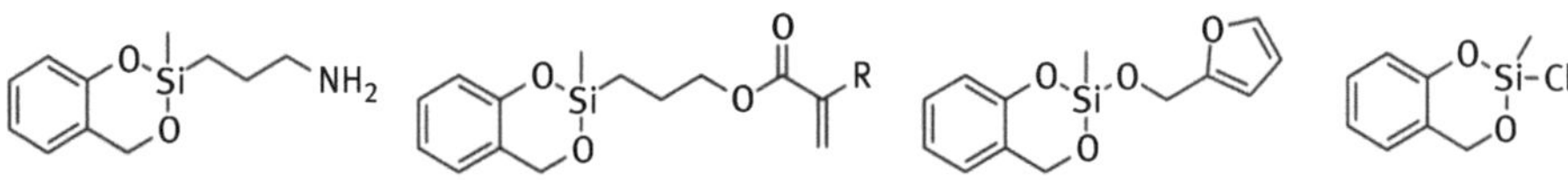

Scheme 2.11: Examples of functional twin monomers that are modified at the silicon atom.

However, the reactivity of the organic fragment can be precisely adjusted by appropriate substituents which is dealt with in detail in Chapter 3.1. It is discussed which kind of catalysis is the most appropriate one with regard to a specific twin monomer species. By a careful inspection of the substituent effects, important mechanistic insights into the course of the twin polymerization are identified and thus, a variety of substituted salicyl alcohol-, furan-, and thiophene derivatives have been prepared and used as reactants to fabricate twin monomers (see Chapter 3.1).

When the functional group is bound at the central atom, the synthesis of corresponding twin monomers is much less expensive if suitable precursor molecules are available as reactants. It is important to note that the silicon–carbon bond of functionalized twin monomers is robust under the conditions of twin polymerization. The incorporation of additional functionalities can serve for various purposes.

For example, reactive organic groups, such as primary amino groups, are covalently bonded and can react with a variety of organic electrophiles, such as epoxides, isocyanates, or carboxylic anhydrides. Those reactive groups are suitable to combine twin polymerization with established step-growth polymerization processes. That important feature is discussed in Chapter 5.2.

Acid or base groups can serve as internal catalyst for the twin polymerization. Thus, the use of additional catalysts which may disturb applications might be avoided. This aspect is demonstrated for the production of siliceous hybrid materials in Chapter 4.1.

The vinyl group can be integrated to copolymerize twin monomers with olefinic monomers. However, this methodology does only work for special monomer combinations and radical initiators because radical polymerization is sometimes inhibited by the twin monomers due to their phenolic character. Several aspects concerning this matter are discussed in Chapter 4.1 and Chapter 5.2.

2.7 Transformation of hybrid materials to porous metal oxides or carbon

The hybrid materials obtained by twin polymerization are composed of interpenetrating organic/inorganic networks that are closely packed on the nanometer scale. In general, these materials are not porous and the specific surface area measured by N_2 adsorption experiments is rather low (< 50 $m^2 \cdot g^{-1}$). The removal of one of the two phases, inorganic or organic, will drastically increase the surface area of the remaining phase if it is dimensionally stable (Fig. 2.4). Pure metal oxides, such as SiO_2, are accessible by simply burning the organic polymer of the hybrid material in air atmosphere. Alternatively, the metal oxide is removed by wet etching from the hybrid material. In the case of SiO_2, this works very well with HF or NaOH solution. In general, there is a carbonization step under inert atmosphere (e.g., argon) and temperatures above 500 °C before the etching. The structure of carbon is mostly much more stable and rigid than the former organic polymer (phenolic resin, polyfurfuryl alcohol). This helps to retain the morphology. Additionally, the final microporous carbon materials are much more attractive for applications which is discussed in more detail in Chapter 5.1.

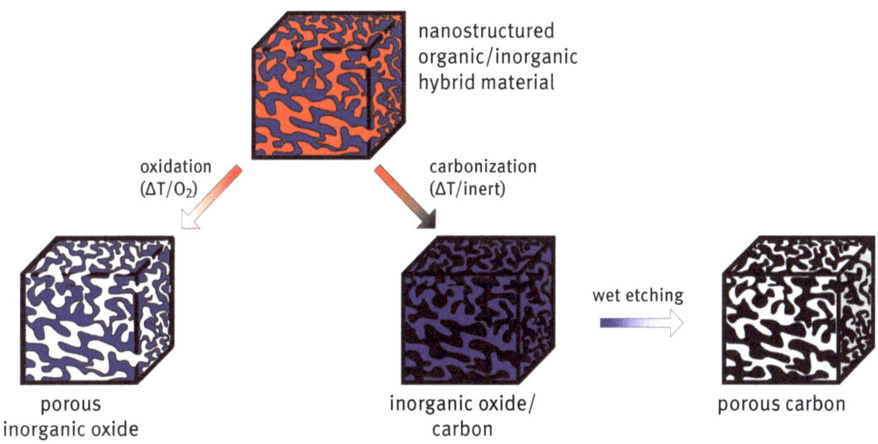

Fig. 2.4: Process steps to convert nanostructured hybrid materials obtained by twin polymerization into porous materials.

2.8 Twin polymerization in combination with additives

Every polymerization process including the simultaneous twin polymerization, STP, or other TP processes can be carried out in the presence of additives. The additive can either serve as a structurally acting template during the polymerization process and influence the morphology of the hybrid material or it can be added under the objective to remain in the hybrid material and enable a required (chemical) feature such as catalytic activity. The use of additives as hard or soft templates which can be removed from the final product by various techniques, is reported for production of both shape-determined silicate hybrid and carbon materials in Chapters 4.1 and 4.2.

The second class of additives does not alter the morphology of the hybrid materials but remains in the final material and adds an additional property. They are usually used in small quantities. A very straightforward approach for the syntheses of this kind of materials is simply by adding the additive to the twin monomer mixture or by melting before the polymerization process. A homogeneous distribution in the hybrid material is expected if the additive is well soluble. Due to the crosslinked nature of the polymers formed, the diffusion of the additive is hindered in most cases. For example, metal (oxide) nanoparticle decorated hybrid materials are accessible by the use of (noble) metal carboxylates as additives. Using a single source methodology, these metal-organic carboxylate complexes decompose directly to metal nanoparticles (e.g., silver or gold) by thermal treatment without the need of reducing agents. Of course, the modified organic/inorganic hybrid material can be transferred into porous carbon or metal oxides as described in Chapter 2.7, and can also be decorated with the nanoparticles. Furthermore, the concept is not limited to one precursor/additive only, it is also allowing the decoration with several metals opening up a wide range of possibilities, such as the production of nanostructured metal alloy particles enclosed in porous carbon (see Chapter 4.4). For instance, tin alloy nanoparticles can be produced as active components for lithium ion batteries, which is demonstrated in Chapter 5.1.

If the additive is not just soluble in the twin monomer mixture but contains a structurally similar organic group contained in the twin monomer, both organic fragments can copolymerize with each other. In certain cases, this can help to improve the homogeneity of the distribution of the additive within the hybrid material due to the formation of covalent bonds.

There are also many pure organic compounds suitable as smart additives. Their selection depends on the application of the resulting hybrid material. For instance, catechol as additive readily undergoes electrophilic substitution reaction during the twin polymerization of Si-Spiro and catechol moieties are incorporated into the organic polymer of the hybrid material. This can significantly improve the adhesion of the hybrid material to metal surfaces such as iron or aluminum. Some examples and aspects regarding the benefits of additives in twin polymerization processes are treated in Chapter 5.2.

References

[1] Spange S, Grund S. Nanostructured Organic-Inorganic Composite Materials by Twin Polymerization of Hybrid Monomers. Adv Mater 2009;21:2111–6. doi:10.1002/adma.200802797.

[2] Seifert A, Spange S. Aus eins mach zwei: die Zwillingspolymerisation. Nachrichten aus der Chemie 2013;61:1028–33. doi:10.1002/nadc.201390319.

[3] Grund S, Kempe P, Baumann G, Seifert A, Spange S. Nanocomposites Prepared by Twin Polymerization of a Single-Source Monomer. Angew Chem Int Ed 2007;46:628–32. doi:10.1002/anie.200504327.

[4] Choura M, Belgacem NM, Gandini A. Acid-Catalyzed Polycondensation of Furfuryl Alcohol: Mechanisms of Chromophore Formation and Cross-Linking. Macromolecules 1996;29:3839–50. doi:10.1021/ma951522f.

[5] Schmitt CR. Polyfurfuryl Alcohol Resins. Polym Plast Technol Eng 1974;3:121–58. doi:10.1080/03602557408545025.

[6] Furanharze, Polyfurfurylalkohol – Stoffliche Nutzung nachwachsender Rohstoffe – page 446 n.d. http://what-when-how.com/Tutorial/topic-270798bpere/Stoffliche-Nutzung-nachwachsender-Rohstoffe-466.html (accessed July 21, 2017).

[7] Odian GG. Principles of Polymerization. 4th ed. Hoboken, N.J: Wiley-Interscience; 2004.

[8] Holleman AF, Wiberg E, Wiberg N. Lehrbuch der Anorganischen Chemie. 102nd ed. Berlin: de Gruyter; 2007.

[9] Brinker CJ, Scherer GW. Sol-Gel Science: The Physics and Chemistry of Sol-Gel Processing. Boston: Academic Press; 1990.

[10] Kickelbick G. Hybrid Materials Synthesis, Characterization, and Applications. Weinheim: WILEY-VCH; 2007.

[11] Müller H, Rehak P, Jäger C, Hartmann J, Meyer N, Spange S. A Concept for the Fabrication of Penetrating Carbon/Silica Hybrid Materials. Adv Mater 2000;12:1671–5. doi:10.1002/1521-4095(200011)12:22<1671::AID-ADMA1671>3.0.CO;2-M.

[12] Mehner A, Pohlers A, Hoyer W, Cox G, Spange S. Cationic Twin Polymerization Versus Simultaneous Polymerization of Titanium Compounds to Fabricate Nanostructured Organic Polymer/TiO_2 Hybrid Materials. Macromol Chem Phys 2013;214:1000–10. doi:10.1002/macp.201200629.

[13] Mehner A, Rüffer T, Lang H, Pohlers A, Hoyer W, Spange S. Synthesis of Nanosized TiO_2 by Cationic Polymerization of (μ4-oxido)-hexakis(μ-furfuryloxo)-octakis(furfuryloxo)-tetra-titanium. Adv Mater 2008;20:4113–7. doi:10.1002/adma.200801376.

[14] Mehner A, Rüffer T, Lang H, Schlesinger M, Mehring M, Spange S. Chirality and Reversible Solid State Phase Transition of a Tetranuclear Titanium Oxido Cluster. New J Chem 2013;37:1290–3. doi:10.1039/C3NJ40775F.

[15] Ozaki H, Hirao A, Nakahama S. Polymerization of monomers containing functional silyl groups. 11. Anionic living polymerization of 3-(tri-2-propoxysilyl)propyl methacrylate. Macromolecules 1992;25:1391–5. doi:10.1021/ma00031a005.

[16] Chung JJ, Jones JR, Georgiou TK. Toward Hybrid Materials: Group Transfer Polymerization of 3-(Trimethoxysilyl)propyl Methacrylate. Macromol Rapid Commun 2015;36:1806–9. doi:10.1002/marc.201500356.

[17] Klapdohr S, Moszner N. New Inorganic Components for Dental Filling Composites. Monatsh Chem 2005;136:21–45. doi:10.1007/s00706-004-0254-y.

[18] Novak BM, Davies C. "Inverse" organic-inorganic composite materials. 2. Free-radical routes into nonshrinking sol-gel composites. Macromolecules 1991;24:5481–3. doi:10.1021/ma00019a044.

[19] Ellsworth MW, Novak BM. Mutually interpenetrating inorganic-organic networks. New routes into nonshrinking sol-gel composite materials. J Am Chem Soc 1991;113:2756–8. doi:10.1021/ja00007a062.

[20] Ellsworth MW, Novak BM. "Inverse" organic-inorganic composite materials. 3. High glass content "nonshrinking" sol-gel composites via poly(silicic acid esters). Chem Mater 1993;5:839–44. doi:10.1021/cm00030a020.

[21] Niederberger M, Garnweitner G, Buha J, Polleux J, Ba J, Pinna N. Nonaqueous synthesis of metal oxide nanoparticles: Review and indium oxide as case study for the dependence of particle morphology on precursors and solvents. J Solgel Sci Technol 2006;40:259–66. doi:10.1007/s10971-006-6668-8.

[22] Niederberger M. Nonaqueous Sol-Gel Routes to Metal Oxide Nanoparticles. Acc Chem Res 2007;40:793–800. doi:10.1021/ar600035e.

[23] Garnweitner G, Niederberger M. Organic Chemistry in Inorganic Nanomaterials Synthesis. J Mater Chem 2008;18:1171–82. doi:10.1039/B713775C.

[24] Pinna N, Niederberger M. Surfactant-Free Nonaqueous Synthesis of Metal Oxide Nanostructures. Angew Chem Int Ed 2008;47:5292–304. doi:10.1002/anie.200704541.

[25] Niederberger M, Garnweitner G. Organic Reaction Pathways in the Nonaqueous Synthesis of Metal Oxide Nanoparticles. Chem Eur J 2006;12:7282–302. doi:10.1002/chem.200600313.

[26] Pinna N, Garnweitner G, Beato P, Niederberger M, Antonietti M. Synthesis of Yttria- Based Crystalline and Lamellar Nanostructures and their Formation Mechanism. Small 2004;1:112–21. doi:10.1002/smll.200400014.

[27] Karmaoui M, Sá Ferreira RA, Carlos LD, Pinna N. Lanthanide-based lamellar nanohybrids: The case of erbium. Mater Sci Eng C 2007;27:1368–71. doi:10.1016/j.msec.2006.08.019.

[28] Karmaoui M, Sá Ferreira RA, Mane AT, Carlos LD, Pinna N. Lanthanide-Based Lamellar Nanohybrids: Synthesis, Structural Characterization, and Optical Properties. Chem Mater 2006;18:4493–9. doi:10.1021/cm060705l.

[29] Karmaoui M, Mafra L, Sá Ferreira RA, Rocha J, Carlos LD, Pinna N. Photoluminescent Rare-Earth Based Biphenolate Lamellar Nanostructures. J Phys Chem C 2007;111:2539–44. doi:10.1021/jp0672609.

[30] Garnweitner G, Tsedev N, Dierke H, Niederberger M. Benzylamines as Versatile Agents for the One-Pot Synthesis and Highly Ordered Stacking of Anatase Nanoplatelets. Eur J Inorg Chem 2008; 890–5. doi:10.1002/ejic.200700995.

[31] Polleux J, Gurlo A, Barsan N, Weimar U, Antonietti M, Niederberger M. Template-Free Synthesis and Assembly of Single-Crystalline Tungsten Oxide Nanowires and their Gas-Sensing Properties. Angew Chem Int Ed 2006;45:261–5. doi:10.1002/anie.200502823.

[32] Polleux J, Antonietti M, Niederberger M. Ligand and Solvent Effects in the Nonaqueous Synthesis of Highly Ordered Anisotropic Tungsten Oxide Nanostructures. J Mater Chem 2006;16:3969–75. doi:10.1039/b607008f.

[33] Polleux J, Pinna N, Antonietti M, Niederberger M. Growth and Assembly of Crystalline Tungsten Oxide Nanostructures Assisted by Bioligation. J Am Chem Soc 2005;127:15595–601. doi:10.1021/ja0544915.

[34] Böttger-Hiller F, Lungwitz R, Seifert A, Hietschold M, Schlesinger M, Mehring M, et al. Nanoscale Tungsten Trioxide Synthesized by *In Situ* Twin Polymerization. Angew Chem Int Ed 2009;48:8878–81. doi:10.1002/anie.200903636.

[35] Ebert T, Seifert A, Spange S. Twin Polymerization – a New Principle for Hybrid Material Synthesis. Macromol Rapid Commun 2015;36:1623–39. doi:10.1002/marc.201500182.

[36] Spange S, Kempe P, Seifert A, Auer AA, Ecorchard P, Lang H, et al. Nanocomposites with Structure Domains of 0.5 to 3 nm by Polymerization of Silicon Spiro Compounds. Angew Chem Int Ed 2009;48:8254–8. doi:10.1002/anie.200901113.

[37] Kempe P, Löschner T, Auer AA, Seifert A, Cox G, Spange S. Thermally Induced Twin Polymerization of 4*H*-1,3,2-Benzodioxasilines. Chem Eur J 2014;20:8040–53. doi:10.1002/chem.201400038.

[38] Ebert T, Cox G, Sheremet E, Gordan O, Zahn DRT, Simon F, et al. Carbon/carbon nanocomposites fabricated by base catalyzed twin polymerization of a Si-spiro compound on graphite sheets. Chem Commun 2012;48:9867–9. doi:10.1039/c2cc34775j.

[39] Löschner T, Seifert A, Lungwitz R, Cox G, Lange A, Hähnle H-J, et al. Simultaneous Twin Polymerization – Controlling the Nanostructure Formation of Hybrid Materials. Macromol Rapid Commun, vol. 32, Freiburg: 2011, p. F60–1.

[40] Löschner T, Mehner A, Grund S, Seifert A, Pohlers A, Lange A, et al. A Modular Approach for the Synthesis of Nanostructured Hybrid Materials with Tailored Properties: The Simultaneous Twin Polymerization. Angew Chem Int Ed 2012;51:3258–61. doi:10.1002/anie.201108011.
[41] Jenkins AD, Kratochvíl P, Stepto RFT, Suter UW. Glossary of basic terms in polymer science (IUPAC Recommendations 1996). Pure Appl Chem 1996;68:2287–311. doi:10.1351/pac199668122287.

3 Structure and Theory

Philipp Kitschke, Michael Mehring

3.1 Structure – reactivity relationships of the precursors for twin polymerization

3.1.1 Introduction

The structural features of the precursors that are capable of twin polymerization (TP), the so-called twin monomers, control several aspects of TP. First, the chemical nature and the specific molecular structure of the macromolecular compounds that are obtained as a result of the polymerization process are dependent of the chemical composition and the molecular structure of the single source precursors. For instance, TP of tetrakis(furan-2-ylmethyl)orthosilicate (tetrafurfuryloxysilane [TFOS]) monomers provides a hybrid material that consists of dense interpenetrating networks of polyfurfuryl alcohol and SiO_2, whereas TP of bis(furan-2-ylmethoxy)dimethylsilane (difurfuryloxydimethylsilane [DFOS]) results in the formation of polyfurfuryl alcohol particles and cyclic oligodimethylsiloxanes (Scheme 3.1) [1]. These precursors are structurally related to each other due to the fact that they both possess a central silicon atom and at least two furan-2-ylmethanolate moieties that are bound to the silicon atom. But their TP, which can be catalyzed by addition of a proton source, gives two different sets of material classes, i.e., a single hybrid material consisting of interpenetrating networks of polyfurfuryl alcohol and SiO_2 (TP of TFOS) and polymeric (polyfurfuryl alcohol) as well as oligomeric (oligodimethylsiloxanes) compounds (TP of DFOS). Thus, a modification of the precursor's molecular structure such as the replacement of two furan-2-ylmethanolate moieties with two methyl groups [$Si(OFur)_4 \rightarrow SiMe_2(OFur)_2$ with $OFur = OCH_2C_4H_3O$] has an enormous impact on the nature of the polymerization products and, thus, on their properties.

Second, as TP is feasible for different classes of molecular compounds, e.g., silicon furfuryl alcoholates or germanium salicyl alcoholates, diverse organic/inorganic hybrid material classes are accessible by TP. In addition to this aspect, morphology of the hybrid materials, e.g., the average phase domain sizes of the single components of the hybrid material, is also affected by the choice of the specific classes of precursors. This is due to the fact that the proceeding of the ensemble of the reactions (triggering, propagation, and termination reactions) that are defined as the TP process is specific for each class of twin monomers. This is best illustrated by comparing proton-catalyzed TP of TFOS with the proton-catalyzed TP of Si-Spiro in solution, respectively (Scheme 3.1). Both compounds represent typical silicon alkoxides and their TP is catalyzed by addition of a proton source to give nanostructured organic/inorganic hybrid materials. However, as the natures

https://doi.org/10.1515/9783110499360-003

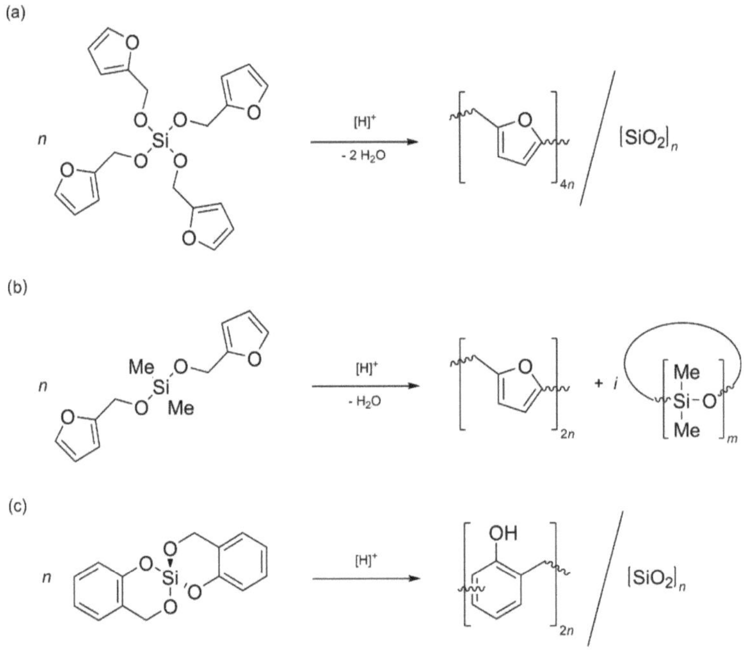

Scheme 3.1: Proton-catalyzed twin polymerization of (a) tetrafurfuryloxysilane (TFOS), (b) difurfuryloxydimethylsilane (DFOS), and (c) 2,2′-spirobi[4*H*-1,3,2-benzodioxasiline] (Si-Spiro). This scheme illustrates the gross conversions for these polymerization reactions.

of the ensemble of the reactions that result in the formation of their TP products differ, the morphology of these nanostructured organic/inorganic hybrid materials (polyfurfuryl alcohol/SiO_2 starting from TFOS and phenolic resin/SiO_2 starting from Si-Spiro) is distinguishable. Due to the fact that TFOS polymerizes *via* condensation reactions to form water upon its TP, larger phase domain sizes of the single macromolecular components (polyfurfuryl alcohol/SiO_2) within the as-prepared nanostructured hybrid materials are present than compared to the phase domain sizes of the hybrid materials (phenolic resin/SiO_2) as-obtained starting from Si-Spiro (Scheme 3.1) [1, 2]. Notably, the latter compound is one of the few examples that undergoes an ideal TP, which is the formation of two macromolecular structures starting from one monomer in just one process step without the formation of any by-products [3].

Noteworthy, three different triggering processes of TP are reported (thermally, addition of an acid, and addition of a base). Regarding this, a general classification of the initial steps in (a) proton-catalyzed and Lewis-acid-catalyzed as well as (b) base-catalyzed and fluoride-initiated TP of the Si-Spiro is given in Scheme 3.2. Thus, the first step in either proton-catalyzed or Lewis-acid-catalyzed TP leads to an addition of the acid (proton or Lewis acid) at the benzylic oxygen atom. Furthermore, a

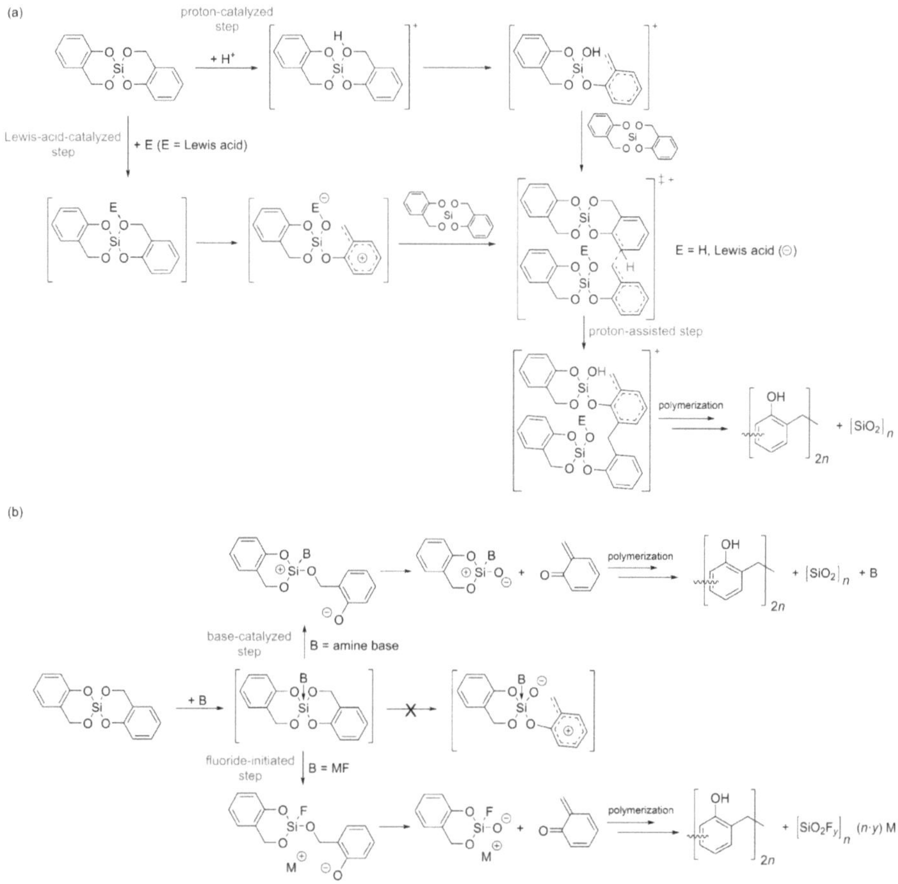

Scheme 3.2: Initial steps in (a) proton-catalyzed and Lewis-acid-catalyzed twin polymerization as well as (b) base-catalyzed (e.g., *tertiary* amine bases) and fluoride-initiated twin polymerization. This scheme illustrates the gross conversions for these polymerization reactions [4–6].

C–O bond cleavage is induced and gives a benzylic cation that reacts with a second monomer whereupon a σ-complex is formed. Independent of the first step, further steps are proton-assisted, while the proton of the σ-complex is transferred to a benzylic oxygen atom and reacts further, resulting in the hybrid material (Scheme 3.2a). As compared to acid-catalyzed TP, it might be expected that a pentacoordinated silicon compound is obtained when the TP process of the Si-Spiro is triggered by addition of a base. Therefore, the Si–B bond energy (B = base) is crucial for the following reaction path. In the case of a low Si–B bond energy, e.g., for amine bases, a catalytic process is expected, whereas a strong Si–B bond, e.g., in case of fluoride, leads to an incorporation of the base within the hybrid material. Thus, the base is an initiator rather than a catalyst in the latter case (Scheme 3.2b). By definition, an initiator is consumed in the polymerization process while the catalyst is neither consumed nor bonded within the formed polymer. Therefore, the terms proton-catalyzed, Lewis-acid-catalyzed, base-catalyzed, and fluoride-initiated have been used in this book.

A third aspect of the relationships between the structural features of the precursors and their TP processes becomes obvious by studying the TP of monomers which only show small modifications in their molecular structure, e.g., possessing different substituents (R = H, Me, Br, ^{t}Bu) at specific positions in their molecular backbone structure. Such precursors usually polymerize by following similar reaction cascades that can be described by a universal reaction mechanism (similar triggering, propagation, and termination reactions). Thus, the properties of the resulting products are expected to be quite similar apart from the discriminable chemical compositions. However, the influence of the different substituents on the polymerization process was determined to be quite pronounced with regard to the properties of the final polymerization products. For instance, studies on a series of precursors that are structurally related to Si-Spiro revealed that this is attributed to the specific electronic and steric features of the substituents. The latter significantly modify the reaction kinetics of the complex reaction cascades proceeding during the TP processes and affect the properties of the resulting hybrid materials eventually [7]. Moreover, if precursors exhibit a similar molecular structure but possess a different chemical composition, e.g., due to substitution of atoms by group homolog elements at specific positions in the molecular structure, the suitability of the precursors for TP may be dramatically altered. For example, studies on the TP of *4H,4′H*-2,2′-spirobi[benzo[*d*][1,3,2]-dioxagermine] [the germanium homolog compound to Si-Spiro (Scheme 3.1)] revealed that this compound is highly suitable for TP, but its sulfur-substituted homolog *4H,4′H*-2,2′-spirobi[benzo[*d*][1,3,2]dithiagermine] is not at all suitable for TP [8]. In addition to this, substitution of the central atom by another group homolog element, e.g., silicon by germanium or tin, may also cause alterations of the general molecular structural motif of the precursors. Hence, twin monomers that are obtained on the basis of salicyl alcoholates but possess different elements (M = Si, Ge, Sn) as central unit were determined to show diverse coordination motifs of these central atoms, e.g., a tetrahedral coordination motif for the spirocyclic silicon salicyl alcoholates or octahedral coordination motifs for molecular germanates as well as for spirocyclic tin salicyl alcoholates, which form coordination polymers (Scheme 3.3) [2, 9, 10]. As a consequence of the differences in their molecular structures and, thus, coordination behavior, the reactivity of these precursors toward TP is also altered. Thus, TP of spirocyclic silicon salicyl alcoholates can be induced thermally, by addition of a proton source, or even by addition of a Lewis base to always give a phenolic resin/SiO_2 hybrid material, whereas TP of the molecular germanates can only be induced thermally providing phenolic resin/GeO_2 hybrid materials [2, 9, 11]. The spirocyclic tin salicyl alcoholates exhibit poor reactivity toward TP. Thus, the synthesis of hybrid materials consisting of phenolic resin and SnO_2 starting from these precursors is only feasible if these compounds are polymerized in the presence of the more reactive spirocyclic

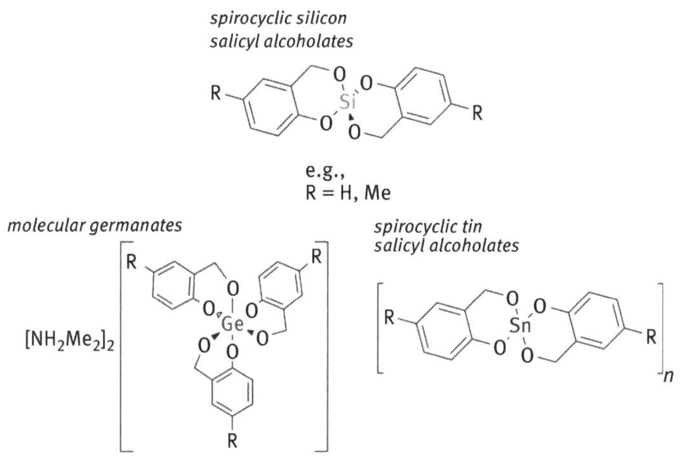

Scheme 3.3: Illustration of the molecular structures of twin monomers based on group 14 salicyl alcoholates.

silicon salicyl alcoholates [10]. The latter polymerization process is called simultaneous twin polymerization (STP) [3].

In the following sections, all these aspects according to the relationships between the structural features of the twin monomers and their specific reactivity with regard to TP will be discussed. We start by providing an overview of the diversity of the molecular structures of precursors that are suitable for TP. A discussion on the precursors' reactivity with regard to their structural features is given. Thus, the reactivity of the specific twin monomer classes and general approaches to modify them are reviewed. The second section focuses on precursors that exhibit a similar molecular structure as compared to Si-Spiro but possess a different chemical composition. The reasons for the modifications in their reactivity that are a result of the substitution of specific atoms (e.g., the atoms that connect the inorganic moiety of the monomers with the organic moieties and the central atom of the spirocyclic precursor) by group homolog elements are revealed. Then, the reactivity of spirocyclic silicon salicyl alcoholates toward TP is summarized. With regard to structural modifications of these twin monomers, the influence of substituents at their aromatic moieties on the reactivity of these compounds in TP and on the properties of the as-obtained nanostructured hybrid materials is given. Afterward, detailed studies on the different triggering processes and, thus, reaction mechanisms according to the TP of Si-Spiro are presented. Please note that this compound was the first representative of a series of spirocyclic salicyl alcoholates for which TP proceeds in an ideal manner. Therefore, this precursor represents an ideal model system to illustrate the key features of the molecular precursors and, thus, of their reactivity to obtain organic/inorganic hybrid materials through TP. This chapter closes by presenting molecular dynamic simulation studies on the TP of Si-Spiro which were performed to gain insight into the formation processes resulting in its nanostructured hybrid material which usually exhibits phase domain sizes of the single components below 3 nm [12].

3.1.2 Precursor classes and their reactivity in TP

Quite a number of twin monomer classes ranging from precursors possessing early transition metals to compounds exhibiting main group metals as central molecular units have been reported so far [3]. The molecular structures of these precursors vary greatly including alkoxido oxido clusters [e.g., $[Ti_4(\mu_4\text{-}O)(\mu_2\text{-}OCH_2C_4H_3O)_6(OCH_2C_4H_3O)_8]$ – Scheme 3.4a], adducts of alkoxides with alcohols [e.g., $[Sn(OCH_2C_4H_3S)_3(\mu_2\text{-}OCH_2C_4H_3S)(HOCH_2C_4H_3S)]_2$ and $[Sn(OCH_2C_6H_4OMe\text{-}2)_3(\mu_2\text{-}OCH_2C_6H_4OMe\text{-}2)(HOCH_2C_6H_4OMe\text{-}2)]_2$ – Scheme 3.4b and c, respectively], oligomeric alkoxides [e.g., $[Ge(O_2C_7H_6)]_n$ with $n = 3$, 4 – Scheme 3.4d], coordination polymers [e.g., $[Sn(O_2C_7H_6)_2]_n$ with $n > 3$ – Scheme 3.4e], and small molecules [e.g., $Ge(OSC_7H_6)_2$ – Scheme 3.4f].

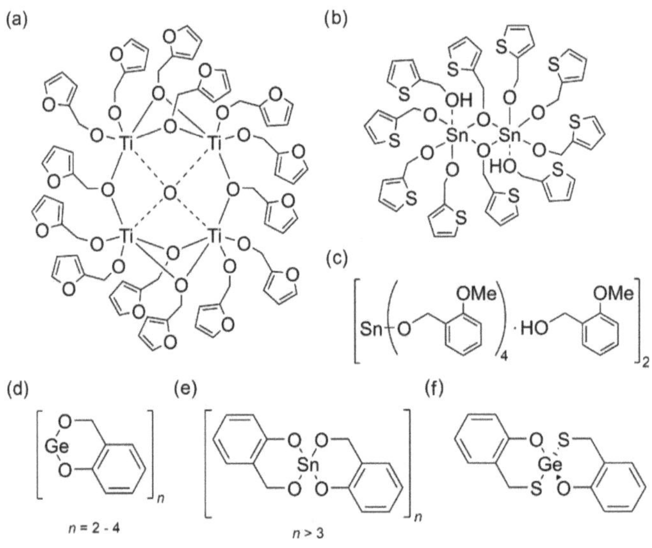

Scheme 3.4: Illustration of the molecular structures of selected representatives of precursors for twin polymerization: (a) (μ_4-oxido)-hexakis(μ_2-thiophene-2-methoxido)-octakis(thiophene-2-methoxido)-tetra-titanium {$[Ti_4(\mu_4\text{-}O)(\mu_2\text{-}OCH_2C_4H_3O)_6(OCH_2C_4H_3O)_8]$}, (b) a dimer of ($\mu_2$-thiophen-2-ylmethoxy)-trikis(thiophen-2-ylmethoxy)(thiophen-2-methanol)stannane {$[Sn(OCH_2C_4H_3S)_3(\mu_2\text{-}OCH_2\text{-}C_4H_3S)(HOCH_2C_4H_3S)]_2$}, (c) a dimer of [$\mu_2$-(2-methoxybenzyl)oxyl]-trikis[(2-methoxybenzyl)oxy)-[(2-methoxyphenyl)methanol]stannane {$[Sn(OCH_2C_6H_4OMe\text{-}2)_3(\mu_2\text{-}OCH_2C_6H_4OMe\text{-}2)\text{-}(HOCH_2C_6H_4OMe\text{-}2)]_2$ – its solid state structure is similar to the structure of $[Sn(OCH_2C_4H_3S)_3\text{-}(\mu_2\text{-}OCH_2C_4H_3S)(HOCH_2C_4H_3S)]_2$ as depicted in (b)}, (d) germanium(II) 2-(oxidomethyl)phenolate {$[Ge(O_2C_7H_6)]_n$ with $n = 3$, 4}, (e) 4*H*,4′*H*-2,2′-spirobi[benzo[*d*][1,3,2]dioxastannine] {$[Sn(O_2C_7H_6)_2]_n$ with $n > 3$}, and (f) 4*H*,4′*H*-2,2′-spirobi[benzo[*d*][1,3,2]oxathiagermine] $[Ge(OSC_7H_6)_2]$.

Regarding these complex molecular structures of the twin monomers, a simpler classification with respect to their organic moieties seems to be reasonable. Thus, five categories of metal-based precursor classes are distinguished up to now. These are twin monomers that are based on furan-2-ylmethanolates (Scheme 3.4a), thiophen-2-ylmethanolates (Scheme 3.4b), benzyl alcoholates (Scheme 3.4c), salicyl alcoholates (Scheme 3.4d and e), and 2-(sulfidomethyl)phenolates (Scheme 3.4f). Notably, all compounds of each category exhibit a similar general reactivity toward their TP. Thus, their specific reactivity is related in a more general sense to the reactivity of their specific organic moieties. Therefore, the relationships between the structural features of the precursors and their reactivity in TP are presented with regard to the classification of the twin monomers toward their organic moieties in the following.

3.1.2.1 Twin monomers based on furan-2-ylmethanolate and thiophen-2-ylmethanolate

The conversion of TFOS into polyfurfuryl alcohol and SiO_2 by an acid-catalyzed polymerization process provided the first example of TP to give an organic/inorganic hybrid material [1]. With regard to the promising features of this "novel" hybrid material, quite a couple of twin monomers that are derivatives of furan-2-ylmethanolate ranging from (μ_4-oxido)-hexakis(μ_2-furan-2-methoxido)-octakis(furan-2-methoxido)-tetratitanium {$[Ti_4(\mu_4\text{-}O)(\mu_2\text{-}OCH_2C_4H_3O)_6(OCH_2C_4H_3O)_8]$} [13, 14], tetrakis(furan-2-ylmethoxy)zirconium [$Zr(OFur)_4$] [15], tetrakis(furan-2-ylmethoxy)hafnium·furan-2-ylmethanol [$Hf(OFur)_4\cdot HOCH_2C_4H_3O$] [15], tri(furan-2-ylmethyl)borate [$B(OFur)_3$] [12], DFOS [$SiMe_2(OFur)_2$] [1], and tetrakis(furan-2-ylmethoxy)stannane [$Sn(OFur)_4$] [10] have been synthesized since 2007. Utilizing TP on these compounds gives nanostructured hybrid materials consisting of the polyfurfuryl alcohol and the respective oxides with the exception of DFOS [$SiMe_2(OFur)_2$] which provides polyfurfuryl alcohol and oligodimethylsiloxanes (Scheme 3.5).

In general, there are three ways to induce TP of these precursors. These are the proton-catalyzed (Scheme 3.6a), the acid-anhydride-catalyzed (Scheme 3.6b and c), and the Lewis-acid-catalyzed triggering of TP (Scheme 3.6d). Notably, if there are traces of water present in the reaction mixture at the time an acid anhydride is added to the precursor's solution, TP of the precursors may also follow a proton-catalyzed process due to the formation of a Brønsted–Lowry acid (Scheme 3.6c).

Please note that self-triggering of Lewis-acid-catalyzed TP will only be feasible if precursors possessing an electrophilic metal atom that is also able to undergo hypercoordination are applied. TP of tetrakis(furan-2-ylmethoxy)stannane {$[Sn(OFur)_4]_n$} at elevated temperature (120 °C) represents an example where the polymerization process is attributed to a Lewis-acid-catalyzed self-triggering of a twin monomer based on furan-2-ylmethanolate [10]. All these initiation processes are related to the ability of the precursors to form furan-2-ylmethylium cations under acidic conditions.

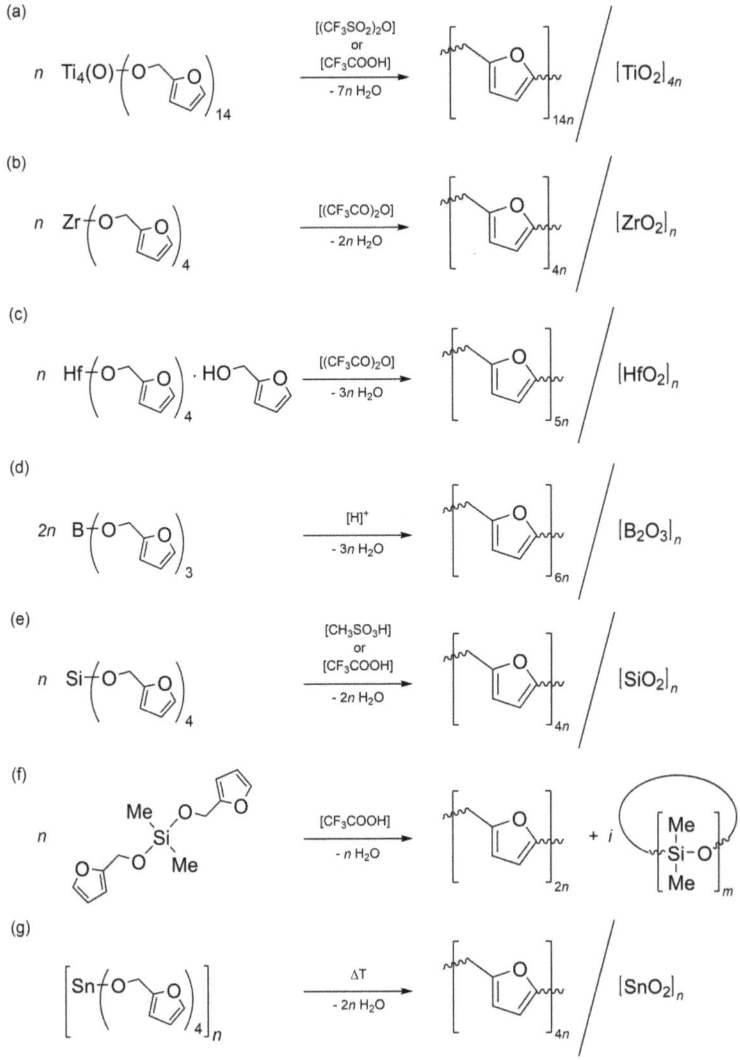

Scheme 3.5: Molecular precursors for twin polymerization: (a) $[Ti_4(\mu_4\text{-}O)(\mu_2\text{-}OCH_2C_4H_3O)_6(OCH_2\text{-}C_4H_3O)_8]$, (b) $Zr(OFur)_4$, (c) $Hf(OFur)_4{\cdot}HOCH_2C_4H_3O$, (d) $B(OFur)_3$, (e) TFOS, and (f) $SiMe_2(OFur)_2$ (DFOS) and $[Sn(OFur)_4]_n$ according to the references [1, 10, 12–16]. Twin polymerization of $B(OFur)_3$ can be catalyzed by Brønsted–Lowry acids in general according to Ref. [12]. The twin polymerization of $[Sn(OFur)_4]_n$ at elevated temperature (≈ 120 °C) is attributed to the formation of a Lewis acid–base adduct of the precursor, which induces the polymerization process [10]. This scheme illustrates the gross conversions for these polymerization reactions.

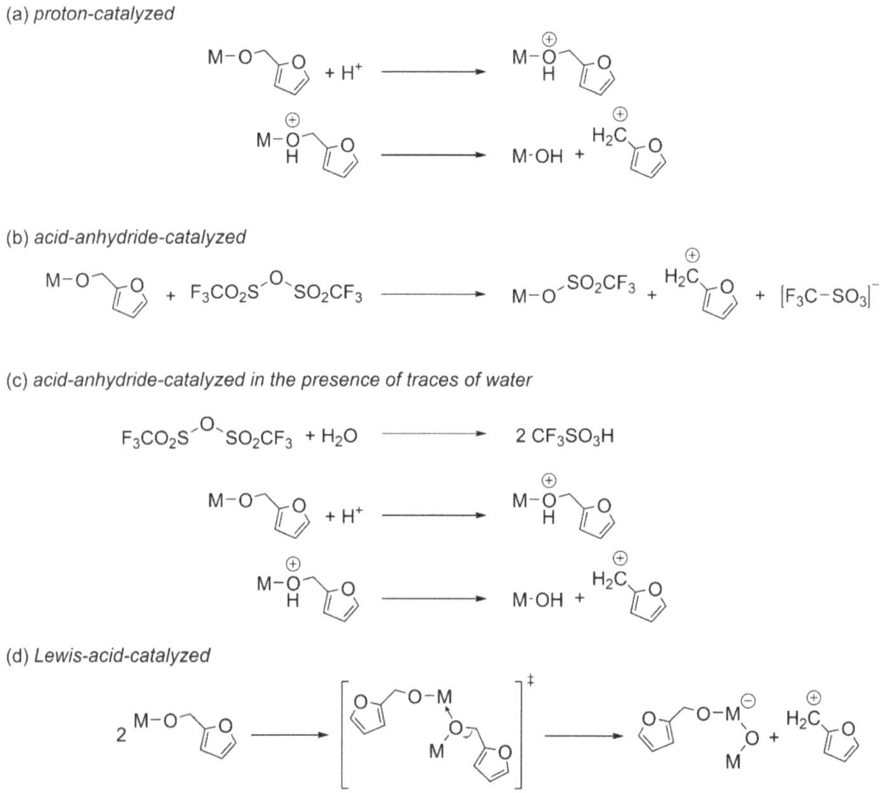

Scheme 3.6: Illustration of initial steps of (a) proton-catalyzed, (b) acid-anhydride-catalyzed, (c) acid-anhydride-catalyzed in the presence of traces of water, and (d) Lewis-acid-catalyzed twin polymerization of twin monomers based on furan-2-ylmethanolates. M denotes the central metal or metalloid atoms of the precursors. In (b) and (c), trifluoromethanesulfonic anhydride was chosen as representative of an acid anhydride due to its common use for the process of twin polymerization.

The formation of furan-2-ylmethylium cations is energetically favored here. Among other facts, the latter is attributed to the ability of this molecular cation to distribute its positive charge over the whole molecular entity as illustrated in Scheme 3.7 by its mesomeric structures.

Once a furan-2-ylmethylium cation is formed, the precursors quickly react with this highly reactive intermediate to give polyfurfuryl alcohol eventually as a result of propagation reactions. The propagation reactions between the precursors, the

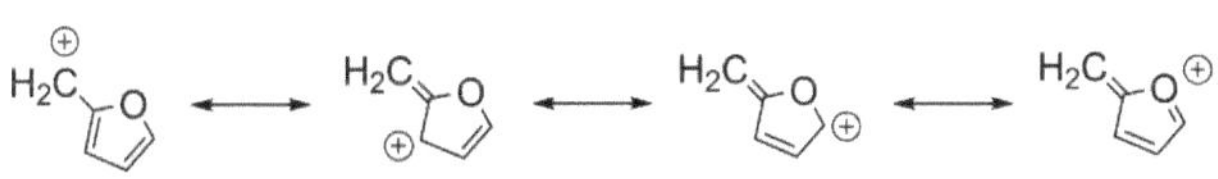

Scheme 3.7: Mesomeric structures of the furan-2-ylmethylium cation (leftmost).

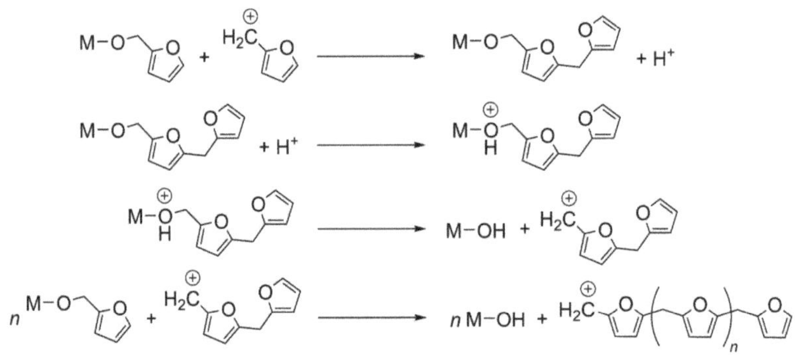

Scheme 3.8: Illustration of the propagation reactions to give polyfurfuryl alcohol starting from precursors based on furan-2-ylmethanolate with furan-2-ylmethylium cations as intermediates. M denotes the central metal or metalloid atoms of the precursors.

furan-2-ylmethylium cation, and the consecutively formed positively charged intermediates continue during the whole TP process (Scheme 3.8).

In addition, M–OH groups (M denotes the central metal or metalloid atoms of the precursors here) are formed as a result of these propagation reactions. The hydroxyl groups consecutively undergo condensation reactions either with other M–OH groups or with furan-2-ylmethanolate moieties of the precursors to give the respective metal oxide/metalloid oxide and water or furan-2-ylmethanol, respectively (Scheme 3.9).

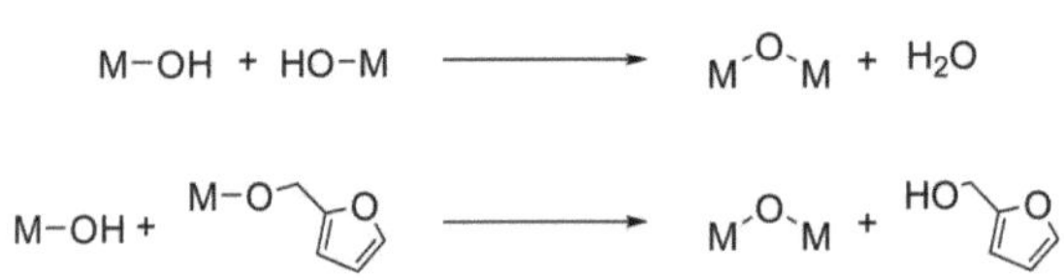

Scheme 3.9: Illustration of the condensation reactions providing metal oxide/metalloid oxide, water, and furan-2-ylmethanol. M denotes the central metal or metalloid atoms of the precursors.

The presence of furan-2-ylmethanol may result in the formation of further furan-2-ylmethylium cations due to its reaction with H^+ as given in Scheme 3.10. These furan-2-ylmethylium cations can induce additional chain propagation reactions to give polyfurfuryl alcohol (Scheme 3.8).

Scheme 3.10: Formation of the furan-2-ylmethylium cation starting from furan-2-ylmethanol under acidic conditions.

At this point, it should be mentioned that the formation of organic/inorganic hybrid materials consisting of polyfurfuryl alcohol and metal oxide/metalloid oxide is only possible if one of the following conditions is complied with (i) the propagation reactions to give polyfurfuryl alcohol and the metal oxide/metalloid oxide take place on approximately the same timescale, or (ii) one of these propagation reactions proceeds so fast that segregation of the two forming phases, e.g., polyfurfuryl alcohol and metal oxide or metalloid oxide, is inhibited (Scheme 3.11) [12]. The latter is due to the quick formation of a dense matrix mainly consisting of the product that is formed by the faster propagation reaction. Notably, if one set of the functional groups of the precursors reacts much slower than the other set of functional groups participating in the formation of the polymer matrices, e.g., furan-2-ylmethanolate to give polyfurfuryl alcohol, the slowly reacting functional groups, e.g., M–OH groups, remain bonded to the residual precursor moieties during the polymerization process in the early reaction stages. Thus, the formation of the other polymeric species takes place as downstream reactions, for instance, due to higher reaction barriers of its propagation reactions. Please note that the condensation reactions of the Si–OH groups as part of a TP process to give a silica network starting from Si-Spiro were determined to be such downstream reactions due to higher reaction barriers as compared to the reactions resulting in the formation of the organic polymer network [4]. The complex reaction kinetic scenarios [either case (i) or (ii)] characterizing the formation reactions to give interpenetrating polymeric structures of, e.g., two polymers A and B, are illustrated in a simplified way in Scheme 3.11. Here, the rate constants k_A and k_B are defined as the rate constants characterizing the whole formation process of the polymeric species $-(A)_n-$ and $-(B)_{m,n}-$, respectively. It is

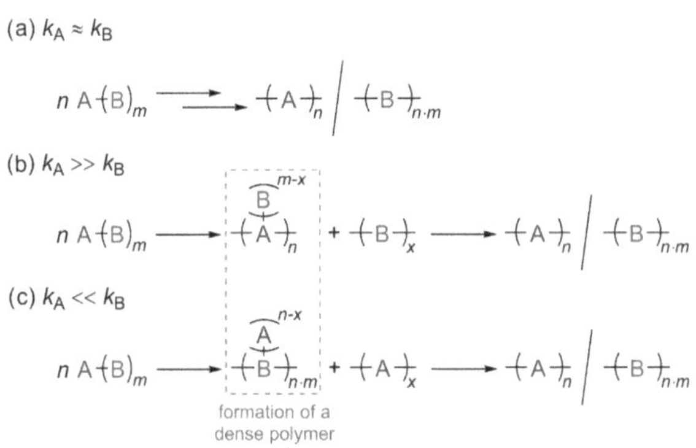

Scheme 3.11: Illustration of possible scenarios of the reaction kinetics for the propagation reactions starting from bifunctional precursors resulting in an interpenetrating network of two polymers. In (a), the rate constants of the propagation reactions are almost equal, whereas in (b) and (c), one of the propagation reactions is much faster than the other one. The rate constants k_A and k_B are defined as the rate constants characterizing the formation process of the polymeric species A and B, respectively. It is assumed that these rate constants are mainly characterized by the rate constants of the propagation reactions to form the polymers $-(A)_n-$ and $-(B)_{m,n}-$, respectively.

assumed that these rate constants are mainly characterized by the rate constants of the propagation reactions to form the polymers $-(A)_n-$ and $-(B)_{m,n}-$, respectively.

TP of furan-2-ylmethanolate derivatives gave organic/inorganic hybrid materials exhibiting phase domain sizes on the nanometer scale of the single components, e.g., polyfurfuryl alcohol and metal oxide or metalloid oxide, with the exception of bis(furan-2-ylmethoxy)dimethylsilane [$SiMe_2(OFur)_2$] [1, 10, 12–14]. Moreover, oxidation of the organic/inorganic hybrid materials that were obtained starting from, e.g., $B(OFur)_3$, $[Ti_4(\mu_4\text{-}O)(\mu_2\text{-}OCH_2C_4H_3O)_6(OCH_2C_4H_3O)_8]$, $Zr(OFur)_4$, $Hf(OFur)_4{\cdot}HOCH_2C_4H_3O$, and $[Sn(OFur)_4]_n$ provided porous materials consisting of metal oxides with BET surface areas (analyzed by method of S. **B**runauer, P. H. **E**mmett, and E. **T**eller) ranging from 20 $m^2{\cdot}g^{-1}$ to 109 $m^2{\cdot}g^{-1}$ and exhibiting crystallite sizes of the respective metal oxides on the nanometer scale [1, 10, 12–14, 17]. In contrast, the twin monomer TFOS provides porous materials consisting of SiO_2 exhibiting BET surface areas ranging up to 518 $m^2{\cdot}g^{-1}$. Thus, TP of such precursors results in the formation of network-like structures of the single polymers (polyfurfuryl alcohol and metal oxide or metalloid oxide) that interpenetrate on the nanometer scale. However, larger phases mainly consisting of one of the two components were also determined within these hybrid materials with phase domain sizes ranging up to the micrometer scale. The latter is attributed to segregation processes that may be caused by the presence of water during the TP process. Since TP of twin monomers that are derivatives of furan-2-ylmethanolate is accompanied by water formation, water phases may be formed with sizes ranging up to the micrometer scale. Intermediates that are soluble in water phase, e.g., M–OH species, cumulate in this phase and further react here to give the respective polymers. The accumulation of such species in the water phase also results in their depleting in other phases, whereby the concentration of intermediates that are poorly soluble in water increases there. Regarding this, the apparent contradiction may be explained by the fact that the hybrid materials as obtained by TP starting from precursors based on furan-2-ylmethanolate exhibit phase domain sizes of single phases ranging from the nanometer as well as up to the micrometer scale. It is noteworthy that TP of Si-Spiro provides nanostructured hybrid materials exhibiting only phase domain sizes of the single polymers, i.e., phenolic resin and SiO_2, on the nanometer scale. As illustrated in Scheme 3.1, its TP does not formally result in the formation of water as additional product.

Twin monomers that are derivatives of thiophen-2-ylmethanolate were obtained starting from thiophen-2-ylmethanol by similar synthesis protocols as reported for precursors on the basis of furan-2-ylmethanolate. Three representatives of this class of compounds have been reported so far. These are (μ_4-oxido)-hexakis-(μ_2-thiophene-2-methoxido)-octakis(thiophene-2-methoxido)-tetra-titanium {$[Ti_4(\mu_4\text{-}O)(\mu_2\text{-}OCH_2C_4H_3S)_6(OCH_2C_4H_3S)_8]$}, tetrakis(thiophen-2-ylmethyl)orthosilicate [$Si(OCH_2C_4H_3S)_4$], and tetrakis(thiophen-2-ylmethoxy)stannane which could be isolated as $[Sn(OCH_2C_4H_3S)_3(\mu_2\text{-}OCH_2C_4H_3S)(HOCH_2C_4H_3S)]_2$ in the solid state (Scheme 3.4) [10, 13, 18, 19]. The reactivity of these compounds toward TP is similar to the reactivity of homolog twin monomers based on furan-2-ylmethanolate. Thus,

TP of these precursors results in the formation of organic/inorganic hybrid materials composed of poly(2-thiophenemethanol) and the respective metal oxides/metalloid oxides. In addition, the initial step of their TP is also acid-catalyzed and water is formed during the polymerization process (Scheme 3.12).

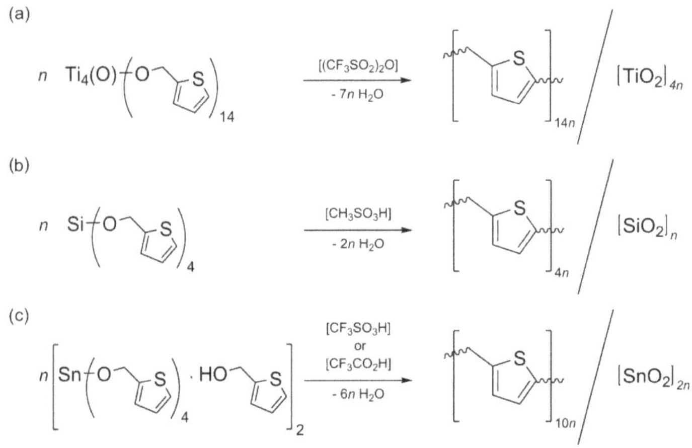

Scheme 3.12: Twin polymerization of (a) $[Ti_4(\mu_4\text{-}O)(\mu_2\text{-}OCH_2C_4H_3S)_6(OCH_2C_4H_3S)_8]$, (b) $Si(OCH_2C_4H_3S)_4$, and (c) $[Sn(OCH_2C_4H_3S)_4(HOCH_2C_4H_3S)]_2$ according to Refs. [10, 13, 18]. This scheme illustrates the gross conversions for these polymerization reactions.

Hybrid materials obtained by TP of thiophen-2-ylmethanolates exhibit phase domain sizes of the single components on the nanometer scale. However, they also show larger phases mainly consisting of one of the two components within the hybrid materials with sizes ranging up to the micrometer scale. Oxidation of the organic/inorganic hybrid materials provided porous metal oxides with BET surface areas ranging from 80 $m^2 \cdot g^{-1}$ to 790 $m^2 \cdot g^{-1}$ and exhibiting crystallite sizes of the respective metal oxide on the nanometer scale [10, 13, 18, 19]. Thus, the phase domain sizes of the as-obtained hybrid materials and the properties of the consecutively obtained porous metal oxides (TiO_2 and SnO_2)/metalloid oxide (SiO_2) are quite similar compared to those materials that were obtained starting from the homolog twin monomers based on furan-2-ylmethanolate. Thus, it is assumed that the TP process for thiophen-2-ylmethanolates is characterized by similar reaction cascades as generally outlined in Schemes 3.6–3.11 for the twin monomers that are derivatives of furan-2-ylmethanolate.

3.1.2.2 Twin monomers based on phenylmethanolates

Twin monomers that are based on phenylmethanolates have been rarely reported till now. There are only two isolated representatives for this class of compounds.

These reactants are $[\{Sn(OCH_2Ar)_3(\mu_2\text{-}OCH_2Ar)(HOCH_2Ar)\}_2]$ with Ar = C_6H_4OMe-2 and $C_6H_3(OMe)_2$-2,4 [10]. Acid-catalyzed TP of $[Sn(OCH_2C_6H_4OMe\text{-}2)_3(\mu_2\text{-}OCH_2C_6H_4OMe\text{-}2)(HOCH_2C_6H_4OMe\text{-}2)]_2$ gave an organic/inorganic hybrid material consisting of polymethoxymethylbenzene and, whereas $[Sn\{OCH_2C_6H_3(OMe)_2\text{-}2,4\}_3\{\mu_2\text{-}OCH_2C_6H_3(OMe)_2\text{-}2,4\}\{HOCH_2C_6H_4(OMe)_2\text{-}2,4\}]_2$ provided a hybrid material consisting of octamethoxy calix[4] resorcinarene and SnO_2 after addition of an acid (Scheme 3.13) [10].

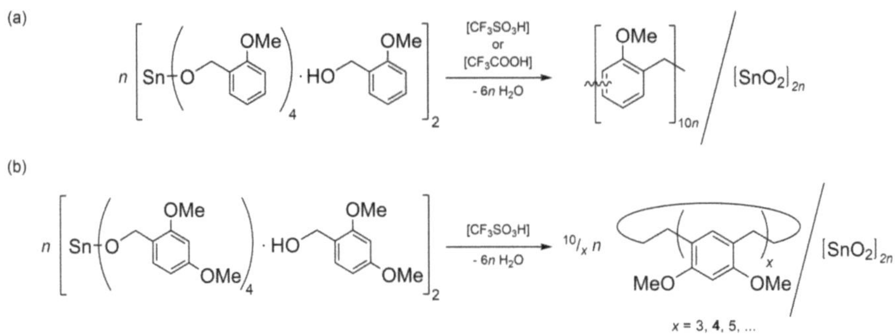

Scheme3.13: Acid-catalyzed twin polymerizations of (a) $[Sn(OCH_2C_6H_4OMe\text{-}2)_3(\mu_2\text{-}OCH_2C_6H_4OMe\text{-}2)\text{-}(HOCH_2C_6H_4OMe\text{-}2)]_2$ and (b) $[Sn\{OCH_2C_6H_3(OMe)_2\text{-}2,4\}_3\{\mu_2\text{-}OCH_2C_6H_3(OMe)_2\text{-}2,4\}\{HOCH_2\text{-}C_6H_3(OMe)_2\text{-}2,4\}]_2$ according to Ref. [10]. Octamethoxy calix[4]resorcinarene was isolated as crystalline solid by sublimation from the polymerization product of $[Sn\{OCH_2C_6H_3(OMe)_2\text{-}2,4\}_3\text{-}\{\mu_2\text{-}OCH_2C_6H_3(OMe)_2\text{-}2,4\}\{HOCH_2C_6H_3(OMe)_2\text{-}2,4\}]_2$. The syntheses of calixarene derivatives were reported, e.g., in Refs. [20, 21]. This scheme illustrates the gross conversions for these polymerization reactions.

The hybrid material as-obtained from the polymerization of $[Sn(OCH_2C_6H_4OMe\text{-}2)_3\text{-}(\mu_2\text{-}OCH_2C_6H_4OMe\text{-}2)(HOCH_2C_6H_4OMe\text{-}2)]_2$ possesses a much lower carbon content (≈ 23.5 %, calculated: 63.9 %) than was expected for an ideal TP. Thus, the degree of polymerization was low and oligomers formed during the polymerization process were removed upon workup. In the case of $[Sn\{OCH_2C_6H_3(OMe)_2\text{-}2,4\}_3\{\mu_2\text{-}OCH_2C_6H_3(OMe)_2\text{-}2,4\}\{HOCH_2C_6H_3(OMe)_2\text{-}2,4\}]_2$, almost complete conversion to give the octamethoxy calix[4]resorcinarene/SnO_2 material was observed. However, the latter calixarene derivative can be removed from this material by sublimation upon heating. Thus, both compounds do not follow an ideal TP. Notably, the low content of organic materials within the polymerization product of $[Sn(OCH_2C_6H_4OMe\text{-}2)_3(\mu_2\text{-}OCH_2C_6H_4OMe\text{-}2)\text{-}(HOCH_2C_6H_4OMe\text{-}2)]_2$ may also be a result of cyclic oligoarene formation during the polymerization process. However, the products of their polymerization reactions are organic/inorganic hybrid materials exhibiting SnO_2 nanoparticles (≈ 2–4 nm) that built agglomerates on the micrometer scale [10]. The latter indicates that during the polymerization process, phase domain sizes of the single components are formed which are on the nanometer scale. This is similar to TP of precursors based on furan-2-ylmethanolates and thiophen-2-ylmethanolates. In addition, the initial step of

TP of $[Sn(OCH_2C_6H_4OMe\text{-}2)_3(\mu_2\text{-}OCH_2C_6H_4OMe\text{-}2)(HOCH_2C_6H_4OMe\text{-}2)]_2$ and $[Sn\{OCH_2\text{-}C_6H_3(OMe)_2\text{-}2,4\}_3\{\mu_2\text{-}OCH_2C_6H_3(OMe)_2\text{-}2,4\}\{HOCH_2C_6H_3(OMe)_2\text{-}2,4\}]_2$ is acid-catalyzed as reported for furan-2-ylmethanolates and thiophen-2-ylmethanolate. Thus, the initial step of their TP is assumed to follow a similar reaction cascade as given for the precursors based on the latter classes of compounds (Scheme 3.14). The kinetics of the whole polymerization process must be characterized by similar reaction conditions as outlined for TP of precursors based on furan-2-ylmethanolates and thiophen-2-ylmethanolates (see Scheme 3.11) to give organic/inorganic hybrid materials exhibiting SnO_2 nanoparticles (≈ 2–4 nm).

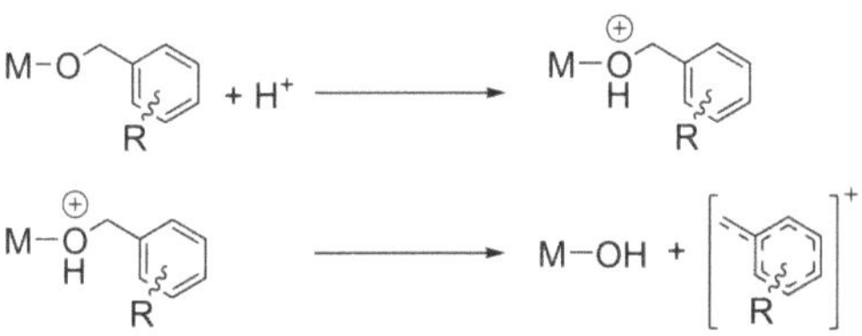

Scheme 3.14: Illustration of the initial steps of acid-catalyzed twin polymerization of twin monomers based on phenylmethanolates. M denotes the metal atom of the precursors, e.g., M = Sn, and R denotes substituents such as methoxy groups.

The formation of benzylic cations and M–OH groups, i.e., Sn–OH groups, upon addition of a proton source is assumed to be energetically favored, again, among other facts, due to the ability of the benzylic cation to distribute its charge over the whole molecular entity. Especially, the +M effect of methoxy substituents will contribute to stabilize the benzylic cations (Scheme 3.15).

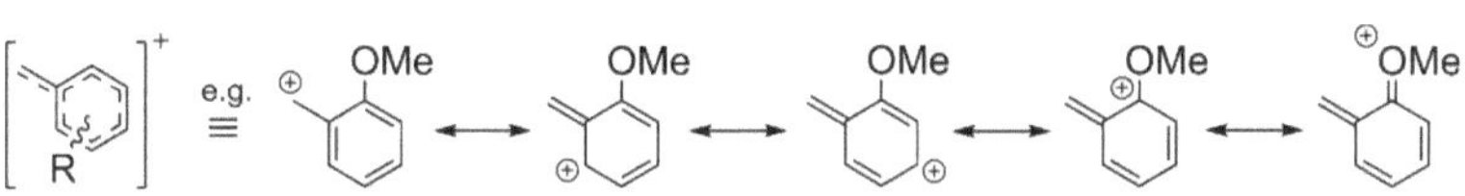

Scheme 3.15: Mesomeric structures of a benzylic cation such as the 2-methoxybenzylic cation.

Once a benzylic cation is formed, the precursors react with this intermediate by electrophilic aromatic substitution reactions (Scheme 3.16).

The propagation of these reactions rather gives oligomeric organic products, e.g., cyclic oligoarene species, than polymethylbenzenes. In the case of TP of $[Sn\{OCH_2\text{-}C_6H_3(OMe)_2\text{-}2,4\}_3\{\mu_2\text{-}OCH_2C_6H_3(OMe)_2\text{-}2,4\}\{HOCH_2C_6H_3(OMe)_2\text{-}2,4\}]_2$, the formation of octamethoxy calix[4]resorcinarene as major organic product may be a result of the electronic (+M effect) and steric features of the methoxy substituents that especially favor electrophilic aromatic substitution reactions in fifth position at the aromatic moiety (being in *ortho-* as well as in *para-*position with regard to the methoxy

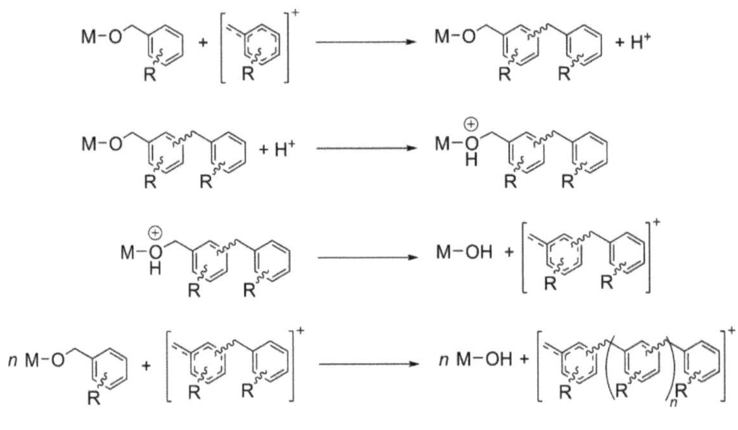

Scheme 3.16: Illustration of electrophilic aromatic substitution reactions of precursors based on phenylmethanolate and being induced by benzylic cations. M denotes the metal atom of the precursors, e.g., M = Sn, and R denotes substituents such as methoxy groups.

substituents). Hence, if four aromatic moieties are connected to each other by methylene groups, a further intramolecular electrophilic aromatic substitution reaction gives octamethoxy calix[4]resorcinarene (Scheme 3.17). In addition, the formation of methoxy calixresorcine species exhibiting other ring sizes cannot be excluded.

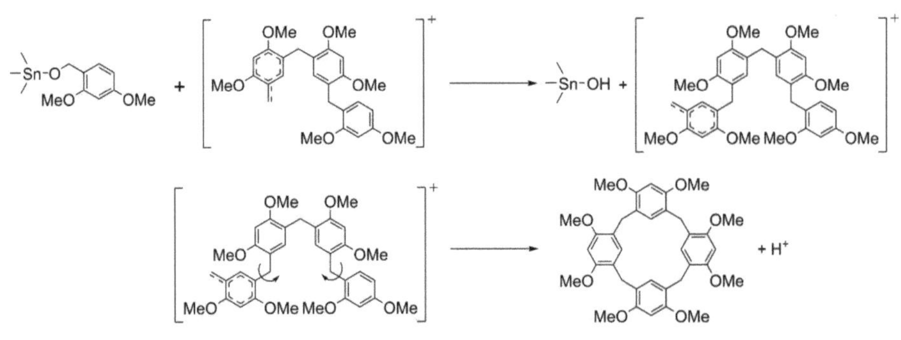

Scheme 3.17: Illustration of octamethoxy calix[4]resorcinarene formation by electrophilic aromatic substitution reactions starting from [Sn{$OCH_2C_6H_3(OMe)_2$-2,4}$_3${μ_2-$OCH_2C_6H_3$-$(OMe)_2$-2,4}{$HOCH_2C_6H_3(OMe)_2$-2,4}]$_2$ that is depicted here only by one Sn{$OCH_2C_6H_3(OMe)_2$-2,4} moiety for simplification.

Condensation reactions of the M–OH groups with either M–OH groups or with phenylmethanolate moieties of the precursors eventually result in the formation of a metal oxide, e.g., SnO_2 (Scheme 3.18).

Again, water is formed as a result of these condensation reactions and upon protonation of benzyl alcohols that may be formed during the polymerization process.

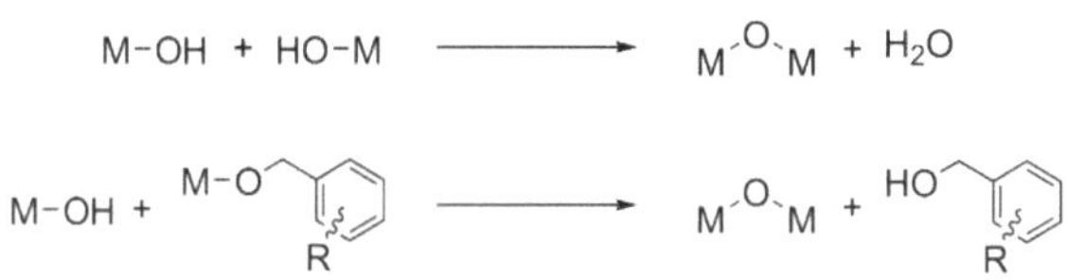

Scheme 3.18: Illustration of the condensation reactions providing metal oxide, e.g., SnO_2, water, and benzyl alcohols. M denotes the metal atom of the precursors, e.g., M = Sn, and R denotes substituents such as methoxy groups.

The latter provides further benzylic cations that may induce new propagation reactions (Scheme 3.19).

Scheme 3.19: Formation of benzylic cations starting from benzyl alcohol under acidic conditions. R denotes substituents such as methoxy groups.

Similar to TP of the compounds that are derivatives of furan-2-ylmethanolate and thiophen-2-ylmethanolate, the presence of water during the polymerization process of $[Sn(OCH_2C_6H_4OMe\text{-}2)_3(\mu_2\text{-}OCH_2C_6H_4OMe\text{-}2)(HOCH_2C_6H_4OMe\text{-}2)]_2$ and $[Sn\{OCH_2C_6H_3(OMe)_2\text{-}2,4\}_3\{\mu_2\text{-}OCH_2C_6H_3(OMe)_2\text{-}2,4\}\{HOCH_2C_6H_3(OMe)_2\text{-}2,4\}]_2$ may explain that their polymerization products exhibit SnO_2 nanoparticles (≈ 2–4 nm) in form of SnO_2 agglomerates ranging up to the micrometer scale.

Please note that studies on the reactivity of $[Sn(OCH_2C_6H_4OMe\text{-}2)_3(\mu_2\text{-}OCH_2C_6H_4OMe\text{-}2)(HOCH_2C_6H_4OMe\text{-}2)]_2$ revealed that catalysis of the TP process by addition of a strong acid such as CF_3SO_3H significantly accelerates the polymerization process compared to its catalysis by a weak acid such as CF_3COOH [10].

It is also noteworthy that addition of an acid such as CH_3SO_3H to a solution of $Ge(OCH_2C_6H_4OMe\text{-}2)_2$ rather gives oligoarene species and $Ge(O_3SCH_3)_2$ than the expected organic/inorganic hybrid material [22, 23]. In addition, studies toward TP of $Sn(OCH_2C_6H_4OMe\text{-}2)_2$ and $Sn[OCH_2C_6H_3(OMe)_2\text{-}2,4]_2$ indicated a similar reactivity for these compounds as observed for $Ge(OCH_2C_6H_4OMe\text{-}2)_2$.

3.1.2.3 Twin monomers based on salicyl alcoholates and homolog thiolates

3.1.2.3.1 General aspects

Several classes of twin monomers are derivatives of salicyl alcoholates and of homolog thiolates. These classes of precursors exhibit a large variety of structural motifs ranging from compounds possessing a low valent metal/metalloid atom to ionic twin monomers consisting of anions exhibiting hypercoordination at their metal/metalloid atom (Scheme 3.20).

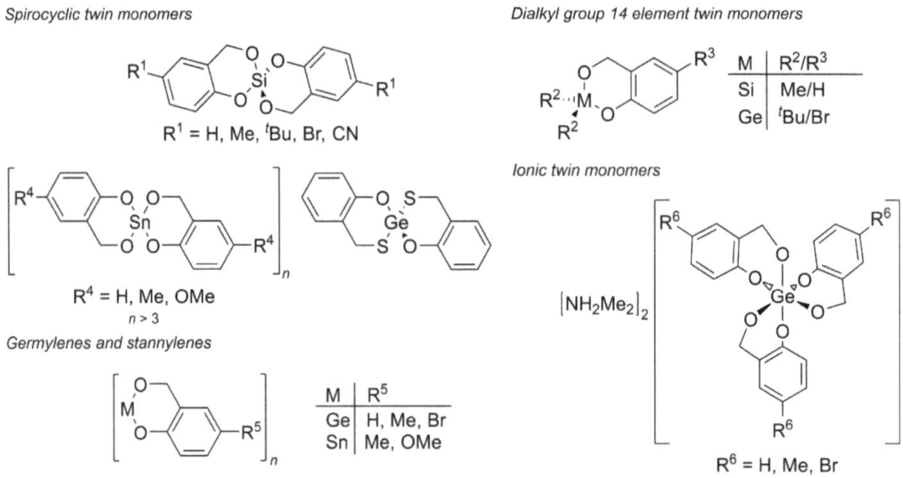

Scheme 3.20: Selected representatives illustrating the structural diversity of twin monomers that are derivatives of salicyl alcoholates or homolog thiolates.

The diversity of these classes of twin monomers allowed comprehensive studies on the relationships between the structure of the precursors and their reactivity toward TP. Notably, considering the molecular structures of neutral twin monomers that are derivatives of salicyl alcoholates, their polymerization should provide products consisting of phenolic resins and the respective metal oxide/metalloid oxide without the formation of water as by-product (Scheme 3.21). Thus by theory, no quantitative amounts of water should influence their polymerization process. This is in contrast to TP of those classes of twin monomers presented in the previous sections (Schemes 3.5, 3.12, and 3.13). However, in the case of TP of ionic precursors based on salicyl alcoholates, equimolar amounts of water are formed during the polymerization process even under the conditions of an ideal polymerization process (Scheme 3.21).

Apart from the general structural features of these classes of twin monomers and the distinguishable characteristics of these compounds, e.g., neutral molecular germanium salicyl alcoholates versus molecular germanates, it is also possible to introduce substituents at the aromatic moiety of the precursors possessing selective features (Scheme 3.20). Hence, twin monomers that are derivatives of salicyl alcoholates represent ideal model compounds to systematically study diverse relationships between their structure and their reactivity in TP.

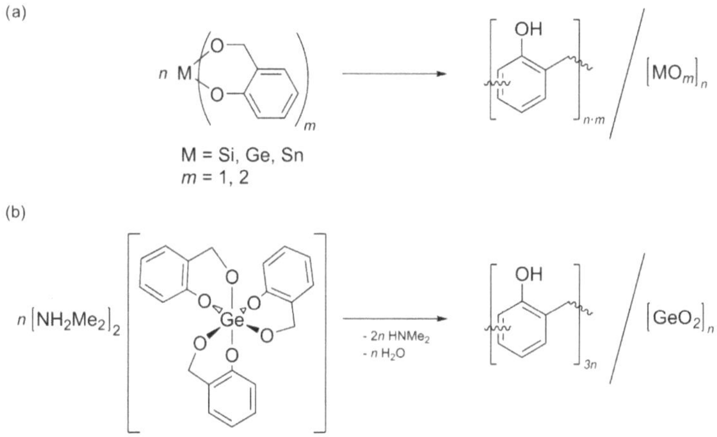

Scheme 3.21: Twin polymerization of (a) neutral twin monomers based on salicyl alcoholates and (b) bis(dimethylammonium) tris[2-(oxidomethyl)phenolate(2-)]germanate as representative of an ionic precursor based on salicyl alcoholate. This scheme illustrates the gross conversions for these polymerization reactions.

3.1.2.3.2 Spirocyclic silicon salicyl alcoholates

The spirocyclic silicon salicyl alcoholate Si-Spiro was the first representative of twin monomers based on a salicyl alcoholate [2]. Its polymerization process can be triggered under acidic conditions, e.g., by addition of a Brønsted–Lowry acid or Lewis acid, upon thermal annealing, and by addition of a Lewis base to give a nanostructured hybrid material [2, 6, 7, 11]. These hybrid materials consist of phenolic resin and SiO_2 which both exhibit phase domain sizes below 3 nm resulting from the formation of polymeric networks of each component interpenetrating on the nanometer scale (Scheme 3.1) [2, 6, 7, 11]. Notably, the type of triggering the TP does not significantly influence the morphology of the resulting hybrid materials on the nanoscale, which is remarkable. Thus, systematic studies on the TP of Si-Spiro and its derivatives exhibiting a modified molecular structure, e.g., possessing substituents at the aromatic moieties, were performed to gain insights into the differently triggered polymerization processes. Please note that, for instance, the morphology of a hybrid material should be altered by modifying the molecular structure of the spirocyclic silicon salicyl alcoholate. An *ortho*/*ortho*- as well as *ortho*/*para*-connectivity of the bridging methylene groups interconnecting the benzene rings in the phenolic resin is favored for the TP of Si-Spiro. In contrast, spirocyclic silicon salicyl alcoholates exhibiting substituents at their aromatic moieties, e.g., in *para*-position, favor a different connectivity pattern of the bridging methylene groups in the phenolic resin, e.g., a prevailing *ortho*/*ortho*-connectivity pattern, with respect to the blocked positions at their aromatic moieties. Thus, the properties of the hybrid materials may be altered at the nanoscale. The results of our studies according to the TP of spirocyclic silicon salicyl alcoholates and the relationships between their structural features and their reactivity in TP, also considering the diverse types of its triggering, are presented in detail within the following sections.

3.1.2.3.2.1 Acid-catalyzed TP

TP of spirocyclic silicon salicyl alcoholates can be triggered under acidic condition, e.g., by addition of CH_3SO_3H to their solution. The formation of a benzylic cation upon protonation at a benzylic oxygen atom of these precursors was determined to be the initial reaction step (Scheme 3.22a) [2, 4, 7]. In the case of a Lewis-acid-catalyzed TP, the formation of a Lewis-acid–base-adduct-like species due to coordination of a benzylic oxygen atom of the spirocyclic silicon salicyl alcoholate to the electrophilic atom of the Lewis acid is assumed to result in the formation of a benzylic cation (Scheme 3.22b).

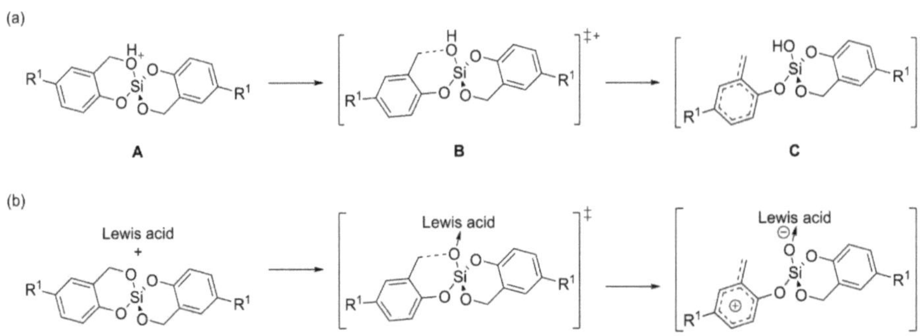

Scheme 3.22: Illustration of the first reaction step of (a) proton-catalyzed and (b) Lewis-acid-catalyzed twin polymerization. In (a), the relative energies of the species that are denoted as A, B, and C are given in Figure 3.2. R^1 denotes H or substituents such as Me, Br, and CN.

Once the benzylic cation is formed, it quickly reacts with another monomer in an electrophilic aromatic substitution reaction to give an intermediate species (**F** in Scheme 3.23a). This intermediate species is able to form another benzylic cation upon its protonation (Scheme 3.23b). Thus, further electrophilic aromatic substitution reactions can propagate.

The latter results in the formation of oligomers and polymers forming a dense matrix. As illustrated, e.g., by species **F** (Scheme 3.23a), there are silanol groups present at these early oligomeric and polymeric species. These silanol groups react in downstream reactions to form a silica network (Scheme 3.24). Please note that the phenolic resin formation is kinetically favored, whereas the downstream condensation reactions to give the silica network exhibit much larger reaction barriers [4]. Due to the formation of a dense organic matrix in the first place, segregation of the arising

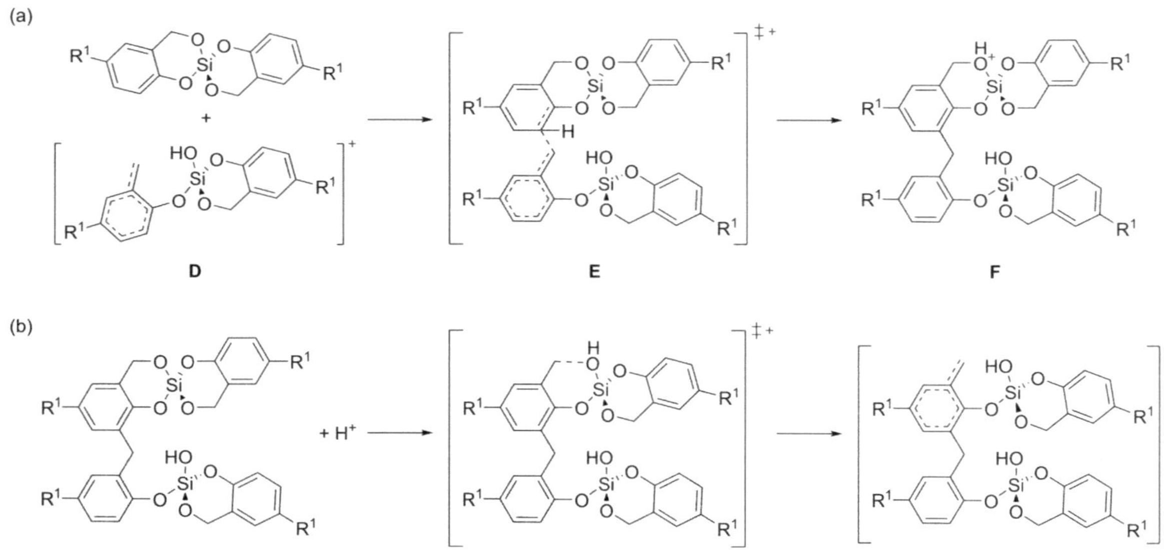

Scheme 3.23: Illustration of (a) the second reaction step of proton-catalyzed and (b) Lewis-acid-catalyzed twin polymerization. In (a), the relative energies of the species that are denoted as **D**, **E**, and **F** are given in Table 3.1. R^1 denotes H or substituents such as Me, Br, and CN.

phenolic resin and silica is inhibited. This eventually leads to the formation of nano-structured hybrid materials consisting of phenolic resin and SiO_2, which exhibit phase domain sizes down to below 3 nm [2, 6, 7, 11].

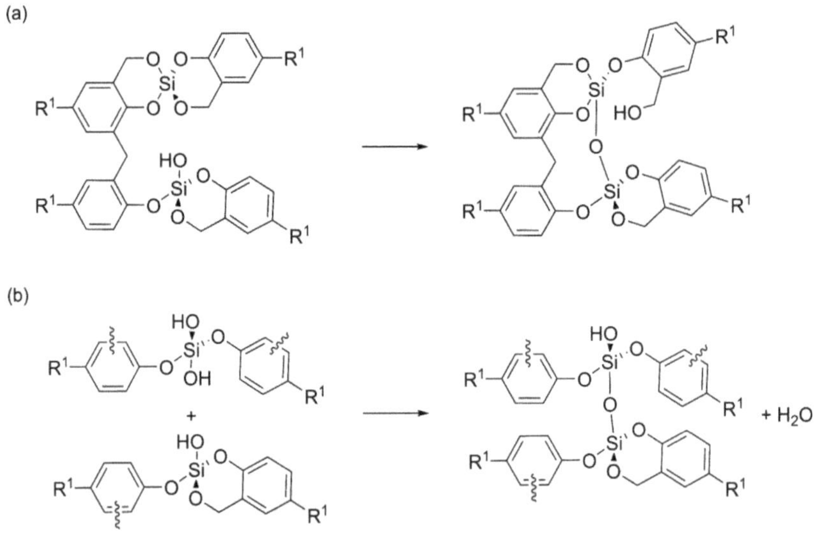

Scheme 3.24: Illustration of (a) intramolecular and (b) intermolecular condensation reactions of silanol groups. R^1 denotes H or substituents such as Me, Br, and CN.

Notably, water that may be formed *in situ* by the condensation of two silanol groups can react with monomers and intermediate silicon alkoxide species to give the respective alcohols and other species containing silanol groups. Protonation of the as-formed alcohols may result in benzylic cationic species and further propagation reactions to finally give a phenolic resin (Scheme 3.25). Thus, the proton-catalyzed TP process of spirocyclic silicon salicyl alcoholates is a composition of diverse types of reactions. Their interplay results in complex reaction cascades providing phenolic resin/SiO_2 hybrid materials eventually.

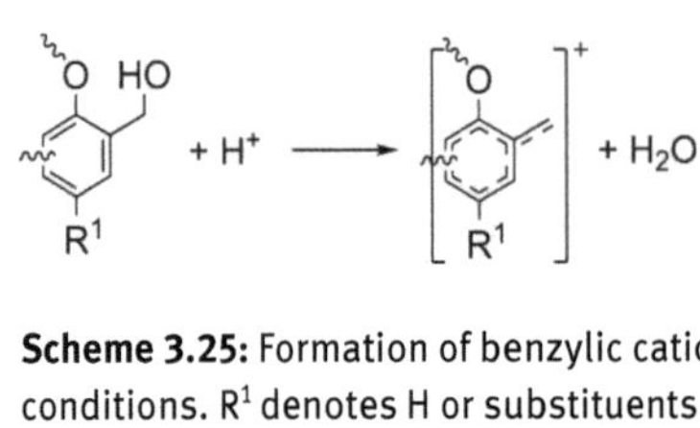

Scheme 3.25: Formation of benzylic cations starting from derivatives of benzyl alcohol under acidic conditions. R^1 denotes H or substituents such as Me, Br, and CN.

Differential scanning calorimetry (DSC) studies on mixtures of a series of structurally related spirocyclic silicon salicyl alcoholates possessing substituents with distinguishable electronic and steric features with trichloroacetic acid revealed that the process of proton-catalyzed TP is influenced by the properties of the substituents (Fig. 3.1). Lower onset temperatures for the triggered polymerization process were determined for the compounds exhibiting electron-density-donating substituents, e.g., an onset temperature of 57 °C for 6,6′-dimethyl-4*H*,4′*H*-2,2′-spirobi[benzo[*d*][1,3,2]-dioxasiline], whereas significantly higher onset temperatures were detected for precursors possessing electron-density-withdrawing substituents, e.g., an onset temperature of 139 °C for 6,6′-dibromo-4*H*,4′*H*-2,2′-spirobi[benzo[*d*][1,3,2]dioxasiline] [7]. The steric features of the substituents seem to be less important regarding the reactivity of the twin monomers since bromine (van der Waals radius: 187 pm [24]) exhibits a lower steric hindrance than a methyl group (van der Waals radius: ≈ 200 pm [25]). These observations were also proven by experiments in CH_2Cl_2 solution at ambient temperature using methanesulfonic acid as the catalyst. Although, for instance, Si-Spiro and 6,6′-dimethyl-4*H*,4′*H*-2,2′-spirobi[benzo[*d*][1,3,2]dioxasiline] rapidly react to give the respective nanostructured hybrid materials as pink and green powder, respectively, 4*H*,4′*H*-2,2′-spirobi[benzo[*d*][1,3,2]dioxasiline]-6,6′-dicarbonitrile does not provide polymeric products under similar conditions. The latter is indicative of a hindered reactivity of this precursor that was related to the features of its substituent, i.e., a cyano group [7].

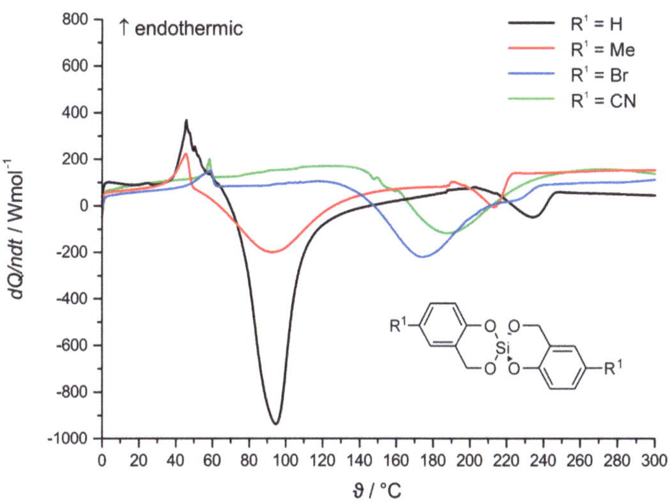

Fig. 3.1: Differential scanning calorimetry of the mixtures of spirocyclic silicon salicyl alcoholates with trichloroacetic acid: mixture of Si-Spiro (black curve) with a molar ratio of $n_{precursor}$:n_{acid} = 1.5, 6,6′-dimethyl-4*H*,4′*H*-2,2′-spirobi[benzo[*d*][1,3,2]dioxasiline] (red curve) with a molar ratio of $n_{precursor}$:n_{acid} = 1.8, 6,6′-dibromo-4*H*,4′*H*-2,2′-spirobi[benzo[*d*][1,3,2]dioxasiline] (blue curve) with a molar ratio of $n_{precursor}$:n_{acid} = 1.5, 4*H*,4′*H*-2,2′-spirobi[benzo[*d*][1,3,2]dioxasiline]-6,6′-dicarbonitrile (green curve) with a molar ratio of $n_{precursor}$:n_{acid} = 4.6; heating rate 10 K·min⁻¹, N_2 atmosphere, N_2 volume flow of 50 mL·min⁻¹. This figure is adapted from Ref. [7] by permission of John Wiley & Sons, Inc.

In order to obtain insights into the energetics of the first reaction steps (see Schemes 3.22a and 3.23a) of the proton-catalyzed TP of this series of spirocyclic silicon salicyl alcoholates and, thus, to disclose the origins of the observed reactivity, density functional theory (DFT) calculations (B3LYP/def2-TZVPP level of theory including conductor-like screening model [COSMO] corrections using the permittivity of CH_2Cl_2) were performed. The calculations revealed that the reaction barrier heights as well as the values of the reaction energies for the first reaction step, the formation of a benzylic cation after protonation of the monomer, follow the ability of the substituents to donate electron density. Thus, the lowest reaction barrier ($E_A = 25.0$ kJ·mol^{-1}) and the largest reaction energy ($\Delta E = -26.9$ kJ·mol^{-1}) for the formation of the benzylic cation were determined for 6,6′-dimethyl-4*H*,4′*H*-2,2′-spirobi[benzo[*d*][1,3,2]dioxasiline], whereas the largest reaction barrier ($E_A = 35.2$ kJ·mol^{-1}) and the lowest reaction energy ($\Delta E = -8.4$ kJ·mol^{-1}) for this type of reaction were calculated for 4*H*,4′*H*-2,2′-spirobi[benzo[*d*][1,3,2]dioxasiline]-6,6′-dicarbonitrile (Fig. 3.2). Both the ordering of the reaction barriers and the ordering of the reaction energies are attributed to the ability of the substituents to donate electron density. Substituents that donate electron density into the aromatic system stabilize the product of this reaction, i.e., a benzylic cation, and they also lower the energy of the transition state. Furthermore, the species formed at the transition state is a cation possessing a similar chemical nature to that of a benzylic cation. On the other hand, electron-withdrawing substituents exhibit a destabilizing effect. Natural bond orbital (NBO) analyses of the reactants, the intermediate species, and the products of this reaction step also support the latter conclusions.

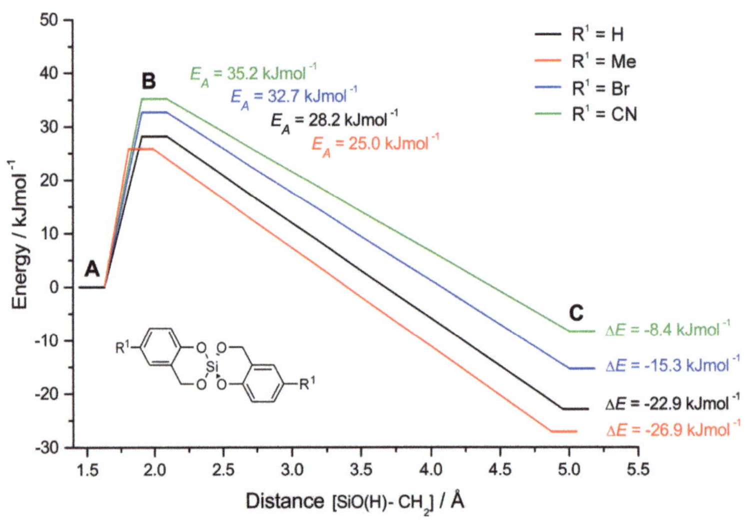

Fig. 3.2: Relative energies for the first reaction step, the ring-opening of the protonated monomer (Scheme 3.22a) of spirocyclic silicon salicyl alcoholates. The distance of the $SiO(H)–CH_2$ bond cleavage was chosen as reaction coordinate. This figure is adapted from Ref. [7] by permission of John Wiley & Sons, Inc.

A similar ordering for the reaction barriers of the second reaction step in proton-catalyzed TP of spirocyclic silicon salicyl alcoholates which is the aromatic substitution reaction of the *ortho*-substituted benzylic cation with another monomer (Scheme 3.23a) was determined (Tab. 3.1). However, a slightly higher reaction barrier for this reaction was calculated for 6,6′-dimethyl-4*H*,4′*H*-2,2′-spirobi[benzo[*d*][1,3,2]-dioxasiline] ($E_A = 20.5$ kJ·mol^{-1}) than for Si-Spiro ($E_A = 17.0$ kJ·mol^{-1}). The change in the ordering was attributed to the steric hindrance of the methyl groups. Nevertheless, the trend that higher reaction barriers were calculated for reactions of monomers that possess substituents donating less electron density is in agreement with the fact that benzylic cation-like species are involved in this reaction step. The determined reaction energies of the aromatic substitution reaction follow an inverse ordering compared to that determined for the first reaction step (see Fig. 3.2 and Tab. 3.1). Notably, the as-calculated products of this reaction are dimers of the spirocyclic silicon salicyl alcoholates that carry the positive charge at a benzylic oxygen atom due to its protonation (Scheme 3.23a). Hence, there is no stabilizing effect of electron-density-donating substituents at the aromatic moiety for these dimers. In addition, the energetics of the protonated dimers may also depend on their specific conformation that always resembled a parallel alignment of the two monomeric moieties for these calculations (Scheme 3.23a).

Tab. 3.1: Relative energies (given in kJ·mol^{-1}) for the second reaction step, the aromatic condensation reaction of the *ortho*-substituted benzylic cation with another monomer (Scheme 3.23 a, **D–F**) of spirocyclic silicon salicyl alcoholates (R^1 = H, Me, Br, and CN) [7].

R^1	E_A	ΔE
H	17.0	−23.3
Me	20.5	−17.5
Br	23.7	−29.8
CN	30.7	−33.2

Generally speaking, the DFT calculations revealed that lower reaction barriers (E_A) were calculated for precursors possessing substituents that exhibit higher electron-density-donation abilities, whereas higher activation energies (E_A) were determined for precursors with electron-density-withdrawing substituents. Thus, the ordering of the reaction barriers as well as the exothermic character of the first two reaction steps in proton-catalyzed TP are in agreement with the ordering of the onset temperatures and the exothermic nature of the polymerization processes as observed by the DSC measurements. Therefore, a qualitative correlation of the reaction barrier heights with the Hammett substituent constants for substituents in *meta*-position (σ_m), given as $\sigma_m = -0.07$ for R^1 = Me, $\sigma_m = 0.00$ for R^1 = H, $\sigma_m = 0.39$ for R^1 = Br, and $\sigma_m = 0.61$ for R^1-= CN, was proposed [7, 26]. On the other hand, please note that the deviation in the onset temperatures (74 °C for $R^1 = H$, 57 °C for $R^1 = H$, 139 °C for $R^1 = Br$, and 152 °C for $R^1 = CN$) of the process of proton-catalyzed TP for this series of spirocyclic silicon salicyl alcoholates as determined by the DSC measurement cannot be fully explained by the quite small differences of the calculated reaction barriers. Thus, its initial steps might be influenced by other processes as well. For instance, proton transfer reactions, the mobility of the formed reactive species in the bulk phase, and the energetic character of consecutive reaction steps, e.g., condensation reactions to form the silica network, may additionally affect the initial process of the spirocyclic silicon salicyl alcoholates in proton-catalyzed TP [7].

In conclusion, these studies suggest that the Hammett substituent constants allow a prediction of the reactivity of any precursors that are structurally related to Si-Spiro in proton-catalyzed TP.

3.1.2.3.2.2 Thermally induced TP

TP of spirocyclic silicon salicyl alcoholates can be induced by thermal annealing of the pristine monomers. Studies on thermally induced TP of diverse types of spirocyclic silicon salicyl alcoholates including Si-Spiro and structurally related compounds possessing either one or two substituents at their aromatic moieties revealed that the formation of *ortho*-quinone methide species always accompanies the polymerization process [6, 7]. With respect to the results of these studies, the formation of the quinone methide species was reported to be an important initial reaction step for the processes of thermally induced TP [7]. A conceivable reaction path that explains both the triggering of TP as well as the formation of quinone methides during the initial process was proposed by Kitschke *et al.* (Scheme 3.26) [7].

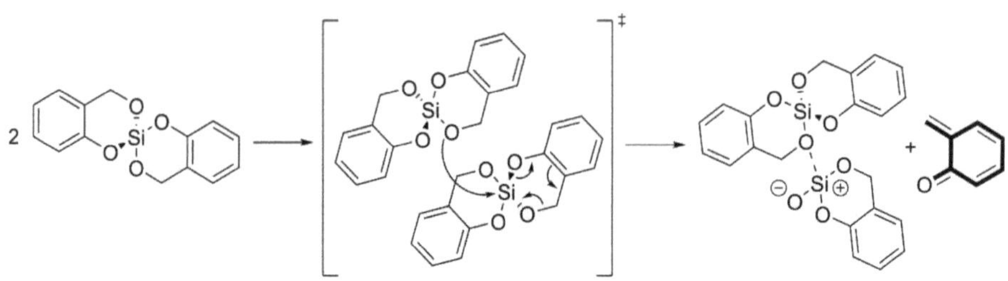

Scheme 3.26: Illustration of a possible reaction path for the process of thermally induced twin polymerization that results in *ortho*-quinone methide (bold structure) formation exemplified for Si-Spiro. This scheme was drawn according to schemes of Ref. [7].

Following this proposed reaction path, a subsequent reaction of the as-formed zwitterionic species with another monomer gives further quinone methide and may explain the formation of the silica network during the polymerization process (Scheme 3.27).

Scheme 3.27: Illustration of a possible propagation reaction path for the process of thermally induced twin polymerization that results in *ortho*-quinone methide (bold structure) formation exemplified for Si-Spiro.

While the formation of the phenolic resin may be explained by the reactions of the highly reactive quinone methide intermediates (Scheme 3.28), it is well known that the latter species do not undergo selective reactions with aromatic systems [27].

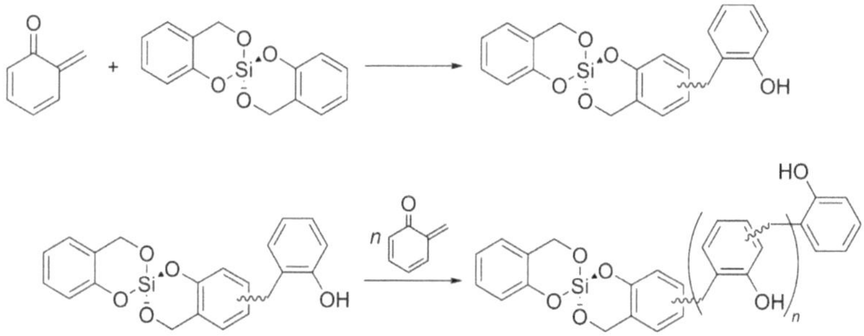

Scheme 3.28: Illustration of possible reaction paths of *ortho*-quinone methide to give phenolic resin in the presence of Si-Spiro.

A modified mechanism for the process of thermally induced TP of Si-Spiro was proposed as a result of theoretical studies by Auer and coworkers [28]. Calculations were performed using DFT including dispersion corrections (D3) and accounting for environmental effects by COSMO [28]. Here, two monomers initially form a dimer exhibiting pentacoordinated silicon atoms (Scheme 3.29). The formation of this dimer was determined to be slightly endothermic ($\Delta E = 12.0$ kJ·mol^{-1}), while this reaction possesses a relatively large reaction barrier ($E_A = 95.9$ kJ·mol^{-1}).

Once the dimer is formed, two reaction paths are conceivable. First, a concerted bond cleavage of the Si–OC_{Aryl} and SiO–CH_2 bonds that belong to the same salicyl alcoholate moiety results in the liberation of a quinone methide (Scheme 3.30). This reaction possessing a reaction barrier of 145.2 kJ·mol^{-1} was calculated to be endothermic by 60.3 kJ·mol^{-1}.

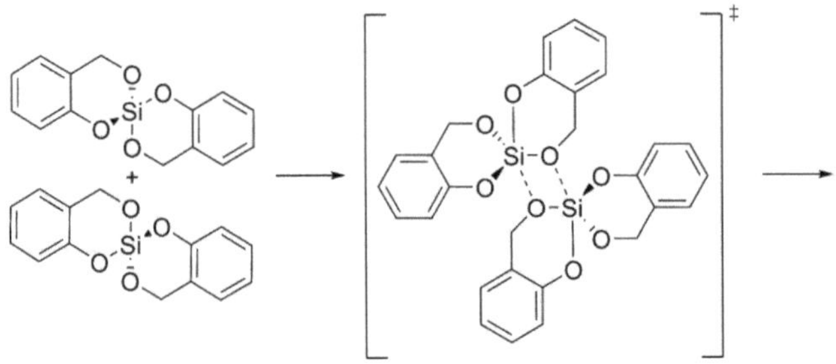

Scheme 3.29: Dimerization of Si-Spiro according to Ref. [28].

The latter reaction providing quinone methide seems to be unlikely due to its high activation energy and its clearly endothermic character. In contrast, the intermediate dimer exhibiting two pentacoordinated silicon atoms can be converted practically without activation barrier to give a species possessing SiO_4 moieties that are bridged by two salicyl alcoholate moieties (Scheme 3.31). The formation of this species was calculated to be slightly exothermic (–3.1 kJ·mol^{-1}) with respect to the van der Waals adduct of two Si-Spiro molecules [28].

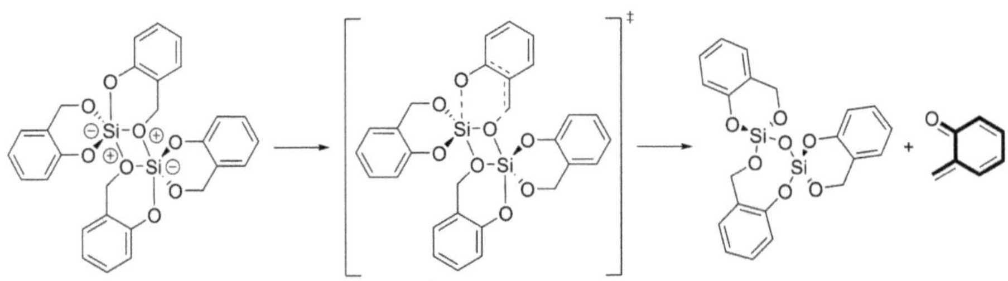

Scheme 3.30: Formation of *ortho*-quinone methide (bold structure) and a species possessing a Si–O–Si unit starting from a dimer of Si-Spiro according to Ref. [28].

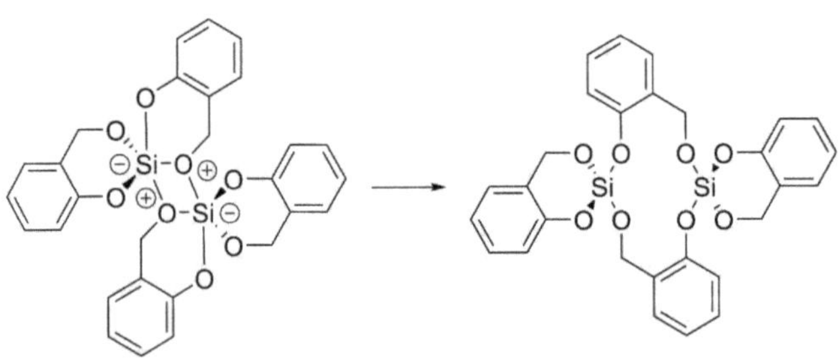

Scheme 3.31: Formation of a silicon salicyl alcoholate possessing two SiO_4 units that are bridged by two salicyl alcoholate moieties starting from a dimer of Si-Spiro according to Ref. [28].

Following the course of the latter reaction path, the formation of different oligomeric species being diverse in size and in the connectivity pattern of the bridging salicyl alcoholate moieties as illustrated in Scheme 3.32 for trimeric species can be easily understood.

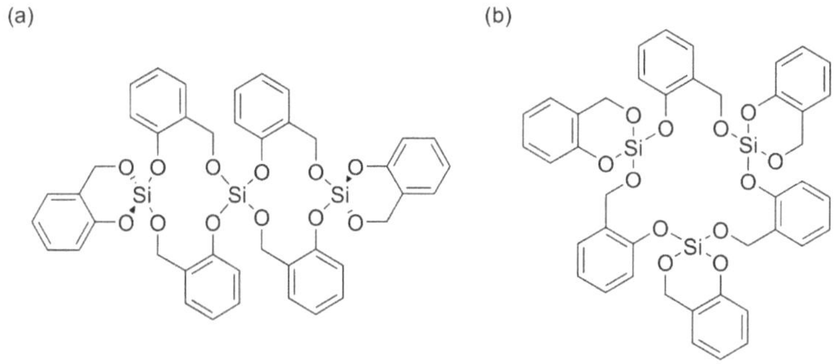

Scheme 3.32: Illustration of trimeric structures possessing three SiO_4 units that are bridged by salicyl alcoholate moieties resembling (a) a linear or (b) a cyclic structure according to Ref. [28].

The formation of the trimeric structures that are illustrated in Scheme 3.32 were both calculated to be endothermic by ≈ 25 kJ·mol^{-1} [28]. The latter indicates that such oligomeric structures may be formed in melts or solutions of Si-Spiro at elevated temperatures being in agreement with experimental data (see below). Subsequent reactions of such oligomeric entities similar to the reactivity of the dimer of Si-Spiro may also result in liberation of reactive quinone methides by simultaneous formation of Si–O–Si bridges (Scheme 3.33). Further reactions of quinone methides as illustrated in Scheme 3.28 and the continuous formation of Si–O–Si bridges as a result of the quinone methide elimination of the oligomeric species lead to the formation of the hybrid material consisting of phenolic resin and silica eventually.

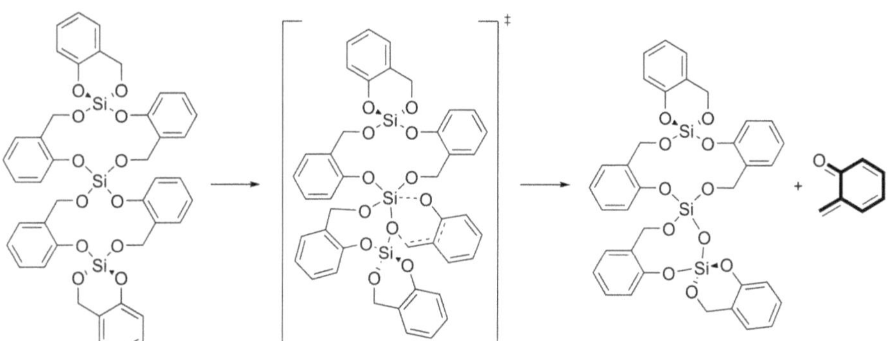

Scheme 3.33: Formation of *ortho*-quinone methide (bold structure) starting from a trimeric species possessing three SiO_4 units that are bridged by salicyl alcoholate moieties.

The formation of such oligomers provides larger molecular entities or even macromolecular structures that exhibit a regular alternating arrangement of salicyl alcoholate moieties in between SiO_4 units. This preorganization at the molecular level in addition with the hindered mobility of the as-formed oligomeric structures may be one of the particular features in the process of TP of spirocyclic silicon salicyl alcoholates that explain the formation of an organic/inorganic hybrid material exhibiting interpenetrating networks of the single components, i.e., phenolic resin and silica, on the nanometer scale. In addition to these complex reaction cascades, it may also be likely that phenolic species are formed as side products during thermal annealing of spirocyclic silicon salicyl alcoholates. The acidity of phenols seems to be strong enough to induce reaction channels similar to the process of proton-catalyzed TP of these precursors [28]. Thus, beneath the reaction paths related to the formation of quinone methides, reaction cascades that result from acidic reaction conditions may be relevant during the process of thermally induced TP of these precursors as well. This makes the process of their thermally induced TP very complex [28]. In order to elucidate the precise reaction cascades that take place during the polymerization process, further experimental and theoretical studies are needed.

Studies on the reactivity of a series of spirocyclic silicon salicyl alcoholates including representatives possessing either one (Fig. 3.3) or two substituents at their aromatic moieties revealed that their TP process is influenced by the electronic and the steric features of their substituents. Hence, lower onset temperatures for the triggering of thermally induced polymerization were determined for compounds exhibiting electron-density-donating substituents, e.g., an onset temperature of 173 °C for 6,6′-dimethyl-4*H*,4′*H*-2,2′-spirobi[benzo[*d*][1,3,2]dioxasiline], whereas higher onset temperatures were detected for precursors possessing electron-withdrawing substituents, e.g., an onset temperature of 235 °C for 6,6′-dibromo-4*H*,4′*H*-2,2′-spirobi[benzo[*d*][1,3,2]-dioxasiline] (Fig. 3.3) [7]. Thus, the triggering of thermally induced TP is facilitated by the ability of substituents to donate electron density into the aromatic moieties of the precursors. The latter observation is in line with the results as observed in the case of their proton-catalyzed TP. However, if precursors possess substituents with similar electron-density-donating abilities, the steric features of the substituents become a decisive factor. Here, sterically demanding substituents seem to hamper the triggering process. For instance, onset temperatures of 173 °C and 180 °C were determined for the thermally induced TP processes of 6,6′-dimethyl-4*H*,4′*H*-2,2′-spirobi[benzo[*d*][1,3,2]dioxasiline] and 6,6′-di-*tert*-butyl-4*H*,4′*H*-2,2′-spirobi[benzo[*d*][1,3,2]dioxasiline], respectively (Fig. 3.3) [7]. Thus, the latter precursor exhibits a higher onset temperature of its polymerization process (180 °C) than the dimethyl-substituted congener (173 °C), although it carries substituents donating slightly more electron density into its aromatic moieties.

Notably, 4*H*,4′*H*-2,2′-spirobi[benzo[*d*][1,3,2]dioxasiline]-6,6′-dicarbonitrile shows a different reaction behavior in thermally induced TP than other spirocyclic silicon salicyl alcoholates. This compound polymerizes prior to melting with an onset temperature of 161 °C (Fig. 3.3) [7]. Regarding the electron-density-withdrawing character of the cyano groups and the low temperature at which TP is triggered for the latter compound, the process of its thermally induced TP might follow a different reaction mechanism than proposed here for the other spirocyclic silicon salicyl alcoholates.

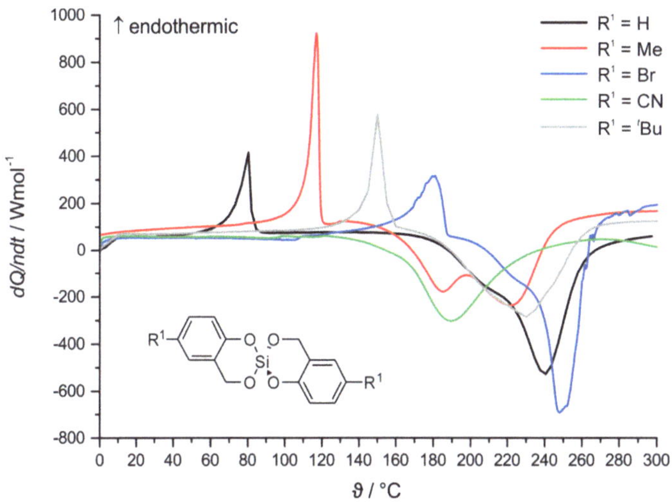

Fig. 3.3: Differential scanning calorimetry of Si-Spiro (black curve), 6,6′-dimethyl-4*H*,4′*H*-2,2′-spirobi[benzo[*d*][1,3,2]dioxasiline] (red curve), 6,6′-dibromo-4*H*,4′*H*-2,2′-spirobi[benzo[*d*][1,3,2]-dioxasiline] (blue curve), 4*H*,4′*H*-2,2′-spirobi[benzo[*d*][1,3,2]dioxasiline]-6,6′-dicarbonitrile (green curve), and 6,6′-di-*tert*-butyl-4*H*,4′*H*-2,2′-spirobi[benzo[*d*][1,3,2]dioxasiline] (gray curve); heating rate 10 $K \cdot min^{-1}$, N_2 atmosphere, N_2 volume flow of 50 $mL \cdot min^{-1}$. This figure is adapted from Ref. [7] by permission of John Wiley & Sons, Inc.

In the case of precursors exhibiting two alkyl substituents at their aromatic moieties, neither steric hindrance accompanied by similar electronic features of the substituents in *ortho*- and *para*-position (with respect to the C_{Aryl}–OR) of the phenolic moiety of the precursors nor the different substitution pattern as compared to precursors such as 6,6′-di-*tert*-butyl-4*H*,4′*H*-2,2′-spirobi[benzo[*d*][1,3,2]dioxasiline] and 6,6′-dimethyl-4*H*,4′*H*-2,2′-spirobi[benzo[*d*][1,3,2]dioxasiline] significantly influences the temperatures (166–178 °C) at which their thermally induced TP is triggered [7]. However, by considering the specific substituent patterns of theses precursors and relating them to the onset temperatures of their thermally induced TP process (R^1/R^2 = Me/Me with 174 °C, R^1/R^2 = Me/^{t}Bu with 178 °C, and R^1/R^2 = ^{t}Bu/Me with 166 °C), the presence of sterically demanding substituents, e.g., *tert*-butyl groups, in *ortho*-position (with respect to the C_{Aryl}–OR) at their aromatic moieties seems to facilitate the initial process of their thermally induced TP.

TP of precursors possessing two substituents at their aromatic moieties gave hybrid materials containing significantly larger mass fractions of soluble by-products (mass losses due to Soxhlet extractions up to ≈ 50 %) such as 2-methylphenols, 1,2-diphenol ethanes, calixarenes, and volatile products of downstreamed reactions of quinone methides than determined for spirocyclic silicon salicyl alcoholates exhibiting only one substituent at their phenolic moieties [mass losses < 10 % (with the exception of 6,6′-dibromo-4*H*,4′*H*-2,2′-spirobi[benzo[*d*][1,3,2]dioxasiline] – mass losses < 30 %)] [6, 7]. Thus, introducing substituents at the aromatic moieties of the precursors results in the formation of larger fractions of oligomeric products during the polymerization process as a result of blocking potential linkage points by the substituents. In addition, the presence of substituents at the phenolic moieties of the precursors also affects the degree of crosslinkages (the number of methylene groups connecting the aromatic moieties with each other) within the as-formed phenolic resin. In general, a larger number of substituents at the aromatic moieties of the precursors lower the degree of crosslinking of the bridging methylene groups within the phenolic resin.

3.1.2.3.2.3 Lewis-base-catalyzed TP

TP of Si-Spiro can be induced by Lewis bases such as pyridine, 1,8-bis(dimethylamino)naphthalene (BDMAN), 1,8-diazabicyclo[5.4.0]undec-7-ene (DBU), 1,4-diazabicyclo[2.2.2]octane (DABCO), and tetra-*n*-butylammonium fluoride (TBAF) [6, 11]. Notably, the difference whether the base is a catalyst or an initiator depends on the Si–B (B = base) bond energy (see Scheme 3.2). The formation of a Lewis acid–base adduct as intermediate species that easily reacts to give quinone methide and a zwitterionic species is assumed to most likely resemble the reaction path inducing its TP (Scheme 3.34). Supporting the latter, Roschke *et al.* were able to isolate the 1,4,7,10,13,16-hexaoxacyclooctadecan stabilized potassium fluorido silicate $[K\{(CH_2)_2O\}_6][Si(O_2C_7H_6)_2F]$ starting from Si-Spiro, 1,4,7,10,13,16-hexaoxacyclooctadecan, and potassium fluoride [5]. This compound exhibiting a pentacoordinated silicon atom polymerizes at much lower temperatures (onset temperature of 134 °C) than Si-Spiro (onset temperature of 196 °C) to give organic/inorganic hybrid materials [5].

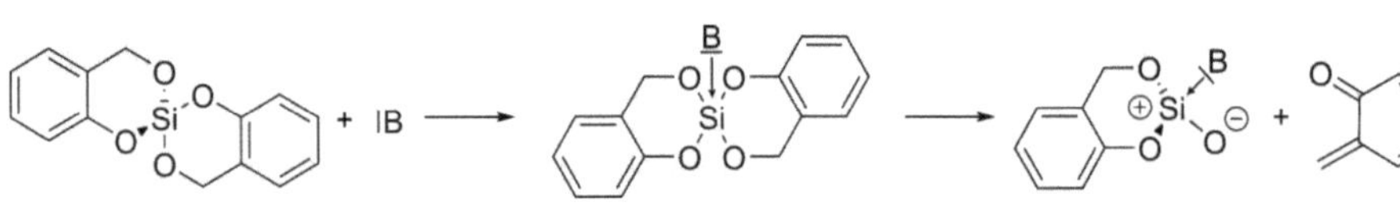

Scheme 3.34: Illustration of the proposed initial reaction path of Lewis-base-catalyzed twin polymerization exemplified for Si-Spiro.

Similar to thermally induced TP of spirocyclic silicon salicyl alcoholates (see Schemes 3.27 and 3.30), subsequent reactions of the as-formed zwitterionic species with other monomers provide further quinone methides and may explain the

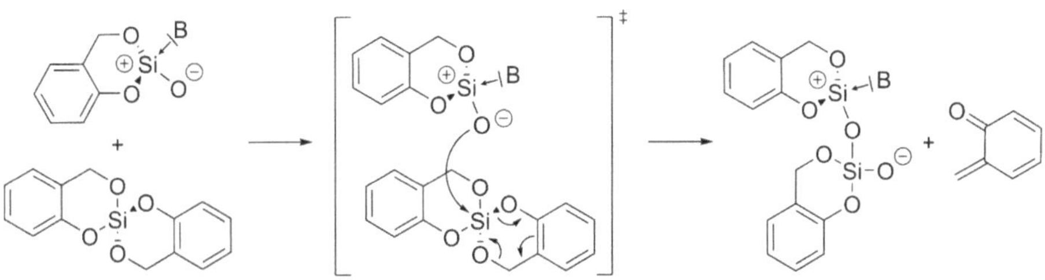

Scheme 3.35: Illustration of a possible propagation reaction path for the process of Lewis-base-catalyzed twin polymerization exemplified for Si-Spiro.

formation of the silica network during the polymerization process (Scheme 3.35). Hence, the formation of the phenolic resin may also be explained by the reactions of the highly reactive quinone methide intermediates (see Scheme 3.28).

So far, systematic studies on the reactivity of different spirocyclic silicon salicyl alcoholates toward Lewis-base-assisted TP are missing. However, there are extensive studies on mixtures of Si-Spiro with different Lewis bases that were performed by DSC [6]. These studies revealed that the chemical nature of the Lewis bases also affects the thermochemical characteristics of the polymerization process. For instance, Lewis bases exhibiting a high nucleophilicity, e.g., BDMAN and DBU, show significantly lower onset temperatures (≈ 110–120 °C) for the TP process (196 °C for pristine Si-Spiro) than Lewis bases that are less nucleophilic, e.g., ε-caprolactam (≈ 180 °C) [6]. Thus, it is assumed that Lewis bases possessing a high nucleophilicity facilitate the triggering of Lewis-base-catalyzed TP by forming more easily reactive Lewis acid–base adducts. The nucleophilicity of the Lewis bases depends, on the one hand, on their ability to donate electron density to the silicon atom that affects the ability of the as-formed Lewis acid–base adduct to form quinone methide by altering the reaction barrier of its formation reaction eventually. On the other hand, the reactivity of a nucleophile regarding its reaction with a specific twin monomer, e.g., Si-Spiro, may also depend on its ability to form a Lewis acid–base adduct with the twin monomer. Moreover, the specific reaction conditions, e.g., the polarity of the solvent, influence the reactivity of the compounds as well. These parameters are considered in the reactivity scale concept of Mayr and coworkers [29]. Here, the reactivity of different Lewis bases is related to distinguishable parameters such as a nucleophilicity parameter (N), a nucleophile-specific sensitivity parameter (S_N), but it also takes into account the specific conditions that are related to the continuous phase, i.e., the solvent in which the reaction is performed [29]. A systematic DSC study on the reactivity of Si-Spiro in the presence of small amounts (5 mol%) of different pyridines supports this assumption (Fig. 3.4). TP can be easily triggered at temperatures as low as 121 °C (onset temperature) by using *N*,*N*-dimethylpyridin-4-amine as Lewis base, whereas a higher onset temperature (135 °C) was determined when pyridine was applied. Notably, *N,N*-dimethylpyridin-4-amine possesses a higher nucleophilicity than pyridine as indicated by Mayr and coworkers, who assigned a higher nucleophilicity parameter for *N,N*-dimethylpyridin-4-amine [$N = 15.8$ (CH_2Cl_2)] compared to pyridine [$N = 12.9$ (CH_2Cl_2)] [29]. The sensitivity parameters of these pyridines are almost identical [$S_N = 0.66$ (CH_2Cl_2) for *N,N*-dimethylpyridin-4-amine versus $S_N = 0.67$ (CH_2Cl_2) for pyridine] [29]. Thus, the different reactivity of the mixtures is attributed to the distinguishable nucleophilicity of these compounds due to the presence of the amine nitrogen atom in *N,N*-dimethylpyridin-4-amine. Lewis-base-catalyzed TP of Si-Spiro in the presence of small amounts (5 mol%) of 2,6-di-*tert*-butylpyridine is triggered at a similar temperature (onset temperature of 136 °C) as determined for the mixture of Si-Spiro with pyridine (onset temperature of 135 °C) (Fig. 3.4). The latter indicates that the initial step of Lewis-base-catalyzed TP of Si-Spiro is not affected by the steric features of the two *tert*-butyl groups in 2,6-di-*tert*-butylpyridine. Unfortunately, there are no nucleophilicity (N) and nucleophile-specific sensitivity (S_N) parameters reported for 2,6-di-*tert*-butylpyridine. However, the nitrogen atom of 2,6-di-*tert*-butylpyridine possesses a quite similar ability to donate electron density to an electrophilic atom to that of the nitrogen atom in pyridine. Thus, their basicity should be similar and thus in agreement with the experimental data.

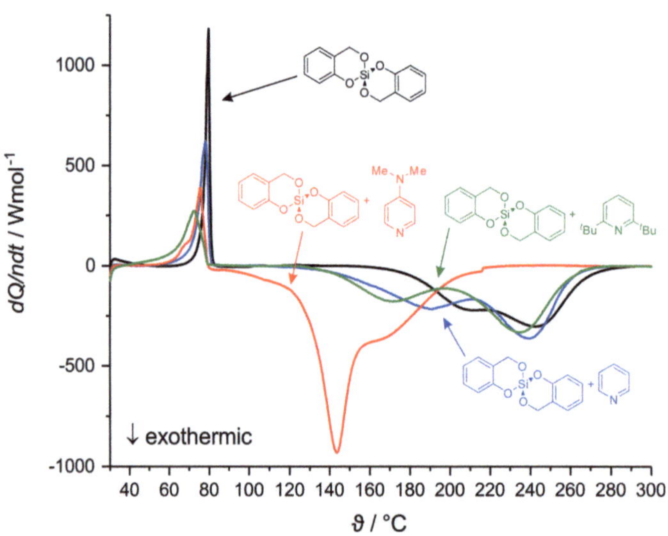

Fig. 3.4: Differential scanning calorimetry of Si-Spiro (black curve) and mixtures of Si-Spiro with pyridine (blue curve), 2,6-di-*tert*-butylpyridine (green curve), and *N,N*-dimethylpyridin-4-amine (red curve) with a molar ratio of $n_{precursor}$:$n_{Lewis\ base} = 20$ (5 mol%), respectively; heating rate 10 K min^{-1}, N_2 atmosphere, N_2 volume flow of 50 mL min^{-1}.

3.1.2.4 Spirocyclic germanium salicyl alcoholates and thiolates

Spirocyclic silicon salicyl alcoholates such as Si-Spiro are precursors that undergo an ideal TP [3]. Thus, extending the concept to spirocyclic germanium salicyl alcoholates that are structurally related to Si-Spiro was expected to provide organic/inorganic hybrid materials by polymerization processes being similar to their spirocyclic silicon salicyl alcoholates. For instance, theoretical studies (B3LYP/def2-TZVPP level of theory including COSMO corrections using the permittivity of CH_2Cl_2) on the proton-catalyzed TP of 4*H*,4′*H*-2,2′-spirobi[benzo[*d*][1,3,2]dioxagermine] revealed a similar reactivity of this compound in comparison to Si-Spiro. Thus, the formation of an *ortho*-substituted benzylic cation starting from its protonated monomer was calculated to be energetically similar to the same reaction of Si-Spiro and, thus, being the most favorable reaction path triggering proton-catalyzed TP of 4*H*,4′*H*-2,2′-spirobi[benzo[*d*][1,3,2]dioxagermine] (Scheme 3.36) [8, 30].

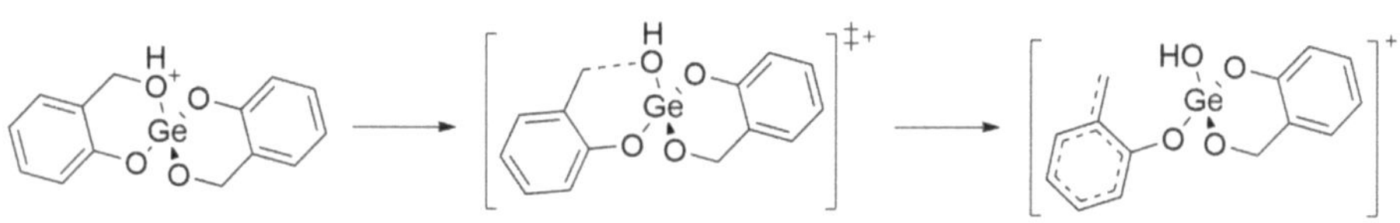

Scheme 3.36: Illustration of the most favorable reaction path triggering proton-catalyzed twin polymerization of 4*H*,4′*H*-2,2′-spirobi[benzo[*d*][1,3,2]dioxagermine]. The given reaction barrier (E_A) and reaction energy (ΔE) are referred to the protonated monomer (left) and were calculated at the B3LYP/def2-TZVPP level of theory including COSMO corrections using the permittivity of CH_2Cl_2 [8].

However, all attempts to synthesize 4*H*,4′*H*-2,2′-spirobi[benzo[*d*][1,3,2]dioxagermine] failed [30, 31]. Notably, molecular germanates were synthesized by reactions of $Ge(NMe_2)_4$ in the presence of various salicyl alcohols (see Chapter 3.1.2.8.3). Single crystal X-ray diffraction analysis revealed that the anions of these molecular germanates possess either a pentacoordinated or a hexacoordinated germanium atom, depending on the steric features of their salicyl alcoholate moieties [9]. On the other hand, introducing sterically demanding groups at the methylene group of the salicyl alcohols, e.g., *tert*-butyl groups, paved a way toward the synthesis of spirocyclic germanium salicyl alcoholates possessing a tetravalent germanium atom and being structurally related to 4*H*,4′*H*-2,2′-spirobi[benzo[*d*][1,3,2]dioxagermine]. Hence, sets of spirocyclic germanium salicyl alcoholates with respect to their stereochemistry were accessible by the reaction of $Ge(NMe_2)_4$ with salicyl alcohols possessing *tert*-butyl groups at their methylene groups (Scheme 3.37) [32].

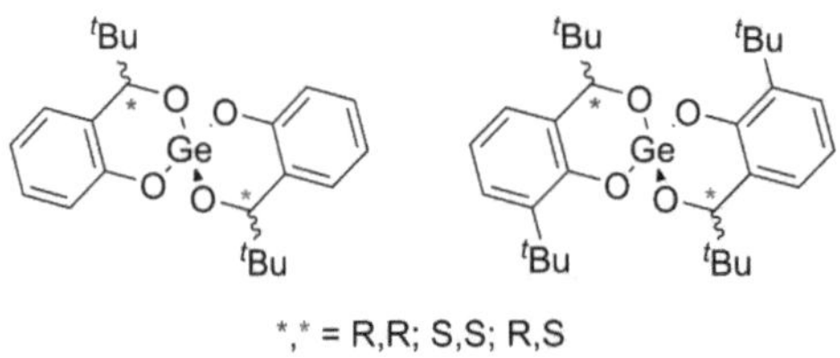

Scheme 3.37: Illustration of the molecular structures of sets of spirocyclic germanium salicyl alcoholates that can in general be denoted as 4,4′-di-*tert*-butyl-4*H*,4′*H*-2,2′-spirobi[benzo[*d*]-[1,3,2]-dioxagermine] (left) and 4,4′,8,8′-tetra-*tert*-butyl-4*H*,4′*H*-2,2′-spirobi[benzo[*d*]-[1,3,2]dioxagermine] (right) by overlooking their stereochemistry.

Again, theoretical studies (B3LYP/def2-TZVPP level of theory including COSMO corrections using the permittivity of CH_2Cl_2) on the proton-catalyzed TP of the stereoisomer (4*R*,4′*R*)-4,4′,8,8′-tetra-*tert*-butyl-4*H*,4′*H*-2,2′-spirobi[benzo[*d*][1,3,2]dioxagermine] qualitatively resembled the results of the compounds 4*H*,4′*H*-2,2′-spirobi[benzo[*d*][1,3,2]dioxagermine] and Si-Spiro. The formation of its *ortho*-substituted benzylic cation ($E_A = 21.5$ kJ·mol^{-1} and $\Delta E = -17.2$ kJ·mol^{-1}) starting from its protonated monomer was calculated to be the energetically most favorable reaction path resulting in a reactive species that may be triggering its proton-catalyzed TP. Based on these results, these spirocyclic germanium salicyl alcoholates should in general be suitable for proton-catalyzed TP. Indeed, addition of methanesulfonic acid and trifluoromethanesulfonic acid to a solution of racemic mixtures (enantiomers exhibiting either a 4*R*,4′*R*- or 4*S*,4′*S*-configuration) of 4,4′-di-*tert*-butyl-4*H*,4′*H*-2,2′-spirobi[benzo[*d*][1,3,2]dioxagermine] and 4,4′,8,8′-tetra-*tert*-butyl-4*H*,4′*H*-2,2′-spirobi[benzo[*d*]-[1,3,2]dioxagermine], respectively, triggered polymerization reactions. But contrary to Si-Spiro, the formation of insoluble products occurred within days instead of few minutes [2, 33]. Analyses of the insoluble products revealed that they mainly consisted of GeO_2, whereas oligomeric products that were identified as larger systems of phenols bridged by methylene groups remained in the solution phase. Thus, the addition of a proton source to the racemic mixtures of 4,4′-di-*tert*-butyl-4*H*,4′*H*-2,2′-spirobi[benzo[*d*][1,3,2]dioxagermine] and 4,4′,8,8′-tetra-*tert*-butyl-4*H*,4′*H*-2,2′-spirobi[benzo[*d*][1,3,2]dioxagermine] triggered decomposition reactions resulting in GeO_2 and oligomeric phenol species rather than their TP [32]. Notably, these decomposition reactions may initially follow a similar reaction cascade as the process of proton-catalyzed TP (Scheme 3.38).

The rapid formation of a dense organic matrix due to electrophilic aromatic substitution reactions of as-formed benzylic cations is hindered by the presence of the sterically demanding *tert*-butyl groups at the methylene groups of the monomers. Please note, addition of methanesulfonic acid to a racemic mixture (enantiomers exhibiting either a 4*R*,4′*R*- or 4*S*,4′*S*-configuration) of the silicon-based precursor 4,4′-di-*tert*-butyl-4*H*,4′*H*-2,2′-spirobi[benzo[*d*][1,3,2]dioxasiline] gave similar results

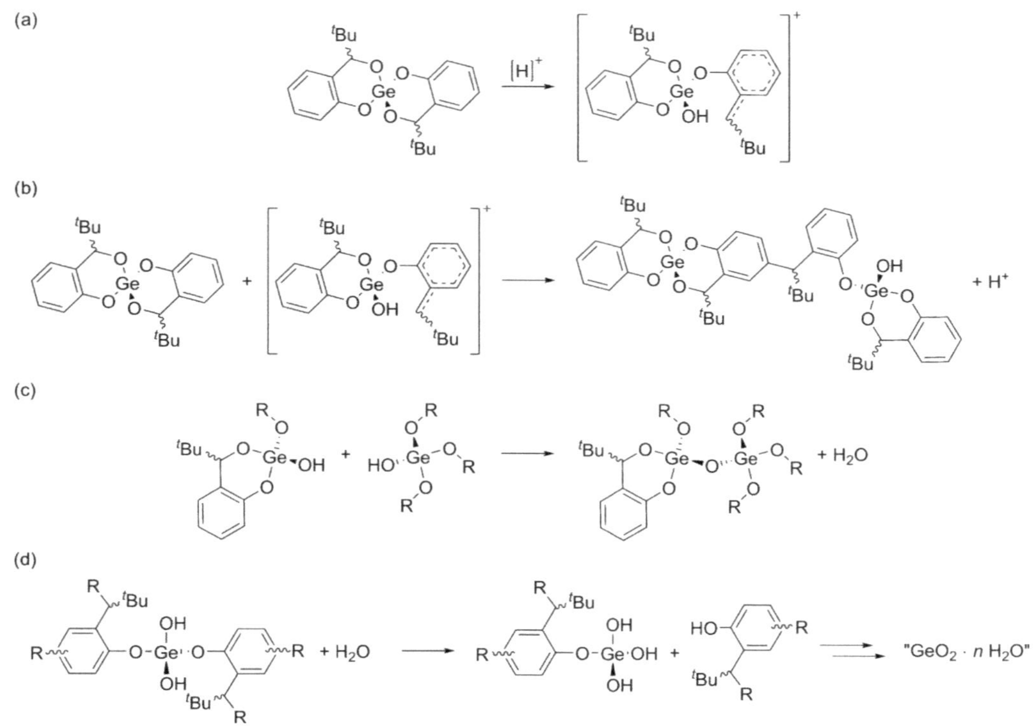

Scheme 3.38: Illustration of general reaction paths leading to the decomposition of spirocyclic germanium salicyl alcoholates and resulting in GeO_2 and oligomeric phenol species eventually. All reactions are exemplified for monomers of the type 4,4′-di-*tert*-butyl-4*H*,4′*H*-2,2′-spirobi[benzo[*d*][1,3,2]dioxagermine]. These reaction paths are in agreement with expected reaction cascades for proton-catalyzed twin polymerization of spirocyclic germanium salicyl alcoholates. However, a much slower reaction rate of the electrophilic aromatic substitution reaction as given in (b) in comparison to a similar reaction of structurally related spirocyclic silicon salicyl alcoholates is expected to cause their decomposition resulting in GeO_2 and oligomeric phenol species rather than their twin polymerization. R denotes organic species such as oligomeric-methylene-group-bridged phenols.

as observed for the racemic mixtures (enantiomers exhibiting either a *4R,4′R*- or *4S,4′S*-configuration) of the germanium based precursors 4,4′-di-*tert*-butyl-4*H*,4′*H*-2,2′-spirobi[benzo[*d*][1,3,2]dioxagermine] and 4,4′,8,8′-tetra-*tert*-butyl-4*H*,4′*H*-2,2′-spirobi[benzo[*d*] [1,3,2]dioxagermine], respectively [33]. The latter observation indicates that the presence of sterically demanding *tert*-butyl groups at methylene groups of the precursors generally causes a slowed down reactivity for proton-catalyzed TP.

Thermal analyses of racemic mixtures of 4,4′-di-*tert*-butyl-4*H*,4′*H*-2,2′-spirobi[benzo[*d*][1,3,2]dioxagermine] and 4,4′,8,8′-tetra-*tert*-butyl-4*H*,4′*H*-2,2′-spirobi[benzo[*d*][1,3,2]dioxagermine] revealed that these precursors are not suitable for thermally induced TP. Both compounds start to decompose directly after melting (onset temperature for decomposition 240 °C for 4,4′-di-*tert*-butyl-4*H*,4′*H*-2,2′-spirobi[benzo[*d*][1,3,2]dioxagermine] and 170 °C for 4,4′,8,8′-tetra-*tert*-butyl-4*H*,4′*H*-2,2′-spirobi[benzo[*d*][1,3,2]dioxagermine]). The decomposition processes are accompanied by high mass losses (> 70 % for 4,4′-di-*tert*-butyl-4*H*,4′*H*-2,2′-spirobi[benzo[*d*][1,3,2]dioxagermine] and > 90 % for 4,4′,8,8′-tetra-*tert*-butyl-4*H*,4′*H*-2,2′-spirobi[benzo[*d*][1,3,2]dioxagermine]) that are attributed mainly to evaporation of the decomposition products [32, 33]. Please note that the silicon-based precursor 4,4′-di-*tert*-butyl-4*H*,4′*H*-2,2′-spirobi[benzo[*d*][1,3,2]dioxasiline] is not suitable for thermally induced TP either. Thermal analysis of this compound revealed that upon its melting (onset temperature 265 °C), further endothermic processes (onset temperature 270 °C) occur that are accompanied by an almost quantitative mass loss (≈ 98 %), which was attributed to decomposition reactions and evaporation of residual amounts of the monomer [33]. Therefore, the presence of the sterically demanding *tert*-butyl groups at the methylene groups of these precursors seems to significantly alter the reactivity of these compounds. The different reactivity of 4,4′-di-*tert*-butyl-4*H*,4′*H*-2,2′-spirobi[benzo[*d*][1,3,2]dioxagermine], 4,4′,8,8′-tetra-*tert*-butyl-4*H*,4′*H*-2,2′-spirobi[benzo[*d*][1,3,2]dioxagermine] (germanium-based precursors), and 4,4′-di-*tert*-butyl-4*H*,4′*H*-2,2′-spirobi[benzo[*d*][1,3,2]-dioxasiline] (silicon-based precursor) was attributed to the steric hindrance of the *tert*-butyl groups that causes decomposition reactions and evaporation of the monomers rather than their thermally induced TP upon their thermal annealing.

Spirocyclic germanium compounds that are structurally related to Si-Spiro are accessible by applying either 2-(mercaptomethyl)benzenethiol, (2-mercaptophenyl)-methanol, or 2-(mercaptomethyl)phenol rather than salicyl alcohol as reactant [8]. TP of this class of compounds was expected to be quite likely regarding the facts that sulfur is a group homolog of oxygen and that the molecular structures of these spirocyclic germanium thiolates do not possess any sterically demanding groups at their aromatic moieties (Scheme 39a–c).

However, experimental studies revealed that none of the spirocyclic germanium thiolates is suitable for proton-catalyzed TP. Only 4*H*,4′*H*-2,2′-spirobi[benzo[*e*][1,3,2]-oxathiagermine] provided an organic/inorganic hybrid material consisting of a phenolic resin, GeO_2, and some unidentified sulfur-containing species upon its thermal annealing by processes that were attributed to its thermally induced TP. Thus,

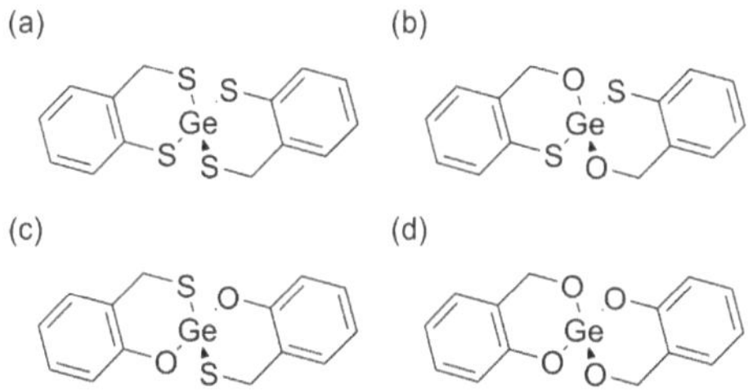

Scheme 3.39: Illustration of the molecular structure of (a) 4*H*,4′*H*-2,2′-spirobi[benzo[*d*][1,3,2]dithiagermine], (b) 4*H*,4′*H*-2,2′-spirobi[benzo[*d*][1,3,2]oxathiagermine], (c) 4*H*,4′*H*-2,2′-spirobi[benzo[*e*][1,3,2]oxathiagermine], and (d) 4*H*,4′*H*-2,2′-spirobi[benzo[*d*][1,3,2]dioxagermine].

systematical studies on the reactivity of this series of spirocyclic germanium(IV) compounds including hypothetical 4*H*,4′*H*-2,2′-spirobi[benzo[*d*][1,3,2]dioxagermine] (Scheme 3.39d) were performed to elucidate the influence of sulfur in comparison to oxygen at the benzylic and the phenylic position of these precursors toward their suitability for TP [8].

First, studies on the compounds' reactivity toward their protonation and possible bond cleavage reactions which should subsequently follow their protonation were carried out by DFT-D calculations (B3LYP-D3/def2-TZVPP level of theory including COSMO corrections using the permittivity of CH_2Cl_2). The formation of a benzylic cation was determined to be the most favorable reaction path for these spirocyclic germanium compounds upon their protonation. This is similar to the results of the studies toward proton-catalyzed TP of the siline Si-Spiro and the germine 4*H*,4′*H*-2,2′-spirobi[benzo[*d*][1,3,2]dioxagermine] (see Chapter 3.1.2.3.2.1 and Scheme 3.22). In contrast to the latter precursors, the reactions of the germanium thiolates were calculated to be endothermic and to exhibit higher reaction barriers (Fig. 3.5) [8].

Detailed NBO analyses of the protonated monomers (**I**-X/Y), the transition state species, and the as-formed benzylic cations (**P**-X/Y) revealed that the presence of a sulfur atom at either the benzylic and/or phenylic position of these precursors causes two effects altering the energetics of these species in comparison to 4*H*,4′*H*-2,2′-spirobi[benzo[*d*][1,3,2]dioxagermine]. (i) The benzylic cations are formed by the bond cleavage of the protonated benzylic X(H)–CH_2 bonds (X = O, S). A S(H)–CH_2 bond possesses an opposite polarity compared to a O(H)–CH_2 bond. Here, the sulfur atom is positively charged and the carbon atom exhibits a negative net charge, whereas the oxygen atom is negatively charged and the carbon atom possesses a positive net charge in the benzylic O(H)–CH_2 bond. Thus, a heteropolar bond cleavage of an X(H)–CH_2 bond to form a benzylic cation is aggravated for X = S due to its polarity, whereas the polarity of the O(H)–CH_2 facilitates its heteropolar bond cleavage. As a consequence, higher reaction barriers were calculated for the X(H)–CH_2 bond cleavage for compounds possessing a sulfur atom in benzylic position. (ii) A phenolic oxygen atom better stabilizes the benzylic cation than a sulfur atom at the same phenylic position due to its enhanced resonance interactions (+M effect). Thus, benzylic

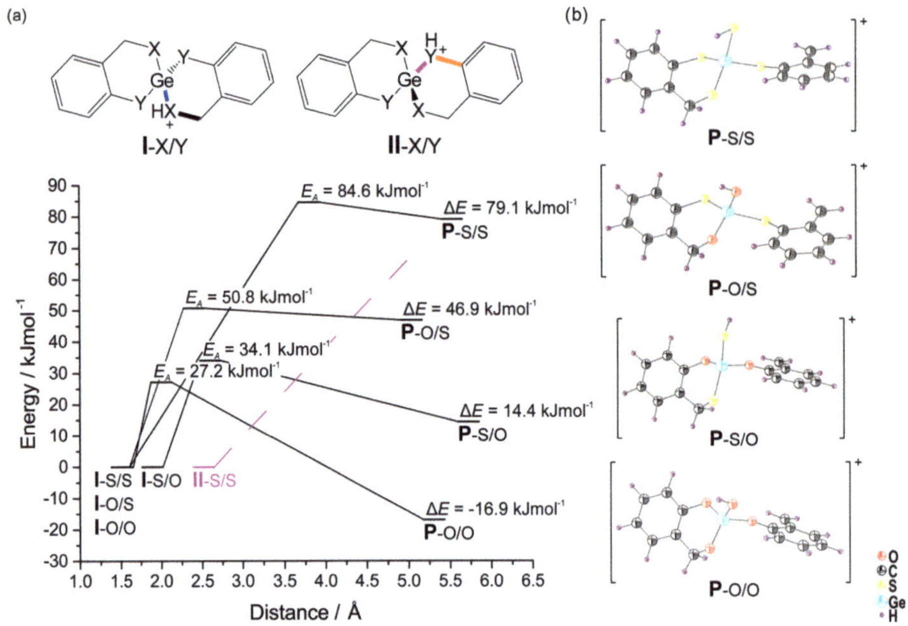

Fig. 3.5: (a) Illustration of the bond cleavage reactions (top) that were studied by variation of the bond lengths after protonation of the monomers of the spirocyclic germanium thiolates and 4*H*,4′*H*-2,2′-spirobi[benzo[*d*][1,3,2]dioxagermine] to give either species I-X/Y or II-X/Y and relative energies for selected bond cleavage reactions [II-S/S → cationic benzenethiol species (pink dashed line), I-S/S → P-S/S, I-O/S → P-O/S, I-S/O → P-S/O and I-O/O → P-O/O] for their protonated monomers. (b) The structures of the products P-S/S–P-O/O that are determined to be local minimum structures are depicted for comparison. No local minimum structure could be determined for the cationic benzenethiol species of the bond cleavage of II-S/S (pink dashed line). This figure is adapted from Ref. [8] by permission of John Wiley & Sons, Inc.

cations carrying an oxygen atom rather than a sulfur atom in *ortho*-position are energetically favored. As a result, lower reaction energies were calculated for the formation of benzylic cations possessing a phenolic oxygen atom rather than a sulfur atom. Furthermore, the reaction barriers of the X(H)–CH_2 bond cleavage reactions (X = O, S) to form the benzylic cations are also affected by the distinguishable resonance interactions (+M effect and –I effect) of the chalcogen atoms. The enhanced +M effect as well as the enhanced –I effect of a phenolic oxygen atom in comparison to a sulfur atom at the same phenylic position better assist the heteropolar X(H)–CH_2 bond cleavage reactions (X = O, S). Thus, the reaction barriers calculated for the X(H)–CH_2 bond cleavages become higher when the precursors possess a sulfur atom at the phenylic position of the aromatic moieties. In conclusion, the presence of sulfur in benzylic and/or phenylic position aggravates the formation of a benzylic cation by increasing the reaction barrier of the X(H)–CH_2 bond cleavage reaction and by destabilizing the as-formed benzylic cation in comparison to an oxygen atom at these positions.

Consequently, the reaction that might induce proton-catalyzed TP of these precursors becomes energetically unfavorable. This might explain that the spirocyclic germanium thiolates are not suitable for proton-catalyzed TP.

The thermochemical behavior of the spirocyclic germanium thiolates was studied by using a variety of thermal analysis techniques such as DSC, thermogravimetric analysis (TGA), and TGA in combination with mass spectrometry (TGA–MS). These studies revealed that in contrast to spirocyclic silicon salicyl alcoholates, each of the spirocyclic germanium thiolates shows a unique thermochemical behavior upon thermal annealing. For instance, endothermic decomposition processes with an onset temperature of 340 °C were observed for *4H,4′H*-2,2′-spirobi[benzo[*d*][1,3,2]-dithiagermine], whereas *4H,4′H*-2,2′-spirobi[benzo[*d*][1,3,2]oxathiagermine] decomposes at much lower temperatures (onset temperature 279 °C) by processes exhibiting an exothermic character (Fig. 3.6) [8]. As mentioned above, both compounds are not suitable for thermally induced TP.

In the case of *4H,4′H*-2,2′-spirobi[benzo[*e*][1,3,2]oxathiagermine], slightly exothermic processes that are attributed to its thermally induced TP were observed (Fig. 3.6). However, its thermally induced TP does not proceed smoothly, which is in contrast to thermally induced TP of spirocyclic silicon salicyl alcoholates. TGA–MS measurements and analysis of the as-obtained hybrid materials revealed the formation of volatile by-products such as SO_2 and *ortho*-cresol during the polymerization process providing an explanation for the observed weight loss of ≈ 32 % and the presence of GeO_2 within the hybrid material. Similar to thermally induced TP of Si-Spiro [6], quinone methide and products derived from it were identified to be formed as a result of decomposition. Thus, with regard to all of these observations, it was assumed that *4H,4′H*-2,2′-spirobi[benzo[*e*][1,3,2]oxathiagermine] undergoes complex reaction cascades upon heating. These reaction cascades are characterized by decomposition reactions but also by the formation of a phenolic resin attributed to the reactions of quinone methide. As a result, a hybrid material is formed eventually. Please note, with regard to the molecular structures of the spirocyclic germanium thiolates, only *4H,4′H*-2,2′-spirobi[benzo[*e*]-[1,3,2]oxathiagermine] is able to form a quinone methide (Scheme 3.40).

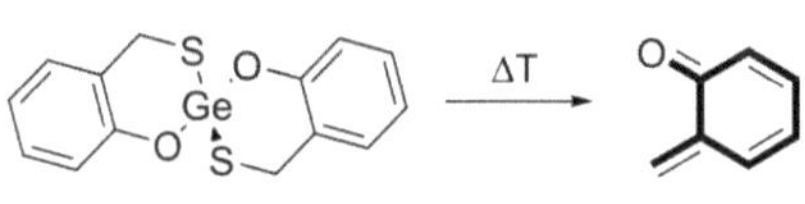

Scheme 3.40: Illustration of the ability of *4H,4′H*-2,2′-spirobi[benzo[e][1,3,2]oxathiagermine] to form *ortho*-quinone methide upon thermal annealing.

Therefore, the ability to form quinone methides upon thermal treatment is expected to be a key feature for spirocyclic compounds structurally related to the spirocyclic siline Si-Spiro to form hybrid materials by processes that are assigned to their thermally induced TP.

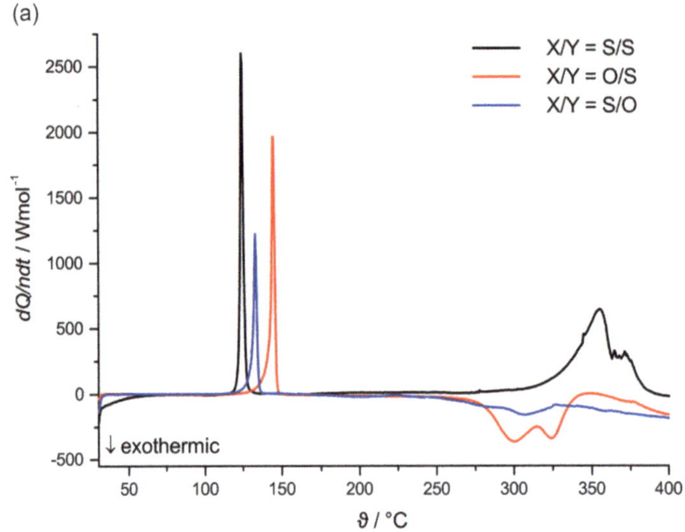

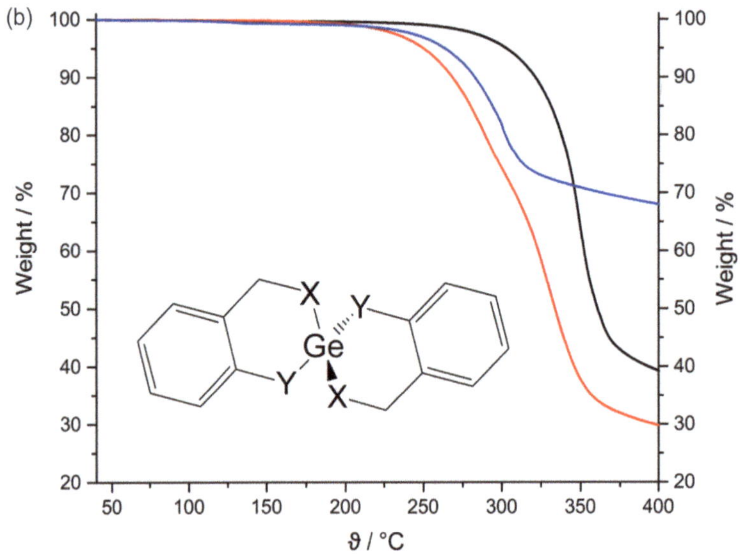

Fig. 3.6: Differential scanning calorimetry given in (a) and thermogravimetric analysis given in (b) of 4*H*,4′*H*-2,2′-spirobi[benzo[*d*][1,3,2]dithiagermine] (black curves), 4*H*,4′*H*-2,2′-spirobi[benzo[*d*][1,3,2]oxathiagermine] (red curves), and 4*H*,4′*H*-2,2′-spirobi[benzo[*e*][1,3,2]oxathiagermine] (blue curves); heating rate 10 K·min^{-1}, N_2 atmosphere with a volume flow of 50 mL·min^{-1} (DSC) and Ar atmosphere with a volume flow of 60 mL·min^{-1} (TGA). This figure is adapted from Ref. [8] by permission of John Wiley & Sons, Inc.

3.1.2.5 Spirocyclic oxazasilines, oxazagermines, diazasilines, and diazagermines

Spirocyclic silicon-containing compounds as well as spirocyclic germanium-containing compounds that are structurally related to Si-Spiro are also accessible by applying either 2-(aminomethyl)phenols, (2-aminophenyl)methanols, or 2-(aminomethyl)anilines rather than salicyl alcohol as reactant. Thus, various oxazasilines, oxazagermines, diazasilines, and diazagermines were synthesized and studied with regard to their suitability for TP (Scheme 3.41) [34, 35].

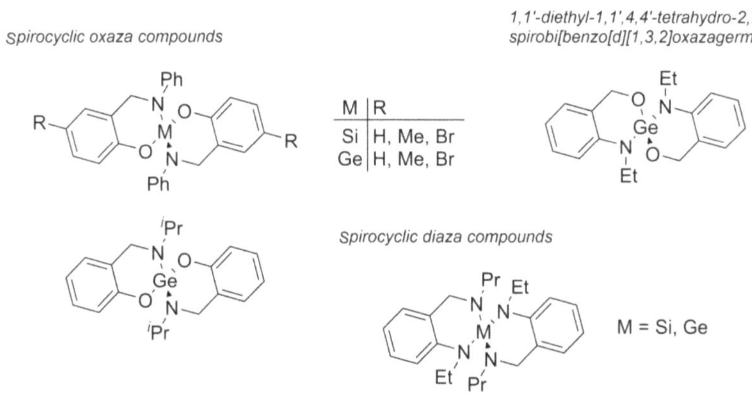

Scheme 3.41: Illustration of the molecular structures of a series of oxazasilines and oxazagermines as well as of spirocyclic diaza compounds that were studied with regard to their suitability for twin polymerization.

However, none of these compounds was suitable for any type of TP. At this point it should be noted that by contrast, spirocyclic diazasiline compounds do polymerize thermally since the alkyl groups at the nitrogens are substituted by hydrogen atoms. This feature is discussed in Chapter 4.1 of this book. With regard to a proton-catalyzed TP, it is obvious that addition of a Brønsted–Lowry acid gives an ammonium species due to the preferred protonation of the amino groups rather than resulting in any reactive species that is able to start a polymerization reaction. DFT-D calculations (B3LYP/def2-TZVPP level of theory including COSMO corrections using the permittivity of CH_2Cl_2) that were performed to screen the reactivity of these compounds are in agreement with the latter thesis.

Studies on the thermal behavior of these compounds revealed that the oxazasilines 3,3′-diphenyl-3,3′,4,4′-tetrahydro-2,2′-spirobi[benzo[*e*][1,3,2]oxazasiline] and 6,6′-dimethyl-3,3′-diphenyl-3,3′,4,4′-tetrahydro-2,2′-spirobi[benzo[*e*][1,3,2]oxazasiline] vaporize almost completely without decomposition at higher temperatures (> 300 °C), whereas 6,6′-dibromo-3,3′-diphenyl-3,3′,4,4′-tetrahydro-2,2′-spirobi[benzo[*e*][1,3,2]oxazasiline] decomposes exhibiting an overall weight loss of ≈ 58 % at these temperatures [34]. In contrast, the oxazagermines decompose stepwise to give porous materials composed of germanium and carbon (Ge@C) eventually. These decomposition reactions can occur at temperatures as low as 120 °C (3,3′-diisopropyl-3,3′,4,4′-tetrahydro-2,2′-spirobi[benzo[*e*][1,3,2]oxazagermine]), but they are induced at temperatures not higher than 217 °C (3,3′-diphenyl-3,3′,4,4′-tetrahydro-2,2′-spirobi[benzo[*e*][1,3,2]oxazagermine]) [34]. Notably, 2-(phenyliminomethyl)phenol and 2-(phenylaminomethyl)phenol as well as 2-(isopropyliminomethyl)phenol and 2-(isopropylaminomethyl)phenol were determined as major volatile decomposition products during the decompositions of the corresponding oxazagermines, respectively [34]. The formation of the 2-(iminomethyl)-phenols was attributed to a β-hydride elimination process occurring during the decomposition reactions. This process might be a result of Ge–N bond cleavages upon thermal annealing of the oxazagermines providing a hydrogenated intermediate species that release the imines later on (Scheme 3.42). The formation of 2-(aminomethyl)phenols was reported to account for reactions of the precursors with traces of water [34].

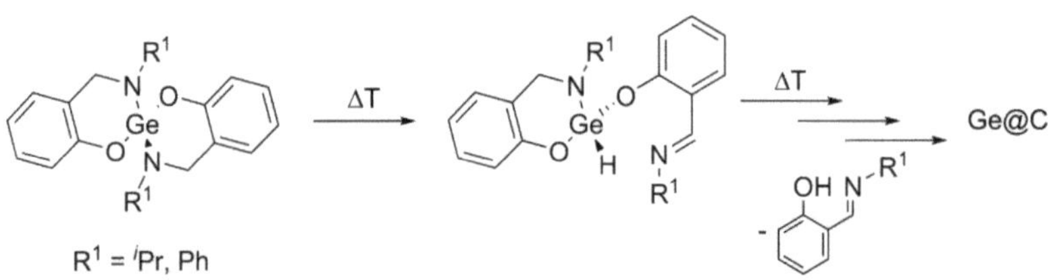

Scheme 3.42: Proposed reaction pathway for a β-hydride elimination to give Ge@C hybrid materials upon thermal annealing of 3,3′-diphenyl-3,3′,4,4′-tetrahydro-2,2′-spirobi[benzo[*e*]-[1,3,2]oxazagermine] (R^1 = Ph) and 3,3′-diisopropyl-3,3′,4,4′-tetrahydro-2,2′-spirobi[benzo[*e*]-[1,3,2]oxazagermine] (R^1 = ^{i}Pr) eventually. This figure is reproduced from Ref. [34] with permission of The Royal Society of Chemistry (RSC) on behalf of the Centre National de la Recherche Scientifique (CNRS) and the RSC.

Please note that with regard to the molecular structure of these oxazagermines, the formation of quinone methides as by-product of decomposition reactions upon their thermal annealing seems to be conceivable. Moreover, any formation of quinone methides might result in reaction cascades similar to the ones reported for thermally induced TP of other spirocyclic precursors. However, the formation of 2-(iminomethyl)phenols seems to be favored under the applied conditions. This may explain that these spirocyclic oxazagermines are not suitable for TP. The oxazagermine 1,1′-diethyl-1,1′,4,4′-tetrahydro-2,2′-spirobi[benzo[*d*][1,3,2]oxazagermine] (see Scheme 3.41) decomposes at much higher temperature (onset temperature 392 °C) also resulting in porous Ge@C materials, whereas the spirocyclic diaza compounds 1,1′-diethyl-3,3′-dipropyl-3,3′,4,4′-tetrahydro-1*H*,1′*H*-2,2′-spirobi[benzo[*d*][1,3,2]diazasiline] and 1,1′-diethyl-3,3′-dipropyl-3,3′,4,4′-tetrahydro-1*H*,1′*H*-2,2′-spirobi[benzo[*d*][1,3,2]diazagermine] vaporize above 274 °C and 281 °C, respectively. Please note that these three compounds are not able to form quinone methides with regard to their molecular structures. Thus, the reactivity observed upon thermal annealing of these compounds is in agreement with the latter [35].

Regarding a Lewis-base-catalyzed TP of these compounds, it is noteworthy that their nitrogen atoms at the amino groups possess intrinsically the features of a Lewis base. However, they do not induce any polymerization reaction as determined by thermal analyses of these compounds. Moreover, experimental studies toward Lewis-base-catalyzed TP of 1,1′-diethyl-1,1′,4,4′-tetrahydro-2,2′-spirobi[benzo[*d*][1,3,2]oxazagermine] and 1,1′-diethyl-3,3′-dipropyl-3,3′,4,4′-tetrahydro-1*H*,1′*H*-2,2′-spirobi[benzo[*d*][1,3,2]diazagermine] using DBU as Lewis base did not induce any polymerization reaction. The fact that the compounds are not suitable for Lewis-base-catalyzed TP may be attributed to two aspects. (i) The presence of the substituents at the amino groups causes steric hindrance aggravating attacks of the Lewis acidic central atom, i.e., Si or Ge, by a Lewis base. (ii) Even if the Lewis acidic group 14 atom can form a Lewis acid–base adduct with a Lewis base, no reactive species will be formed which is able to induce any polymerization reaction. The latter may also hold for the oxazasilines and oxazagermines possessing an oxygen atom in phenolic position at their aromatic moieties with regard to the observation that these compounds rather form 2-(iminomethyl)phenols than quinone methides upon thermal annealing.

3.1.2.6 Spirocyclic tin salicyl alcoholates

Spirocyclic tin salicyl alcoholates being structurally related to Si-Spiro are suitable for TP [10]. In contrast to the spirocyclic silicon salicyl alcoholates, these tin-containing precursors, e.g., 4*H*,4′*H*-2,2′-spirobi[benzo[*d*][1,3,2]dioxastannine], 6,6′-dimethyl-4*H*,4′*H*-2,2′-spirobi[benzo[*d*][1,3,2]dioxastannine], and 6,6′-dimethoxy-4*H*,4′*H*-2,2′-spirobi[benzo[*d*][1,3,2]dioxastannine] (Scheme 3.43), are not monomeric in the solid state. As a consequence, these precursors are poorly soluble in common organic solvents such as diethyl ether, dichloromethane, or toluene. However, a higher solubility was observed for the compounds exhibiting substituents at the aromatic moieties, e.g., 6,6′-dimethyl-4*H*,4′*H*-2,2′-spirobi[benzo[*d*][1,3,2]dioxastannine] and 6,6′-dimethoxy-4*H*,4′*H*-2,2′-spirobi[benzo[*d*][1,3,2]dioxastannine]. All spirocyclic tin salicyl alcoholates exhibit chemical shifts of their tin atoms in the range between $\delta = -683$ ppm and -705 ppm as determined by solid state $^{119}Sn\{^1H\}$-CP-MAS NMR spectroscopy ($\{^1H\}$ means ^1H decoupling during data aquisition, CP means cross polarization, MAS means magic angle spinning of the sample) for the pristine powders of these precursors. This is indicative of a hexacoordination of the tin atoms by six oxygen-containing donors [10]. With this in mind, the solid state structure of a Lewis acid–base adduct of 8,8′-di-*tert*-butyl-6,6′-dimethyl-4*H*,4′*H*-2,2-spirobi[benzo[*d*][1,3,2]-dioxastannine] with two molecules of dimethyl sulfoxide, determined by single crystal X-ray diffraction, confirms that its tin atom prefers hexacoordination (Scheme 3.43b) [32]. Please note that 8,8′-di-*tert*-butyl-6,6′-dimethyl-4*H*,4′*H*-2,2-spirobi[benzo[*d*][1,3,2]-dioxastannine] is not suitable for TP regarding its sterically demanding substituents that hamper any polymerization process.

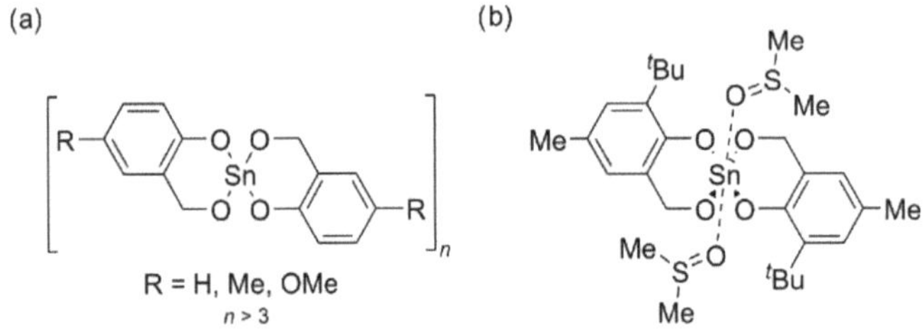

Scheme 3.43: Illustration of (a) the coordination polymeric structural motif of spirocyclic tin salicyl alcoholates such as 4*H*,4′*H*-2,2′-spirobi[benzo[*d*][1,3,2]dioxastannine] (R = H), 6,6′-dimethyl-4*H*,4′*H*-2,2′-spirobi[benzo[*d*][1,3,2]dioxastannine] (R = Me), and 6,6′-dimethoxy-4*H*,4′*H*-2,2′-spirobi[benzo[*d*][1,3,2]dioxastannine] (R = OMe), and (b) the molecular structure motif of the Lewis acid–base adduct of 8,8′-di-*tert*-butyl-6,6′-dimethyl-4*H*,4′*H*-2,2′-spirobi[benzo[*d*][1,3,2]-dioxastannine] with two dimethyl sulfoxide molecules as determined by single crystal X-ray diffraction analysis [32].

In addition to these experimental results, DFT calculations (at the B3LYP/SVP level of theory) were performed to elucidate the solid state structures of the spirocyclic tin-containing compounds on a reliable theoretical basis. These studies revealed that spirocyclic tin salicyl alcoholates tend to form larger oligomeric units such as tetramers exhibiting tin atoms being hexacoordinated by the oxygen atoms of the salicyl alcoholate moieties (Fig. 3.7) [30]. Moreover, with regard to the relative energies of diverse optimized tetrameric structures, oligomers exhibiting only additional coordination of the phenolic oxygen atoms to their tin atoms are less favorable in comparison to oligomers of the spirocyclic tin salicyl alcoholates that also possess additional coordination of the benzylic oxygen atoms to their tin atoms (Fig. 3.7) [30]. Thus, considering all experimental and theoretical data, it is assumed that spirocyclic tin salicyl alcoholates form Lewis acid–base adducts in the solid state to give at least oligomeric entities. In the case of 4*H*,4′*H*-2,2′-spirobi[benzo[*d*][1,3,2]dioxastannine], the formation of coordination polymers was proposed [30].

TP of the spirocyclic tin salicyl alcoholates is carried out in substance rather than in solution phase due to their low solubility. Notably, the spirocyclic tin salicyl alcoholates show a much lower reactivity toward processes that are assigned to their TP in comparison to their homolog spirocyclic silicon salicyl alcoholates. This may be a result of the distinguishable molecular structures of these precursors. For instance, TP of 4*H*,4′*H*-2,2′-spirobi[benzo[*d*][1,3,2]dioxastannine] was achieved by mixing this precursor with a Brønsted–Lowry acid such as 4-methylbenzene-1-sulfonic acid and upon heating the mixture to 280 °C. Thus, elevated temperatures are a prerequisite to polymerize this precursor by a proton-catalyzed polymerization process. The latter is in agreement with results of DFT calculations that were performed in order to study the reactivity of a monomer and dimers of 4*H*,4′*H*-2,2′-spirobi[benzo[*d*][1,3,2]dioxastannine] upon their protonation. Similar to the spirocyclic silicon salicyl alcoholates, the formation of *ortho*-substituted benzylic cations starting from the protonated species was calculated to be the most favorable reaction paths being able to trigger proton-catalyzed TP

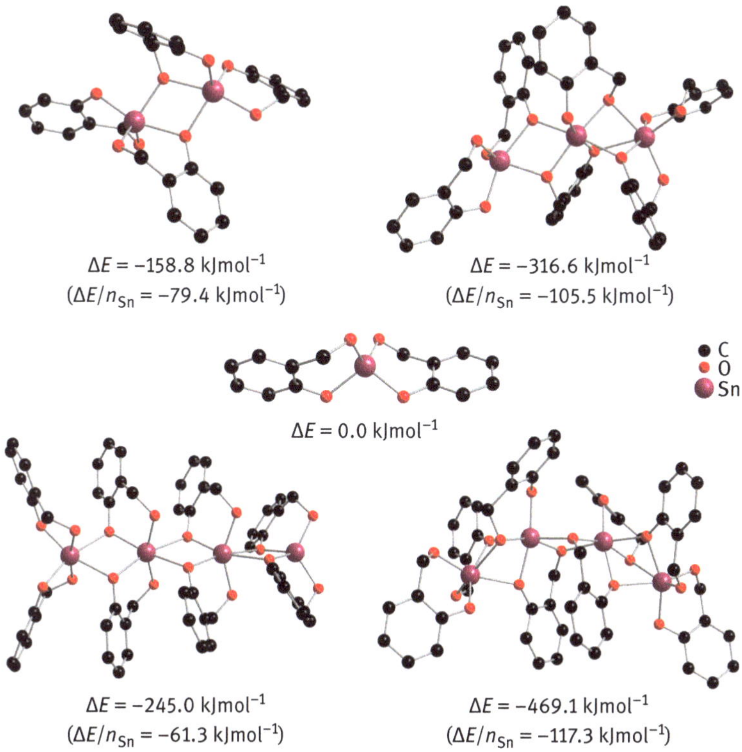

Fig. 3.7: Illustration and relative energies (ΔE) of optimized oligomeric structures of 4*H*,4′*H*-2,2′-spirobi[benzo[*d*][1,3,2]dioxastannine] that were determined by DFT calculations at the B3LYP/SVP level of theory. The relative energies related to one monomeric unit ($\Delta E/n_{Sn}$) are given in parentheses. Hydrogen atoms have been omitted for clarity. The molecular structures and the relative energies are presented according to results reported in Ref. [30].

of 4*H*,4′*H*-2,2′-spirobi[benzo[*d*][1,3,2]dioxastannine]. However, the formation of a benzylic cation starting from the protonated monomer was determined to be an exothermic reaction, whereas the formation of a benzylic cation starting from protonated dimers of 4*H*,4′*H*-2,2′-spirobi[benzo[*d*][1,3,2]dioxastannine] is either an endothermic process or an almost thermally neutral process depending on the specific benzylic cation that is formed (Scheme 3.44) [30]. Moreover, significantly larger reaction barriers were calculated for the formation of the benzylic cations starting from the protonated dimers of 2,2′-spirobi[benzo[*d*][1,3,2]dioxastannine] (Scheme 3.44) [30]. Please note that a downstream reaction of the energetically favored benzylic cation that was obtained starting from a protonated dimer of 2,2′-spirobi[benzo[*d*][1,3,2]dioxastannine] may be aggravated by steric hindrance. Thus, based on these results and regarding the experimental data, it has been assumed that the energetics of the initial step of proton-catalyzed TP are altered to be endothermic in the case of spirocyclic tin salicyl alcoholates due to their formation of oligomeric or even polymeric molecular structures in the solid state.

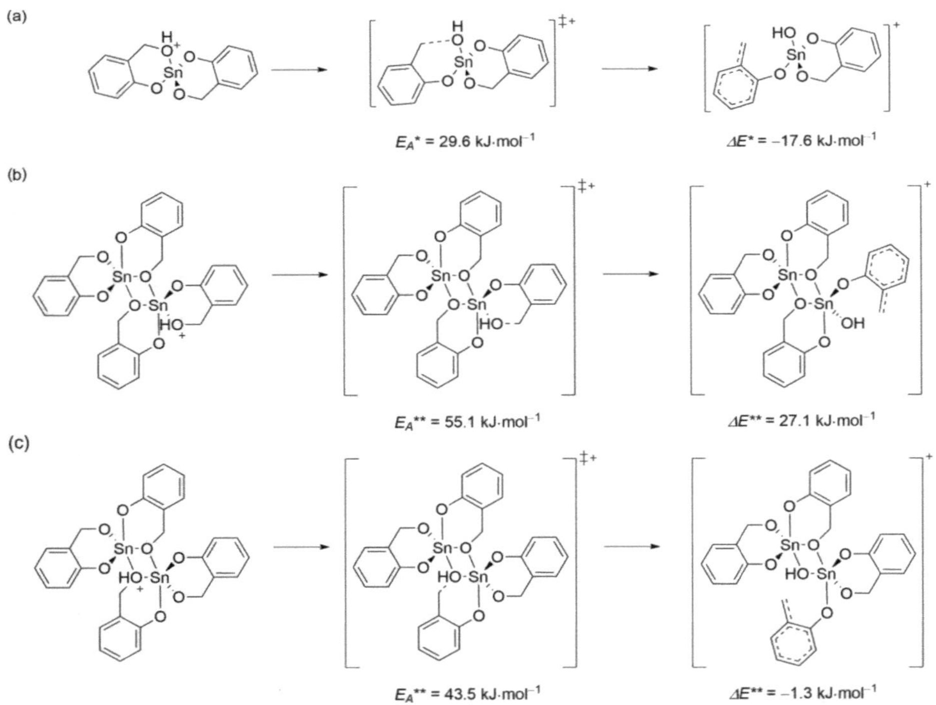

Scheme 3.44: Illustration of the most favorable reaction paths that were determined by DFT calculations starting from (a) a protonated monomer or (b) and (c) starting from a protonated dimer of 4*H*,4′*H*-2,2′-spirobi[benzo[*d*][1,3,2]dioxastannine] to give *ortho*-substituted benzylic cations being able to trigger proton-catalyzed twin polymerization of this precursor. In (a), the reaction barrier (E_A*) and reaction energy (Δ*E**) are referred to the protonated monomer of 4*H*,4′*H*-2,2′-spirobi[benzo[*d*][1,3,2]dioxastannine] (left) and were calculated at the B3LYP/def2-TZVPP level of theory. In (b) and (c), the reaction barriers (E_A**) and reaction energies (Δ*E***) are referred to protonated dimers of 4*H*,4′*H*-2,2′-spirobi[benzo[*d*][1,3,2]dioxastannine] (left) and were calculated at the B3LYP/def2-SVP level of theory.

Heating of the 2,2′-spirobi[benzo[*d*][1,3,2]dioxastannine] results in its decomposition to give undefined species [10]. Thus, this precursor is not suitable for thermally induced TP. In contrast, 6,6′-dimethoxy-4*H*,4′*H*-2,2′-spirobi[benzo[*d*][1,3,2]dioxastannine] is suitable for thermally induced TP. Heating of this precursor results in its almost quantitative conversion into an organic/inorganic hybrid material consisting of a phenolic resin and SnO_2 [32]. The latter may result from two effects that are caused by the presence of methoxy groups in *para*-position with respect to the phenolic oxygen atoms at the aromatic moieties of this spirocyclic tin salicyl alcoholate. First, these groups are sterically demanding. Therefore, oligomeric entities rather than coordination polymers of this compound are formed in the solid state as indicated by its higher solubility in common organic solvents than observed for 4*H*,4′*H*-2,2′-spirobi[benzo[*d*][1,3,2]dioxastannine]. This may affect its reactivity. Second, the electronic features of the methoxy groups further enhance its reactivity toward its thermally induced TP. Here, the triggering of the polymerization process is facilitated by the ability of its substituents to donate more electron density into the aromatic moieties than compared to hydrogen atoms. With this in mind, a similar reactivity scale of these precursors according to the electronic features of their substituents as determined for the spirocyclic silicon salicyl alcoholates is proposed.

Another way to overcome the poor reactivity of the spirocyclic tin salicyl alcoholates, e.g., 4*H*,4′*H*-2,2′-spirobi[benzo[*d*][1,3,2]dioxastannine] and 6,6′-dimethyl-4*H*,4′*H*-2,2′-spirobi[benzo[*d*][1,3,2]dioxastannine], toward their TP is provided by applying the concept of STP. The latter is defined as the polymerization of at least two types of precursors simultaneously in a single process step [3]. Here, mixtures of the spirocyclic tin salicyl alcoholates with a spirocyclic silicon salicyl alcoholate such as Si-Spiro can be polymerized without addition of a catalyst or an initiator at temperatures below 100 °C to give an organic/inorganic hybrid material consisting of phenolic resin, SiO_2 and SnO_2 (Scheme 3.45) [10, 36]. The polymerization processes are self-catalyzed due to the Lewis acidic nature of the spirocyclic tin salicyl alcoholates. As mentioned in Chapter 3.1.2.3.2.1, the formation of a Lewis acid–base adduct due to coordination of a benzylic oxygen atom of the spirocyclic silicon salicyl alcoholate to a Lewis acidic tin atom is assumed to result in the formation of a benzylic cation triggering the polymerization process (see Scheme 3.22b in Chapter 3.1.2.3.2.1). Notably, better conversions of the precursors into the hybrid material were observed for mixtures of Si-Spiro with spirocyclic tin salicyl alcoholates possessing electron-density-donating substituents at their aromatic moieties such as 6,6′-dimethoxy-4*H*,4′*H*-2,2′-spirobi[benzo[*d*][1,3,2]-dioxastannine] [10].

3.1.2.7 Precursors based on dialkyl group 14 salicyl alcoholates

Dialkyl group 14 salicyl alcoholates represent another type of cyclic precursors that are structurally related to Si-Spiro. Two representatives, namely, 2,2-dimethyl-4*H*-benzo[*d*]-[1,3,2]dioxasiline and 6-bromo-2,2-di-*tert*-butyl-4*H*-benzo[*d*][1,3,2]dioxagermine, of such

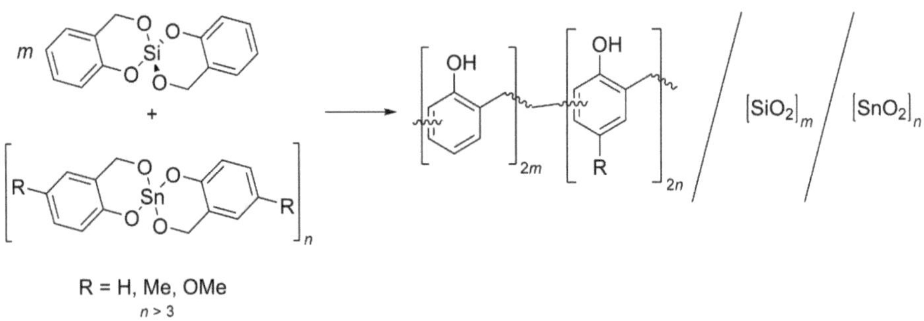

Scheme 3.45: Illustration of simultaneous twin polymerization of mixtures of Si-Spiro with spirocyclic tin salicyl alcoholates, e.g., *4H,4′H*-2,2′-spirobi[benzo[*d*][1,3,2]dioxastannine] (R = H), 6,6′-dimethyl-*4H,4′H*-2,2′-spirobi[benzo[*d*][1,3,2]dioxastannine] (R = Me), and 6,6′-dimethoxy-*4H,4′H*-2,2′-spirobi[benzo[*d*][1,3,2]dioxastannine] (R = OMe).

compounds have been reported so far [37–39]. Analyses of the thermochemical behavior of these precursors revealed that these compounds are not suitable for thermally induced TP due to their high vapor pressure. However, *ortho*-quinone methide formation was detected by gas chromatography in combination with mass spectrometry (GC–MS) studies on 6-bromo-2,2-di-*tert*-butyl-4*H*-benzo[*d*][1,3,2]dioxagermine at elevated temperatures (Fig. 3.8) [38]. Therefore, thermally induced TP may generally be feasible for this type of compounds on the condition that they possess lower vapor pressures.

TP of these precursors is achieved upon addition of a Brønsted–Lowry acid such as trifluoromethanesulfonic acid or trichloroacetic acid to solutions of the compounds or to their melts [38, 39]. Thus, the catalysis process of the polymerization may follow a similar reaction cascade as reported for the spirocyclic silicon salicyl alcoholates, which starts with the formation of an *ortho*-benzylic cation upon protonation of the monomer. Electrophilic aromatic substitution reactions first propagate the formation of a phenolic resin, whereupon downstream condensation reactions of the as-formed metalloid hydroxyl groups give either silicone or germocane species (see Schemes 3.22–3.24 in Chapter 3.1.2.3.2.1). Analysis of the as-obtained hybrid materials revealed that the silicone and germocane products mainly consist of cyclic oligomeric species that can be extracted from the phenolic resin matrix. Thus, proton-catalyzed TP of the dialkyl group 14 salicyl alcoholates 2,2-dimethyl-4*H*-benzo[*d*][1,3,2]dioxasiline and 6-bromo-2,2-di-*tert*-butyl-4*H*-benzo[*d*][1,3,2]dioxagermine results in hybrid materials consisting of phenolic resin and cyclic oligomers of either dimethyl silicones or di-*tert*-butyl germocanes (Scheme 3.46), respectively.

Studies on the reactivity of mixtures of 2,2-dimethyl-4*H*-benzo[*d*][1,3,2]dioxasiline with Lewis bases such as DBU and TBAF revealed that this precursor is also suitable for Lewis-base-catalyzed TP [39]. The onset temperatures (≈ 130–140 °C) of the polymerization processes are slightly higher than the onset temperatures (≈ 115–125°C)

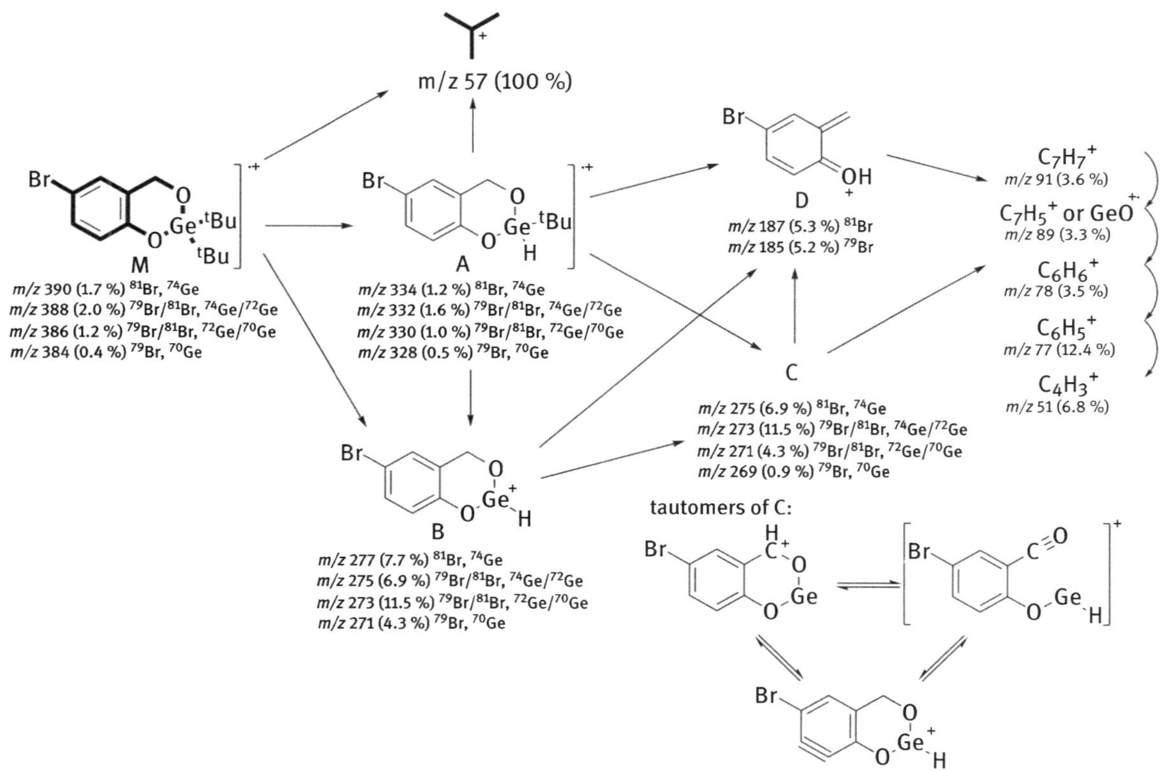

Fig. 3.8: Fragmentation scheme of 6-bromo-2,2-di-*tert*-butyl-4*H*-benzo[*d*][1,3,2]dioxagermine as obtained by gas chromatography in combination with mass spectrometry (GC–MS) studies at elevated temperatures (injection temperature: 230 °C). The proposed structure of the molecular ion (M) and further species (A–C) occurring upon fragmentation of 6-bromo-2,2-di-*tert*-butyl-4*H*-benzo[*d*][1,3,2]dioxagermine to finally result in quinone methide formation (D) are presented. This figure is adapted from Ref. [38] with permission from Elsevier Limited, Oxford (United Kingdom), Copyright 2014.

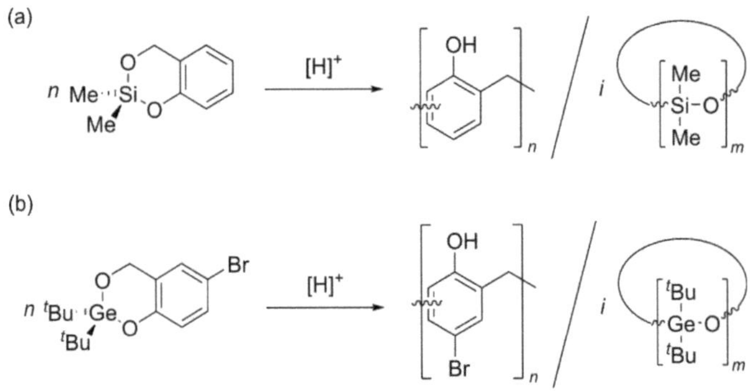

Scheme 3.46: Proton-catalyzed twin polymerization of (a) 2,2-dimethyl-4*H*-benzo[*d*][1,3,2]dioxasiline and (b) 6-bromo-2,2-di-*tert*-butyl-4*H*-benzo[*d*][1,3,2]dioxagermine.

that were determined for the Lewis-base-catalyzed TPs of Si-Spiro using the same Lewis bases [39]. Notably, the latter compound possesses a higher stress envelop of its heterocycles (the [Si–O–CH_2–C_{Aryl}=C_{Aryl}–O–] cycles). As a consequence, Si-Spiro is expected to show a higher reactivity toward nucleophilic reactions with a Lewis base than 2,2-dimethyl-4*H*-benzo[*d*][1,3,2]dioxasiline being in agreement with the experimental observations. With this in mind, it is assumed that the initial processes of their Lewis-base-catalyzed TP follow a similar reaction cascade starting from the formation of *ortho*-quinone methide resulting from the reaction of the monomers with the Lewis base (Scheme 3.47).

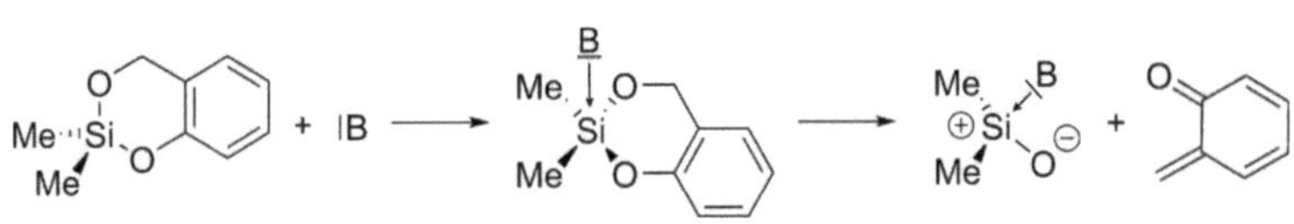

Scheme 3.47: Illustration of the proposed reaction path triggering the process of Lewis-base-catalyzed twin polymerization of 2,2-dimethyl-4*H*-benzo[*d*][1,3,2]dioxasiline.

Please note that STP of 2,2-dimethyl-4*H*-benzo[*d*][1,3,2]dioxasiline with Si-Spiro provides hybrid materials consisting of phenolic resin and covalently linked SiO_2-dimethylsiloxane phases [39, 40]. The formation of volatile cyclic dimethyl siloxanes is inhibited due to the reaction of the dimethyl silanol intermediates with silanol groups that are concurrently formed as a result of the polymerization process of Si-Spiro. It is also noteworthy that STP of the latter precursors can be induced either thermally, by addition of a Brønsted–Lowry acid, or even by addition of a Lewis base, respectively [39, 40].

3.1.2.8 Germylenes and stannylenes

Compounds that are derivatives of salicyl alcoholates and exhibit a low valent group 14 element are more easily accessible from precursors possessing heavier group 14 elements such as germanium and tin than starting from precursors possessing a lighter group homolog, e.g., silicon, due to the more pronounced ability of the heavier elements to feature formally low oxidation states, e.g., +II. Therefore, only syntheses of germylenes, namely, germanium(II) 2-(oxidomethyl)phenolate, germanium(II) 4-methyl-2-(oxidomethyl)phenolate, and germanium(II) 4-bromo-2-(oxidomethyl)-phenolate, and stannylenes, namely, tin(II) 4-methyl-2-(oxidomethyl)phenolate, tin(II) 4-methoxy-2-(oxidomethyl)phenolate, and tin(II) 5-methoxy-2-(oxidomethyl)-phenolate, being derivatives of salicyl alcoholates, have been reported up to now [36, 41–43]. These precursors are suitable for thermally induced TP attributed to their ability to form quinone methides upon thermal annealing [44]. Addition of a Brønsted–Lowry acid to their solutions results in decomposition reactions of the compounds, whereas Lewis acid–base adducts are formed upon addition of a Lewis base such as *N,N*-dimethylpyridin-4-amine to their solutions [41].

3.1.2.8.1 Germanium(II) salicyl alcoholates

Germanium(II) salicyl alcoholates form dimers, trimers, and tetramers in the solid state as shown by single crystal X-ray diffraction analysis (Scheme 3.48) [41, 42]. These oligomeric species are in a chemical equilibrium in solution phase [41, 42], and

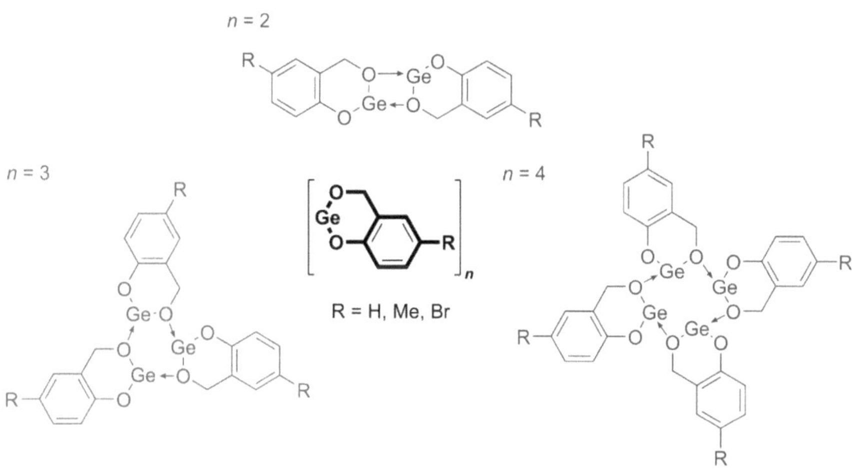

Scheme 3.48: Illustration of the molecular structure of germanium(II) 2-(oxidomethyl)phenolate (R = H), germanium(II) 4-methyl-2-(oxidomethyl)phenolate (R = Me), and germanium(II) 4-bromo-2-(oxidomethyl)-phenolate (R = Br) and their oligomeric structures as determined by single crystal X-ray diffraction analyses. A dimer, trimer, or tetramer is depicted in violet, green, and blue, respectively.

thus, different solid state forms of the germylenes are accessible. For instance, germanium(II) 2-(oxidomethyl)phenolate was obtained as crystals of its trimer, as crystals of its tetramers incorporating one molecule of *n*-pentane per two tetrameric units, and as amorphous powder (Scheme 3.48) [42].

DSC studies of amorphous powders of the germylenes did not reveal a melting point. As a consequence, their TP proceeds in the solid state. The temperatures at which their TP is triggered cannot be related to the electronic and/or to steric features of the substituents at their aromatic moieties (Fig. 3.9). This is in contrast to the results of similar studies toward the reactivity of the spirocyclic silicon salicyl alcoholates (see Chapter 3.1.2.3.2.2). Notably, thermally induced TP of the spirocyclic silicon salicyl alcoholates was always performed in liquid phase, e.g., in melt. Thus, a reactivity scale based on the electronic and steric features of the substituents at the aromatic moieties of precursors based on salicyl alcoholates may only be applicable for polymerization processes carried out in liquid phase resolving the latter apparent contradiction.

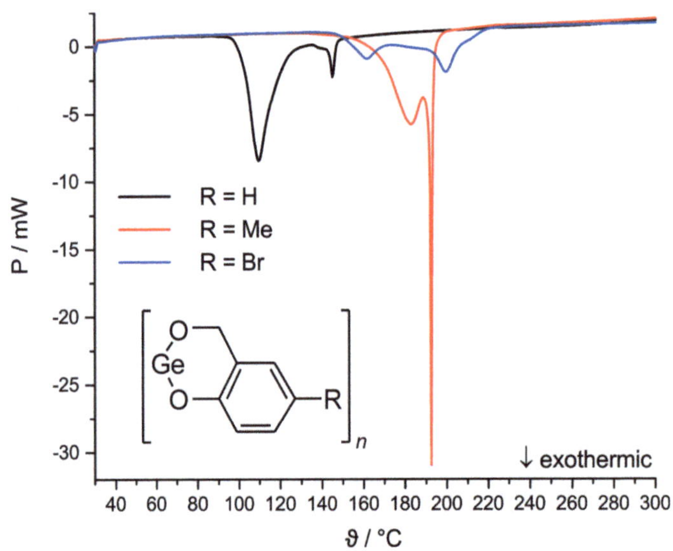

Fig. 3.9: Differential scanning calorimetry of the amorphous materials of germanium(II) 2-(oxidomthyl)phenolate (black curve), germanium(II) 4-methyl-2-(oxidomethyl)phenolate (red curve), and germanium(II) 4-bromo-2-(oxidomethyl)phenolate (blue curve); heating rate 10 $K{\cdot}min^{-1}$, N_2 atmosphere, N_2 volume flow 50 $mL{\cdot}min^{-1}$. This figure is reproduced from Ref. [42] with permission of The Royal Society of Chemistry.

However, a significant dependency on the solid state form of the germylenes, e.g., amorphous versus crystalline material, was determined by DSC studies of germanium(II) 2-(oxidomethyl)phenolate (amorphous powder vs. crystalline trimers vs. crystals of tetramers) (Fig. 3.10). Higher onset temperatures and narrower temperature ranges of the polymerization processes were observed for the crystalline materials rather than for its amorphous powder, whereas the reaction enthalpies ($\Delta H \approx 70$ kJ·mol^{-1}) and the temperature ranges of the second minima that are indicative of a downstream reaction process are comparable for all distinguishable solid state forms (Fig. 3.10). With this in mind, it was concluded that amorphous materials possess smaller reaction barriers for the initial step of thermally induced TP, whereas the initial step is aggravated for crystalline material presumably due to a regularly densely packed structure. On the other hand, once the polymerization process is triggered, a preorganized arrangement of the precursors facilitates the polymerization process.

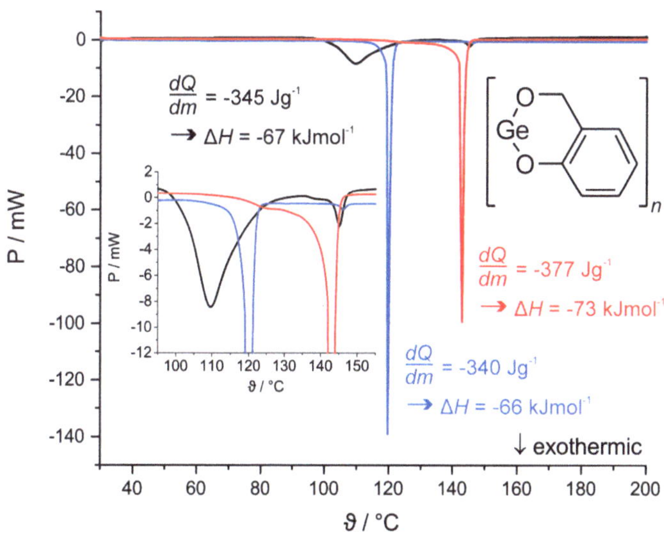

Fig. 3.10: Differential scanning calorimetry of amorphous germanium(II) 2-(oxidomethyl)phenolate (black curve), crystals of trimers of germanium(II) 2-(oxidomethyl)phenolate (red curve), and crystals of tetramers of germanium(II) 2-(oxidomethyl)phenolate incorporating one molecule of *n*-pentane per two tetramers of germanium(II) 2-(oxidomethyl)phenolate (blue curve); heating rate 10 K·min^{-1}, N_2 atmosphere, N_2 volume flow 50 mL·min^{-1}. The inlet magnifies the range between 95 °C and 155 °C. The determined d*Q*/d*m* values and the as-calculated molar reaction enthalpies (Δ*H*) that refer to one monomeric unit of germanium(II) 2-(oxidomethyl)phenolate of the exothermic processes are depicted in black for amorphous germanium(II) 2-(oxidomethyl)phenolate, red for crystals of trimers of germanium(II) 2-(oxidomethyl)phenolate, and blue for crystals of tetramers of germanium(II) 2-(oxidomethyl)phenolate incorporating one molecule of *n*-pentane per two tetramers of germanium(II) 2-(oxidomethyl)phenolate within the crystals. This figure is reproduced from Ref. [42] with permission of The Royal Society of Chemistry.

Thermally induced TP of germanium(II) salicyl alcoholates provided hybrid materials consisting of phenolic resin and germanium-containing compounds such as germanium oxide. This was expected by theory (see Scheme 3.21 in Chapter 3.1.2.3.1). However, in addition to the resonance signals that are typical for phenolic resins of such hybrid materials, further resonance signals of low intensity with chemical shifts centered at δ = 16 ppm were detected by solid state $^{13}C\{^1H\}$-CP-MAS NMR spectroscopy of these hybrid materials (Fig. 3.11). The latter resonance signals were assigned to small portions of species exhibiting Ge–CH_2 connectivity (Fig. 3.11). The presence of these Ge–CH_2 groups is indicative of side reactions occurring during the polymerization processes. Thus, TP of the germylenes does not follow an ideal course, even when their polymerization processes proceed smoothly.

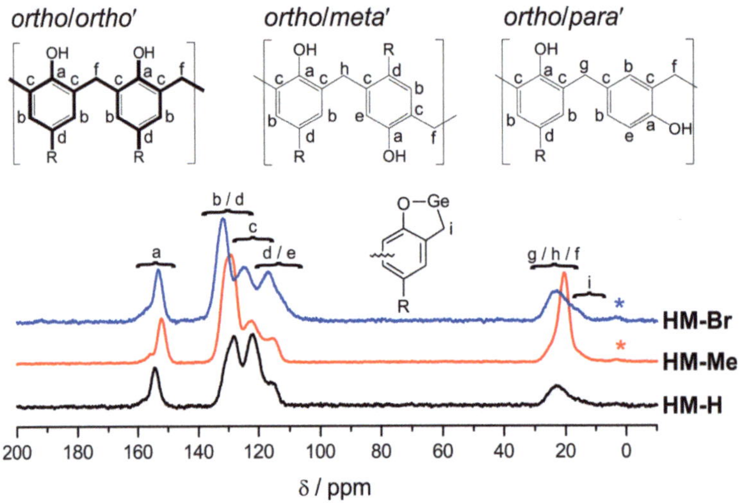

Fig. 3.11: Solid state $^{13}C\{^1H\}$-CP-MAS NMR spectra of the germanium-containing phenolic resin hybrid materials as obtained by thermally induced twin polymerization of germanium(II) 2-(oxidomethyl)-phenolate (HM-H – black curve), germanium(II) 4-methyl-2-(oxidomethyl)phenolate (HM-Me – red curve), and germanium(II) 4-bromo-2-(oxidomethyl)phenolate (HM-Br – blue curve). The asterisks mark spinning side bands. Assignment of the resonance signals is given for the assigned connectivity patterns of the bridging methylene groups (*ortho*/*para'* only for HM-H). The connectivity pattern of the bridging methylene groups that is assumed to be prevailing (*ortho/ortho'*) is shown in bold. This figure is reproduced from Ref. [42] with permission of The Royal Society of Chemistry.

Studies on the reactivity of germanium(II) 2-*tert*-butyl-4-methyl-6-(oxidomethyl)-phenolate provided an explanation for the formation of the by-products exhibiting the Ge–CH_2 groups and being incorporated in the hybrid materials during the polymerization processes. Notably, germanium(II) 2-*tert*-butyl-4-methyl-6-(oxidomethyl)phenolate (Scheme 3.49) is not suitable for TP due to its sterically demanding substituents at its aromatic moiety. Its substituents hamper the formation of a phenolic resin; thus, oligomers rather than polymers are obtained upon thermal annealing of this germylene [41]. Germanium(II) 2-*tert*-butyl-4-methyl-6-(oxidomethyl)phenolate is not stable in solution and it is slowly converted into

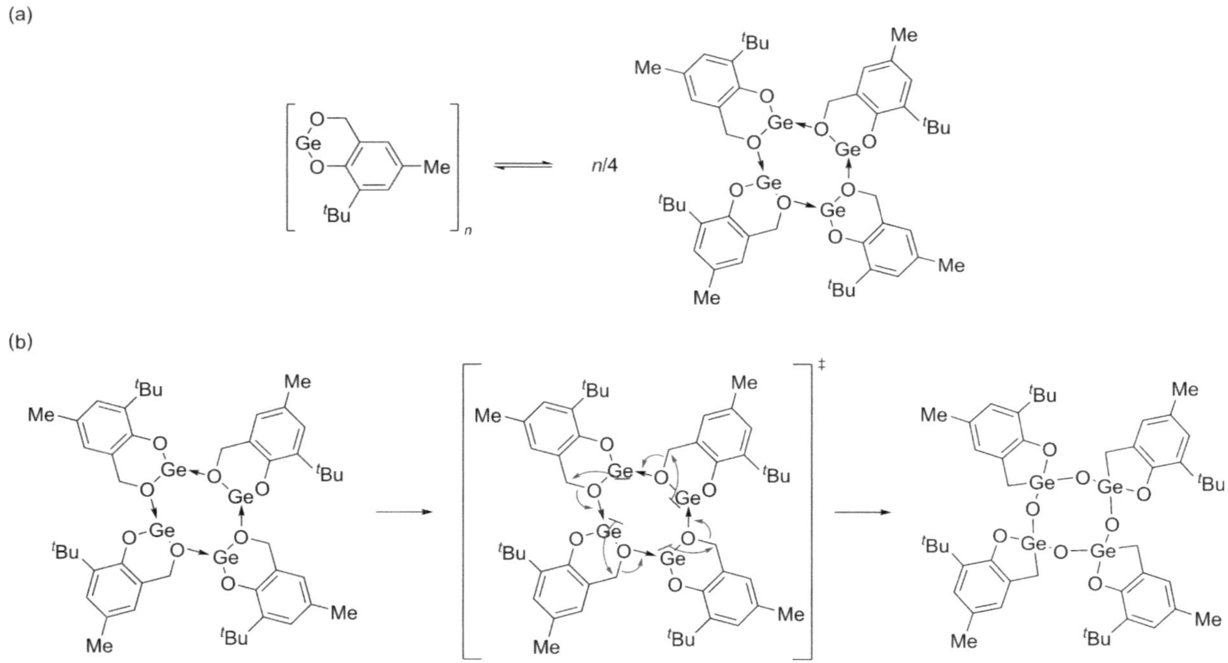

Scheme 3.49: Illustration of (a) the equilibrium being present in solution phases of germanium(II) 2-*tert*-butyl-4-methyl-6-(oxidomethyl)phenolate providing tetramers of this germylene and (b) proposed rearrangement process explaining the formation of Ge–CH_2 groups by an intramolecular C–O insertion reaction to give a cyclic germocane starting from the tetramer of germanium(II) 2-*tert*-butyl-4-methyl-6-(oxidomethyl)phenolate. This scheme was drawn according to schemes of Ref. [41].

2,4,6,8-tetra-(3-*tert*-butyl-5-methyl-2-oxidophenyl)methanide-1,3,5,7,2,4,6,8-tetraoxidogermocane in solution phase (Scheme 3.49) [41]. NMR spectroscopy studies and DFT-D calculations (at the B3LYP-D3/def2-TZVPP level of theory) revealed that this germylene quickly forms equilibriums between its dimers, trimers, and tetramers in solution phase (Scheme 3.49a) [41]. These oligomers undergo downstream intramolecular C–O insertion reactions to give cyclic germocanes exhibiting Ge–CH_2 groups (Scheme 3.49b) [41].

Please note that the formation of tetramers of germanium(II) 2-*tert*-butyl-4-methyl-6-(oxidomethyl)phenolate was determined to be slightly energetically favored by the DFT-D calculations. Thus, the singular isolation of 2,4,6,8-tetra-(3-*tert*-butyl-5-methyl-2-oxidophenyl)methanide-1,3,5,7,2,4,6,8-tetraoxidogermocane may be a result of a preferred formation of this compound starting from tetramers of this germylene. However, regarding the energetics of the oligomers of all of the presented germanium(II) salicyl alcoholates and the energetics of germocanes that may be formed starting from the different oligomers of germanium(II) 2-*tert*-butyl-4-methyl-6-(oxidomethyl)phenolate [41, 42], it is assumed that the formation of Ge–CH_2 groups *via* intramolecular C–O insertion reactions represents a reaction path being generally feasible for all oligomers of germanium(II) salicyl alcoholates. Moreover, there is no apparent reason why the oligomers of these germylenes should not undergo such intramolecular C–O insertion reactions in the solid state. Therefore, it is assumed that the formation of by-products exhibiting Ge–CH_2 connectivity during the polymerization process of the germanium(II) salicyl alcoholates results from intramolecular C–O insertion reactions of their oligomers occurring upon their thermal annealing as a side reaction.

3.1.2.8.2 Tin(II) salicyl alcoholates

There have only been reported few studies on tin(II) salicyl alcoholates up to now. These precursors, e.g., tin(II) 4-methyl-2-(oxidomethyl)phenolate, tin(II) 4-methoxy-2-(oxidomethyl)phenolate, and tin(II) 5-methoxy-2-(oxidomethyl)phenolate, are not monomeric in the solid state (Scheme 3.50). The chemical shifts of their tin atoms determined by solid state $^{119}Sn\{^1H\}$-CP-MAS NMR spectroscopy are in the range between $\delta = -420$ ppm and $\delta = -445$ ppm, indicating a tetrahedral coordination of the tin atoms by four oxygen atoms. The fourfold coordination of the low valent tin atoms was attributed to the presence of coordination polymers in the solid state resulting from Lewis acid–base adduct formation (Scheme 3.50) [43]. The coordination polymers are expected to resemble the typical polymeric chain structure that was reported for tin alkoxides possessing low valent tin atoms (Scheme 3.50) [43, 45]. Notably, these stannylenes are poorly soluble in nonpolar solvents and they are just moderately soluble in polar solvents such as dimethyl sulfoxide. However, ^1H NMR spectroscopy studies of a solution of tin(II) 4-methyl-2-(oxidomethyl)phenolate in dimethyl sulfoxide revealed that these precursors form oligomers in solution phase that are in an equilibrium with each other [43].

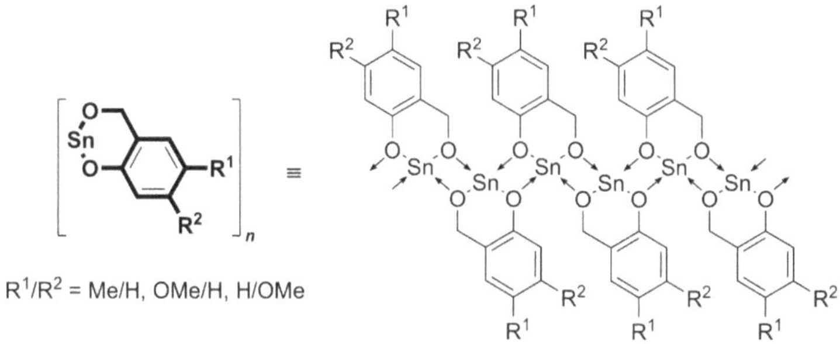

Scheme 3.50: Illustration of the molecular structure of tin(II) 4-methyl-2-(oxidomethyl)phenolate (R^1/R^2 = Me/H), tin(II) 4-methoxy-2-(oxidomethyl)phenolate (R^1/R^2 = OMe/H), and tin(II) 5-methoxy-2-(oxidomethyl)phenolate (R^1/R^2 = H/OMe) and their proposed structures in the solid state according to Ref. [43].

Studies on the reactivity of the tin(II) salicyl alcoholates were exemplarily performed on tin(II) 4-methoxy-2-(oxido-methyl)phenolate and tin(II) 5-methoxy-2-(oxidomethyl)phenolate by thermal analyses. These stannylenes show complex thermochemical behaviors upon thermal annealing. Bulk phase experiments revealed that they do not melt and decompose visibly at temperatures above 250 °C [36]. However, thermogravimetric analyses already detected minor weight losses of ≈ 4 % occurring continuously at temperatures below 250 °C (Fig. 3.12). The weight losses are indicative of decomposition reactions that might result from the formation of low-molecular weight compounds such as quinone methides. The thermochemical character of the decomposition reactions is endothermic at temperatures below 150°C, but it becomes highly exothermic at higher temperatures (Fig. 3.12). The exothermic processes were assigned to their thermally induced TP. In addition, the exothermic processes possess clearly distinguishable onset temperatures being 157 °C starting from tin(II) 4-methoxy-2-(oxidomethyl)phenolate and 244 °C starting from tin(II) 5-methoxy-2-(oxidomethyl)phenolate (Fig. 3.12). Thus, a different reactivity of these precursors upon their thermal annealing is obvious. The molecular structure of these stannylenes differs only by the position of the methoxy group at their aromatic moieties. Thus, on the condition that the specific arrangement of the coordination polymers formed in the solid state does not or does just slightly modify their reactivity, the apparent difference in their reactivity must be attributed to the specific position of their methoxy groups at their aromatic moieties. Consequently, a methoxy group in *para*-position of the phenolic oxygen atom facilitates the triggering of TP of tin(II) salicyl alcoholates, whereas a methoxy group in *para*-position of the benzylic methylene group (being in *meta*-position to the phenolic oxygen atom) does not seem to facilitate its initial step.

Notably, STP of mixtures of tin(II) 4-methoxy-2-(oxidomethyl)phenolate and tin(II) 5-methoxy-2-(oxidomethyl)phenolate with Si-Spiro, respectively, is triggered at temperatures below 100 °C due to the Lewis acidity of the tin atoms of the tin(II) salicyl alcoholates and the reactivity of the spirocyclic silicon compound [36].

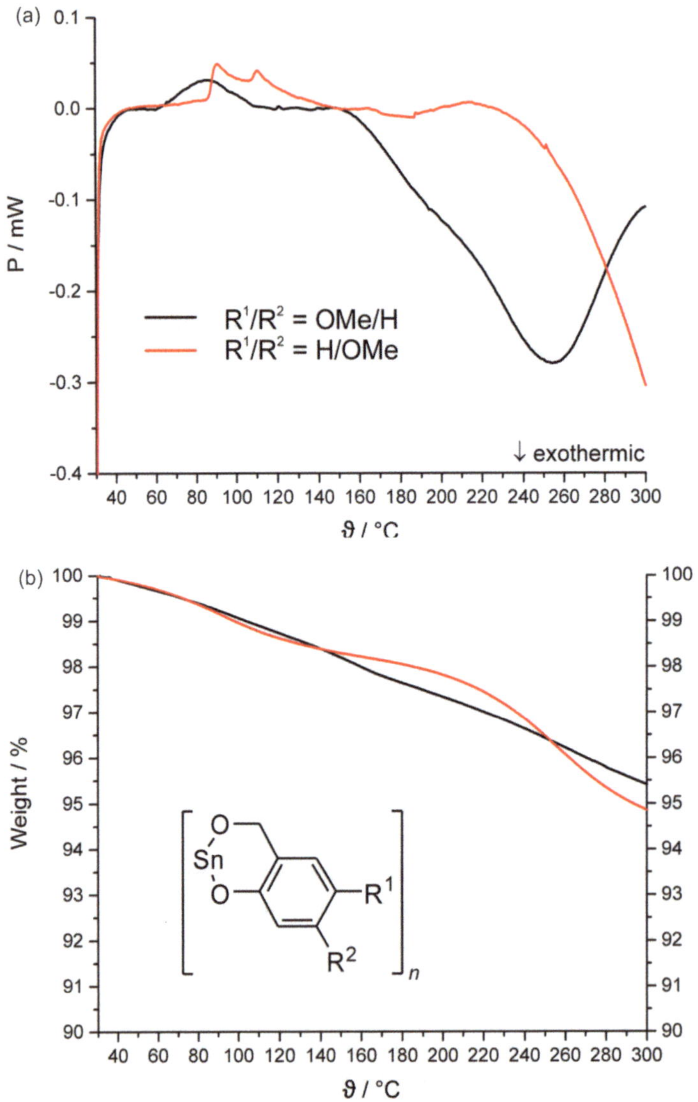

Fig. 3.12: Differential scanning calorimetry given in (a) and thermogravimetric analysis given in (b) of tin(II) 4-methoxy-2-(oxidomethyl)phenolate (black curve), tin(II) 5-methoxy-2-(oxidomethyl)phenolate (red curve); heating rate 10 K·min^{-1}, N_2 atmosphere, N_2 volume flow 50 mL·min^{-1}.

3.1.2.8.3 Molecular germanates based on salicyl alcoholates

Molecular germanates are a particular type of precursors that are suitable for TP. These compounds, e.g., bis(dimethylammonium) tris[2-(oxidomethyl)phenolate(2-)]germanate, bis(dimethylammonium) tris[4-methyl-2-(oxidomethyl)phenolate(2-)]germanate, and bis(dimethylammonium) tris[4-bromo-2-(oxidomethyl)phenolate(2-)]germanate, are composed of dimethylammonium cations and germanate dianions consisting of three salicyl alcoholate moieties and a central germanium atom [9]. They are accessible by the reaction of $Ge(NMe_2)_4$ with salicyl alcohols that do not exhibit steric hindrance near their hydroxyl groups. Single crystal X-ray diffraction analysis revealed that the dimethylammonium cations and the dianions of these molecular germanates form one-dimensional, infinite hydrogen bridged networks to give cation–anion chains [9]. The germanium atoms of the dianions are hexacoordinated by the oxygen atoms of the salicyl alcoholate moieties with benzylic and phenolic oxygen atoms being opposite to each other in the solid state [9]. However, NMR spectroscopic analyses of these molecular germanates revealed that they are in an equilibrium in solution between their dianionic structure as determined in the solid state and species possessing pentacoordinated anions (Scheme 3.51) [9].

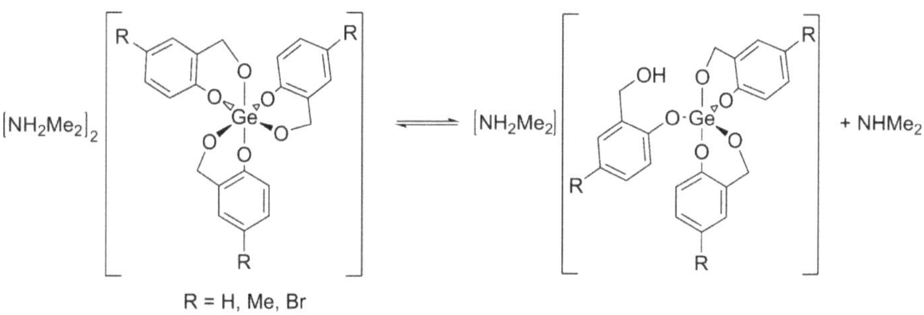

Scheme 3.51: Equilibrium between molecular germanates that were determined by ^{1}H NMR spectroscopy of bis(dimethylammonium) tris[2-(oxidomethyl)phenolate(2-)]germanate, bis(dimethylammonium) tris[4-methyl-2-(oxidomethyl)phenolate(2-)]germanate, and bis(dimethylammonium) tris[4-bromo-2-(oxidomethyl)phenolate(2-)]germanate. This scheme was drawn according to schemes of Ref. [9].

Dimethylammonium bis[2-*tert*-butyl-4-methyl-6-(oxidomethyl)phenolate(2-)][2-*tert*-butyl-4-methyl-6-(hydroxymethyl)phenolate(1-)]germanate represents a molecular germanate exhibiting fivefold coordination of its germanium atom by two phenolic and three benzylic oxygen atoms of its salicyl alcoholate moieties in the solid state. This germanate does not form an equilibrium in solution due to the steric hindrance of its *tert*-butyl groups in *ortho*-position at the salicyl alcoholate moieties [9]. Moreover, dimethylammonium bis[2-*tert*-butyl-4-methyl-6-(oxidomethyl)phenolate(2-)] [2-*tert*-butyl-4-methyl-6-(hydroxymethyl)phenolate(1-)]-germanate is not suitable for TP. It shows a complex thermochemical behavior upon its thermal annealing attributed to decomposition reactions (Fig. 3.13 violet curves). The latter is in contrast to the reactivity of bis(dimethylammonium)-tris[2-(oxidomethyl)phenolate(2)]germanate, bis(di-methylammonium) tris[4-methyl-2-(oxidomethyl)phenolate(2-)]germanate, and bis(dimethylammonium) tris[4-bromo-2-(oxidomethyl)phenolate(2-)]germanate providing hybrid materials upon their thermally induced TP. These precursors also exhibit complex thermochemical behaviors upon their thermal annealing, but they show dominant exothermic processes with onset temperatures in the range between 173 °C and 180 °C that are assigned to their thermally induced TP (Fig. 3.13). Regarding the molecular structure of the anions of the germanates, it was assumed that dimethylammonium bis[2-*tert*-butyl-4-methyl-6-(oxidomethyl)phenolate(2-)][2-*tert*-butyl-4-methyl-6-(hydroxymethyl)phenolate(1-)]germanate is not suitable for TP due to the sterically demanding substituents at its aromatic moieties that hamper the formation of a phenolic resin [9]. Notably, TP of the molecular germanates can only be triggered thermally which is attributed to their ability to form quinone methides upon thermal annealing. Addition of a Brønsted–Lowry acid, Lewis acid, or Lewis base to their solutions does not result in TP, which is easily comprehensible regarding their molecular nature.

Bulk phase experiments revealed that the molecular germanates do not melt prior to polymerization. However, morphologic changes of their powders were visible at temperatures below the temperature range in which their TP was achieved, which indicates some decomposition reactions and is in agreement with thermal analysis data (Fig. 3.13). The temperatures at which TP of these molecular germanates is triggered cannot be related to the electronic and/or steric features of the substituents at their aromatic moieties. This is in contrast to the results of similar studies on the reactivity toward thermally induced TP of the spirocyclic silicon salicyl alcoholates. But it is in accordance with studies on thermally induced TP of germanium(II) salicyl alcoholates. TP of these germylenes was also performed in the solid state, whereas TP of the spirocyclic silicon salicyl alcoholates was carried out in liquid phases, e.g., in melt. Thus, these observations further support the conclusion previously drawn in Chapter 3.1.2.8.1. If TP of precursors based on salicyl alcoholates is performed in the solid state, the influence of the electronic and steric features of the substituents at their aromatic moieties on their reactivity toward the initial step of their TP is negligible.

Thermogravimetric analyses determined weight losses up to 34 % occurring during the polymerization process of germanates being suitable for TP. The latter is in agreement with bulk phase experiments. It is noteworthy that, for instance, a weight loss of only 20.7 % was expected for TP of bis(dimethylammonium) tris[2-(oxidomethyl)phenolate(2-)]germanate due to the formation of $HNMe_2$ and water during the polymerization process as indicated in Scheme 3.21b (see Chapter 3.1.2.3.1). However, other volatile by-products such as 2-[(dimethylamino)-methyl]phenol are additionally formed during its polymerization process providing an explanation for the higher weight loss that was observed. Notably, solid state $^{13}C\{^{1}H\}$-CP-MAS NMR spectroscopy of the as-obtained hybrid material that mainly consists of a phenolic resin and GeO_2 revealed resonance signals that were assigned to small portions of terminating CH_2NMe_2 groups presumably resulting from partial incorporation of 2-[(dimethylamino)methyl]phenol into the phenolic resin matrix

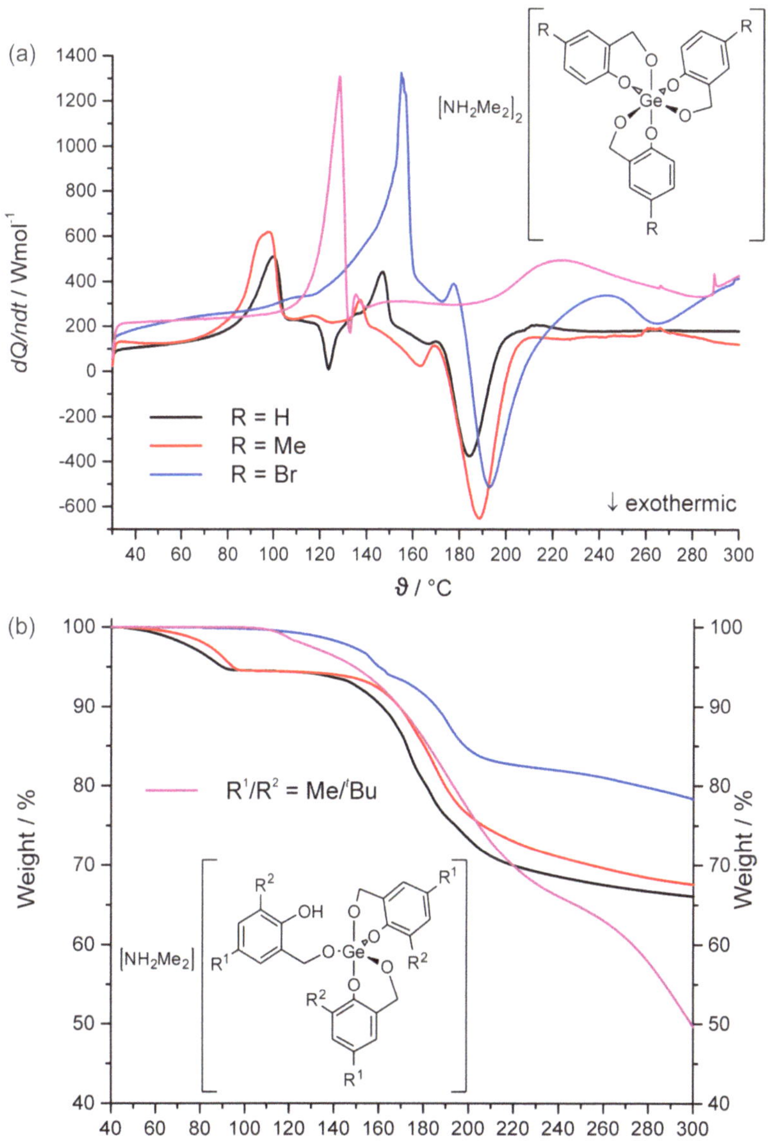

Fig. 3.13: Differential scanning calorimetry in (a) and thermogravimetric analysis given in (b) of bis(dimethylammonium) tris[2-(oxidomethyl)phenolate(2-)]germanate (black curve), bis(dimethylammonium) tris[4-methyl-2-(oxidomethyl)phenolate(2-)]germanate (red curve), bis(dimethylammonium) tris[4-bromo-2-(oxidomethyl)phenolate(2-)]germanate (blue curve), and dimethylammonium bis[2-*tert*-butyl-4-methyl-6-(oxidomethyl)phenolate(2-)][2-*tert*-butyl-4-methyl-6-(hydroxymethyl)phenolate(1-)]germanate (violet curve); heating rate 10 K·min^{-1}, N_2 atmosphere, N_2 volume flow 50 mL·min^{-1}. The figure is drawn according to Ref. [9] with permission of The Royal Society of Chemistry.

during the polymerization process (Fig. 3.14). Thus, TP of the molecular germanates does not proceed in an ideal manner.

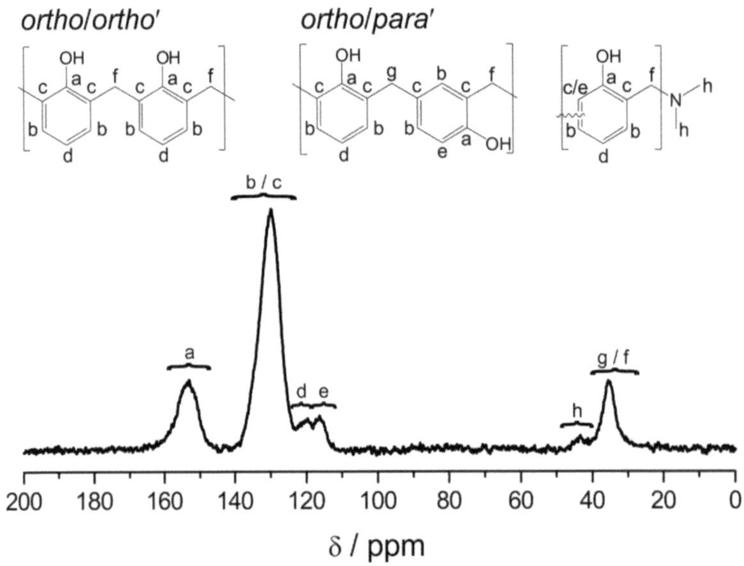

Fig. 3.14: Solid state $^{13}C\{^{1}H\}$-CP-MAS NMR spectrum of germanium-containing phenolic resin hybrid material as obtained by thermally induced twin polymerization of bis(dimethylammonium) tris[2-(oxidomethyl)phenolate(2-)]germanate given with assignment of the resonance signals. This figure is reproduced from Ref. [9] with permission of The Royal Society of Chemistry.

3.1.3 Summary

In this chapter, relationships between the structural features of the precursors being suitable for TP and their reactivity in the polymerization process were discussed. In the following, a brief summary of the main aspects of these relationships is presented.

- Precursors that are suitable for TP are based on metal alkoxides and metalloid alkoxides being derivatives of furan-2-ylmethanolate, thiophen-2-ylmethanolate, phenylmethanolates, salicyl alcoholates, and 2-(mercaptomethyl)phenolate.
- TP of precursors that are derivatives of furan-2-ylmethanolate, thiophen-2-ylmethanolate, and phenylmethanolates is catalyzed by addition of acids and acid anhydrides to their solutions, whereas TP of precursors that are based on salicyl alcoholates can be catalyzed by either addition of acids, acid anhydrides, and Lewis bases to their liquid phases or even thermally induced.
- All precursors that are suitable for thermally induced and Lewis-base-catalyzed TP are able to form *ortho*-quinone methides with respect to their molecu-

lar structure. These reactive intermediates were detected either directly during the polymerization processes or indirectly by the presence of their reaction products within the as-obtained hybrid materials. Thus, it is supposed that the ability to form quinone methides is a prerequisite for precursors based on metal alkoxides or metalloid alkoxides to be suitable for thermally induced and/or Lewis-base-catalyzed TP.

- A reactivity scale based on the Hammett constants of the substituents at the aromatic moieties of spirocyclic silicon salicyl alcoholates is proposed. Thus, the reactivity of these precursors can be estimated with respect to the reactivity of Si-Spiro by comparison of the Hammett constants of their substituents at the salicyl alcohol moiety. Generally speaking, the triggering of TP for this class of compounds is facilitated by substituents that are electron density donating.
- Studies on the reactivity of germanium(II) salicyl alcoholates and molecular germanates that are based on salicyl alcoholates indicate that the influence of the electronic and steric features of substituents at their aromatic moieties on the initial step of their thermally induced TP is negligible. In contrast to the spirocyclic silicon salicyl alcoholates, the germanium(II) compounds do not melt prior to polymerization, which might explain the different performance.
- The nature of the metal or metalloid atom regarding its Lewis acidity, its formal oxidation state, and/or its ability to show hypercoordination influences the molecular structure of the precursors. For instance, spirocyclic silicon salicyl alcoholates are molecular entities possessing a tetrahedral coordination of the silicon atom and being precursors that smoothly undergo TP proceeding in an ideal manner to give nanostructured organic/inorganic hybrid materials. Spirocyclic germanium salicyl alcoholates that exhibit similar structures as their silicon homologs are not accessible, whereas spirocyclic tin salicyl alcoholates form coordination polymers possessing hexacoordinated tin atoms in the solid state. The latter compounds react sluggishly in TP to give hybrid materials exhibiting much larger sizes of each of the product domains, i.e., phenolic resin and SnO_2, in comparison to the phase domain sizes observed within the hybrid materials that were obtained starting from the respective spirocyclic silicon salicyl alcoholates. However, DFT-D calculations indicate that monomeric precursors differing only in the nature of their metal and/or metalloid atom but otherwise possessing the same molecular structure shall exhibit a similar reactivity toward their TP.
- Finally, hybrid materials exhibiting interpenetrating networks with phase domain sizes of the organic and inorganic polymers on the nanometer scale were only accessible starting from precursors that do not generate volatile byproducts upon TP. Such nanostructured hybrid materials are best obtained by TP of Si-Spiro and some of its derivatives. For these precursors, TP proceeds in an ideal manner.

Alexander A. Auer, Giovanni Bistoni

3.2 Theory of the TP – computational studies on the mechanism of the proton-catalyzed and thermally induced TP

3.2.1 Introduction

In this part, the mechanistic studies on the cationic and thermally induced twin polymerization (abbreviated as TP in the following) will be discussed, which have been carried out using quantum chemical methods – mostly DFT – in order to elucidate mechanistic details of the TP.

In the beginning, a word of caution should be given – studying polymerization processes by quantum chemistry is by no means an easy task. These types of reactions start in liquid phase and yield a "soft," mostly disordered, inhomogeneous solid. Hence, spectroscopic data being essential to benchmark computations are rare. Another thing to keep in mind is that the quality of computational results is also limited in many ways, which has to be acknowledged by experimentalists in order to assess how far statements made from simulations can carry. Hence, we will elaborate on this a little more in the next paragraph.

The basic limitations of electronic structure calculations can roughly be cast into three categories – approximations within the electronic structure theory applied to the molecule, approximations concerning the molecule's surroundings, like in solution or in the solid, and approximations to the complexity of the system as a whole, like disregarding impurities, assuming only a limited number of reaction paths, or focusing only on certain structures, etc.

The accuracy of the electronic structure method itself is usually not too difficult to assess, as most methods can be benchmarked with respect to experiment or close-to-exact methods. For DFT, this means that results for geometries and force constants are usually in very good agreement with molecular structures from single crystal X-ray diffraction analysis and IR-spectra [46–50]; relative energies have error bars of approximately 10–20 $kJ \cdot mol^{-1}$ and barriers of roughly 20 $kJ \cdot mol^{-1}$ [51–54]. While rates will be very difficult to predict with this accuracy, it usually suffices to assess whether a reaction path is plausible under the given reaction conditions.

Disregarding or approximating the solvent or packing effects in the solid can introduce significant errors if the solvent binds strongly to the molecule or large intramolecular interactions are present in the solid. However, approximations like polarizable continuum models [55–63] for implicit inclusion of solvation effects usually work well and errors should not be larger than 20 $kJ \cdot mol^{-1}$. While entropy contributions to reaction enthalpies can be computed on the basis of molecule's vibrational, rotational, and translational partition function, the estimate of entropy contributions in the case of oligomerization or fragmentation reactions is more difficult to compute. Hence, some uncertainty remains, for which only the direction of the correction can be estimated.

Last but not least, quantum chemical studies rely to some extend on the hypothesis made about a mechanism by chemists. In principle, electronic structure calculations will yield the energy for a given molecular geometry. Using the gradient of the energy with changes in the geometry, schemes for finding minimum structures and transition states can be devised, but most studies start from chemical intuition of what is possible and what is plausible if further experimental evidence for species involved is not available. As a consequence, it might turn out that perfectly valid results are useless, as a major reaction path has been overlooked, a critical intermediate was not included in the simulation, or a possible product was not taken into account.

Having phrased these concerns and keeping the limitations of computational methods in mind, one also has to acknowledge the power of electronic structure methods as tools in chemistry. These methods can nowadays not only routinely be applied to large systems and yield results with good accuracy, but they also yield essential insight that allows us to understand and rationalize structure–property relationships and aid in the design or improvement of processes, compounds, and materials.

So, while computational results alone may be difficult to relate to experiment, they can provide a deeper understanding of mechanistic details. Here, the essence is the close collaboration of experimentalists and theoreticians in order to be able to assess mechanistic hypothesis, plan computational studies, and to discuss the fundamental influences that govern the outcome of a synthesis.

In the following sections, we will describe the recent progress on understanding the mechanistic details of the proton-catalyzed and thermally induced TP. It should be noted that the studies mainly – also due to the high complexity – focus on the "ideal twin monomer" (see Fig. 3.15) and its polymerization, but that up to date many more compounds have been investigated in the framework of the TP. In principle, each of these will require separate detailed mechanistic studies; however, often basic mechanistic schemes are transferable.

This chapter is arranged as follows: In the first part, we focus on the proton-catalyzed TP of the "ideal twin monomer" (Si-Spiro, see Fig. 3.15), which has been used as

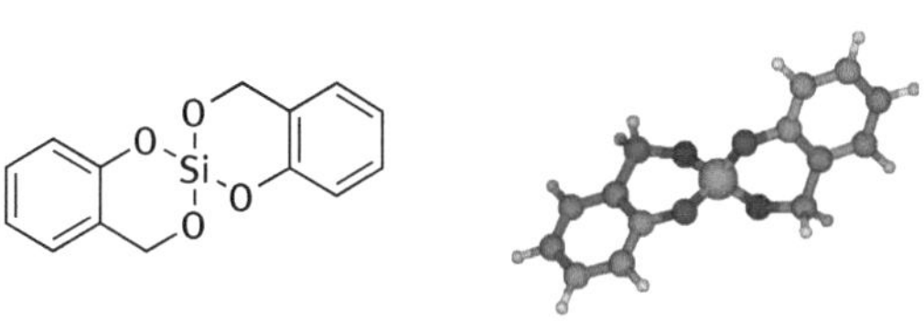

Fig. 3.15: Si-Spiro – 2,2′-spirobi[4*H*-1,3,2-benzodioxasiline] (left: Lewis structure, right: 3D molecular structure as obtained from a geometry optimization), that has been used throughout most studies discussed in the following.

prototype example and yields a nanostructured phenolic resin/SiO_2 hybrid material with unique features. In the second part, we review work on the thermally induced TP, which was found to be much more complex in its initial stages. We conclude the chapter by summarizing the basic insight obtained from the computational studies carried out so far.

3.2.2 Mechanistic studies on the proton-catalyzed TP

3.2.2.1 Proton catalysis and active species

One of the important developments in the pioneering work on TP by Spange *et al.* was that, at a very early stage, one prototype compound was identified that proved to be an ideal object of fundamental studies. Si-Spiro (Fig. 3.15) has a simple molecular structure, yet exhibits all features of an ideal twin monomer: Its polymerization yields the desired nanostructured hybrid material in principle without by-products; it can be obtained in large amounts and good purity and is also simple to handle apart from its tendency to undergo hydrolysis [2]. Hence, the study of this system has led to the deeper understanding of the structure formation process of the unique hybrid materials that are obtained by the TP.

In the beginning, when it was discovered that nanostructured hybrid materials of two materials that are actually immiscible can be obtained from one monomer in a single step [1, 3, 12], the first task was to unravel the mechanism and to identify possible intermediates. Note that Si-Spiro was explicitly designed to stoichiometrically yield SiO_2 and phenolic resin hypothetically without any by-products [2]. Starting from the monomer and the knowledge that after catalysis by small amounts of acid, the hybrid material formation proceeds with an initial gelation and, after some time, solidification, a computational study was carried out in the search for possible reaction paths.

As the monomer exhibits two pairs of equivalent oxygen atoms that are accessible for protonation, and for each oxygen, two bonds can be cleaved, a search along

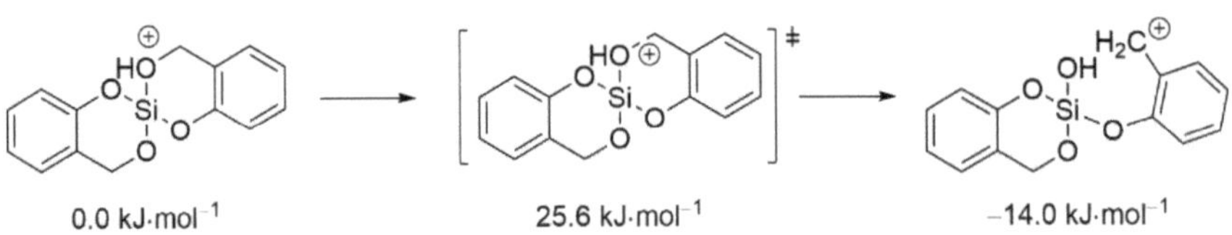

Scheme 3.52: Energetics for the formation of the active species in the proton-catalyzed twin polymerization of Si-Spiro calculated at the B3LYP-D3/TZVPPD + COSMO level of theory. For a comparison of the results obtained at various levels of DFT-D, see Tab. 3.2.

all possible protonation/activation coordinates of this type is trivial. In this case, it yielded results very much in agreement with chemical intuition. After protonation, the most labile bond is actually the Si–O–CH_2 carbon–oxygen bond and its cleavage yields a good electrophile in which the positive charge is stabilized by mesomeric effects (see Scheme 3.52).

The reaction energy calculated for this process is in the order of –10 kJ·mol^{-1} to –20 kJ·mol^{-1} with a low barrier in the order of 20 kJ·mol^{-1} to 30 kJ·mol^{-1}. While solvent effects may alter the values for the stability of the intermediate and the barrier for its formation, the rough number is in line with what is known for the formation of cationic intermediate species in the formation of phenolic resins [64–73].

Note that while the computational results are unambiguous for this case and also spectroscopic data like IR frequencies and NMR chemical shifts could be calculated (for B3LYP/def2-TZVPP computed results, see Fig. 3.16), there is probably not sufficient concentration of this intermediate to observe it experimentally. However, studies by Kitschke and Mehring support the hypothesis that the intermediate resulting from protonation is the bottleneck of the TP process as substituents in the monomer which would stabilize or destabilize the cationic active species have the corresponding effect in the polymerization process [7].

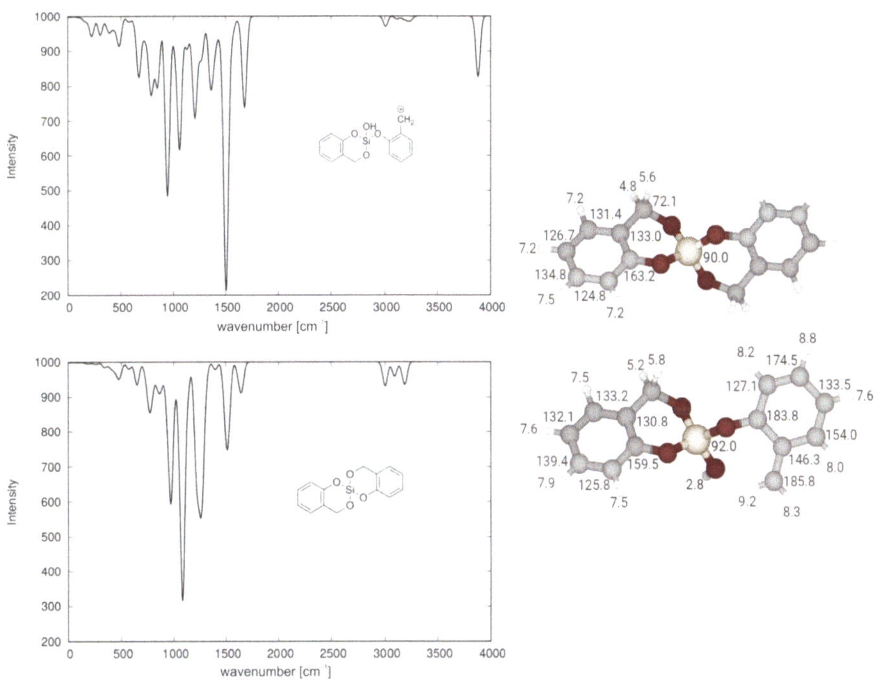

Fig. 3.16: Computed IR spectra (left) and NMR chemical shifts (right, with respect to TMS) of Si-Spiro and its active cationic species.

Hence, by estimating the stability of the active species using computational methods, one can already assess the ability of a given monomer to serve as a starting compound in the TP.

Coming back to the argument of the accuracy of the computational results, one may ask whether the standard DFT methods yield results of sufficient accuracy in this case. While accurate benchmark calculations on larger systems like this are inherently difficult due to the steep scaling of methods like CCSD(T) with system size, in recent years efficient approximations have been developed that allow to obtain near-quantitative results like the DLPNO-CCSD(T) approach [74, 75]. This has proven to be an ideal tool not only to obtain very accurate results in applications but also to benchmark DFT methods in the framework of larger studies [76–80].

In Table 3.2, the results for calculations of the relative energies of the protonated monomer and the active species are given. For this purpose, the geometries of all molecules were optimized at the B3LYP-D3/TZVPPD + COSMO level of theory. The transition state for the ring-opening reaction has been obtained by stretching the C–O bond while relaxing all other degrees of freedom. Based on these geometries, DLPNO-CCSD(T)/cc-pVQZ (TightPNO setting) calculations as well as a variety of DFT/TZVPPD calculations have been carried out. From the results, it becomes obvious that DFT has its deficiencies as it underestimates all barriers by 5–20 kJ·mol^{-1} and also overestimates the reaction energy by roughly the same amount. However, as stated before, while this might seem a large error at first, all results are roughly in qualitative agreement and also fall within the error bars known for the corresponding DFT methods.

Tab. 3.2: Benchmark results for a variety of DFT-D3 (using Becke-Johnson damping) methods in comparison with DLPNO-CCSD(T) results using triple zeta basis sets. (Note that the difference of the B3LYP-D3 results given in the fourth row in comparison to the values given in Scheme 3.52 is the solvation energy as obtained by the COSMO model.)

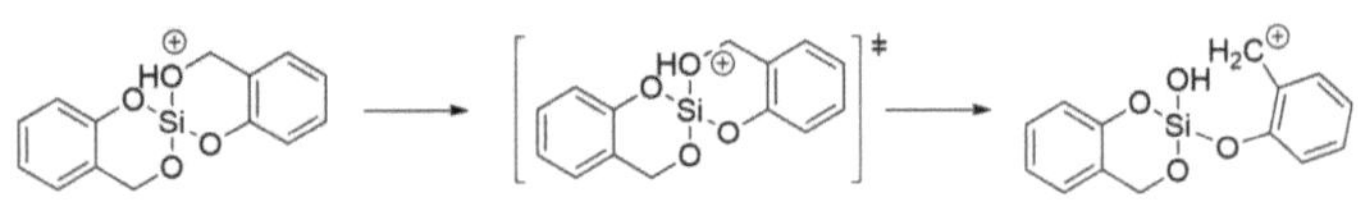

BLYP	0.0 kJ/mol	8.4 kJ/mol	−30.6 kJ/mol
PBE	0.0 kJ/mol	21.0 kJ/mol	−7.5 kJ/mol
BP86	0.0 kJ/mol	18.0 kJ/mol	−10.0 kJ/mol
B3LYP	0.0 kJ/mol	22.6 kJ/mol	−18.3 kJ/mol
PBEO	0.0 kJ/mol	36.8 kJ/mol	4.1 kJ/mol
PW6895	0.0 kJ/mol	32.5 kJ/mol	−4.8 kJ/mol
DLPNO-CCSD(T)	0.0 kJ/mol	41.1 kJ/mol	0.8 kJ/mol

One important question in the formation of the active species is the role of the counter ion or the actual acid molecule used as proton source, respectively. This is the focus of the computational study of Tchernook, Prehl, and Friedrich, who investigated the initial steps of the proton-catalyzed TP of Si-Spiro upon explicit inclusion of a trifluoroacetic acid (TFA) molecule (see Chapter 3.3). For the overall process, they find very similar results (using BP86-D3 + COSMO/SVP geometries and PW6B95-D3 + COSMO/TZVPP single point energies) [81, 82].

For the monomer activation to obtain an active cationic species, they find a reaction energy of 105 kJ·mol^{-1} and a barrier of 106 kJ·mol^{-1} for a monomer + acid model. For a dimer + acid model, they obtain a reaction energy of 109 kJ·mol^{-1} and a barrier of 129 kJ·mol^{-1}. While these values are large in comparison to the data given in Table 3.2, it has to be noted that these include the dissociation of the acid in an organic solvent. Estimates for the pK_a value of organic acids in organic solvents like acetic or picric acid in dichloroethane or acetonitrile usually have values between 10 and 20, which allows to estimate the dissociation free energies to approximately 60–120 kJ·mol^{-1} [83, 84]. These values roughly fit the difference observed between the reaction starting from the monomer plus proton and the reaction starting from the monomer plus acid molecule. This already hints at another important fact: If a certain acid is to be used as catalyst in the proton-catalyzed TP, the acidity has to be assessed in terms of pK_a values in the given organic solvent, and – as experience shows – water as possible impurity of acids like in lactic acid should also be regarded.

In their study, Tchernook *et al.* also obtain *ortho*-quinone methide as a possible by-product of the monomer fragmentation, which will play an important role in the further discussion. Most importantly, Tchernook *et al.* find an almost barrierless, exothermic process for the formation of a dimer from the active species in agreement with all previous studies. This reaction step is essential for the successive polymerization process, as will be discussed in the next paragraph.

3.2.2.2 Progress of the polymerization process

After the initial active species has been formed, the monomer solution contains a potent electrophile that is able to attack aromatic systems in a chain reaction of electrophilic substitutions, regenerating the initial proton. At the same time, hydroxyl groups at the silicon center lend themselves for initiation of the sol-gel-like process, ultimately yielding the SiO_2 component of the nanostructured hybrid material.

While mapping out the full structure formation process is impossible using electronic structure methods, as length- and timescales of the process are both enormous from an atomistic point of view, disassembling the process into its basic steps allows to at least qualitatively explore the potential energy landscape. The decisive reaction paths are essentially the formation of C–C bonds by electrophilic substitution and the formation of Si–O bonds by condensation. Furthermore, the initial steps in the polymerization process consist of activated monomers reacting with their surroundings; so focusing on monomer reactivity can also be viewed as the simulation of the very first steps from monomers toward oligomers.

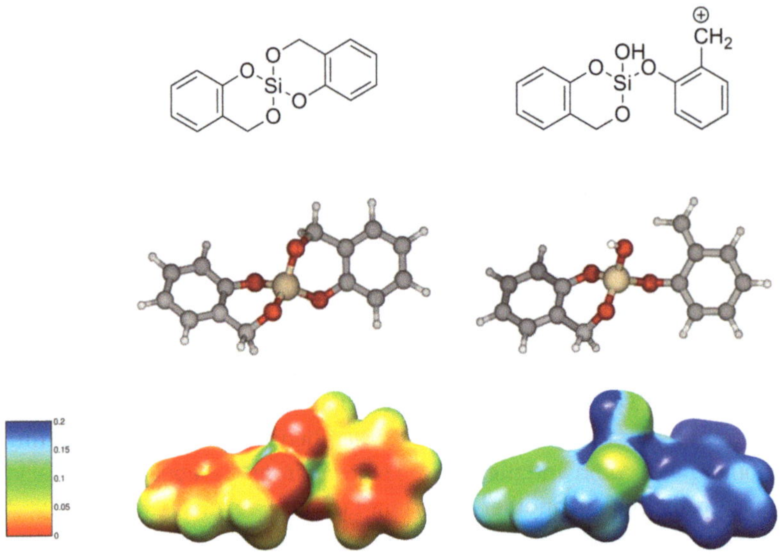

Fig. 3.17: Si-Spiro and its activated cationic form including the electrostatic potential as obtained at the DFT level of theory. Note how the positive charge is stabilized by delocalization in the aromatic system (bottom right).

The next steps to discuss are the electrophilic substitution reaction and the Si–O–Si bond formation of a monomer and an activated monomer. Already the very first computational studies of these compounds revealed a clear difference in the reaction energies and especially in the barriers of the two possible reaction paths. On the one hand, the electrophilic substitution proceeds with very low barriers, as it mostly involves shifts of protons in the organic part of the molecular framework. The restoration of the aromatic system in the cationic species renders this process thermodynamically favorable. On the other hand, the Si–O bond formation is hindered by the fact that an electron-rich oxygen atom approaching the central silicon has to penetrate the coordination sphere of four oxygen atoms, which are partially negatively charged themselves [4]. This is illustrated by the electrostatic potential given at the bottom of Figure 3.17. The plausibility of these results is also confirmed by other computational studies on related reactions [70–73, 85].

Based on these first results obtained for a very simple model system, one can already formulate a hypothesis on how the TP proceeds, as the energetics clearly show that resin formation and silica formation are barely competing processes – the much larger barriers for silica formation render this step probably orders of magnitudes slower. Thus, in the first stage of the polymerization, the organic network will quickly form by successive electrophilic substitutions, impeding demixing of other molecular fragments resulting from this process. This is also in line with the observation of a quick gelation of the reaction mixture. The formation of the silica component

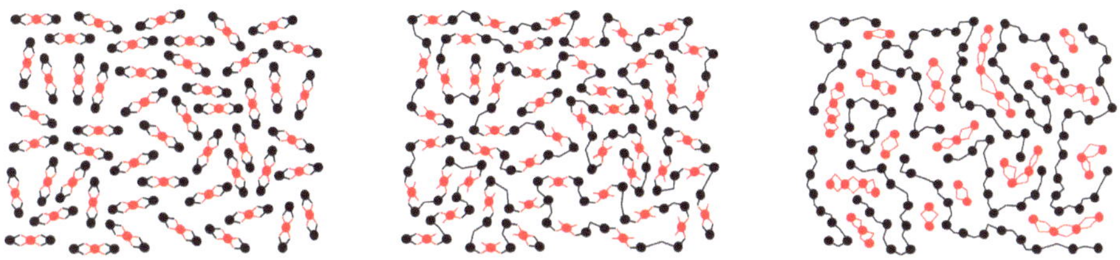

Fig. 3.18: Model for the structure formation process of Si-Spiro. Monomers are depicted as symbols consisting of an organic (black) and inorganic (red) part. Starting from the monomer precursor (left) in solution or melt, first the organic network is quickly formed in a process with small barriers (middle). At this stage, segregation of the two components in the organic network is impeded. Furthermore, the two components are likely still covalently connected to some extent. Only at a later stage, the formation of the inorganic network proceeds (right) to yield the final nanostructured hybrid material.

of the hybrid material would then proceed, although on a much larger timescale – this corresponds to the observed slow curing of the final polymer. Note that this hypothesis already includes a possible explanation for why nanostructured hybrid materials with remarkably small phase domains (< 3 nm) are obtained: The quick formation of the organic network hinders diffusion and also prohibits demixing of the components. At the same time, the cleavage of a SiO_2-precursor unit from the organic moiety of the monomer – which quickly becomes a part of the phenolic resin network – only occurs at later stages of the structure formation process, which also will make segregation of the inorganic part unlikely. Figure 3.18 illustrates this process, and Figure 3.19 depicts a molecular model of how the hybrid material structure with phase domains in the order of a few nm could be structured in atomistic detail.

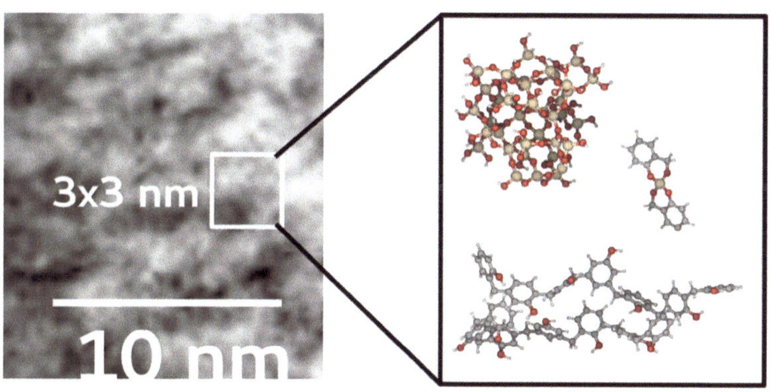

Fig. 3.19: Typical elemental distribution of silicon as obtained by HAADF-STEM for a nanostructured hybrid material obtained from Si-Spiro (left; [2] Copyright John Wiley & Sons, Inc. Reproduced with permission) and an insert of 3 × 3 nm illustrating the length scales of SiO_2 and phenolic resin phase domains as well as a single monomer on the atomic scale.

Since, apart from the monomer described above, furfuryl alcoholates are also a prominent class of monomers for which several TP studies have been published [1, 12], we compare the reactivity for the polymerization of the phenolic resin and polyfurfuryl alcohol in Scheme 3.53. Note that although structurally only remotely similar to Si-Spiro, the furfuryl alcoholate DFOS exhibits very similar energetics in terms of the initiation process of the TP. The proton-catalyzed process starts with the formation of a cationic active species, which is the rate determining step. After this, the formed electrophile attacks another monomer in an exothermal reaction that also represents the first in a series of successive reactions ultimately yielding the organic component of the hybrid material. If one assumes that the reactivity of the silicon moiety in DFOS exhibits the same features as in Si-Spiro, it appears very likely that the structure formation processes are very similar in both cases. Note that also experimentally, both monomers are found to yield hybrid materials with similar morphology [1–3, 12].

A question that remains is whether the information obtained about the simple monomer reactivity can be transferred to all stages of the polymerization. Note that at later points in time, the polymer building blocks will consist of units that can be very different from the monomer itself. The phenyl units will be parts of substituted oligomers and the $Si(OR)_4$ will partly be converted to $Si(OH)_m(R)_n$ or XO–Si–O–Si–OX fragments of the silica network. While the full complexity of all possible motifs and their reactivity is impossible to cover using electronic structure calculations, work of Richter *et al.* has been dedicated to at least exploring the most important influences on the reactivity of the different building blocks of the hybrid material at later stages in the polymerization [4, 85].

For this purpose, the basic reaction steps have been separated into the formation of the organic network and the formation of the silica network. The chain propagation in the formation of the organic network involves mostly the phenyl moiety and can be expected to proceed along more or less the same reaction path

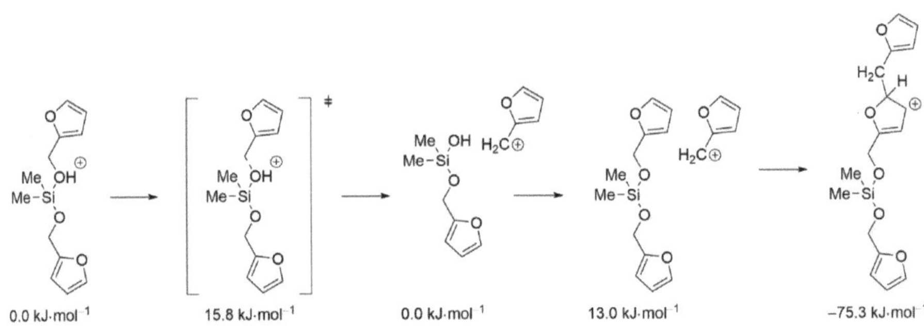

Scheme 3.53: Energetics of the proton-catalyzed formation of the active species for furfuryl alcohol-based systems (difurfuryloxydimethylsilane, denoted as DFOS) as obtained at the B3LYP/TZVPP level of theory. In comparison to the spirocyclic compound Si-Spiro (Fig. 3.15 and Scheme 3.52), the same features can be observed in this case – the formation of an active, cationic species with low barrier and a quick exothermic formation reaction to yield the organic network.

with the electrophilic attack being the rate determining step. Here, just as for the formation of phenol resins, the major question is the substitution pattern in the final polymer. While Richter *et al.* found little preference for any of the two possible substitution patterns, Tchernook *et al.* conclude that there might be a preference for an *ortho*-substitution in subsequent reactions. But they also stress the fact that this should not be overemphasized, as the differences lie within the error bars of the approximations applied [81, 82, 85].

For the formation of the silica network, which appears to be sluggish in comparison to the formation of the organic network, the question arises whether this can change during the TP. During the formation of the nanostructured hybrid material, the organic component is successively split from the silica precursor. Possible candidates for intermediates that can be derived from a $Si(OR)_4$-type monomer would be $HOSiR_3$, $(HO)_2SiR_2$, $(HO)_3SiR$, or even $Si(OH)_4$. Especially, the condensation reactions involving the motif $SiR_3{-}OH...HO{-}SiR_3 \rightarrow R_3{-}Si{-}O{-}Si{-}R_3 + H_2O$ should exhibit a high thermodynamic driving force due to the good leaving group water. At least, this could be expected to be energetically different from the proton-catalyzed reaction of two monomers to form a Si–O–Si bridge upon regeneration of the proton.

B3LYP/TZVP level of theory calculations for a set of prototypical reactions studied on a dimer exhibits negative reaction energies in the order of more than 20 $kJ{\cdot}mol^{-1}$ (see Fig. 3.20). The lowest transition state found in any of those species was in the order of 30 $kJ{\cdot}mol^{-1}$, the highest reaction energy in the order of –50 $kJ{\cdot}mol^{-1}$. Hence, even regarding later steps in the reaction mechanism, one has to conclude that any step in the formation of the silica network will still exhibit considerably higher barriers than the formation of the organic network.

Note that besides these reactions, the work of Richter *et al.* also included calculations on the hydrolysis reaction of the monomer, as this is also a relevant reaction in practice, where traces of water cannot be excluded. First results suggest that a possible hydrolysis mechanism includes attack of the water molecule at the silicon atom to yield a pentacoordinated intermediate, which then cleaves the $Si{-}O{-}CH_2$ bond to yield two hydroxyl groups. Reaction energies and barriers are estimated to be in the order of –10 $kJ{\cdot}mol^{-1}$ and 40 $kJ{\cdot}mol^{-1}$, respectively (B3LYP/TZVP) [85].

3.2.2.3 Larger time and length scales

In order to extend the modeling approaches to larger time and length scales, it is necessary to apply more approximate methods for the simulation of the TP process. This can be achieved, e.g., by scale bridging approaches in which parameters obtained from electronic structure methods are used in more approximate models like classical molecular dynamics or coarse-grained approaches.

In cooperation with A. Berezkin on a coarse-grained model (see Fig. 3.21) for the TP, it was found that if the parameters of the electronic structure calculation were used as a basis for a dissipative particle dynamics (DPD) simulation including

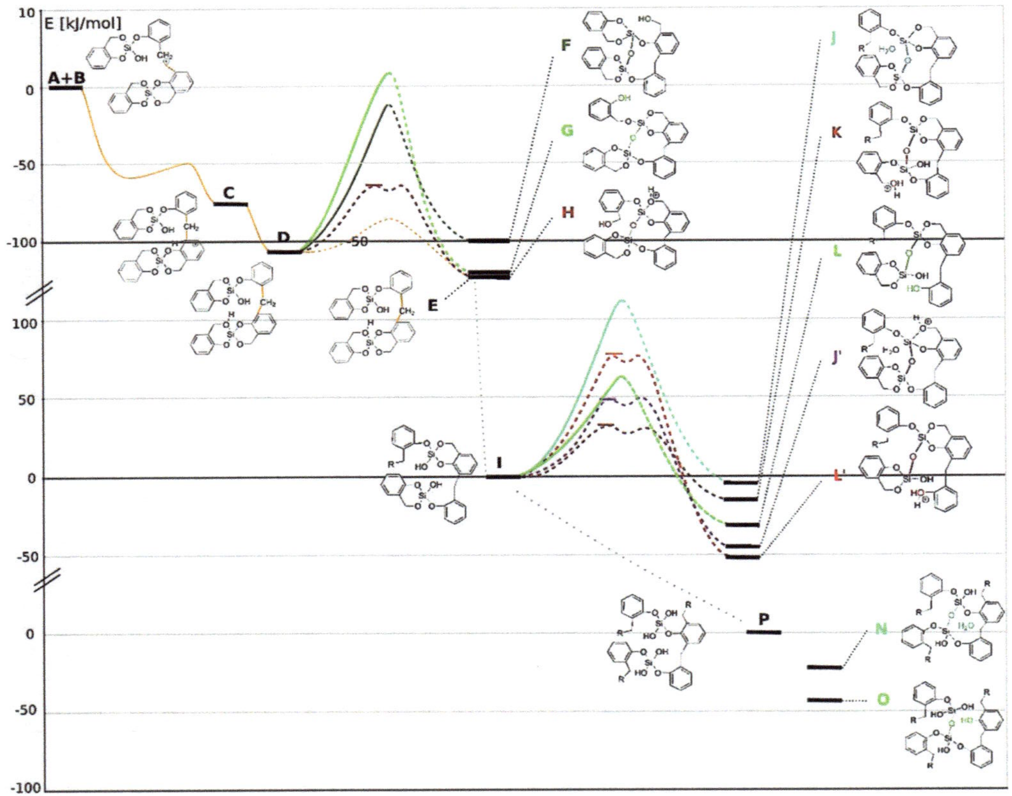

Fig. 3.20: Computational modeling of the proton-catalyzed TP from the monomeric species to oligomers. Note that successive steps of the formation of the organic and the inorganic network are compared including estimates for the reaction barrier height. All calculations have been carried out at the B3LYP/TZVP level of theory. Reproduced from Ref. [85] by the permission of Antje Richter.

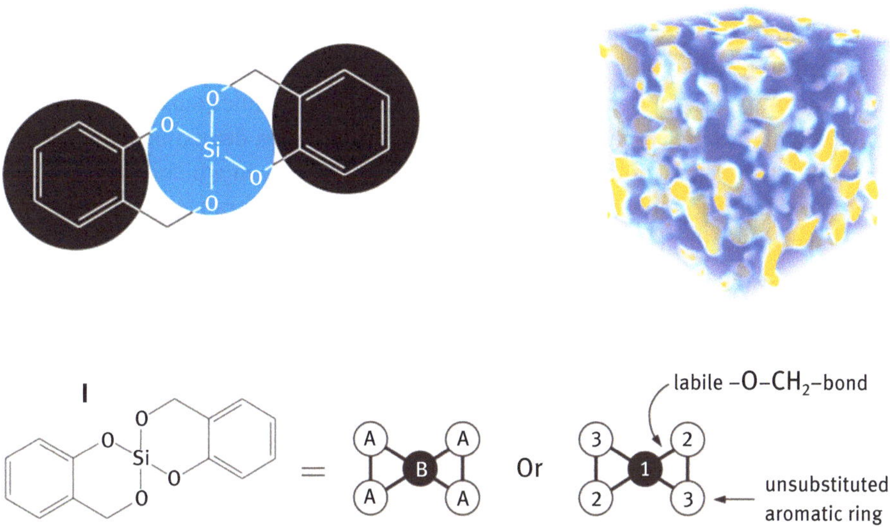

Fig. 3.21: A coarse-grained approach has been applied by Berezkin *et al.* in which Si-Spiro has been reduced to a simple bead-loop representing the building blocks of the organic and inorganic network (left and bottom). For this purpose, a dissipative particle dynamics model has been extended to include the reaction rates estimated from electronic structure calculations. The obtained morphology (top right) agrees well with what is found in experiment [3, 4]. Copyright John Wiley & Sons, Inc. Reproduced with permission.

reactivity, the overall morphology of the hybrid material, the timescales of structure formation, and the experimental values for the scattering structure factor as obtained from small angle X-ray scattering could be well reproduced [4].

While the small dimensions of the structure in the resulting nanostructured hybrid material and the timescale on which the fastest processes occur might be considered outside the scope of the applicability of the DPD simulation, it still demonstrates that in this process, the formation of a network dominates over diffusion and demixing processes which are well described by DPD.

Another approach that is more promising to yield insight into the atomistic details is being pursued in the groups of Friedrich and Prehl. Here, electronic structure calculations are applied in order to parameterize a ReaxFF reactive force field that can then be used to carry out simulations on much longer time- and length scales (see Chapter 3.3) [86].

3.2.2.4 Summary of the proton-catalyzed TP

In summary, all results obtained so far support the hypothesis illustrated in Figure 3.19: In the proton-catalyzed process, all steps involved in the formation of the silica network will have roughly at least twice the barrier heights as the electrophilic substitution involved in the organic network formation, resulting in rates that will differ by approximately an order of magnitude. Hence, after an active species has been formed by protonation, the formation of the organic network quickly proceeds. Only at later stages, the separation of the organic and inorganic components and the formation of the silica network will occur. As a consequence, demixing of immiscible components is impeded and a composite can be obtained, in which phase domain sizes of organic and inorganic structures are in the nanometer range.

However, while it appears that the basic mechanism of the proton-catalyzed TP can be rationalized by means of reactivity in the formation of the two networks, several open questions remain. First of all, it would be highly desirable to put the hypothesis formulated based on computational results on more solid ground by experimental evidence. This would require further *in situ* spectroscopic investigation including time-resolved methods like optical-, IR-, or NMR spectroscopy. Furthermore, a thorough experimental quantification of reaction enthalpies would provide a better framework of basic information on the overall process and would also be relevant for industrial applications of the process. Last but not the least, the recent work on combined experimental and theoretical investigations of further monomers in the TP should proceed. This will not only deepen the existing knowledge but also broaden the scope of compounds it has been obtained for. While the extensive studies on the "ideal twin monomer" (Si-Spiro) provide a blueprint for a hypothesis that may be applied to related TPs as well, mechanistic details and the resulting hybrid material structures will differ, and certainly each requires separate investigations.

3.2.3 Mechanistic studies on the thermally induced TP

The thermally induced TP was first reported in 2013, when the group of S. Spange carried out pioneering work on whether a nanostructured hybrid material can be obtained without the addition of acid as catalyst in the TP [6, 39]. This process is especially appealing when any impurities in the reaction mixture would alter the property of the final polymer or if the technical application of the polymer prohibits an additional mixing step. An example would be if the polymer is used as a bonding agent. Here, simply heating the compound until it melts, then applying it and heating it further to cure the material is a simple and efficient procedure (see Chapter 5.2). However, while the thermally induced TP has been developed quite far for applications (see Chapter 4.1), the mechanistic studies are far less complete than for example for the proton-catalyzed TP.

The mechanistic details obtained from experimental studies are based on work by Kempe *et al.* [6] and Kitschke *et al.* [7] and can be summarized as follows:

While the thermally induced TP can yield a nanostructured hybrid material very similar to those obtained by the proton-catalyzed TP, the thermal process appears to yield several side products that could at least partly be analyzed. An investigation of the influence of the substituents shows that the thermally induced TP is facilitated by electron-donating groups. However, obtaining more detailed mechanistic insight by spectroscopic methods is just as difficult as for any other type of initiation. Nevertheless, in the framework of the first experimental studies, there were also some suggestions made for possible reaction mechanisms. These include, e.g., speculations about radical or concerted mechanisms.

From the fact that the thermally induced polymerization yields practically the same hybrid material morphology as the proton-catalyzed reaction, one might conclude that the mechanism should not be very different. Hence, a starting hypothesis would be that at higher temperatures, the monomer is converted into an active species which will then undergo similar reactions as the cationic active species assumed for the proton-catalyzed process. However, already the analysis of the side products discussed by Kempe *et al.* indicates that the overall process might be much more complex. In the following, we will discuss recent work on computational studies of reactivity in the thermally induced TP, for which the focus was – again – put on the ideal prototype compound Si-Spiro [28].

3.2.3.1 Thermally induced polymerization – computational study in the search for an active species derived from the monomer

In search for an active species that might result from the monomer by a thermal process like bond cleavage, rearrangement reactions, or fragmentation, all plausible reaction coordinates were scanned. Furthermore, open shell configurations were explored in order to assess whether radical species might play a role in possible mechanisms. However, in spite of applying diverse strategies and exploring larger parts of the potential energy surface of the monomer, no viable pathways for the formation of such a species were found. This confirms what chemical intuition might imply – the monomer compound itself is rather stable thermally and obtaining an active species – for example of zwitterionic nature – involves breaking fairly stable C–O, C–C, or Si–O bonds and is very unlikely.

The only pathway yielding a – more or less – stable product is the formation of *ortho*-quinone methide. This product is extremely high in energy but has also been observed as a – possibly abundant – side product in the experimental studies [6, 7]. Nevertheless, with a reaction energy of around 300 kJ·mol^{-1} (B3LYP-D3/TZVPP + COSMO, see Fig. 3.22), it seems very unlikely that the *ortho*-quinone methide is formed from the monomer directly [28].

Already in the first study, Kempe *et al.* suggested an alternative to the "active monomer species" hypothesis – the possibility of a monomer to undergo a concerted reaction with another monomer. Based on an extensive reaction path sampling study on dimeric species, many of the suggested mechanisms could actually be excluded. The reason why many mechanisms that appear plausible at first sight are found to have prohibitive reaction barriers or energies is almost always the same – all Si–O, C–O, and C–C bonds in the system are fairly stable and will not easily undergo cleavage or rearrangement without activation.

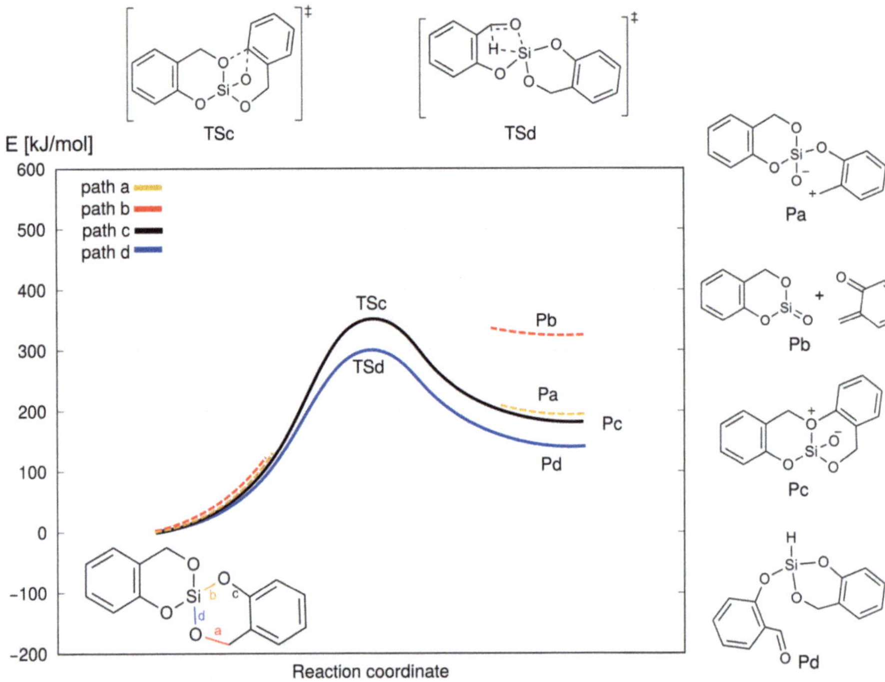

Fig. 3.22: A rough sketch of typical results of reaction path scans for bond-stretching coordinates as obtained by DFT. Extensive studies using various approaches and levels of theory yield similar results – all possible activation/fragmentation reactions are associated with very unfavorable reaction energies. Hence, the monomer itself is very unlikely to be the source of an active species or the start of a possible polymerization reaction. Note that reaction energies are depicted [28]. Copyright John Wiley & Sons, Inc. Reproduced with permission.

A viable reaction path that was discovered during these studies is depicted in Figure 3.23. A low-energy transition state/intermediate was identified in which the five-membered heterocycles of two monomers merge, yielding larger heterocyclic structures. If this process successively connects all monomers in the melt, an organic/inorganic polymer would be obtained. Its structure, however, differs from the hybrid material, as it is a hybrid polymer in which no phase domains of phenolic resin and silica are present, but the two building blocks alternate in loops of aggregated structures. Whether this structure plays a role in reality is to be investigated in more detailed experimental studies in further work. DFT calculations indicate that the overall reaction is thermodynamically unfavorable, however, within the error bars of the computational method.

Nevertheless, it has to be noted that all previous results obtained for the TP indicate that any source of protons would immediately lead the reaction into the fast, strongly exothermic polymerization process of the organic part as it is observed for the proton-catalyzed TP.

Another result obtained from the investigation of larger molecular aggregates is a pathway that yields *ortho*-quinone methide as a possible side product (see TS2/P2 in Fig. 3.23). In contrast to the fragmentation of the monomer, for which the *ortho*-quinone methide is only formed with a prohibitive reaction energy of 300 kJ·mol^{-1}, path TS2/P2 in Fig. 3.23 exhibits a barrier of less than 150 kJ·mol^{-1} and a much more favorable reaction energy.

Let us summarize the computational results obtained so far in conjunction with the experimental evidence, in order to make clear how a consistent hypothesis for the

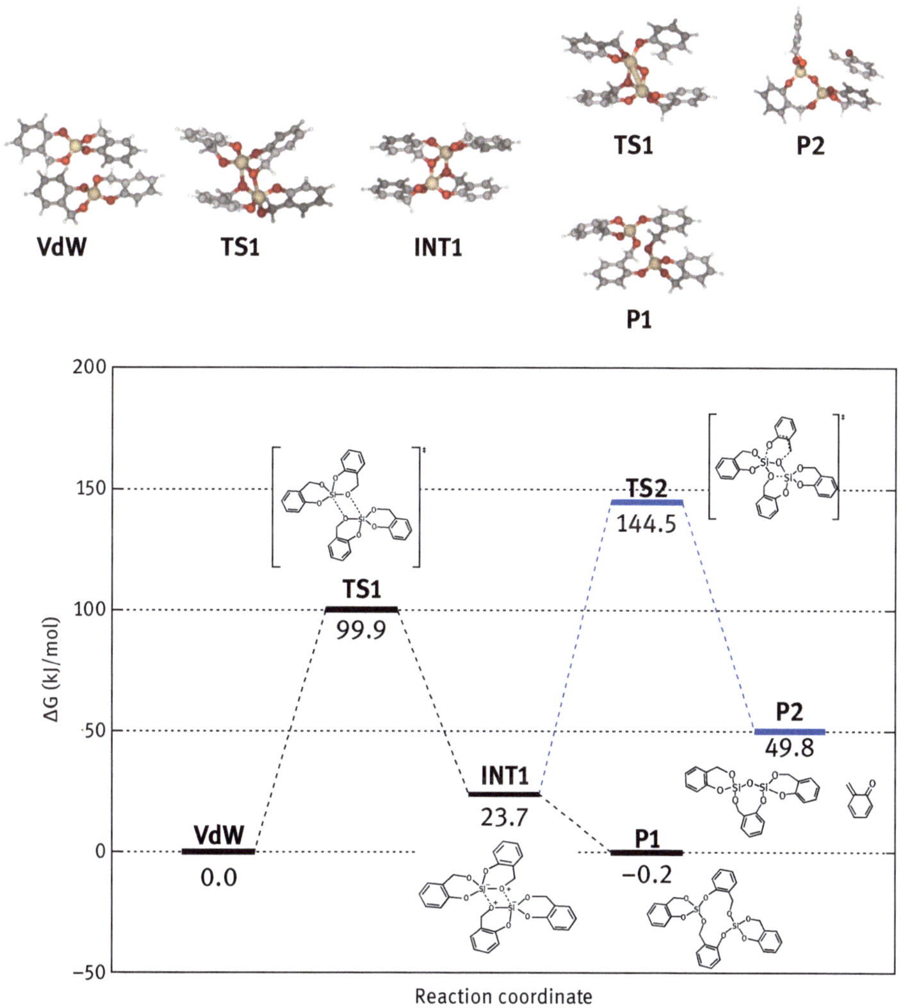

Fig. 3.23: Viable reaction pathways for a concerted reaction of two monomers computed at the B3LYP-D3+COSMO/TZVPP level of theory from Ref. [28]. Copyright John Wiley & Sons, Inc. Reproduced with permission. Note that here, free energies are given.

thermally induced TP can be devised. The experimental evidence indicates that the thermally induced TP is not so different from the proton-catalyzed mechanism – a fairly exothermic reaction yielding a hybrid material very similar in morphology. The computational studies, however, exclude an active species based on the monomer and indicate that there is no direct pathway to the formation of a separate organic or inorganic network. The only pathway that is obtained in agreement with experiment is the formation of *ortho*-quinone methide, for which a viable pathway has been found which is not highly favorable, but at least a plausible side reaction at elevated temperatures.

The fact that there is an exothermal process observed in experiment and there seems to be no viable reaction path from the monomer or dimer raises a central question: How is the system able to enter a downhill slope reaction channel into the formation of SiO_2 and phenolic resin phase domains? While the phenolic resin/SiO_2 hybrid material clearly is the product that should be favored thermodynamically, in absence of protons no direct reaction path seems to be accessible.

One possibility would be to take into account protons from impurities in the reaction melt. This could be residual water, impurities from the monomer synthesis, or degradation products that are formed during heating. From the proton-catalyzed TP, however, it is known that the process requires fairly high concentrations of acid in order to initiate the polymerization process. Furthermore, preliminary studies have been carried out using very low concentrations of acid, as they might arise from impurities in the precursors: Microcalorimetric experiments on a mixture of Si-Spiro and methanesulfonic acid indicate that monomer/initiator ratios of more than 1/500 are required in order for the polymerization to proceed [87]. Hence, the "acidic impurities" hypothesis for the thermally induced TP seems unlikely to be true.

Another hypothesis that offers an answer to the question raised above was already formulated by Kitschke and Mehring (see Chapter 3.1) and includes the *ortho*-quinone methide as essential component. If this species – which might accumulate during the first stage of the reaction – attacks the benzene moiety of another monomer, a quick concerted reaction takes places, yielding phenol derivative structures. This was confirmed in our computational study – all reaction steps included starting from the *ortho*-quinone methide have sufficiently low barriers and reaction energies [28]. The key feature of this reaction path is the resulting functional group, which exhibits protons of increased acidity. Hence, the species that is formed might act as a weak acid, which could replace an acid in initiating the polymerization process similar to the proton-catalyzed TP. However, the acidity of the functional group would have to be sufficient in the given organic solvent or melt. In Table 3.3, the acidity of a variety of functional groups is listed for organic solvents. Indeed, compounds like phenol do exhibit some degree of acidity if compared to other weak acids like acetic acid or benzoic acid (pK_a values of 20–30 in acetonitrile or 10–20 in methanol).

Tab. 3.3: Typical pK_a values of organic alcohols and acids in various solvents, taken from Refs. [83, 84, 88, 89].

Acid	pK_a (water)	pK_a (solvent)
Phenol	10.0	29 (MeCN) 20 (1,2-Dichloroethane) 18 (Dimethylsulfoxide) 14 (Methanol)
Acetic acid	4.76	24 (MeCN) 16 (1,2-Dichloroethane)
Picric acid	0.4	11 (MeCN) 0 (1,2-Dichloroethane)
Trifluoromethanesulfonic acid	−12	0.7 (MeCN) −11 (1,2-Dichloroethane)
Benzoic acid	4.22	20 (MeCN) 11 (Dimethylsulfoxide) 9 (Methanol)
2-Nitrophenol		23 (MeCN)
2,4-Dinitrophenol	4.09	17 (MeCN) 5 (1,2-Dichloroethane)
2-Bromophenol	8.45	27 (MeCN)
Pentafluorophenol	5.5	20 (MeCN)

From these values, one might conclude that the resulting hydroxyl group formed during the first stage of the reaction could just be acidic enough to act as a source of protons, thus opening the downhill reaction path to the hybrid material.

This might be the answer to the question raised in the beginning. Note that already Kempe *et al.* stated that "A significant occurrence of quinone methide adducts indicates that the thermal mechanism proceeds not only by a ring-opening at the oxymethylene position, but also with the *ortho*-quinone methide as a central or alternative intermediate" in their original work [6].

Certainly, a counter argument would be that the formed hydroxyl groups might still not be acidic enough to act as efficient proton sources and further studies are necessary to support this hypothesis. However, it is the first mechanistic proposition that is consistent with the experimental evidence and the results of the computational studies.

This hypothesis is summarized schematically in Figure 3.24 and basically includes three phases. Putting this into perspective by contrasting it to the proton-catalyzed TP, the mechanisms can be described as follows:

The ***proton-catalyzed TP*** can be structured as follows:

(1) addition of catalyst and formation of active species
(2) formation of organic network
(3) formation of the inorganic network/curing to yield the final hybrid material.

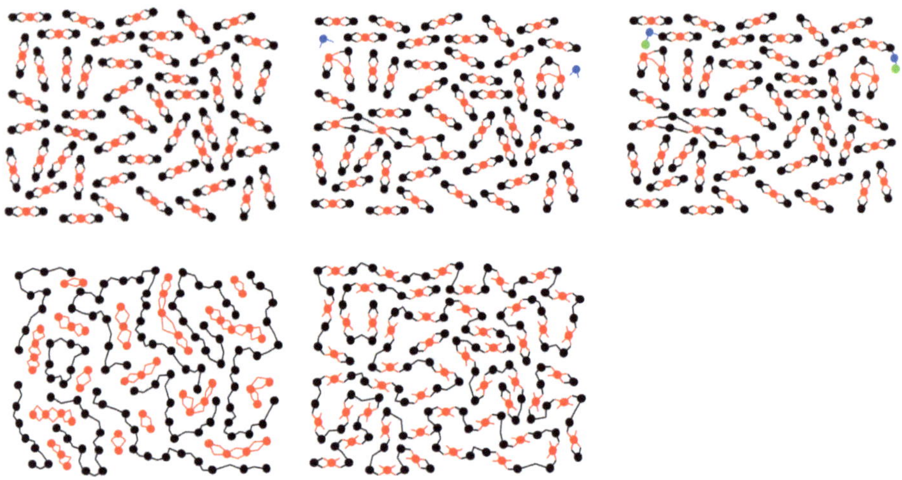

Fig. 3.24: Schematic representation of the hypothesis for the mechanism of the thermally induced twin polymerization. During the first stage of the reaction (top left and middle), a series of alternative oligomerization reactions might proceed which enable cleavage of *ortho*-quinone methide (represented by blue dots). This will react further to generate acidic protons (denoted as green dots on the top right) which will eventually lead to the same mechanism as assumed for the proton-catalyzed twin polymerization (bottom left and right). Computational results indicate that the first three steps require elevated temperatures as no protons that catalyze the polymerization are available. As soon as these are present, the reaction will be exothermic throughout and yield the expected nanostructured hybrid material.

The ***thermally induced TP*** can be structured as follows:

(1) heating and melting of monomer, oligomerization of monomers, and accumulation of *ortho*-quinone methide which, in further reactions, generates acidic protons
(2) if sufficient protons are available, formation of organic network
(3) successive formation of the inorganic network/curing to yield the final hybrid material.

Hence, while very different in the first stages of the reaction, the thermally induced TP yields the same product in an overall exothermal process which resembles the proton-catalyzed TP in its later stages.

This summarizes the current knowledge of the mechanism of the thermally induced TP. Note that this does not so much mark the end of the formation of a hypothesis, but rather the start of more detailed investigations. Further work has to be directed toward how, e.g., adding excess *ortho*-quinone methides (or precursors of it) influences the reaction and whether the molecular structure of initial oligomeric species can be obtained.

3.2.4 Final remarks

Since its invention and the beginning of the joint research efforts on TP, a lot of progress has been achieved. By now, several publications document how the knowledge about the details and possible applications of the TP has been extended in recent years. What is also evident from looking at how progress has been achieved is that the close cooperation of experts from synthesis, analysis, and theory has been a key component of every step.

But while mechanistically many important aspects have been identified – like the importance of covalently bonding together otherwise immiscible precursors or the difference in timescales for the polymerization of organic and inorganic networks – the results reviewed in this chapter only represent the starting point for more detailed investigations. Here, especially *in situ* spectroscopies will yield insights that allow to refine our hypotheses, derive further mechanistic details, and ultimately aid the rational design of novel materials.

Janett Prehl, Benjamin Fiedler, Joachim Friedrich

3.3 Modeling and simulation of nanostructure formation of TP

3.3.1 Introduction

The previous Chapters 3.1 and 3.2 presented the basic concepts of the reaction mechanism regarding acid-catalyzed and thermally induced TPs as well as possible twin monomers and their electronic structures. Based on the mechanistic studies in Chapter 3.2, we present three approaches on different levels of detail to investigate the nanostructure formation process of the (acid-catalyzed) TP.

In general, there are different modeling approaches on various length scales known from literature to investigate a structure formation process and its consequences on the resulting morphology. For instance, quantum chemical approaches [90–93], as presented in Chapter 3.2, can provide detailed geometries and energies of single reaction pathways at an atomic level. For larger systems with 100–1,000 atoms, where also dynamical aspects play an important role, the computational costs of these methods are too high. Thus, for a significant enlargement of the system size, one possibility is a stepwise reduction of the accuracy of the modeling approach. Typical modeling approaches are, with increasing coarsening, molecular dynamics simulations [94–97], coarse-grained molecular dynamics simulations [98, 99], Monte Carlo simulations [95, 100, 101], and lattice-based Monte Carlo methods [102–104]. By comparing the obtained results from these simulation methods at different accuracy levels with experimental results, one can validate the utilized modeling approach.

Molecular dynamics simulations usually refer to atomic length scales, where the time-dependent dynamics of each atom is modeled by taking important atom–atom interactions *via* force fields into account. This leads to detailed results on this length scale. But the calculation of the forces for all atoms with respect to the time requires much computational effort. On a more coarse-grained length scale, i.e., in the case of lattice-based Monte Carlo methods, groups of atoms that are known not to alter their steric structure over time are merged to coarse-grained "super atoms" that can only move on a grid with predefined probabilities. Due to this modeling scheme, the numerical effort per simulation step is kept small, but the knowledge about the underlying structure of the super atom needs to be derived *a priori*.

In the framework of TP, several modeling approaches [4, 86, 105, 106] have been utilized or are newly developed to investigate the corresponding structure formation process. In this section, we will focus on three models from simple to complex:

(1) a coarse-grained lattice-based reactive bond fluctuation model (rBFM),
(2) an atomistic simulation by reactive molecular dynamics (rMD) simulation, and
(3) an extended quantum chemical modeling approach *via* DFT.

We describe how these modeling approaches are adapted and applied to the complex reaction mechanism of TP. Typical results will be discussed and validated by theoretical predictions and experimental measurements. Note that the extended quantum chemical modeling approach is additionally also applied to the base-catalyzed TP. The base-catalyzed TP provides viable advantages for the TP synthesis process. Amongst others, this mechanism needs less amount of catalyst [107, 108] (see Chapter 4.1) than the acid-catalyzed TP; the formation of the silica network is already completed at $T = 160°C$ instead of $T = 220°C$ for the acid-catalyzed TP (with additional post thermal treatment), and furthermore, there are several bases, like amine [109] and imine bases [110], than can be directly built in the reaction process.

This section is structured as follows: First, we give a short review of the basic ideas of the acid-catalyzed TP as shown in Chapter 3.2. Then, we introduce our three modeling approaches for the investigation of the structure formation process based on the concept from "simple to complex." Thus, we first present the rBFM, second the rMD simulation, and third the extended quantum chemical modeling approach. In all three cases, typical results are discussed and compared to experimental results if available. Finally, a brief summary is given.

3.3.2 Basic concept of the proton-catalyzed reaction mechanism of TP

The twin monomer 2,2′-spirobi[4*H*-1,3,2-benzodioxasiline] (Si-Spiro, **1**), as introduced in Chapters 3.1 and 3.2 and depicted in Figure 3.25a, represents an ideal twin monomer to examine the TP and its resulting morphologies [1, 2, 12]. It can be polymerized either catalyzed by an acid [2, 37] or a base [11], as well as thermally induced [6].

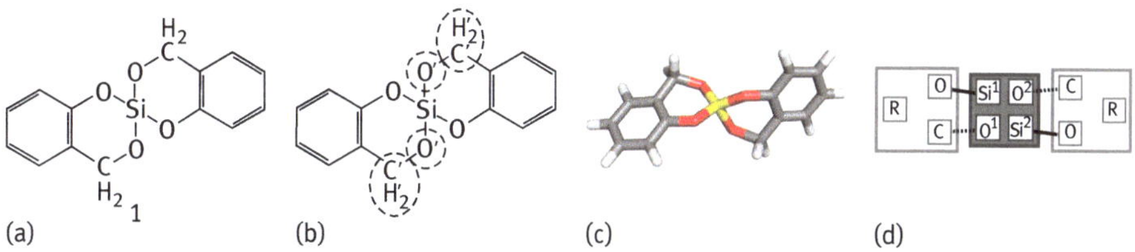

Fig. 3.25: The scheme of 2,2′-spirobi[4*H*-1,3,2-benzodioxasiline] (Si-Spiro, **1**) is given in (a). In (b), the scheme of Si-Spiro with different atom types is depicted as explained in Section 3.3.4; in (c), the molecular structure of Si-Spiro is shown (Si – yellow, O – red, C – gray, H – white); and in (d), the representation of Si-Spiro in the context of a coarse-grained modeling approach is given (organic part – white, inorganic part – gray).

Thus, correlations between the different reaction mechanisms and the final morphology can be investigated in detail. Furthermore, various derivatives of Si-Spiro are known [7, 40]. This allows us to analyze the influence of structural changes of the twin monomer on the resulting material. Therefore, it is convenient to perform the theoretical and numerical investigations of the twin monomer Si-Spiro to enable the comparison of the obtained results with experimental data [2, 4, 81, 86, 105, 106].

First quantum chemical calculations of the acid-catalyzed TP [2, 4, 81, 85] revealed a quite complex and still not fully understood reaction mechanism based on the peculiar structure of the twin monomers, as described in Chapters 3.1 and 3.2. Prior to the modeling of the structure formation process *via* coarse-grained models and the evaluation of the results from atomistic modeling approaches, we briefly review the known acid-catalyzed reaction mechanism of Si-Spiro **1**. The resulting reaction scheme of the TP, yielding the phenolic resin **7** and silica **8**, is given in detail in Scheme 3.54. Note that for these calculations, the acid is reduced to pure protons and thus we will call this the proton-catalyzed reaction mechanism in the following. First, the proton attacks the methylene bond ($O{-}CH_2$) of Si-Spiro **1** leading to an activated monomer **2**. This process is called the bridge-opening process. Next, **2** reacts with the benzyl moiety of a second twin monomer **1** and gives a more stable σ-complex. The subsequent rearomatization by dissociation of a proton leads to stable **3**, which can then initiate the next bridge-opening process. These steps (dashed arrow in Scheme 3.54) represent the pathway for the organic network formation and take place during the whole reaction process, as long as methylene bonds are available.

From this point, the inorganic reaction pathway is accessible as depicted by the dotted arrow in Scheme 3.54. An available hydroxyl group at the Si atom is connected with a neighboring Si atom and forms a siloxane bridge (Si–O–Si) as can be observed in structure **4**. This leads to the silica network formation. At the same time, the bond between the Si atom and the phenolic oxygen is cleaved and a phenol group is formed due to a hydrogen transfer, which is the final step of the organic network formation.

This proton-catalyzed reaction mechanism will be utilized for the next two modeling approaches to investigate the occurring structure formation process of the TP exemplarily for the twin monomer **1** (Si-Spiro).

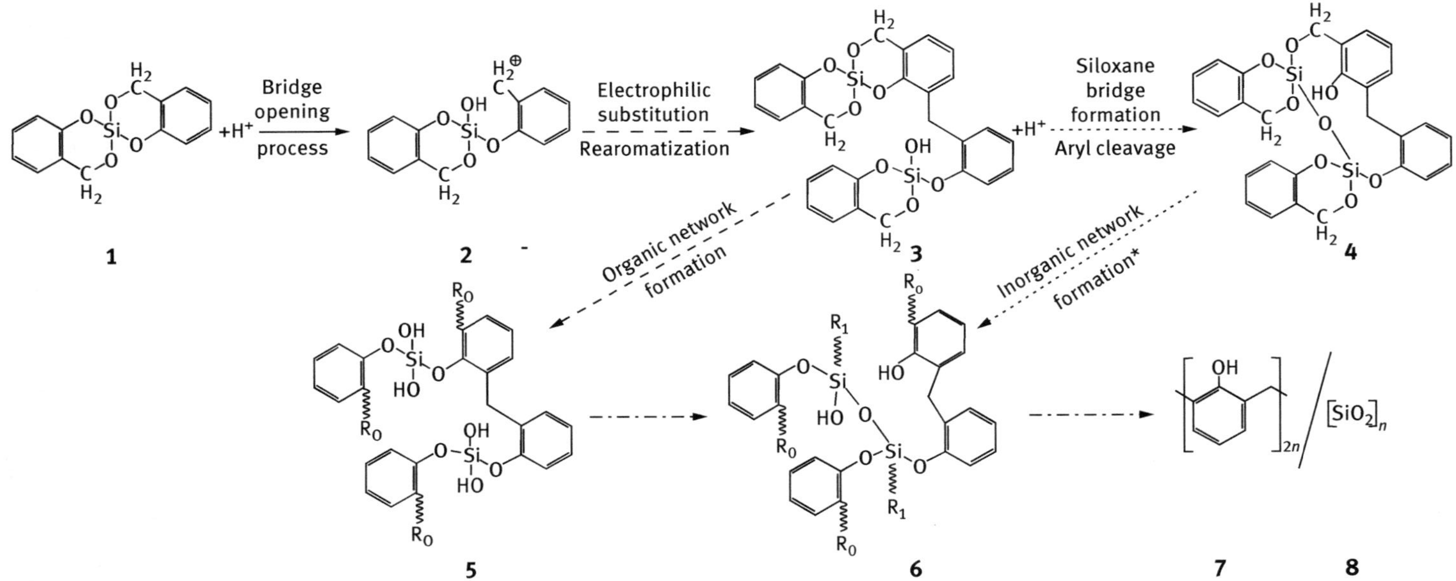

Scheme 3.54: Main features of the reaction scheme of the proton-catalyzed twin polymerization (TP) are depicted. The dashed arrows indicate the organic, the dotted arrows the inorganic reaction pathway, and the combination of both pathways is shown by the dashed-dotted arrows, representing the TP reaction mechanism.

3.3.3 Reactive bond fluctuation model

Based on the reaction mechanism of the proton-catalyzed TP, we first introduce a coarse-grained modeling approach to investigate the time-dependent overall structure formation process of TP. With such an approach, many twin monomers can be considered and their dynamical behavior can be analyzed over the reaction time with reasonable numerical effort. This will allow us to investigate the reaction process itself over time and the final morphology, depending on the initial conditions.

The developed rBFM for TP is based on the bond fluctuation model (BFM) of Carmesin and Kremer [102]. It combines the strength of the BFM, i.e., of being a realistic simulation of the dynamical behavior and confirmation of complex polymer systems, like polymer brushes [111], dilute long polymer chains [112], or semiflexible polymers [113], with the additional information of the structure formation process by including the underlying chemical reaction mechanisms.

There are various extensions of the BFM for reactions given in the literature. Typical examples are the modeling of reactive binary mixtures [114], reactions at interfaces [115, 116], film growth processes [117], reactivities within polymer melts [117, 118], as well as for the representation of enzyme reactions [119]. In all these extensions, two structural units are connected by only one bond. However, the peculiarity of the twin monomers is that all structural units within a twin monomer are connected *via* two bridges exhibiting different activation barriers. Therefore, the mentioned BFMs are not reasonable for the TP. In order to include these properties and the reaction mechanism of the TP, a more generic BFM for reactions is needed.

The rBFM for TP starts from the basic idea of the BFM [102, 103] that a polymer is represented by beads, which are connected *via* bond vectors with each other. The beads represent the monomers and the bond vectors refer to the chemical bonds. The beads can move on a cubic lattice with lattice constant a, where each bead occupies one lattice site. A representative example of a polymer in the BFM is shown in Figure 3.26, where each bead is represented by a square.

Similar to the chemical bonds, the bond vector lengths can also "fluctuate" between predefined finite lengths. According to the implementation of Shaffer [103], we only allow bond vectors of length 1, $\sqrt{2}$, or $\sqrt{3}$. This definition leads to a set of 26 possible bond vectors on a three-dimensional lattice.

The BFM includes the property of excluded volume by enforcing that each lattice site can only be occupied by one bead at the same time and by forbidding the crossing of bond vectors.

The dynamical behavior of the beads over time is defined by Monte Carlo steps (MCSs). A MCS is defined as an attempted movement of a bead from its current position to a neighboring one on the lattice. This movement is only accepted if (1) the selected lattice site is unoccupied, (2) the resulting bond vectors do not cross, and (3) all bond vectors are allowed. Otherwise, the move is rejected. In Figure 3.26, an exemplary MCS with an allowed and a rejected move is depicted.

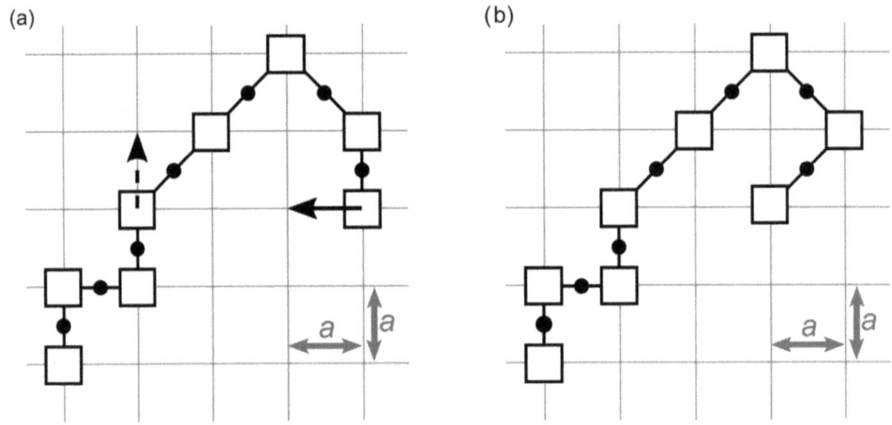

Fig. 3.26: A polymer in the bond fluctuation model representation. Each square represents a bead (one monomer) and the connecting black lines with dot are the bond vectors representing the chemical bonds. In (a), the black arrow indicates a possible movement of a bead, whereas the dashed arrow is not allowed according to the implemented Monte Carlo step (MCS). In (b), the polymer is given after the MCS.

The extension of the BFM to the rBFM for TP includes the following three aspects: (1) different bead types that represent the different structural units of the twin monomers; (2) different bond vector types, representing the different chemical bonds; and (3) a reactive Monte Carlo step (rMCS), representing the reaction mechanism of the TP.

Each bead type is defined by a unique set of reaction centers. A reaction center represents a reactive site of a bead, where it can connect to another bead *via* a bond vector. According to the underlying chemical structure, there are different reaction center types, which can take values, like bonded, nonbonded, or blocked.

The different bond vector types are defined according to the reaction centers they connect. The underlying reaction mechanism specifies the combinations of reaction centers that are allowed.

The rMCS defines how the chemical reactions, the bond formations, and cleavages are represented in the rBFM. In an rMCS, two beads (one random bead of the system and one of its possible reaction partners) as well as one allowed reaction between these beads are chosen randomly. The reaction is allowed if (1) the required reaction center of each bead has the appropriate value for the chosen reaction, (2) the resulting bond vector does not cross another one, and (3) the corresponding reaction probability is larger than a random number chosen from the interval [0,1). Otherwise, the rMCS is rejected.

According to the chemical structure of the twin monomer Si-Spiro we define two bead types. The one bead type (ORG bead) represents the organic part. It has three reaction centers, i.e., the O for the **o**xygen in the aryl bond, the C for the **c**arbon in the methylene bridge, and the R for the *ortho*/*para*-substitution at the benzene **r**ing. The other bead type (INO bead) represents the inorganic O–Si–O part of the twin monomer with four reaction centers. There are two Si^x reaction centers representing the two possible bonds of **s**ilicon with neighboring oxygen atoms and two O^x reaction centers for two possible bonds of the **o**xygen with the benzene ring or silicon ($x = 1, 2$). Additionally, there are four different bond vector types: the methylene bond (C–O^x – dotted black), the phenolic Si–O bond (Si^x–O – solid black), the organic bond (C–C – solid gray), and the siloxane bond (Si^x–O^x – dashed black). The formation of the phenol group is represented by a blocked (gray O-square) reaction center. The mapping of the twin monomer Si-Spiro to the rBFM is depicted in the left-hand side of Figure 3.27 as well as in Figure 3.25d.

From the proton-catalyzed TP described in Section 3.3.2, four main reaction steps are needed: (1) the opening of the methylene (C–O^x) bond, (2) the electrophilic substitution and rearomatization leading to the organic network (C–R) formation, (3) the cleavage of the Si–O bond (Si^x–O) with the phenolic oxygen, and (4) the inorganic network (Si^x–O^x) formation in combination with the formation of the phenol group. On the right-hand side of Figure 3.27, all possible reactions being part of the rMCS for the TP are given with the corresponding reaction probabilities.

Note that the reaction probabilities are closely related to the Boltzmann factors of the activation energies for the single reaction steps, determined by quantum chemical calculations [4, 85] in analogy to the DPD model [4] mentioned in Chapter 3.2. Furthermore, we want to point out that within this model the protons are not modeled explicitly, but that they are included indirectly within the reaction probability of the bridge-opening process. For further technical details to the numerical implementation of the rBFM, we refer to Huster *et al.* [106, 120]

Applying and implementing the rBFM for TP, there are several aspects that can be investigated: How does the amount of bond vectors change over time? Do the reaction probabilities influence the reaction process? Is there a change of diffusivity of the beads over reaction time? What are typical chain lengths? Which radial distribution function (RDF) of beads (similar to pore-size distribution) can be observed? Or does phase separation take place?

Here, we focus on the analysis of the time development of the structure formation process and on the validation of the obtained simulated morphologies with experimental results.

First, we investigate the time development of the structure formation process by tracking the bond fraction over time. The bond fraction is the ratio between the amount of bonds and the maximum possible number of bonds per bond vector type. We also determine the analog bond fraction for the bonded, nonbonded, and blocked reaction centers. In Figure 3.28, the resulting bond fractions over logarithmic time are shown for the reaction probabilities given in Figure 3.27. The simulation time is given in reactive Monte Carlo cycles (rMCCs). One rMCC is a combined reactive and nonreactive MCS.

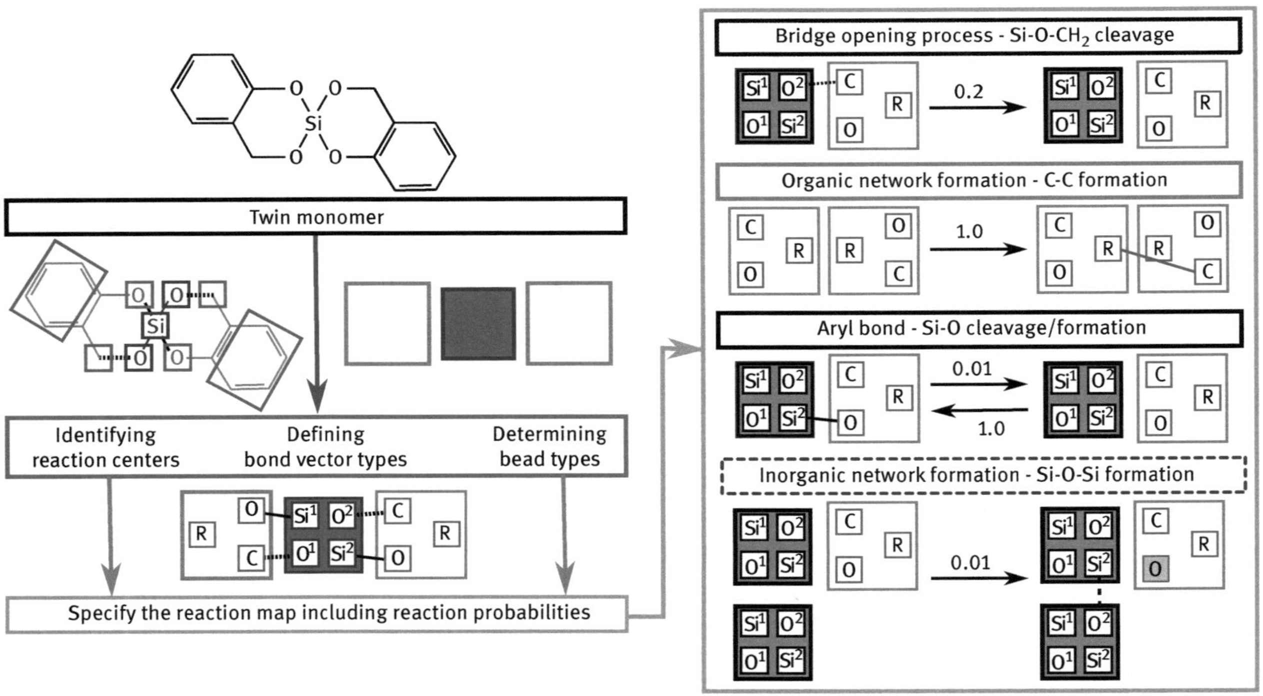

Fig. 3.27: A schematic picture of the mapping process from the twin monomer Si-Spiro to the reactive bond fluctuation model (rBFM). On the left-hand side, the identification of bead types, reaction centers, and bond vector types for the twin monomer is given, and on the right-hand side, the reaction mechanism including the reaction probabilities for the proton-catalyzed twin polymerization for rBFM is depicted.

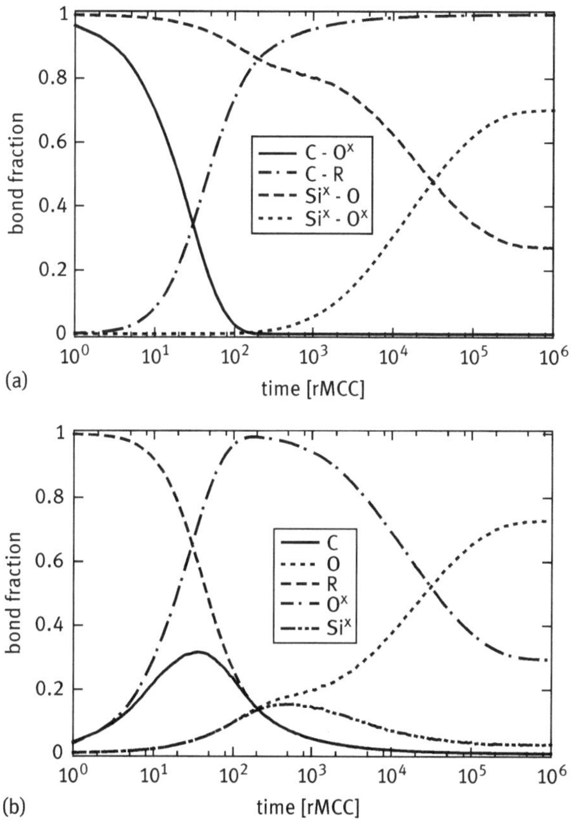

Fig. 3.28: The bond fraction is given over logarithmic time. In (a), it is shown for the four bond vector types methylene bond (C–O^x), the organic bond (C–R), the aryl bond (Si^x–O), and the siloxane bond (Si^x–O^x) and in (b), for different nonbonded reaction centers.

We observe that the reaction process can be divided into three separate phases: The first phase, which takes place for $t < 10$ rMCC, is characterized by only a strong decrease of the number of C–O^x bonds and thus can be identified with the ring-opening process. The second phase occurs from 10 rMCC $< t < 10^3$ rMCC. The most prominent aspect is a further decrease of the number of C–O^x bonds together with a drastic increase of the number of C–R bonds, as the nonbonded C connect to the available R. This phase reflects the organic network formation. Note that at $t \approx 50$ rMCC, the first cleavages of the Si^x–O bonds (i.e., with phenolic O) take place, but the siloxane formation starts at $t \approx 10^3$ rMCC, which refers to the start of the third phase. In this phase, the number of Si^x–O^x bonds mainly increases, which is accompanied by a strong decrease of the number of phenolic Si^x–O bonds.

Note that the crossover from the first to the second phase is indicated by the maximum of nonbonded C reaction centers, whereas the maximum of nonbonded

Si^x reaction centers represents the crossover from the second to the third phase. First investigations regarding the variation of the reaction probabilities showed that these phases can be identified in all cases. However, the time span of each phase may vary.

Second, we use the obtained structures of the rBFM to derive information about the morphology of the model in comparison to experimental results. Therefore, we utilize data about the pore size distribution [7] from gas adsorption and connectivity information between the polymers from NMR spectroscopy [2, 12].

To analyze the pore size distribution of the simulated material, we determine the RDF $g_{\alpha,\alpha}(r)$. The RDF gives the number of beads of one type α (with α = ORG or INO) in a distance r to a reference bead of the same type in comparison to the corresponding number of beads in an ideal gas. This gives a measure of the typical domain sizes of type α. We find that the INO beads form phase domains of typical sizes from 3.5 Å to 13 Å exhibiting a narrow peaked distribution over distance, whereas the ORG beads are nearly homogeneously distributed over the whole material. This fits well with the experimental pore size distribution given in Kitschke *et al.* [7], where the removed organic domains in the inorganic polymer are homogeneously distributed, showing a broad flat distribution, and the removed inorganic phase domains in the organic polymer had pores in the range of 3–11 Å.

Furthermore, one can compare the final connectivity between the organic and the inorganic network to the bond fractions of corresponding bonded and non-bonded reaction centers. From NMR spectroscopy [2, 12] (see also Chapter 4.1), we can derive the amounts of Si atoms that are connected either to two (Q_2), three (Q_3), or four (Q_4) oxygen atoms. The ratio $Q_4/(Q_2+Q_3)$ gives a measure for the connectivity within the inorganic network. For the twin monomer Si-Spiro, this ratio is ≈ 1 [2]. From the rBFM model, we obtain 1.2 which is reasonably close for our coarse-grained model.

From this, we find that the rBFM is a powerful tool for connecting theoretical predictions of an occurring reaction mechanism with experimental data from the synthesis process. The discussed findings validate the proton-catalyzed TP as presented in Chapter 3.2 and provide the possibility to investigate the influences of the reaction probabilities on the morphology and the structure formation process.

However, as the rBFM depends on the *a priori* knowledge of the reaction mechanism, it is important to make sure that all important reaction pathways are known and included. Therefore, we additionally investigate the TP on a more detailed level *via* rMD simulations in the next section.

3.3.4 rMD simulation

With the rMD approach, the movements of all atoms are determined based on the underlying reactive force field. Employing an appropriate force field, the rMD approach can provide in principle all possible reaction paths. This allows us to verify the reaction mechanism, to point out additional reaction paths, and to identify the structure formation processes taking place on the atomic length scale.

A reactive force field that has already been applied for a wide range of materials is the ReaxFF force field introduced by van Duin *et al.* [97]. Amongst others, it has been utilized for hydrocarbons [121], carbon nanotubes [122], polydimethylsiloxane [123], silicon/silicon oxide [124], metal/metal oxides, and metal hydrides [125–128]. This made it convenient for the application to TP [86, 105].

ReaxFF is a reactive force field of the class of bond order potential. The bond orders are defined as single, double, triple, or nonbonded, and they are directly related to the interatomic distance r_{ij} between two atoms i and j. Based on a bond length to bond order relationship, where the bond length is updated in every time step, a smooth transition between different bond orders is possible. As the connectivity is thus developing dynamically, there is no further need for defining reactive sites as it is necessary in the rBFM.

Instead, important force field parameters are determined during a training process. Therefore, one needs to set up a data set for each kind of reaction to approximate the underlying chemical reaction mechanism. These data sets are based on quantum chemical calculations of the desired level of accuracy and thus allow the reproduction of energies close to the accuracy of quantum chemical calculations at a lower computational cost.

A special feature of the classical ReaxFF is the basic philosophy that each chemical element can be described in principle by one parameter set including many-atom interactions as the angle between three and the torsions between four atoms. In the supplementary information of Chenoweth *et al.* [121], a good description of the mathematical background is given.

In the case of the TP of Si-Spiro, the most reasonable ReaxFF parametrizations are given by Chenoweth [121], Nielson [122], or Newsome [129]. Their direct application allowed the ring-opening process but did not lead to a further proceeding of the reaction mechanism [86].

Thus, we developed a new ReaxFF parametrization for TP, called TP-ReaxFF [86]. Therefore, a new training data set has been developed in the context of TP based on a reduced data set of Chenoweth *et al.* [121, 123] utilizing the ReaxFF parameters for C/H/O [121] and Si/O [123] interactions. These data are extended by quantum chemical calculations for the reactions of the aromatic compounds, where we focused on the C/H/O/Si interactions present in the twin monomer as well as during the electrophilic substitution and the rearomatization. Further details about the training data set can be found in Schönfelder *et al.* [86].

In Figure 3.25c, the molecular structure of Si-Spiro in the TP-ReaxFF representation is given. The resulting TP-ReaxFF parametrization captures the electrophilic substitution of the initial organic network formation of the complex reaction mechanism, depicted in Figure 3.29. However, applying the TP-ReaxFF parametrization for larger systems to simulate the whole reaction mechanism does not give satisfactory results.

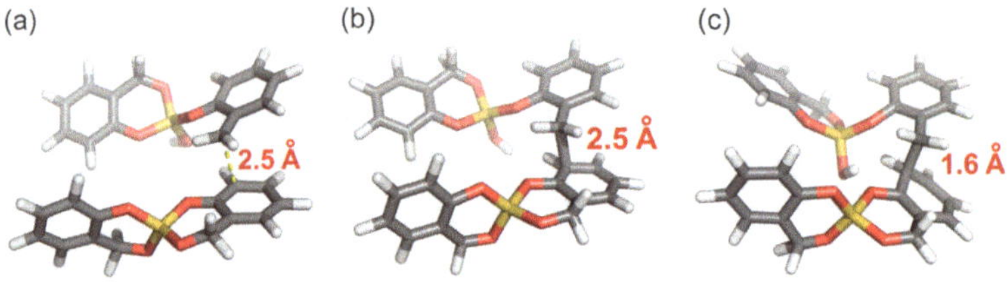

Fig. 3.29: The sequence of snapshots shows the formation of the C–C bond between two Si-Spiro molecules: (a) is taken after 0.05 ps with a distance of 2.5 Å between the reacting carbon atoms (dashed line), (b) presents the moment of binding (after 0.202 ps) at a distance of 2.5 Å, and in (c), the covalent bond with a bond length (distance) of 1.6 Å is given after 2.5 ps. This represents the σ-complex. Reprinted from Ref. [86], Copyright 2014, with permission from Elsevier.

Detailed investigations show that the occurring problems result from different hybridization types of the carbon atoms. The sp^3-hybridized carbon atoms of CH_2 groups are important for the reaction mechanism of the TP (responsible for the electrophilic substitution), whereas the sp^2-hybridized carbon atoms of the benzyl moiety do not take part in the relevant TP reactions. However, this is not fully captured in the atom type description of ReaxFF. Instead of extending the ReaxFF force field description by adding many-body interactions, we decided to overcome this limitation of TP-ReaxFF by leaving the basic philosophy of ReaxFF of having only one parameter set per atom type and to allow multiple atom types per element. This allows us to keep the numerical and mathematical simplicity of the force field.

In the case of Si-Spiro, three new atom types are introduced. In this site-dependent atom type representation (SDAT) of ReaxFF for TP (SDAT-ReaxFF) [105], the atom types depend on the position of the atom within the monomer. Thus, there are aromatic as well as nonaromatic carbons and hydrogens (C, C′, H, H′), phenolic oxygens (O), and dioxide-related oxygens (O′). The resulting structure of Si-Spiro is given in Figure 3.25b.

Due to the additional atom types, new bond-, angle-, and torsion-relations need to be parameterized in the training process of the SDAT-ReaxFF, which enables more flexibility to adjust the force field parametrization to the complex reaction mechanism of the TP. All details of the SDAT-ReaxFF parametrization are given by Prehl *et al.* [105].

Applying the SDAT-ReaxFF parametrization to Si-Spiro allows the modeling of all reaction steps of the proton-catalyzed reaction mechanism of the TP. In Figure 3.30, a sequence of snapshots of a typical SDAT-ReaxFF simulation run for a small system of twin monomers taken from Prehl *et al.* [105] is shown. The presented system contains three twin monomers Si-Spiro and one proton. The simulation starts with an activated twin monomer **2**, given in Figure 3.30a. In Figure 3.30b, the first electrophilic substitution (C′–C bond formation) has taken place, leading to the rearomatization and a further bridge-opening process of the next monomer, as depicted in Figure 3.30c. With ongoing simulation time, first available O′H groups occur

(Fig. 3.30d), leading to first Si–O′H–Si bridges (Fig. 3.30e) and thus to the initial inorganic Si–O'_2 network formation, as shown in Figure 3.30f). At the same time, the organic network formation proceeds *via* a second C′–C bond and formation of an OH group at a benzyl moiety (Fig. 3.30f). For the numerical details of this simulation, we refer to Prehl *et al.* [105].

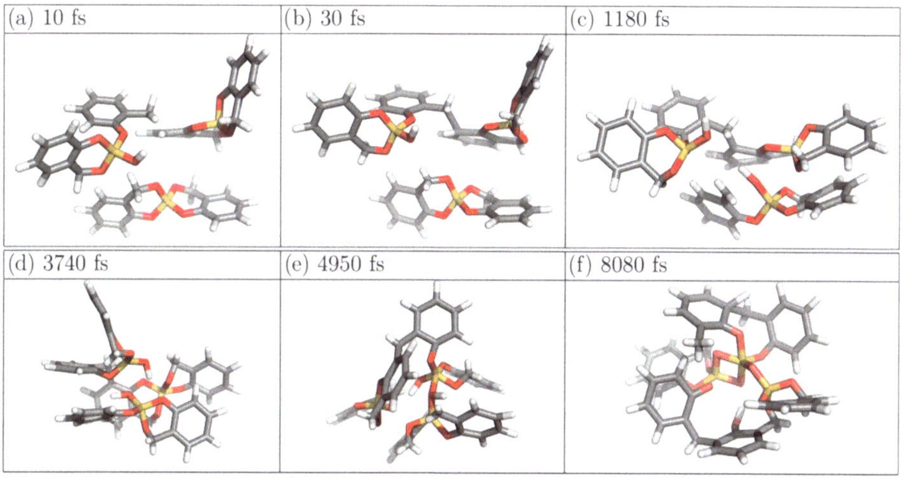

Fig. 3.30: Sequence of snapshots of a typical SDAT-ReaxFF simulation run of a system containing three twin monomers Si-Spiro and one proton. Each snapshot shows one aspect of the proton-catalyzed reaction scheme as presented in Scheme 3.54: (a) starting structure of the system, (b) formation of first C′–C bond, (c) rearomatization of the σ-complex and subsequent bridge-opening of a second monomer, (d) occurrence of available O′H groups, (e) first Si–O′H–Si formation leading to Si–O'_2 network formation as shown in (f), and (f) formation of second C′–C bond and of an OH-group at a benzene ring [105]. Reprinted with permission. Copyright 2017 American Chemical Society.

Overall, this typical SDAT-ReaxFF simulation shows all reaction steps of the proton-catalyzed reaction mechanism, as described in Section 3.3.2. To investigate possible additional reaction pathways, it is necessary to analyze larger systems including more monomers and the full (neutral) acid molecule needs to be taken into account to capture the Coulomb force appropriately, instead of a positively charged proton. From the discussion given in Chapter 3.2, the influence of the anion on the different reaction steps as well as on the overall reaction mechanism is still an open issue. However, this information needs to be known for the appropriate training data set to parameterize the reactive force field. Thus, in the next section, we will present an extended DFT investigation of the TP process. Hereby, we do not only examine the influence of full acid catalyst molecules but also consider the base-catalyzed TP.

3.3.5 Extended quantum chemical modeling

In Chapter 3.2, the reaction mechanism for the thermally induced and proton-catalyzed TP of the Si-Spiro monomer has already been presented and important findings from quantum chemistry have been discussed. In these calculations, a proton was used instead of the full acid molecule, giving a simpler and therefore computationally less demanding model. For the investigation of the nanostructure formation process of TP, there are various reasons to extend this analysis to uncharged molecules, using also the counter anion. The three main reasons are as follows:

(1) The reaction energies and activation barriers should be as close to reality as possible. Beside an appropriate method to optimize the geometrical and electronic structure as well as suitable descriptions of solvation, temperature effects, etc., also the definition of the model size is an essential key point in computational chemistry. Thus, a bare proton in the gas phase provides an excessive acidity compared to the real uncharged system, leading to a less exact description of the electronic structure. As a strong positive charge (high acidity) will better stabilize the species involved in the ring-opening reaction, the activation barriers will be too low in this crude model.
(2) Beside the electronic structure, the counter anion may have a non-negligible steric effect.
(3) In the previously discussed rMD simulations with ReaxFF, the forces of the anion are not taken into account appropriately.

In summary, the inclusion of the counter anion is required for a proper simulation of the nanostructure formation. Therefore, we introduce an extended quantum chemical approach including the interactions of the twin monomer with a full acid. As acid, we use trifluoroacetic acid (TFA).

Beside the extensive research about the thermally induced and acid-catalyzed TP [2, 6], there are also efforts to use bases as catalysts, as for instance DABCO or DBU [11]. Hereby, DBU has been found especially suitable to trigger the base-catalyzed TP of Si-Spiro, because it provides a complete silica network formation. In order to reveal differences to the acid-catalyzed reaction mechanism, we also apply the extended quantum chemical approach for the base-catalyzed TP, employing the base DBU.

The modeling approach utilizes DFT as a rather low-scaling *ab initio* method. The calculations are executed with RI-DFT [130–133] as implemented in TURBOMOLE V6.5 [134, 135]. All stationary points of the potential energy surfaces as well as the corresponding vibrational frequencies and the zero-point vibrational energies are obtained by geometry optimizations and subsequent analyses of the Hessian matrix at the BP86-D3/SVP [136–141] level including COSMO [142] using the dielectricity parameter $\varepsilon = \infty$. In some cases, more accurate electronic energies are needed. In the case of the acid-catalyzed TP, these energies are calculated with PW6B95-D3/def2-TZVPP

[139, 140, 143–145] and $\varepsilon = 10$, where ε is chosen close to the realistic value for TFA $\varepsilon_{TFA} = 8.55$, and for the base-catalyzed TP, PW6B95-D3/def2-QZVPP [139, 140, 143, 146] and $\varepsilon = \infty$ is utilized. In all calculations, the thermal effects are included by means of statistical thermodynamics (freeh module [134, 135]), yielding Gibbs free energies. Starting geometries for the transition state structures were obtained *via* relaxed scans along the reaction coordinate. Further technical details regarding the acid-catalyzed case can be found in Tchernook *et al.* [81, 82].

Note that in each of the following two subsections, an independent numbering of the chemical structures is applied in order to keep the numbering scheme as simple as possible.

3.3.5.1 Acid-catalyzed TP

Modeling the TP *via* DFT, one has always to keep the computational effort in mind, which increases with the system size. Therefore, it is important to find small systems, containing only few atoms that show all important features of the reaction of interest. In the case of the acid-catalyzed TP, it is known that the mechanism for the formation of the organic polymer network can be described by two essential steps: (1) The protonation of the monomer facilitates the O–CH_2 bond cleavage, which leads to a ring-opening with a stabilized benzylic cation as intermediate product. A supposable Si–O cleavage has already been excluded within the previous work of Auer *et al.* [4]. This part of the reaction mechanism will be described by the 1M1T model, i.e., one monomer (M) and one TFA (T) molecule are involved (Scheme 3.55). (2) The propagation step is characterized by an electrophilic aromatic substitution, i.e., an attack of the previously generated reactive benzylic cation on another monomer as well as the subsequent proton transfer from the σ-complex to a neighboring trifluoroacetate anion. Therefore, at least one additional monomer is necessary, giving a 2M1T model (Scheme 3.56). However, further acid molecules may also participate, providing a stabilization of the intermediate counter anion *via* hydrogen bonds as well as a higher steric flexibility. Thus, a 2M2T model was utilized (Scheme 3.57), where the first acid molecule protonates the monomer and the second, stabilizing one, acts as a proton acceptor in the decomposition of the σ-complex. Reaction step (1), i.e., the ring-opening, has been examined in much detail within the 1M1T model and the resulting energies (in kcal·mol^{-1}) are presented in Scheme 3.55.

Note that in the case of TP, typically an acid exhibiting hydrogen bond added to a nonpolar solvent may form droplets instead of the acid in the solvent instead of a homogeneous mixture. Thus, the above described model is suitable for the catalytic process.

At first, a hydrogen bond between the OH group of the TFA and the benzylic oxygen of the Si-Spiro monomer activates the latter for the subsequent ring-opening reaction. Depending on the orientation of the TFA molecule relative to the monomer, different reaction pathways are obtained.

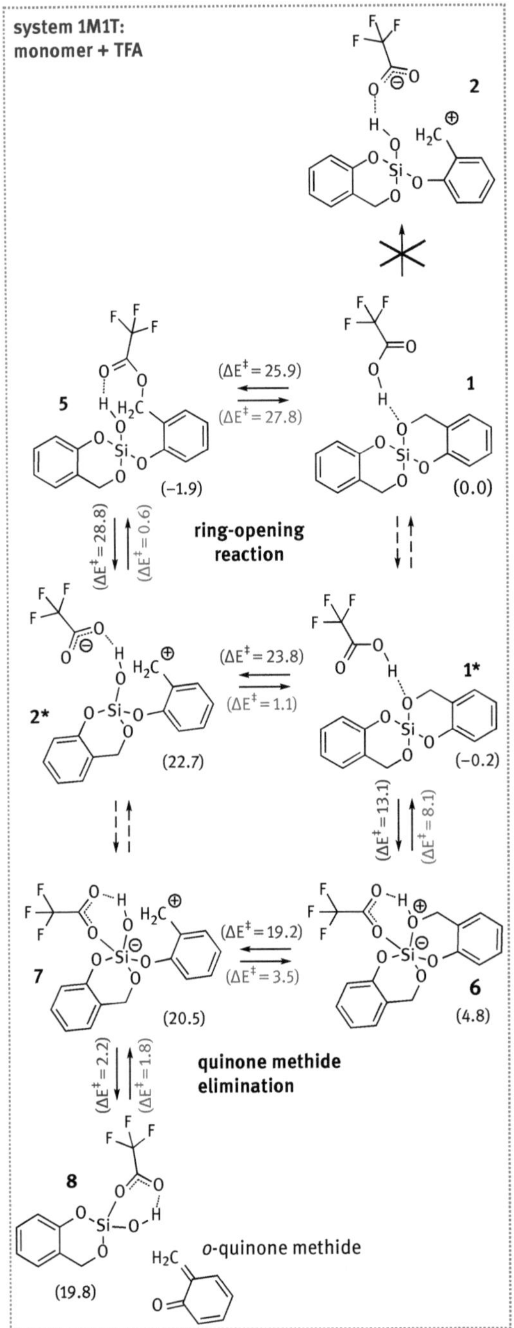

Scheme 3.55: Reaction scheme for the 1M1T model. Energies (relative to **1**) for all products (in parenthesis) and activation barriers (blue, reverse reaction: red) are given in kcal·mol^{-1} and include electronic contribution (PW6B95-D3/def2-TZVPP/ε = 10) as well as ZPVE (BP86/SVP/ε = ∞). Reprinted from Ref. [81], Copyright 2015, with permission from Elsevier.

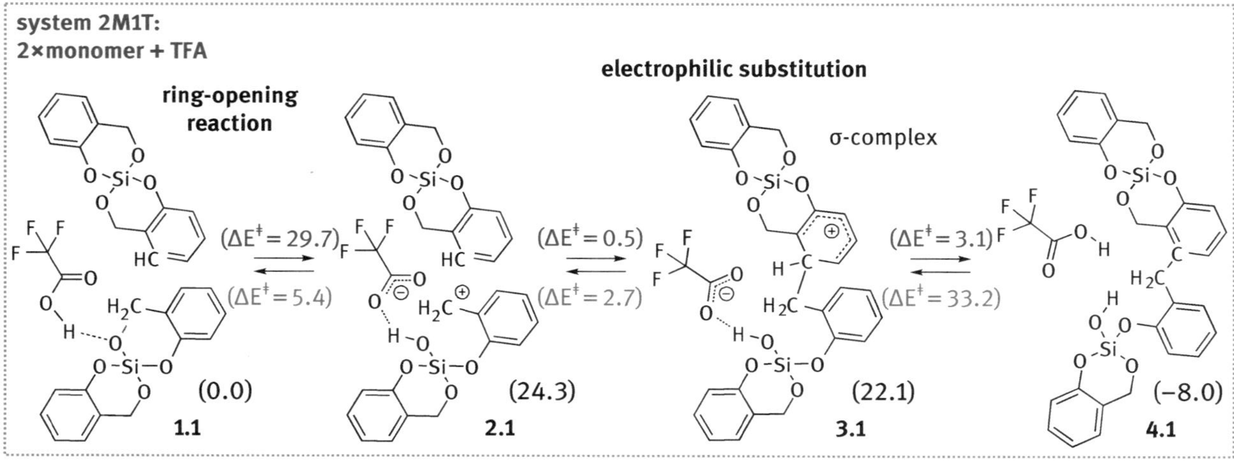

Scheme 3.56: Reaction scheme for the 2M1T model (*ortho*-substitution). Energies (relative to **1.1**) for all products (in parenthesis) and activation barriers (blue, reverse reaction: red) are given in kcal·mol^{-1} and include electronic contribution (PW6B95-D3/def2-TZVPP/$\varepsilon = 10$) as well as ZPVE (BP86/SVP/$\varepsilon = \infty$). Reprinted from Ref. [81], Copyright 2015, with permission from Elsevier.

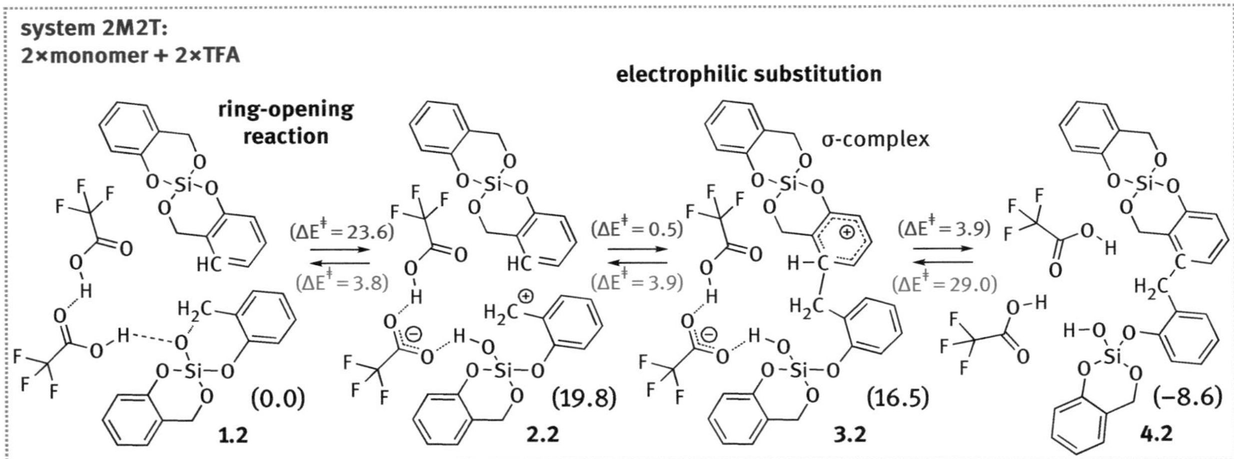

Scheme 3.57: Reaction scheme for the 2M2T model (*ortho*-substitution). Energies (relative to **1.1**) for all products (in parenthesis) and activation barriers (blue, reverse reaction: red) are given in kcal·mol^{-1} and include electronic contribution (PW6B95-D3/def2-TZVPP/$\varepsilon = 10$) as well as ZPVE (BP86/SVP/$\varepsilon = \infty$). Reprinted from Ref. [81], Copyright 2015, with permission from Elsevier.

Considering the protonated monomer conformation **1**, the O–CH_2 bond cleavage does not lead to the expected electrophile **2**, as the emerging positive (benzylic cation) and negative (Si–O^-) charges are very close and therefore not stabilized. In contrast, an adduct of the benzyl group with the free O atom of the TFA molecule (**5**) is formed. Only a subsequent dihedral rotation of the acid leads to the stable intermediate electrophile **2***. Both steps of this pathway feature activation barriers of about 28 kcal·mol^{-1}.

The electrophile **2*** can also be generated directly from conformer **1***, which already exhibits such a dihedral rotation. Thus, the charges that emerge during the O–CH_2 bond cleavage are spatially separated, which leads to the local energy minimum **2***. Additionally, the activation barrier is slightly lower (≈ 25 kcal·mol^{-1}).

The steric orientation of **1*** also provides the possibility of an intermediate coordination of TFA's free O atom at the Si atom, leading to a pentavalent species **6**. This step also supports the subsequent O–CH_2 bond cleavage, as it features an activation energy of only about 20 kcal·mol^{-1}. Due to the pentavalent coordination, the elimination of an *ortho*-quinone methide is feasible. Such a reaction even needs hardly any activation and leads to a quite stable intermediate **8** (≈ 20 kcal·mol^{-1} relative to educt **1***), compared to the electrophile **2*** (≈ 23 kcal·mol^{-1}).

In general, **2***, **7**, and **8** are possible electrophiles for the following reaction steps, but they are less stabilized and feature reverse reactions with very low barriers; so they are expected to be very short-living. Thus, a model with a close second monomer is necessary to explain the electrophilic aromatic substitution. This additional molecule might also have a steric or electronic influence.

Due to the addition of a second monomer (2M1T model, results see Scheme 3.56), an electrophile **2.1**, which is equivalent to the fictitious structure **2** in the 1M1T model, is energetically stabilized and represents a local minimum now. The corresponding direct ring-opening reaction of **1.1**–**2.1** proceeds *via* an activation barrier of about 31 kcal·mol^{-1}.

By adding a second acid molecule (2M2T model, results see Scheme 3.57), only a barrier of about 25 kcal·mol^{-1} remains for the ring-opening of **2.1**, leading to the electrophile **2.2**. This promotion of the initial bond cleavage is caused by the stabilization of the trifluoroacetate anion that emerges during this step *via* a hydrogen bond to the second TFA molecule.

Considering the second important process, almost no differences between the electrophilic aromatic substitutions in the *ortho*- and *para*-positions have been found by Auer *et al.* [4]. Thus, for simplification, only the *ortho*-substitution is in the focus at first. However, the *ortho*/*para*-selectivity gets considered later within the 2M2T model. Both steps of the electrophilic substitution feature lower barriers in comparison to the ring-opening reaction, which is clearly the rate-determining step. While the electrophilic attack, i.e., the formation of the σ-complex **3.1** (**3.2**), occurs almost barrierless with activation energies of 0.7 kcal mol^{-1} (0.4 kcal·mol^{-1}) within the 2M1T (2M2T) model, the barriers are 2.7 kcal·mol^{-1} (3.7 kcal·mol^{-1}) for the final H^+ transfer and are therefore also very low. The last step is very exothermic, as the aromaticity is retrieved, and the product **4.1** (**4.2**), which is by 7.2 kcal·mol^{-1} (7.8 kcal·mol^{-1}) more stable than the educt, is obtained. Concerning the *ortho*/*para*-selectivity, the energetics of the electrophilic attack are nearly the same for both cases. However, the last step indicates a preference for the *ortho*-substitution, as the H^+ transfer proceeds *via* a 3.4 kcal·mol^{-1} lower activation barrier and the *ortho*-product is by 5.6 kcal·mol^{-1} more stable than the *para* one.

The extended DFT analysis of the acid-catalyzed TP shows that the generation of the electrophile by a ring-opening reaction is the rate-determining step of the organic network formation, while the subsequent steps (electrophilic attack, H^+ transfer) feature low barriers. In summary, the addition of a further monomer and an acid leads to the stabilization of the electrophile, a decreased activation barrier for the initial rate-determining O–CH_2 cleavage, and the deactivation by the counter anion is hindered.

3.3.5.2 Base-catalyzed TP

In a first approach for the base-catalyzed TP of the Si-Spiro (M) with the base DBU (B), a simple 1M1B model is considered. Similar to the extended acid-catalyzed TP, the description is limited to the initiation process in this simple model. In order to treat also the propagation steps, an extension to the 2M1B model is carried out.

The formation of pentacoordinated atoms by addition of a nucleophile is well known for silicon [147]. As a deprotonation of the monomer is not feasible, the only reasonable role of DBU is the coordination at the Si atom by its most basic nitrogen atom. For the resulting pentacoordinated complex, different conformers with (more or less) distorted trigonal-bipyramidal geometries are obtained. The most stable one features the base molecule as well as the benzylic oxygen atoms of the monomer in equatorial and its phenolic oxygen atoms in axial positions. The formation of this complex **1.2** requires a rather small activation barrier of 2.6 kcal·mol^{-1} and is clearly exothermic with –11.7 kcal·mol^{-1}.

In general, the TP of the Si-Spiro monomer formally includes the polymerizations of a quinone methide (organic network) as well as an O–Si–O moiety (inorganic network). Therefore, the Si–O (phenolic O) as well as the O–C (benzylic O) bond must be broken during the reaction mechanism and thus a scan of the potential energy surface with respect to the two corresponding bond lengths enables further insights into the initiation (see Fig. 3.31). In contrast to the acid-catalyzed mechanism with the initial protonation of the benzylic oxygen, the O–C bond is not activated any more in case of the base-catalyzed TP. This is also observable in Figure 3.31, as the exclusive separation of the O–C bond continuously increases the energy and no local minimum is received. However, the Si–O cleavage is now possible, which is due to the increased electron density in the monomer, leading to a stabilization of the free phenolate. From the resulting structure **1.3**, the subsequent cleavage of the O–C bond is now feasible, yielding a (less stabilized) quinone methide **1.4**. In this initiation process, the Si–O cleavage is the rate-determining step with an activation energy of 25.1 kcal·mol^{-1}.

Next, the propagation process is investigated by means of the 2M1B model as shown in Scheme 3.58. Within the 1M1B model, the formation of the pencoordinated

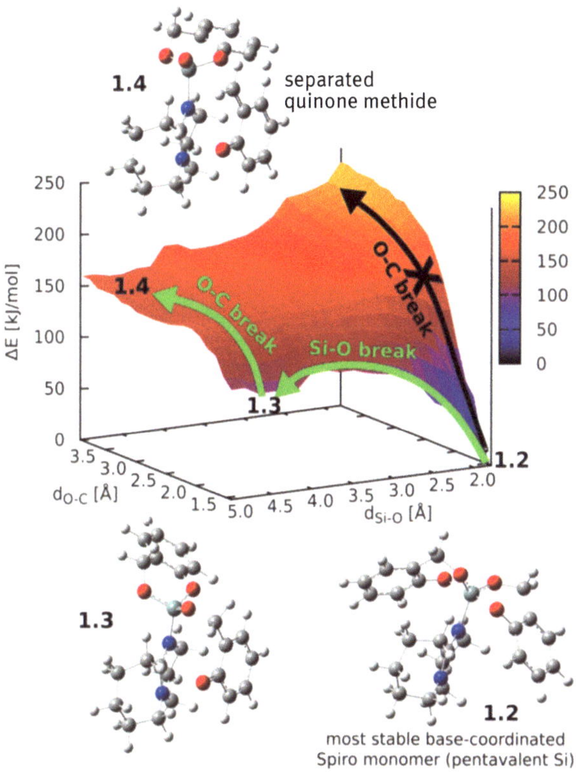

Fig. 3.31: Potential energy surface with respect to the relevant O–C and Si–O bond-lengths for the quinone methide formation (see **1.4**) from the DBU-coordinated Si-Spiro **1.2** *via* intermediate **1.3**. Optimizations of the presented molecular structures and calculations of electronic energies have been performed with BP86-D3/SVP and the COSMO model ($\varepsilon = \infty$).

Si atom as well as the cleavage of the Si–O bond turned out to be the mandatory initiation steps and the influence of a second monomer on these two steps is assumed to be not determining. Therefore, they are not taken into account within the 2M1B model and structure **2.3**, i.e., **1.3** with an additional monomer molecule, is the starting point. In contrast, the separation of the quinone methide is very endothermic and might be avoided by a concerted mechanism or at least stabilized by the second monomer.

At first, the O–C cleavage is again considered within the 2M1B model, depicted in Scheme 3.58, leading to a separated quinone methide in structure **2.4**.

The activation and reaction energies for this formation are not clearly different from those in the 1M1B case (**1.3**–**1.4** in Fig. 3.31), i.e., no significant stabilization is recognizable. Beside the quinone methide, this cleavage generates also a free silanolate as a very reactive moiety. Therefore, two possibilities occur for the following step: (1) electrophilic attack of the methylene group within the quinone methide at the aromatic ring of the second monomer, leading to a σ-complex (organic

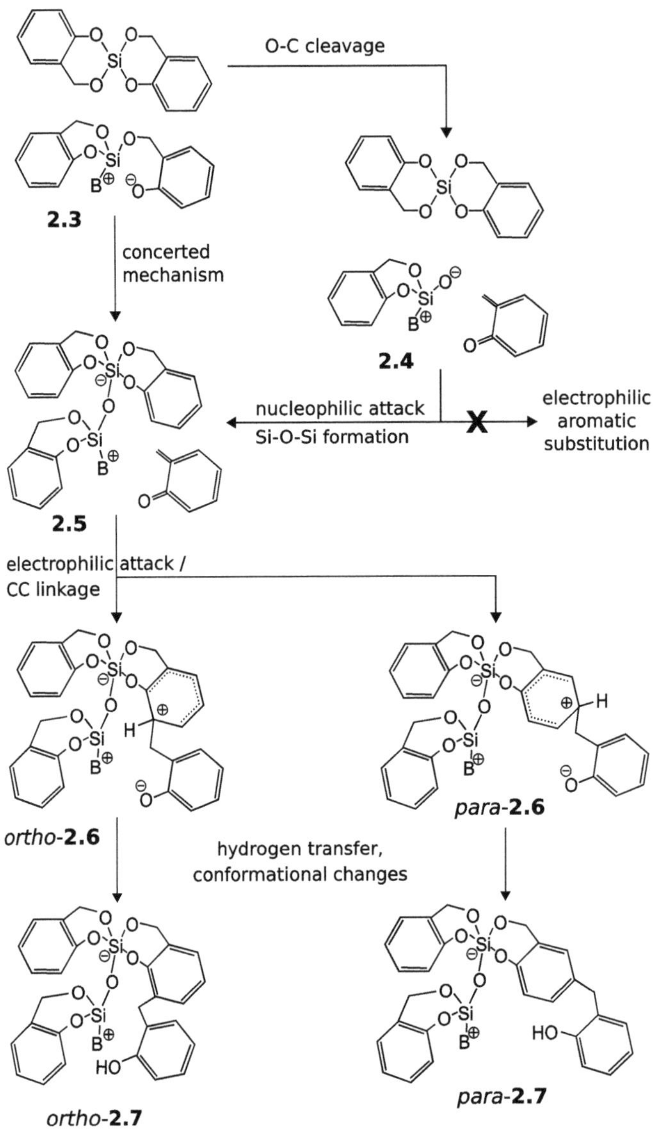

Scheme 3.58: Schematic representation of the considered mechanistic steps within the 2M1B model, where B represents the catalytic base molecule.

propagation step), or (2) a nucleophilic attack of the silanolate at the Si atom of the second monomer, leading to a Si–O–Si bridge (inorganic propagation step).

For possibility (1), relaxed scans of the energy with respect to the formed C–C bond regarding the *ortho*- and the *para*-position exhibit no local minimum, representing the formation of a stable σ-complex. Therefore, the direct organic propagation step from **2.4** is not possible. In contrast, reaction (2) is feasible and leads to the clearly exothermic formation of the Si–O–Si bridge (**2.5**) *via* a rather low activation barrier (≈ 6.6 kcal·mol^{-1}). However, within the obtained mechanism, the separation of the quinone methide by the O–C cleavage and the approach of the silanolate toward the Si atom of the second monomer occur consecutively, i.e., in a two-step mechanism. As shown in Scheme 3.58, an alternative might be the concerted proceeding of these two processes, i.e., a direct transfer of the silanolate *via* a transition state **2.TS$_{35}$**. For the more accurate PW6B95-D3/def2-QZVPP scheme, as presented in Figure 3.32, the two-step mechanism provides a somewhat smaller activation barrier of its rate-determining step (≈ 17.5 kcal·mol^{-1}) than the concerted mechanism (≈ 20.3 kcal·mol^{-1}) and is therefore preferred. This comparison shows the importance of evaluating and choosing a sufficiently accurate DFT method, as the order of preference is the other way around for the inferior BP86-D3/SVP method, i.e., the barrier for the concerted mechanism here is about 3.9 kcal·mol^{-1} smaller.

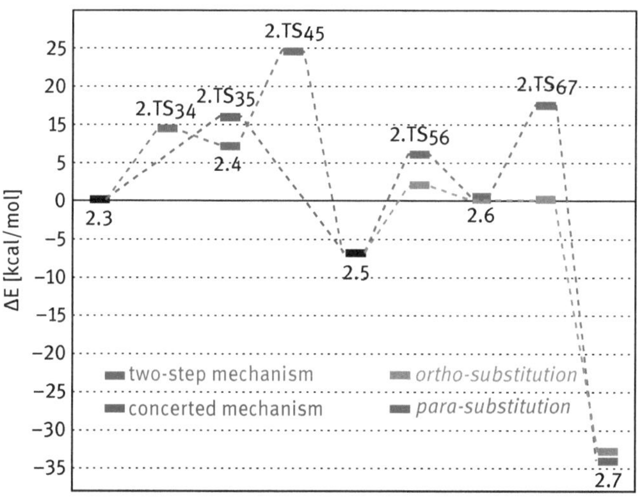

Fig. 3.32: Energy profile for the propagation process of the base-catalyzed twin polymerization of Si-Spiro with DBU in a 2M1B model. Geometry optimizations and calculations of ZPVEs as well as COSMO (ε = ∞) solvation contributions were performed at the BP86-D3/SVP level. Electronic energy contributions were calculated with PW6B95-D3/def2-QZVPP.

Starting with structure **2.5**, which has been formed by the discussed two-step mechanism, an electrophilic aromatic substitution by the quinone methide now appears to be possible again. Indeed, the approach of the quinone methide's methylene group leads to a stable σ-complex **2.6**, characterized as local minimum, for both the *ortho*- and the *para*-position. The energy profile for this electrophilic aromatic substitution is also presented in Figure 3.32. In the final step, hydrogen transfer and conformational changes lead to the strongly exothermic formation of product **2.7**. When considering the *ortho*/*para*-selectivity, the *ortho*-σ-complex is by ≈ 1.6 kcal·mol^{-1} more stable and especially the activation energy for its formation is by 3.1 kcal·mol^{-1} smaller. Furthermore, the final step occurs almost barrierless, while the *para*-σ-complex requires a significant activation for the product formation. Therefore, the electrophilic aromatic substitution in *ortho*-position is kinetically preferred, although the final *para*-product features almost the same thermodynamical stability.

In conclusion, the presented modeling for the Si-Spiro monomer with DBU revealed several insights into the reaction mechanism of the base-catalyzed TP. This process is initiated by the coordination of the base at the Si atom, giving a pentacoordinated species, and the rate-determining cleavage of the Si–O bond, i.e., to the benzylic oxygen atom. A quinone methide may occur as a short-lived intermediate. During the final product formation, the *ortho*-position is kinetically preferred within the electrophilic aromatic substitution. However, a main difference to, e.g., the acid-catalyzed case is the formation of the Si–O–Si bridge prior to the electrophilic aromatic substitution, referring to the organic propagation. Thus, the latter appears only due to a previous inorganic propagation step, which demonstrates the coupling of the two polymerization processes. This is a special feature of the TP and leads to the formation of nanostructured hybrid materials.

3.3.6 Summary

In this chapter, we presented three different modeling approaches to investigate the nanostructure formation process of TP. We started with a very simple rBFM, where extended steric structures that do not alter over reaction time are coarse-grained to beads. With this model, we were able to simulate the complete reaction mechanism of TP for large system sizes and to investigate the dynamical structure formation process over time. Our model reflected the reaction mechanism derived from first quantum chemical calculations. Additionally, we compared the obtained morphologies with experimentally obtained pore size distributions and connectivity information between the twin polymer networks. The structures from the rBFM and experiments showed very good to reasonable agreements.

In a second step, we increased the complexity of our modeling approach to an atom-based rMD simulation, where all atoms of the system and their interactions are taken into account. After developing a new SDAT ReaxFF force field representation, we were able to simulate the complete reaction mechanism for a small system. Extending this method to the full acid instead of only taking the proton into account, we can utilize this method to search for additional reaction pathways and to identify structure formation effects on the atomic length scale.

From these results, we find that an extended quantum chemical calculation of the reaction mechanism of TP is important for further simulations. In this third approach, we applied DFT to determine accurate structures, energies, and reaction pathways for the interactions of twin monomers with the acid or base molecules.

Thus, we did not only examine the acid-catalyzed TP but also the base-catalyzed TP. The most important finding is that although the acid- and the base-catalyzed TP process show similar final morphologies, we observe two clearly distinct reaction processes. The acid-catalyzed TP is determined by the initial electrophilic ring-opening leading to a prior organic network formation. In contrast to this, the base-catalyzed TP is rate-determined by the cleavage of the Si–O bond, which leads to a prior inorganic network formation. These results are in complete agreement with first experimental results as shown in Chapter 4.1 and it will be a promising starting point for further investigations on the structure formation process of TP.

References

[1] Grund S, Kempe P, Baumann G, Seifert A, Spange S. Nanocomposites Prepared by Twin Polymerization of a Single-Source Monomer. Angew Chem Int Ed 2007;46:628–32. doi:10.1002/anie.200504327.

[2] Spange S, Kempe P, Seifert A, Auer AA, Ecorchard P, Lang H, et al. Nanocomposites with Structure Domains of 0.5 to 3 nm by Polymerization of Silicon Spiro Compounds. Angew Chem Int Ed 2009;48:8254–8. doi:10.1002/anie.200901113.

[3] Ebert T, Seifert A, Spange S. Twin Polymerization – a New Principle for Hybrid Material Synthesis. Macromol Rapid Commun 2015;36:1623–39. doi:10.1002/marc.201500182.

[4] Auer AA, Richter A, Berezkin AV, Guseva DV, Spange S. Theoretical Study of Twin Polymerization – From Chemical Reactivity to Structure Formation. Macromol Theory Simul 2012;21:615–28. doi:10.1002/mats.201200036.

[5] Roschke F, Rüffer T, Seifert A, Nagel K, Spange S, Lang H, et al. Chiral Molecular Fluoridosilicates and their Twin Polymerization for the Preparation of Fluorine-doped Mesoporous Silica and Microporous Carbon. Inorg Chem Front 2018; accepted. doi: 10.1039/C8QI00533H.

[6] Kempe P, Löschner T, Auer AA, Seifert A, Cox G, Spange S. Thermally Induced Twin Polymerization of *4H*-1,3,2-Benzodioxasilines. Chem Eur J 2014;20:8040–53. doi:10.1002/chem.201400038.

[7] Kitschke P, Auer AA, Löschner T, Seifert A, Spange S, Rüffer T, et al. Microporous Carbon and Mesoporous Silica by Use of Twin Polymerization: An Integrated Experimental and Theoretical Approach to Precursor Reactivity. ChemPlusChem 2014;79:1009–23. doi:10.1002/cplu.201402029.

[8] Kitschke P, Rüffer T, Lang H, Auer AA, Mehring M. Chiral Spirocyclic Germanium Thiolates – An Evaluation of Their Suitability for Twin Polymerization based on A Combined Experimental and Theoretical Study. ChemistrySelect 2016;1:1184–91.

[9] Kitschke P, Walter M, Rüffer T, Lang H, Kovalenko MV, Mehring M. From molecular germanates to microporous Ge@C via twin polymerization. Dalton Trans 2016;45:5741–51. doi:10.1039/C6DT00049E.

[10] Leonhardt C, Brumm S, Seifert A, Cox G, Lange A, Rüffer T, et al. Tin Oxide Nanoparticles and SnO_2/SiO_2 Hybrid Materials by Twin Polymerization Using Tin(IV) Alkoxides. ChemPlusChem 2013;78:1400–12. doi:10.1002/cplu.201200242.

[11] Ebert T, Cox G, Sheremet E, Gordan O, Zahn DRT, Simon F, et al. Carbon/carbon nanocomposites fabricated by base catalyzed twin polymerization of a Si-spiro compound on graphite sheets. Chem Commun 2012;48:9867–9. doi:10.1039/c2cc34775j.

[12] Spange S, Grund S. Nanostructured Organic-Inorganic Composite Materials by Twin Polymerization of Hybrid Monomers. Adv Mater 2009;21:2111–6. doi:10.1002/adma.200802797.
[13] Mehner A, Pohlers A, Hoyer W, Cox G, Spange S. Cationic Twin Polymerization Versus Simultaneous Polymerization of Titanium Compounds to Fabricate Nanostructured Organic Polymer/TiO_2 Hybrid Materials. Macromol Chem Phys 2013;214:1000–10. doi:10.1002/macp.201200629.
[14] Mehner A, Rüffer T, Lang H, Pohlers A, Hoyer W, Spange S. Synthesis of Nanosized TiO_2 by Cationic Polymerization of (μ_4-oxido)-hexakis(μ-furfuryloxo)-octakis(furfuryloxo)-tetra-titanium. Adv Mater 2008;20:4113–7. doi:10.1002/adma.200801376.
[15] Schliebe C, Gemming T, Noll J, Mertens L, Mehring M, Seifert A, et al. Zirconium and Hafnium Twin Monomers for Mixed Oxides. ChemPlusChem 2015;80:559–67. doi:10.1002/cplu.201402338.
[16] Böttger-Hiller F, Kempe P, Cox G, Panchenko A, Janssen N, Petzold A, et al. Twin Polymerization at Spherical Hard Templates: An Approach to Size-Adjustable Carbon Hollow Spheres with Micro- or Mesoporous Shells. Angew Chem Int Ed 2013;52:6088–91. doi:10.1002/anie.201209849.
[17] Grund S. Synthese anorganisch-organischer Polyfurfurylalkohol-Nanokomposite durch die Zwillingspolymerisation. Dissertation. Technische Universität Chemnitz, 2010.
[18] Böttger-Hiller F, Mehner A, Anders S, Kroll L, Cox G, Simon F, et al. Sulphur-doped porous carbon from a thiophene-based twin monomer. Chem Commun 2012;48:10568–70. doi:10.1039/C2CC35112A.
[19] Mehner A, Rüffer T, Lang H, Schlesinger M, Mehring M, Spange S. Chirality and reversible solid state phase transition of a tetranuclear titanium oxido cluster. New J Chem 2013;37:1290–3. doi:10.1039/C3NJ40775F.
[20] Falana OM, Al-Farhan E, Keehn PM, Stevenson R. High yield synthesis of the parent C-unsubstituted calix[4]resorcinarene octamethyl ether. Tetrahedron Lett 1994;35:65–8. doi:10.1016/0040-4039(94)88163-4.
[21] Li D, Kusunoki T, Yamagishi T-A, Nakamoto Y. Synthesis of C-unalkylated calix[4]resorcinarene from 1,3-dimethoxybenzene-formaldehyde condensation. Polym Bull 2002;47:493–9. doi:10.1007/s002890200013.
[22] Preda A-M, Kitschke P, Rüffer T, Lang H, Mehring M. Synthesis and Characterization of the Germanium Sulfonate $Ge(CH_3SO_3)_2$ – a 3D Coordination Network Solid. Z Anorg Allg Chem 2016;642:467–71. doi:10.1002/zaac.201600023.
[23] Preda A-M, Kitschke P, Rüffer T, Lang H, Mehring M. Characterization of $Ge(CH_3SO_3)_2$, Tagungsheft "Innovative Materialsynthesen" / "Innovative Syntheses of Materials." Z Anorg Allg Chem 2014;640:2309–17. doi:10.1002/zaac.201490027.
[24] Mantina M, Chamberlin AC, Valero R, Cramer CJ, Truhlar DG. Consistent van der Waals Radii for the Whole Main Group. J Phys Chem A 2009;113:5806–12. doi:10.1021/jp8111556.
[25] Glusker JP, Lewis M, Rossi M. Crystal structure analysis for chemists and biologists. VCH; 1994.
[26] Shorter J. Die Hammett-Gleichung – und was daraus in fünfzig Jahren wurde. Chem unserer Zeit 1985;19:197–208. doi:10.1002/ciuz.19850190604.
[27] Stevens MP. Polymer chemistry: an introduction. 3rd ed. New York: Oxford University Press; 1999.
[28] Auer AA, Bistoni G, Kitschke P, Mehring M, Ebert T, Spange S. Electronic Structure Calculations and Experimental Studies on the Thermal Initiation of the Twin Polymerization Process. ChemPlusChem 2017;82:1396–407. doi:10.1002/cplu.201700358.
[29] Mayr H, Ofial AR. Do general nucleophilicity scales exist? J Phys Org Chem 2008;21:584–95. doi:10.1002/poc.1325.
[30] Kitschke P. Theoretische und experimentelle Untersuchungen zur Zwillingspolymerisation an neuen Monomeren auf der Basis vierwertiger Metallalkoxide. Diploma Thesis. Technische Universität Chemnitz, 2011.

[31] Kitschke P, Schulze S, Hietschold M, Mehring M. Synthesis of germanium dioxide nanoparticles in benzyl alcohols – a comparison. Main Group Met Chem 2013;36:209–14. doi:10.1515/mgmc-2013-0038.
[32] Kitschke P, Preda A-M, Auer AA, Scholz S, Rüffer T, Lang H, et al. Spirocyclic tin salicyl alcoholates - A combined experimental and theoretical study on their structures, ^{119}Sn NMR chemical shifts and reactivity in Twin Polymerization. Submitted 2018.
[33] Schäfer A. Synthese, Charakterisierung und Studien zur Zwillingspolymerisation von neuen spirocyclischen Präkursoren. Bachelor Thesis. Technische Universität Chemnitz, 2016.
[34] Dannenberg F, Thiele G, Dornsiepen E, Dehnen S, Mehring M. Synthesis, structure and thermolysis of oxazagermines and oxazasilines. New J Chem 2017;41:4990–7. doi:10.1039/C7NJ00894E.
[35] Roschke F. Alkoxid- und Amidverbindungen der Gruppe-14-Elemente Germanium und Silicium als molekulare Präkursoren zur Herstellung kohlenstoffhaltiger Kompositmaterialien. Dissertation. Technische Universität Chemnitz, 2018.
[36] Leonhardt C, Brumm S, Seifert A, Lange A, Csihony S, Mehring M. Tin Nanoparticles in Carbon/Silica Hybrid Materials by the Use of Twin Polymerization. ChemPlusChem 2014;79:1440–7. doi:10.1002/cplu.201402137.
[37] Kempe P, Löschner T, Adner D, Spange S. Selective ring opening of *4H*-1,3,2-benzodioxasiline twin monomers. New J Chem 2011;35:2735–9. doi:10.1039/C1NJ20654K.
[38] Kitschke P, Auer AA, Seifert A, Rüffer T, Lang H, Mehring M. Synthesis, characterization and Twin Polymerization of a novel dioxagermine. Inorg Chim Acta 2014;409, Part B:472–8. doi:10.1016/j.ica.2013.09.052.
[39] Löschner T. Synthese nanostrukturierter, organisch-anorganischer Hybridmaterialien über Zwillingspolymerisation. Dissertation. Technische Universität Chemnitz, 2013.
[40] Löschner T, Mehner A, Grund S, Seifert A, Pohlers A, Lange A, et al. A Modular Approach for the Synthesis of Nanostructured Hybrid Materials with Tailored Properties: The Simultaneous Twin Polymerization. Angew Chem Int Ed 2012;51:3258–61. doi:10.1002/anie.201108011.
[41] Kitschke P, Rüffer T, Korb M, Lang H, Schneider WB, Auer AA, et al. Intramolecular C–O Insertion of a Germanium(II) Salicyl Alcoholate: A Combined Experimental and Theoretical Study. Eur J Inorg Chem 2015; 5467–79. doi:10.1002/ejic.201501109.
[42] Kitschke P, Walter M, Rüffer T, Seifert A, Speck F, Seyller T, et al. Porous Ge@C materials via twin polymerization of germanium(II) salicyl alcoholates for Li-ion batteries. J Mater Chem A 2016;4:2705–19. doi:10.1039/C5TA09891B.
[43] Leonhardt C. Zinnalkoxide als Präkursoren für zinnhaltige Nanokomposite. Dissertation. Technische Universität Chemnitz, 2016.
[44] Kitschke P. Experimental and theoretical studies on germanium-containing precursors for twin polymerization. Dissertation. Technische Universität Chemnitz, 2016.
[45] Ewings PFR, Harrison PG. Derivatives of bivalent germanium, tin, and lead. Part X. Tin(II) bis(phenoxides), bis(O-methyl dithiocarbonate), and bis(diethyidithiocarbamate). J Chem Soc, Dalton Trans 1975:2015–7. doi:10.1039/DT9750002015.
[46] Grimme S, Brandenburg JG, Bannwarth C, Hansen A. Consistent structures and interactions by density functional theory with small atomic orbital basis sets. J Chem Phys 2015;143:054107. doi:10.1063/1.4927476.
[47] Schröder H, Hühnert J, Schwabe T. Evaluation of DFT-D3 dispersion corrections for various structural benchmark sets. J Chem Phys 2017;146:044115. doi:10.1063/1.4974840.
[48] Howard JC, Enyard JD, Tschumper GS. Assessing the accuracy of some popular DFT methods for computing harmonic vibrational frequencies of water clusters. J Chem Phys 2015;143:214103. doi:10.1063/1.4936654.
[49] Alecu IM, Zheng J, Zhao Y, Truhlar DG. Computational Thermochemistry: Scale Factor Databases and Scale Factors for Vibrational Frequencies Obtained from Electronic Model Chemistries. J Chem Theory Comput 2010;6:2872–87. doi:10.1021/ct100326h.

[50] Andersson MP, Uvdal P. New Scale Factors for Harmonic Vibrational Frequencies Using the B3LYP Density Functional Method with the Triple-ζ Basis Set 6-311+G(d,p). J Phys Chem A 2005;109:2937–41. doi:10.1021/jp045733a.
[51] Goerigk L, Grimme S. A General Database for Main Group Thermochemistry, Kinetics, and Noncovalent Interactions – Assessment of Common and Reparameterized (*meta*-)GGA Density Functionals. J Chem Theory Comput 2010;6:107–26. doi:10.1021/ct900489g.
[52] Boese AD, Martin JML. Development of density functionals for thermochemical kinetics. J Chem Phys 2004;121:3405–16. doi:10.1063/1.1774975.
[53] Zhao Y, Truhlar DG. The M06 suite of density functionals for main group thermochemistry, thermochemical kinetics, noncovalent interactions, excited states, and transition elements: two new functionals and systematic testing of four M06-class functionals and 12 other functionals. Theor Chem Account 2008;120:215–41. doi:10.1007/s00214-007-0310-x.
[54] Sousa SF, Fernandes PA, Ramos MJ. General Performance of Density Functionals. J Phys Chem A 2007;111:10439–52. doi:10.1021/jp0734474.
[55] Sure R, Grimme S. Comprehensive Benchmark of Association (Free) Energies of Realistic Host–Guest Complexes. J Chem Theory Comput 2015;11:3785–801. doi:10.1021/acs.jctc.5b00296.
[56] Sure R, Grimme S. Correction to Comprehensive Benchmark of Association (Free) Energies of Realistic Host–Guest Complexes. J Chem Theory Comput 2015;11:5990. doi:10.1021/acs.jctc.5b01016.
[57] Wilson S. Advances in quantum chemistry. Volume 32, Part II Volume 32, Part II. San Diego: Academic Press; 1999.
[58] Andzelm J, Kölmel C, Klamt A. Incorporation of solvent effects into density functional calculations of molecular energies and geometries. J Chem Phys 1995;103:9312–20. doi:10.1063/1.469990.
[59] Diedenhofen M, Eckert F, Klamt A. Prediction of Infinite Dilution Activity Coefficients of Organic Compounds in Ionic Liquids Using COSMO-RS. J Chem Eng Data 2003;48:475–9. doi:10.1021/je025626e.
[60] Tomasi J, Mennucci B, Cammi R. Quantum Mechanical Continuum Solvation Models. Chem Rev 2005;105:2999–3094. doi:10.1021/cr9904009.
[61] Marenich AV, Cramer CJ, Truhlar DG. Universal Solvation Model Based on Solute Electron Density and on a Continuum Model of the Solvent Defined by the Bulk Dielectric Constant and Atomic Surface Tensions. J Phys Chem B 2009;113:6378–96. doi:10.1021/jp810292n.
[62] Takano Y, Houk KN. Benchmarking the Conductor-like Polarizable Continuum Model (CPCM) for Aqueous Solvation Free Energies of Neutral and Ionic Organic Molecules. J Chem Theory Comput 2005;1:70–7. doi:10.1021/ct049977a.
[63] Isegawa M, Neese F, Pantazis DA. Ionization Energies and Aqueous Redox Potentials of Organic Molecules: Comparison of DFT, Correlated ab Initio Theory and Pair Natural Orbital Approaches. J Chem Theory Comput 2016;12:2272–84. doi:10.1021/acs.jctc.6b00252.
[64] Tieke B. Makromolekulare Chemie: eine Einführung. 2., vollst. überarb. und erw. Aufl., 4. Nachdr. Weinheim: Wiley-VCH; 2012.
[65] Elias H-G. Makromoleküle: Struktur, Eigenschaften, Synthesen, Stoffe. Basel, Heidelberg: Hüthig & Wepf; 1971.
[66] Lee Y-K, Kim D-J, Kim H-J, Hwang T-S, Rafailovich M, Sokolov J. Activation energy and curing behavior of resol- and novolac-type phenolic resins by differential scanning calorimetry and thermogravimetric analysis. J Appl Polym Sci 2003;89:2589–96. doi:10.1002/app.12340.
[67] Braun D, Cherdron H, Ritter H. Polymer Synthesis: Theory and Practice. Berlin, Heidelberg: Springer Berlin Heidelberg; 2001. doi:10.1007/978-3-662-04573-2.
[68] Sultania M, Rai JSP, Srivastava D. A study on the kinetics of condensation reaction of cardanol and formaldehyde, part I. Int J Chem Kinet 2009;41:559–72. doi:10.1002/kin.20411.
[69] Devi A, Srivastava D. Cardanol-based novolac-type phenolic resins. I. A kinetic approach. J Appl Polym Sci 2006;102:2730–7. doi:10.1002/app.24363.

[70] Elanany M, Selvam P, Yokosuka T, Takami S, Kubo M, Imamura A, et al. A Quantum Molecular Dynamics Simulation Study of the Initial Hydrolysis Step in Sol–Gel Process. J Phys Chem B 2003;107:1518–24. doi:10.1021/jp026816z.
[71] Schaffer CL, Thomson KT. Density Functional Theory Investigation into Structure and Reactivity of Prenucleation Silica Species. J Phys Chem C 2008;112:12653–62. doi:10.1021/jp066534p.
[72] Okumoto S, Fujita N, Yamabe S. Theoretical Study of Hydrolysis and Condensation of Silicon Alkoxides. J Phys Chem A 1998;102:3991–8. doi:10.1021/jp980705b.
[73] Okumoto S, Yamabe S. Water-trimer clusters as nucleophilic reagents in hydrolyses of substrates for metal–alkoxides – a computational study. J Non-Cryst Solids 2001;291:167–75. doi:10.1016/S0022-3093(01)00808-0.
[74] Hansen A, Liakos DG, Neese F. Efficient and accurate local single reference correlation methods for high-spin open-shell molecules using pair natural orbitals. J Chem Phys 2011;135:214102. doi:10.1063/1.3663855.
[75] Liakos DG, Sparta M, Kesharwani MK, Martin JML, Neese F. Exploring the Accuracy Limits of Local Pair Natural Orbital Coupled-Cluster Theory. J Chem Theory Comput 2015;11:1525–39. doi:10.1021/ct501129s.
[76] Neese F, Hansen A, Liakos DG. Efficient and accurate approximations to the local coupled cluster singles doubles method using a truncated pair natural orbital basis. J Chem Phys 2009;131:064103. doi:10.1063/1.3173827.
[77] Riplinger C, Sandhoefer B, Hansen A, Neese F. Natural triple excitations in local coupled cluster calculations with pair natural orbitals. J Chem Phys 2013;139:134101. doi:10.1063/1.4821834.
[78] Riplinger C, Neese F. An efficient and near linear scaling pair natural orbital based local coupled cluster method. J Chem Phys 2013;138:034106. doi:10.1063/1.4773581.
[79] Liakos DG, Neese F. Is It Possible To Obtain Coupled Cluster Quality Energies at near Density Functional Theory Cost? Domain-Based Local Pair Natural Orbital Coupled Cluster vs Modern Density Functional Theory. J Chem Theory Comput 2015;11:4054–63. doi:10.1021/acs.jctc.5b00359.
[80] Bistoni G, Auer AA, Neese F. Understanding the Role of Dispersion in Frustrated Lewis Pairs and Classical Lewis Adducts: A Domain-Based Local Pair Natural Orbital Coupled Cluster Study. Chemistry 2017;23:865–73. doi:10.1002/chem.201604127.
[81] Tchernook I, Prehl J, Friedrich J. Quantum chemical investigation of the counter anion in the acid catalyzed initiation of 2,2′-spirobi[4*H*-1,3,2-benzodioxasiline] polymerization. Polymer 2015;60:241–51. doi:10.1016/j.polymer.2015.01.042.
[82] Tchernook I, Prehl J, Friedrich J. Corrigendum to 'Quantum chemical investigation of the counter anion in the acid catalyzed initiation of 2,2'-spirobi[4H-1,3,2-benzodioxasiline] polymerization' [Polymer 60 (2015) 241–251]. Polymer 2015;74:272–4. doi:10.1016/j.polymer.2015.07.058.
[83] Raamat E, Kaupmees K, Ovsjannikov G, Trummal A, Kütt A, Saame J, et al. Acidities of strong neutral Brønsted acids in different media: ACIDITIES OF STRONG NEUTRAL BRØNSTED ACIDS. J Phys Org Chem 2013;26:162–70. doi:10.1002/poc.2946.
[84] Chemical Rubber Company, Lide DR, editors. CRC handbook of chemistry and physics: a ready-reference book of chemical and physical data. 84th ed. Boca Raton: CRC Press; 2003.
[85] Richter A. Quantenchemische Berechnungen zu Spirosiliciumzwillingsmonomeren. Diploma Thesis. Technische Universität Chemnitz, 2010.
[86] Schönfelder T, Friedrich J, Prehl J, Seeger S, Spange S, Hoffmann KH. Reactive force field for electrophilic substitution at an aromatic system in twin polymerization. Chem Phys 2014;440:119–26. doi:10.1016/j.chemphys.2014.06.003.
[87] Weißhuhn J. Simultane Zwillingspolymerisation von Silizium- und Borzwillingsmonomeren. Dissertation. Technische Universität Chemnitz, 2018.
[88] Rossini E, Netz RR, Knapp E-W. Computing p K_a Values in Different Solvents by Electrostatic Transformation. J Chem Theory Comput 2016;12:3360–9. doi:10.1021/acs.jctc.6b00446.

[89] Eckert F, Leito I, Kaljurand I, Kütt A, Klamt A, Diedenhofen M. Prediction of acidity in acetonitrile solution with COSMO-RS. J Comp Chem 2009;30:799–810. doi:10.1002/jcc.21103.

[90] Cornil J, dos Santos DA, Crispin X, Silbey R, Brédas JL. Influence of Interchain Interactions on the Absorption and Luminescence of Conjugated Oligomers and Polymers: A Quantum-Chemical Characterization. J Am Chem Soc 1998;120:1289–99. doi:10.1021/ja973761j.

[91] Pople JA. Quantum Chemical Models (Nobel Lecture). Angew Chem Int Ed 1999;38:1894–902. doi:10.1002/(SICI)1521-3773(19990712)38:13/14<1894::AID-ANIE1894>3.0.CO;2-H.

[92] Heine DR, Grest GS, Curro JG. Structure of Polymer Melts and Blends: Comparison of Integral Equation Theory and Computer Simulations. In: Dr. Holm C, Prof. Dr. Kremer K, editors. Advanced Computer Simulation, vol. 173, Berlin, Heidelberg: Springer Berlin Heidelberg; 2005, p. 209–52. doi:10.1007/b99431.

[93] Grimme S, Steinmetz M, Korth M. How to Compute Isomerization Energies of Organic Molecules with Quantum Chemical Methods. J Org Chem 2007;72:2118–26. doi:10.1021/jo062446p.

[94] Kremer K, Grest GS. Dynamics of entangled linear polymer melts: a molecular-dynamics simulation. J Chem Phys 1990;92:5057–86. doi:10.1063/1.458541.

[95] Madkourt TM, Barakat AM. Computer simulation of polymers. Comput Theor Polym Sci 1997;7:35–46.

[96] Starr FW, Schrøder TB, Glotzer SC. Molecular Dynamics Simulation of a Polymer Melt with a Nanoscopic Particle. Macromolecules 2002;35:4481–92. doi:10.1021/ma010626p.

[97] van Duin ACT, Dasgupta S, Lorant F, Goddard WA. ReaxFF: A Reactive Force Field for Hydrocarbons. J Phys Chem A 2001;105:9396–409. doi:10.1021/jp004368u.

[98] Rudd RE, Broughton JQ. Coarse-grained molecular dynamics: Nonlinear finite elements and finite temperature. Phys Rev B 2005;72:144104-1–32. doi:10.1103/PhysRevB.72.144104.

[99] Karimi-Varzaneh HA, Qian H-J, Chen X, Carbone P, Müller-Plathe F. IBIsCO: A Molecular Dynamics Simulation Package for Coarse-Grained Simulation. J Comput Chem 2010;32:1475–87. doi:10.1002/jcc.21717.

[100] Zeng QH, Yu AB, Lu GQ. Multiscale modeling and simulation of polymer nanocomposites. Prog Polym Sci 2008;33:191–269. doi:10.1016/j.progpolymsci.2007.09.002.

[101] Binder K. Applications of Monte Carlo methods to statistical physics. Rep Prog Phys 1997;60:487–559.

[102] Carmesin I, Kremer K. The bond fluctuation method: a new effective algorithm for the dynamics of polymers in all spatial dimensions. Macromolecules 1988;21:2819–23. doi:10.1021/ma00187a030.

[103] Shaffer JS. Effects of chain topology on polymer dynamics: Bulk melts. J Comput Phys 1994;101:4205–13. doi:10.1063/1.467470.

[104] Kremer K, Binder K. Monte Carlo simulation of lattice models for macromolecules. Comput Phys Rep 1988;7:259–310. doi:10.1016/0167-7977(88)90015-9.

[105] Prehl J, Schönfelder T, Friedrich J, Hoffmann KH. Site Dependent Atom Type ReaxFF for the Proton-Catalyzed Twin Polymerization. J Phys Chem C 2017;121:15984–92. doi:10.1021/acs.jpcc.7b03219.

[106] Huster C, Prehl J. Reactive Bond Fluctation Model: An approach for structure formation analysis of twin polymerization. In Preparation 2018.

[107] Windberg T, Ebert T, Uhlig D, Schulze S, Spange S. Hierarchically structured carbon/carbon nanocomposites with adjustable porosity fabricated by twin polymerization. Microporous Mesoporous Mater 2017;246:62–71. doi:10.1016/j.micromeso.2017.03.012.

[108] Weißhuhn J, Mark T, Martin M, Müller P, Seifert A, Spange S. Ternary organic-inorganic nanostructured hybrid materials by simultaneous twin polymerization. Polym Chem 2016;7:5060–8. doi:10.1039/C6PY00903D.

[109] Göring M, Seifert A, Schreiter K, Müller P, Spange S. A non-aqueous procedure to synthesize amino group bearing nanostructured organic–inorganic hybrid materials. Chem Commun 2014;50:9753–6. doi:10.1039/C4CC03640A.

[110] Oschatz S, Lange A, Csihony S, Cox G, Gronwald O, Seifert A, et al. Functional twin monomers and twin macro monomers as components for the synthesis of hierarchically nanostructured hybrid materials. J Polym Sci Part A: Polym Chem 2016;54:2312–20. doi:10.1002/pola.28102.
[111] Jentzsch C, Sommer J-U. Polymer brushes in explicit poor solvents studied using a new variant of the bond fluctuation model. J Chem Phys 2014;141:104908. doi:10.1063/1.4895555.
[112] i Serra RS, Torregrosa-Cabanilles C, Duenas JMM, Ribelles JLG, Molina-Mateo J. Conformation and dynamics of a diluted chain in the presence of an adsorbing wall: a simulation with the bond fluctuation model. J Non-Cryst Solids 2014;402:7–15. doi:10.1016/j.jnoncrysol.2014.05.009.
[113] Hsu H-P, Paul W, Binder K. Estimation of Persistence Lengths of Semiflexible Polymers: Insight from Simulations. Polym Sci Ser C 2013;55:39–59. doi:10.1134/S1811238213060027.
[114] Glotzer SC, Stauffer D, Jan N. Monte Carlo simulations of phase separation in chemically reactive binary mixtures. Phys Rev Lett 1994;72:4109–12. doi:10.1103/PhysRevLett.72.4109.
[115] Müller M. Reactions at Polymer Interfaces: A Monte Carlo Simulation. Macromolecules 1997;30:6353–7. doi:10.1021/ma9619086.
[116] Berry H. Monte Carlo Simulations of Enzyme Reactions in Two Dimensions: Fractal Kinetics and Spatial Segregation. Biophys J 2002;83:1891–901. doi:10.1016/S0006-3495(02)73953-2.
[117] Yang S, Pandey RB. Monte Carlo simulation of a film growth with reactive hydrophobic, polar, and aqueous components by a covalent bond fluctuating model. J Chem Phys 2007;126:164708-1–8. doi:10.1063/1.2722264.
[118] He X, Nagel J, Lehmann D, Heinrich G. Interface Structure between Immiscible Reactive Polymers under Transreaction: a Monte Carlo Simulation. Macromol Theory Simul 2005;14:305–11. doi:10.1002/mats.200500006.
[119] Zhdanov VP, Kasemo B. Monte Carlo simulation of the kinetics of rapid reactions on nanometer catalyst particles. Surf Sci 1998;405:27–37. doi:10.1016/S0039-6028(97)01078-9.
[120] Hoffmann KH, Prehl J. Modeling the structure formation process of twin polymerization. Reac Kinet Mech Cat 2018;123:367–83. doi:10.1007/s11144-017-1303-y.
[121] Chenoweth K, van Duin ACT, Goddard WA. ReaxFF Reactive Force Field for Molecular Dynamics Simulations of Hydrocarbon Oxidation. J Phys Chem A 2008;112:1040–53. doi:10.1021/jp709896w.
[122] Nielson KD, van Duin ACT, Oxgaard J, Deng W-Q, Goddard III WA. Development of the ReaxFF Reactive Force Field for Describing Transition Metal Catalyzed Reactions, with Application to the Initial Stages of the Catalytic Formation of Carbon Nanotubes. J Phys Chem A 2005;109:493–9. doi:10.1021/jp046244d.
[123] Chenoweth K, Cheung S, van Duin ACT, Goddard WA, Kober EM. Simulations on the Thermal Decomposition of a Poly(dimethylsiloxane) Polymer Using the ReaxFF Reactive Force Field. J Am Chem Soc 2005;127:7192–202. doi:10.1021/ja050980t.
[124] van Duin ACT, Strachan A, Stewman S, Zhang Q, Xu X, Goddard WA. $ReaxFF_{SiO}$ Reactive Force Field for Silicon and Silicon Oxide Systems. J Phys Chem A 2003;107:3803–11. doi:10.1021/jp0276303.
[125] Huang L, Joshi KL, Duin ACT van, Bandosz TJ, Gubbins KE. ReaxFF molecular dynamics simulation of thermal stability of a $Cu_3(BTC)_2$ metal–organic framework. Phys Chem Chem Phys 2012;14:11327. doi:10.1039/c2cp41511a.
[126] Krishnasamy Bharathi A, Kamat A, van Duin ACT. Study of effect of water vapor and mechanical strain on thermal conductivity of zinc oxide using the ReaxFF reactive force field. Comput Theor Chem 2012;987:71–6. doi:10.1016/j.comptc.2012.02.004.
[127] Narayanan B, van Duin ACT, Kappes BB, Reimanis IE, Ciobanu CV. A reactive force field for lithium–aluminum silicates with applications to eucryptite phases. Modell Simul Mater Sci Eng 2012;20:015002. doi:10.1088/0965-0393/20/1/015002.
[128] Keten S, Chou C-C, van Duin ACT, Buehler MJ. Tunable nanomechanics of protein disulfide bonds in redox microenvironments. J Mech Behav Biomed Mater 2012;5:32–40. doi:10.1016/j.jmbbm.2011.08.017.

[129] Newsome DA, Sengupta D, Foroutan H, Russo MF, van Duin ACT. Oxidation of Silicon Carbide by O_2 and H_2O: A ReaxFF Reactive Molecular Dynamics Study, Part I. J Phys Chem C 2012;116:16111–21. doi:10.1021/jp306391p.
[130] Treutler O, Ahlrichs R. Efficient molecular numerical integration schemes. J Chem Phys 1995;102:346–54. doi:10.1063/1.469408.
[131] Eichkorn K, Treutler O, Öhm H, Häser M, Ahlrichs R. Erratum – Auxiliary basis sets to approximate Coulomb potentials. Chem Phys Lett 1995;242:652–60. doi:10.1016/0009-2614(95)00838-1.
[132] Eichkorn K, Weigend F, Treutler O, Ahlrichs R. Auxiliary basis sets for main row atoms and transition metals and their use to approximate Coulomb potentials. Theor Chem Acc 1997;97:119–24.
[133] Von Arnim M, Ahlrichs R. Performance of parallel TURBOMOLE for density functional calculations. J Comput Chem 1998;19:1746–57. doi:10.1002/(SICI)1096-987X(19981130)19:15$<$1746::AID-JCC7$>$3.0.CO;2-N.
[134] TURBOMOLE V6.5 2013, a development of University of Karlsruhe and Forschungszentrum Karlsruhe GmbH, 1989–2007, TURBOMOLE GmbH, since 2007; available from www.turbomole.com. 2013.
[135] Ahlrichs R, Bär M, Häser M, Horn H, Kölmel C. Electronic structure calculations on workstation computers: The program system turbomole. Chem Phys Lett 1989;162:165–9. doi:10.1016/0009-2614(89)85118-8.
[136] Perdew JP. Density-functional approximation for the correlation energy of the inhomogeneous electron gas. Phys Rev B 1986;33:8822–4. doi:10.1103/PhysRevB.33.8822.
[137] Perdew JP. Erratum: Density-functional approximation for the correlation energy of the inhomogeneous electron gas. Phys Rev B 1986;34:7406. doi:10.1103/PhysRevB.34.7406.
[138] Becke AD. Density-functional exchange-energy approximation with correct asymptotic behavior. Phys Rev A 1988;38:3098–100. doi:10.1103/PhysRevA.38.3098.
[139] Grimme S, Antony J, Ehrlich S, Krieg H. A consistent and accurate *ab initio* parametrization of density functional dispersion correction (DFT-D) for the 94 elements H-Pu. J Chem Phys 2010;132:154104-1–19. doi:10.1063/1.3382344.
[140] Grimme S, Ehrlich S, Goerigk L. Effect of the Damping Function in Dispersion Corrected Density Functional Theory. J Comput Chem 2011;32:1456. doi:10.1002/jcc.21759.
[141] Schäfer A, Horn H, Ahrlichs R. Fully optimized contracted gaussian basis sets for atoms Li to Kr. J Chem Phys 1992;97:2571–7. doi:10.1063/1.463096.
[142] Klamt A, Schüürmann G. COSMO: a new approach to dielectric screening in solvents with explicit expressions for the screening energy and its gradient. J Chem Soc, Perkin Trans 2 1993:799–805. doi:10.1039/P29930000799.
[143] Zhao Y, Truhlar DG. Design of Density Functionals That Are Broadly Accurate for Thermochemistry, Thermochemical Kinetics, and Nonbonded Interactions. J Phys Chem A 2005;109:5656–67.
[144] Weigend F, Häser M, Patzelt H, Ahlrichs R. RI-MP2: optimized auxiliary basis sets and demonstration of efficiency. Chem Phys Lett 1998;294:143–52. doi:10.1016/S0009-2614(98)00862-8.
[145] Weigend F, Furche F, Ahlrichs R, Weigend F, Ahlrichs R. Balanced basis sets of split valence, triple zeta valence and quadruple zeta valence quality for H to Rn: Design and assessment of accuracy. Phys Chem Chem Phys 2005;7:3297–305. doi:10.1039/B508541A.
[146] Weigend F, Furche F, Ahlrichs R. Gaussian basis sets of quadruple zeta valence quality for atoms H-Kr. J Chem Phys 2003;119:12753–62. doi:10.1063/1.1627293.
[147] Chuit C, Corriu RJP, Reye C, Young JC. Reactivity of penta- and hexacoordinate silicon compounds and their role as reaction intermediates. Chem Rev 1993;93:1371–448. doi:10.1021/cr00020a003.

4 Materials

Stefan Spange, Mandy Göring, Lysann Kaßner, Kevin Nagel and Andreas Seifert

4.1 Silicon-containing hybrid materials

In this subchapter, the chemical synthesis and physical characterization of organic/inorganic hybrid materials, which are produced from twin monomers based on ortho-silicic acid ester derivatives and several nitrogen-containing analogs, are presented. Hybrid materials by twin polymerization are produced *via* multistage chemical syntheses which involve the following steps: design and synthesis of the twin monomer; twin polymerization procedure as well as characterization of molecular structure, morphology, and topology of the final product. The availability of appropriate twin monomers is an essential requirement for this synthetic concept. Therefore, synthesis of specific silicon-containing twin monomers is briefly discussed. Mechanistic aspects are not considered in this chapter because this matter is already treated in Chapter 3 of this book. The focus of this subchapter is to demonstrate the synthetic possibilities of twin polymerization for the production of various silicon-containing nanostructured hybrid materials. Generally, polymer/silica hybrid material fabrication is an important issue in the field of materials chemistry and the reader is referred to several reviews [1–10].

4.1.1 History

The development of the scientific landscape of organic/inorganic hybrid materials has been strongly inspired by the chemistry of silicon compounds [1–4]. Especially, the aqueous and non-aqueous sol-gel process and the production of polysiloxane-based hybrid materials are important milestones in this research area [5–10]. The sol-gel process of tetraalkoxysilanes and its relation to and distinction from the twin polymerization has been given as instructive explanation in Chapter 2, which explained the general aspects of twin polymerization.

In the year 1953, Hahn and Staudinger already found out that a portion of silica (SiO_2) was generated when silicon tetrachloride (tetrachlorosilane) was treated with several 1,2-diols to target polymers which contain –Si–O–R–O–Si–O–R–O–Si–O linkages (R = alkyl) along the backbone [11]. The synthetic attempts were only of partial success because silica and mixtures of ethers and cyclic compounds were sometimes obtained. Nevertheless, these studies demonstrated that complex reaction cascades take place when diols react with SiX_4 (X = halogen, alkoxy) compounds. Transesterification reactions of silicic esters are therefore important for the understanding of the formation of the so-called "twin monomers." At this point, it should be noted that twin monomers are not always accessible by transesterification of 1,2-diols

https://doi.org/10.1515/9783110499360-004

or 1,3-diols with appropriate silicon compounds. For example, the reaction of tetraalkoxysilanes with lactic acid does not yield spirocyclic compounds; silica and product mixtures of several *bis*-lactides are formed instead of spirocyclic compounds [12].

4.1.2 Silicon-containing twin monomers

Silicon-containing twin monomers are usually synthesized by transesterification reactions of commercially available alkoxysilanes with polymerizable arylmethanols such as furfuryl alcohol, 2-hydroxybenzyl alcohol (salicyl alcohol), substituted benzyl alcohols, and related compounds [13–18]. The use of corresponding silicon chloro-compounds as reactants is also suitable, but the removal of the formed hydrogen chloride is sometimes experimentally challenging. For that reason, the utilization of alkoxysilane agents is of advantage for the development of silicon-containing twin monomers [19–21]. Schemes 4.1 and 4.2 show typical synthetic routes for the production of established twin monomers.

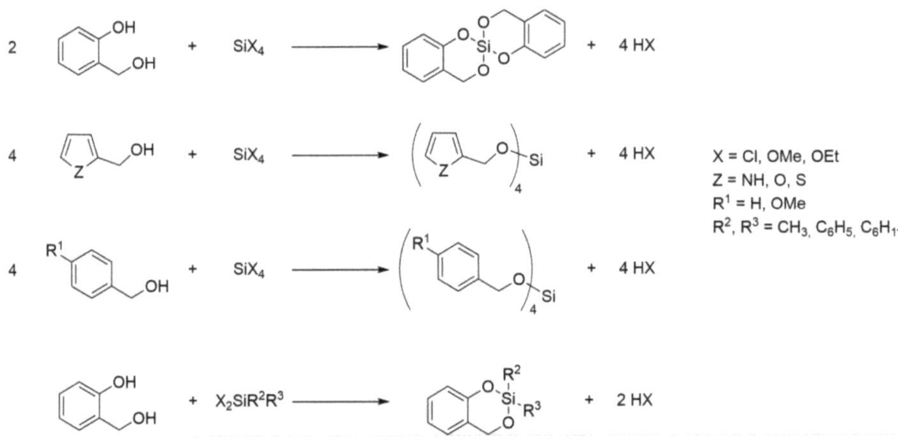

Scheme 4.1: General scheme for production of important silicon-containing twin monomers. Base catalysts and reactants for trapping HX are omitted for clarity.

The twin monomer 2,2'-spirobi[4*H*-1,3,2-benzodioxasiline], briefly called Si-Spiro, should be accentuated, because it is a textbook example for the principle of twin polymerization [22]. Si-Spiro is a white crystalline solid with a melting point of 82 °C (Scheme 4.1 above and Fig. 4.1). It is sensitive to moisture and cannot be kept in air for a longer time period due to hydrolysis. Single crystal X-ray structure analysis verifies the tetracoordination of the silicon atom by oxygen atoms which is also confirmed by the corresponding ^{29}Si NMR spectrum. It should be noted that disodium tris(2-oxymethylphenolate) silicate $[Na_2(C_7H_6O_2)_3]Si$ is a white solid that does neither melt nor dissolve, and is unsuitable as monomer for hybrid materials synthesis, which is most likely a result of its double negatively charged hexacoordinated silicon atom [23].

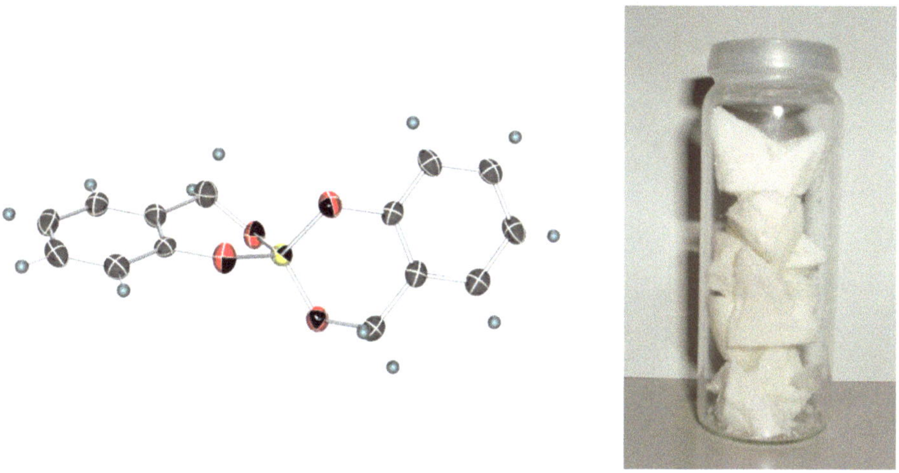

Fig. 4.1: Molecular structure of 2,2'-spirobi[4*H*-1,3,2-benzodioxasiline] (Si-Spiro) and photograph of the crystalline substance [15]. Copyright John Wiley & Sons, Inc. Reproduced with permission.

Silicon-based twin monomers can also be constructed in such a way that additional functional groups can be introduced [23–25]. Two different synthesis variants are possible.

One typical example is the production of 2-chloro-2-methyl-4*H*-1,3,2-benzoxasiline, which is readily available from methyltrichlorosilane and 2-hydroxybenzyl alcohol (Scheme 4.2). The chloro-substituent at the silicon atom of 2-chloro-2-methyl-4*H*-1,3,2-benzoxasiline, which remains after the initial synthesis step, is easily replaceable by other twin polymerizable groups such as furfuryloxy *via* nucleophilic substitution reaction [26]. Furthermore, it can be used to produce several new twin monomers which contain additional reactive groups such as epoxy. An alternative route for the production of functional twin monomers starts from tri- or dialkoxysilanes, which already bear the desired functional group covalently linked by a Si–C bond to the silicon atom. Again, the transesterification reaction with an appropriate polymerizable alcohol gives the twin monomer as depicted in Scheme 4.2 [24, 25].

The chemistry of such silicon-containing monomers, which have additional functional groups such as amino, acrylate, or epoxy, is treated in the course of this chapter and in more detail in Chapter 5.2. These functional groups can polymerize or react with other reagents independently of the twin monomer moiety. Thus, they do not fit in the category “triplet monomer” [27], because the additional function of the twin

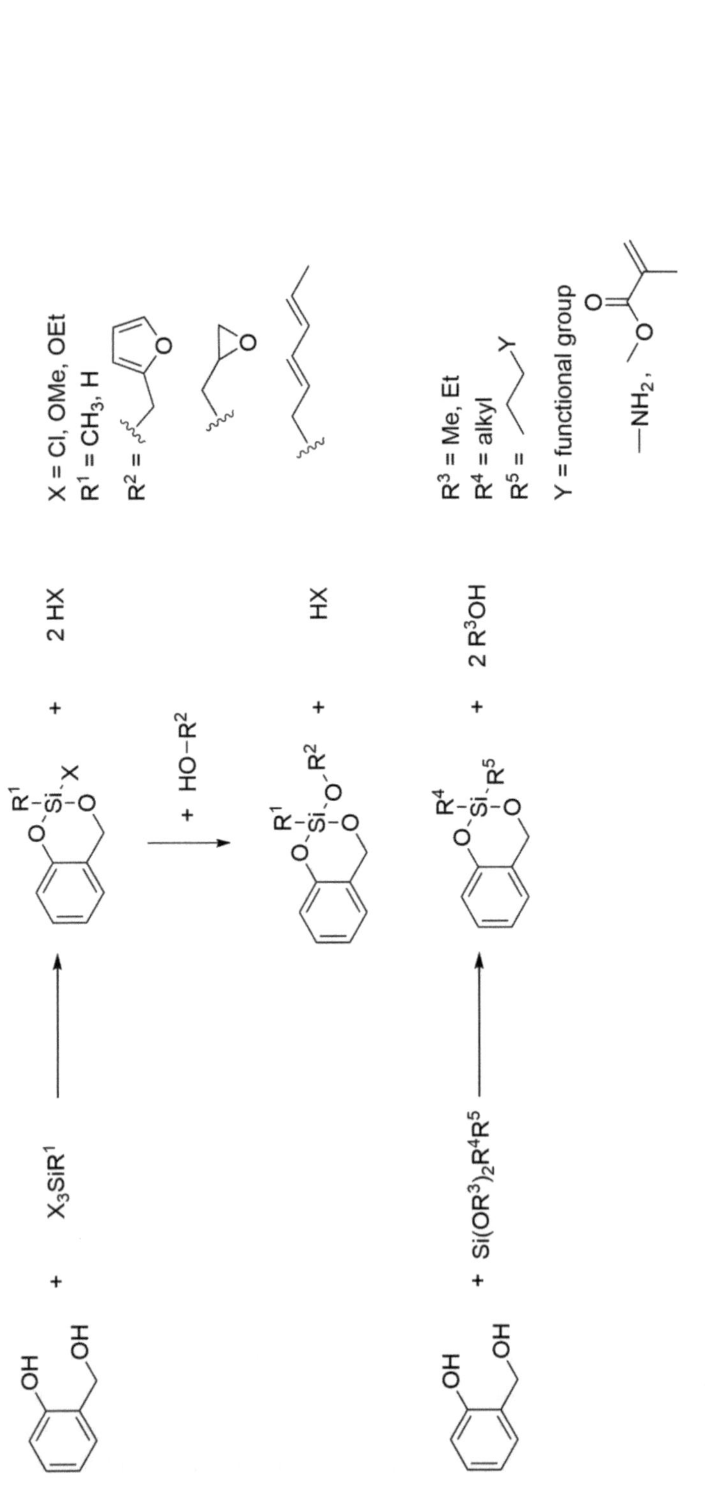

Scheme 4.2: Synthesis of monomers consisting of different twin polymerizable moieties (top), or having additional functional groups Y covalently bonded to the silicon atom (bottom). Base catalysts and reactants for trapping HX are omitted for clarity.

monomer moiety is unable to undergo self-polymerization within one process as the Si–C bond remains intact [26, 27].

4.1.3 Polymer/Silica hybrid materials by twin polymerization

As discussed in Chapter 2, it has to be distinguished between condensative twin polymerization and ideal twin polymerization. Twin polymerization of Si-Spiro has been established as representative example for ideal twin polymerization. A typical instance of the condensative twin polymerization is the cationic polycondensation reaction of furfuryloxysilane derivatives as shown in Scheme 4.3. This aspect is discussed in Chapter 3 of this book.

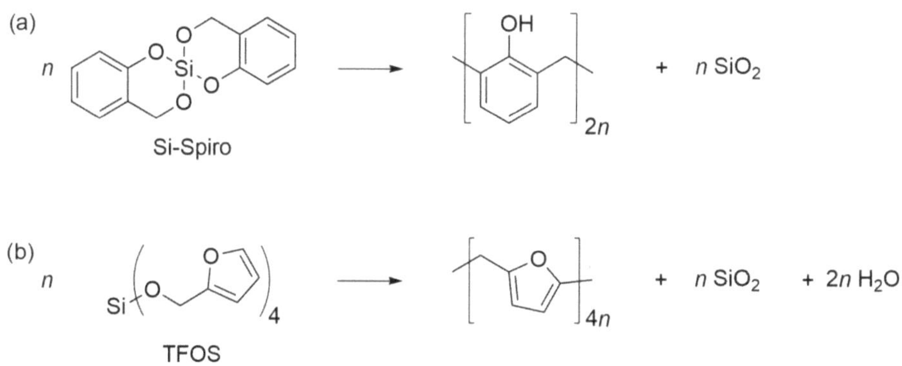

Scheme 4.3: Chemical reactions for the production of nanostructured hybrid materials from (a) Si-Spiro and (b) TFOS.

Product formation and structural characterization of the materials which are obtained from the twin monomers Si-Spiro and tetrafurfuryloxysilane (TFOS) are demonstrated in detail to present the concept of synthesis of the polymer/silica hybrid materials by twin polymerization.

Applied technical polymerization procedure is an important factor for the performance of twin polymerization. For completeness, there are several analytic tools necessary to verify the molecular structure and morphology of the final product. Thus, analytical methods to determine the overall structure of the hybrid materials are briefly discussed in this chapter. Usually, the hybrid materials are solid in the form of monoliths, thin layers on a templating surface, or powders. The form depends on the applied synthetic procedure, with a high concentration of twin monomers being necessary to obtain monolithic materials.

A twin monomer can be dissolved in an appropriate solvent, or it can be melted to be used in an applicable liquid format at least for a given length of time. This is possible because in most cases the melting temperature is much lower than the temperature needed to start a polymerization, in particular without the use of additional catalysts. Twin polymerization of melted twin monomers is especially useful for the following reason: the pure monomer or monomer compositions can be used in a casting mold to produce monolithic hybrid materials of desired shape and size. Twin polymerization of Si-Spiro alone or Si-Spiro together with 2,2-dimethyl-4*H*-1,3,2-benzodioxasiline in a simultaneous twin polymerization (STP) is ideally suited for this purpose because no by-products are released during the polymerization process. Figure 4.2 shows an owl composed of the resulting homogeneous phenolic resin/silica hybrid material that was fabricated using a casting mold originally considered for liquid tin.

Fig. 4.2: Photographs of hybrid materials which show owl shapes (right) produced in a casting mold (left) by STP. The hybrid materials were obtained by lactic acid-catalyzed polymerization and annealing at 200 °C under argon (yellow owl) as well as 175 °C under air (red owl). For comparison, a cast tin owl (gray) is shown [28]. Reproduction with kind permission of Tina Mark.

This possibility is a big advantage in terms of shape retention compared to technologies which need water (sol-gel process) or release water (i.e., polycondensation reaction).

4.1.3.1 Analytical aspects of hybrid materials derived from twin polymerization

The following explanations demonstrate the series of operations that are necessary for the analysis of the molecular structure and morphology of the resulting hybrid materials. For a comprehensive analysis of the structural moieties in the hybrid material, solid state spectroscopy (^{13}C and ^{29}Si NMR) was found to be very helpful in order to prove appropriately the structure of the organic polymer and the corresponding silica species (see Fig. 4.3). The $^{13}C\{^1H\}$-CP-MAS NMR spectroscopy shows the formation of the organic network. Individual molecular structure elements can be clearly assigned as indicated by numbers for the polyfurfuryl alcohol and by letters for the phenolic resin component, respectively, as depicted in Figure 4.3. The twin polymerization of Si-Spiro catalyzed by trifluoroacetic acid (TFA) provides a phenolic resin that shows both possible molecular structure elements: *ortho*/*ortho*'- and *ortho*/*para*'-connections

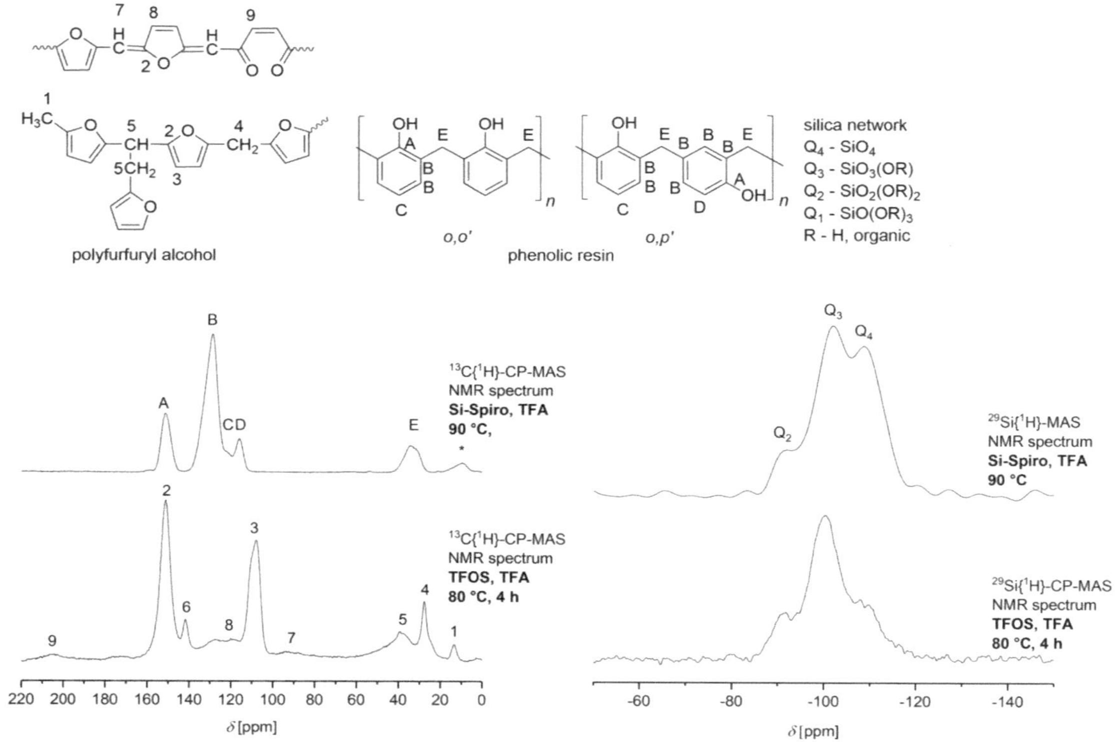

Fig. 4.3: Solid state $^{13}C\{^1H\}$-CP-MAS NMR (left) and $^{29}Si\{^1H\}$-(CP)-MAS NMR spectra (right) of polyfurfuryl alcohol/SiO_2 and phenolic resin/SiO_2 hybrid materials. The molecular structure elements are shown on top. MAS rotational frequencies were 15 kHz for ^{13}C NMR measurements and 12 kHz for ^{29}Si NMR measurements.

according to the molecular structure inset depicted in Figure 4.3. When Si-Spiro is thermally polymerized at 220 °C, several by-products resulting from reactions of the quinone methide intermediate are formed within the phenolic resin network structure which can also be detected by solid state ^{13}C NMR spectroscopy (see [16]).

Solid state ^{29}Si NMR spectra of the same hybrid materials measured under the condition of fast sample spinning (magic angle spinning, MAS) clearly indicates a silica formation by appearance of characteristic Q_2, Q_3, and Q_4 signals that represent different states of condensation of the silicon atoms (see inset Fig. 4.3) [7,8]. The Q_4 signal denotes fully condensed Si-species with four siloxane bridges (Si–O–Si) connecting them to the inorganic network, whereas Q_3 species have one unreacted Si–OR or Si–OH bond left, and Q_2 species have two left. The intensities of the different structural elements measured by solid state ^{29}Si NMR spectroscopy point toward nanometer sized silica phase domains, because high portions of Q_2 and Q_3 signals are a qualitative indication that silicon atoms are present as interfacial components rather than as volume atoms (Q_4). It is important to note that the intensities of the signals do not allow quantitative statements on the fractions of the respective molecular structure if the cross polarization (CP) technique is used to enhance the sensitivity of solid state ^{29}Si NMR spectroscopy. This technique drastically reduces measuring time, but overestimates Q_2 and Q_3 over Q_4 because signals of Si-species with ^1H in close proximity are amplified. For selected samples, quantitative ^{29}Si NMR spectroscopy with direct excitation of ^{29}Si nuclei was used to obtain the real distribution of Q-signals. The disadvantage of this technique results from ^{29}Si longitudinal relaxation, or spin-lattice relaxation (T_1), lasting several minutes. Thus, only a few scans can be collected per hour and several thousand are required for NMR spectra with acceptable signal to noise ratio. The quantitative $^{29}Si\{^1H\}$-MAS NMR spectroscopy of the hybrid material obtained by twin polymerization of Si-Spiro shows that the ratio of volume (Q_4) to interfacial Si atoms (Q_2 and Q_3) is about 1:1 [15]. In comparison, the intensity of the Q_4 signal in a "routine" $^{29}Si\{^1H\}$-CP-MAS NMR spectrum of a material made by polymerization of TFOS under quite similar conditions (80 °C, TFA) is reduced, even though the monomer is more reactive and the degree of condensation should be at least at the same level (Fig. 4.3). As completing and supporting measuring methods, FTIR, Raman, and UV/Vis spectroscopy can be used. Especially, IR spectroscopy is useful when twin polymerization is carried out on metal particle surfaces.

The distribution of the decisive elements (oxygen, silicon, carbon) within the generated solid hybrid materials has a huge impact on its properties and can be studied using analytical procedures which work on different length scales. When we look at the owls depicted in Figure 4.2, it is obvious that there is no macroscopic phase separation between the phenolic resin and the silica formed by the polymerization of Si-Spiro as the material is fully transparent. This is a hint that both phase domain sizes must be below the wavelength of the visible light. Scanning electron microscopy (SEM) and energy-dispersive X-ray (EDX) analysis can be used to analyze the material in more detail and to receive images and element distributions at about micrometer resolution. In general, the materials obtained by twin polymerization are very homogeneous in this dimension. Special TEM techniques like high-angle annular dark-field scanning transmission electron microscopy (HAADF-STEM) and energy-filtered TEM (EFTEM) further increase the resolution in the range of a few nanometers, and can give high-resolution images of the element distributions.

In comparison to Si-Spiro, TFOS is usually polymerized in an organic solvent because released water during this process can disturb the production of monoliths. Therefore, the polymerization is carried out in organic solvents such as toluene or dichloromethane which are suitable for cationic polymerization. The choice of an appropriate acid catalyst is of great importance to guarantee a successful polymerization process with complete conversion of the twin monomer into a hybrid material. Weak acids, such as acetic acid, are not suitable and too strong acids, such as trifluoromethane sulfonic acid, catalyze the polymerization too fast. Then, removal of heat during exothermic reaction is hard to control and a runaway of twin polymerization proceeds. The following rule has been established for silicon-containing twin monomers: the catalyst is suitable when it works appropriately for the corresponding polymerizable alcohol [27, 29–33]. Usually, twin polymerization of TFOS is carried out in organic solvents as precipitation polymerization. SEM and EDX for as-obtained PFA/SiO_2 particles are shown in Figure 4.4.

EFTEM (energy-filtered transmission electron microscopy) clearly demonstrates a homogeneous distribution of the elements C, O, and Si in a hybrid material down to the nanometer scale. EFTEM and HAADF-STEM pictures of nanostructured phenolic resin/SiO_2 materials obtained by polymerization of Si-Spiro are shown in Figure 4.5. The homogeneous distribution of carbon and silicon on the nanoscale is an important feature for hybrid materials obtained by twin polymerization of Si-Spiro. This nanostructure formation takes place in a very robust manner independent of the kind of catalysis or polymerization temperature.

As shown in Figure 4.5, it makes no difference whether the phenolic resin/SiO_2 hybrid material is produced from Si-Spiro by polymerization in the melt, catalyzed by bases or acids, or in solution of organic solvents. The nanostructure formation is independent of the experimental procedure. Thus, the use of Si-Spiro as co-reactant in combined processes, such as STP, is a guarantee for nanostructure formation.

Some aspects of the engineering of twin polymerization will be discussed in the following. It should be emphasized that Si-Spiro can readily be polymerized when it is heated up to 220 °C in melt or in solution [16]. This is very advantageous for such applications where a base or acid catalyst is obstructive because it may remain encapsulated in the final material. But for solution polymerization, the situation becomes more intricate because a high boiling solvent is necessary for that procedure. It has been found that an isomer mixture of di-*iso*-propylnaphtalene (DIPN; bp > 300 °C) is suitable for this purpose. Si-Spiro is readily soluble in DIPN and polymerization starts by heating the solution to 220 °C. Then, the hybrid material precipitates and a fine powder is obtained.

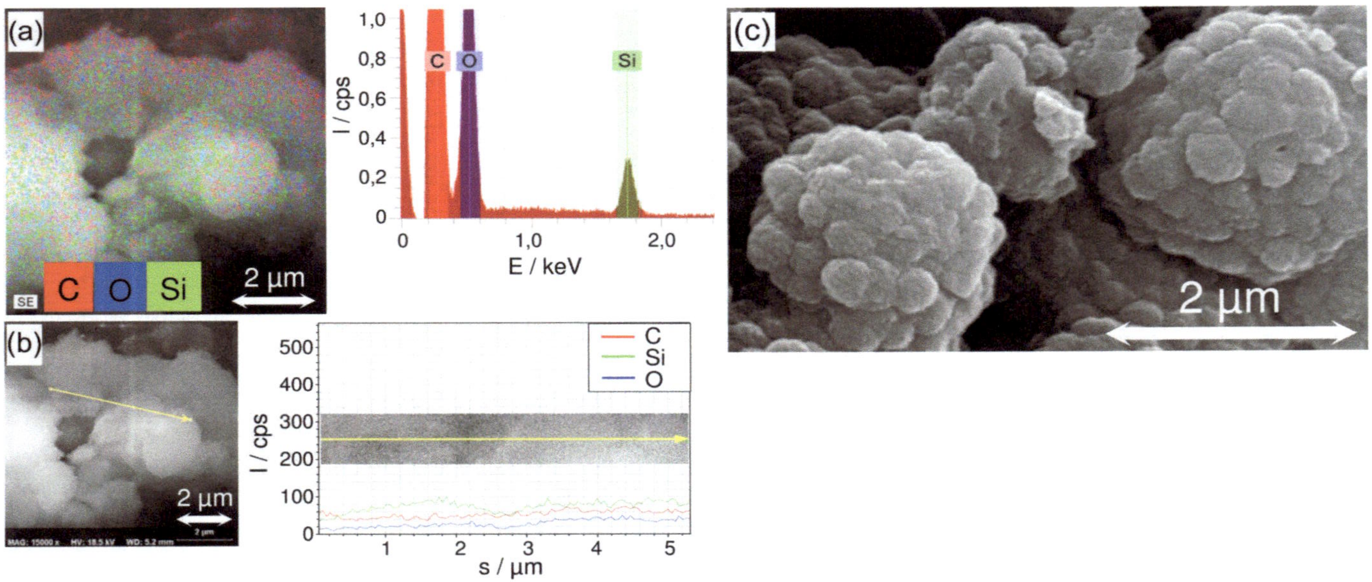

Fig. 4.4: (a) EDX imaging, (b) EDX linescan analysis, and (c) SEM picture of PFA/SiO_2 [13]. Copyright John Wiley & Sons, Inc. Reproduced with permission.

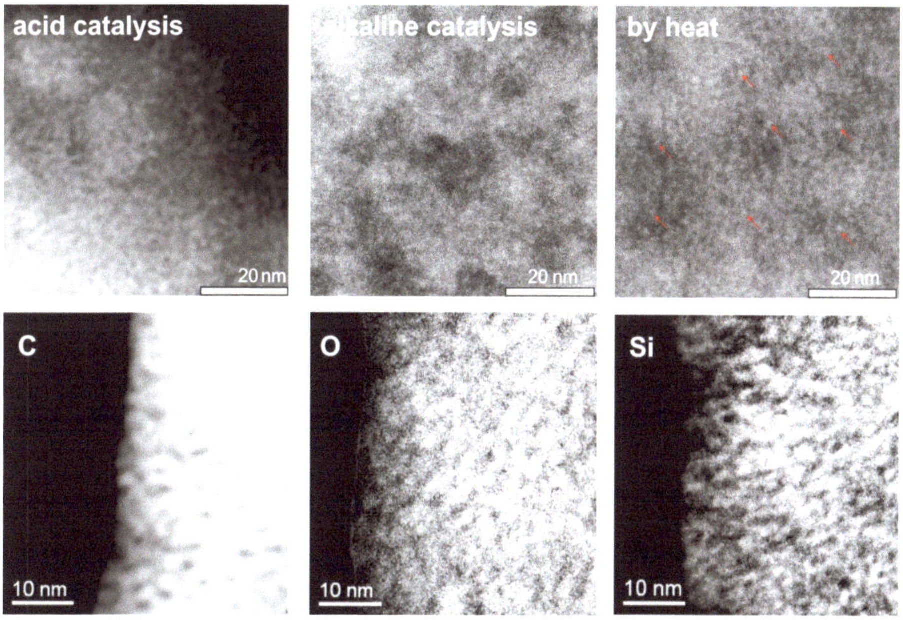

Fig. 4.5: Above: HAADF-STEM images of the silicon distribution of phenolic resin/SiO_2 hybrid material as function of the catalytic process. Left: acid catalysis (CF_3COOH) [15]. Copyright John Wiley & Sons, Inc. Reproduced with permission. Middle: base catalysis with 1,8-diazabicyclo[5.4.0]undec-7-ene (DBU). Reproduced from [32] with permission of The Royal Society of Chemistry. Right: Si-Spiro monomer polymerized at 230 °C in the melt. Reproduced from [16] with permission of the Royal Society of Chemistry. Below: EFTEM images of phenolic resin/SiO_2 sample synthesized by acid catalysis (CF_3COOH) showing the fine nanostructure on a length scale of about 2–3 nm for the elements C, O, and Si [15]. Copyright John Wiley & Sons, Inc. Reproduced with permission.

4.1.3.2 Transformation of hybrid materials into porous silica and carbon

The nanostructured organic/inorganic hybrid materials as obtained are nonporous. This result is independent of the physical shape being a powder or a monolith. There is no measurable BET (*Brunauer-Emmett-Teller*) surface detectable by nitrogen adsorption measurements in either case. Both the nanostructured PFA/silica and the phenolic resin/silica hybrid materials can be chemically transformed into carbon/silica hybrid materials because PFA as well as phenolic resin are well established carbon precursors (see also Chapter 4.2 of this book) [34, 35]. For this purpose, the hybrid material is thermally treated under inert atmosphere (argon) at 800 °C to convert the organic resin component into carbon. Then, the silica component can be removed by treating the C/SiO_2 composite with aqueous HF or with aqueous NaOH in a wet etching process. The general pore texture of the resulting carbon monolith depends on the chemical composition of the former twin monomer. In short, Si-Spiro mainly delivers microporous carbon, whereas TFOS is suitable to produce mesoporous carbon. The fabrication of miscellaneous carbon hybrid materials by combination of hard template-assisted twin polymerization using various twin monomers is reported in detail in Chapter 4.2.

Complementary, the production of highly porous silica is possible when a PFA/silica or a phenolic resin/silica hybrid material is thermally treated in oxygen atmosphere. Then, the organic polymer is removed by oxidation. The integrated synthetic approach for phenolic resin/silica hybrid materials is illustrated in Figure 4.6.

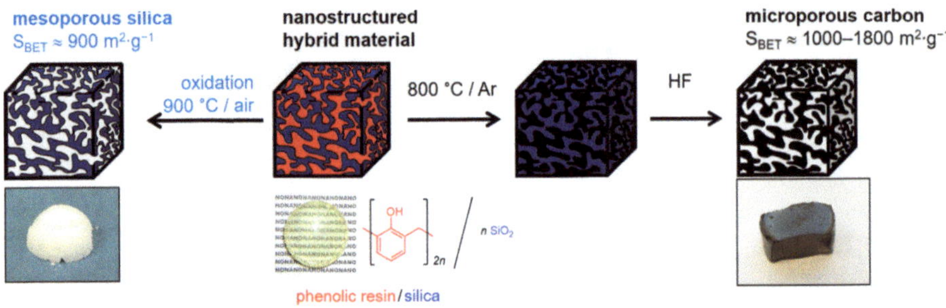

Fig. 4.6: General routes to convert polymer/silica hybrid materials alternatively into microporous carbon (by aq. HF) or mesoporous silica (by oxidation) [36]. Copyright John Wiley & Sons, Inc. Reproduced with permission.

The resulting porous silica and carbon materials show high surface areas of about 1000 $m^2 \cdot g^{-1}$ and narrow pore size distribution. Detailed description of the resulting carbon materials is given in Chapter 4.2.

4.1.4 Multiple polymerizations

From the idea of twin polymerization, it can be deduced that "triple" or "quadruple" polymerizations, leading simultaneously to three or four polymers in one mechanistically coupled process, are also theoretically possible, as long as such types of monomers can be produced. According to the hitherto known twin monomers, the combination of the furfuryloxy, benzoxasiline, and siloxane moiety has been employed for this objective. The synthesis of the monomer is illustrated in Scheme 4.2. Anticipated triple polymerization of 2-furfuryloxy-2-methyl-4*H*-1,3,2-benzoxasiline is demonstrated in Scheme 4.4 [26, 27].

Solid state ^{29}Si and ^{13}C NMR spectroscopies in combination with TEM convincingly show that triple polymerization takes place under specific conditions. The nanostructure formation depends on the catalysis cascade because acids and bases work in different ways for this monomer. In contrast to the benzoxasiline moiety, the furfuryloxy moiety is only polymerizable by acid catalysis. Adding a basic catalyst to this monomer results in twin polymerization of the benzoxasiline moiety, but the

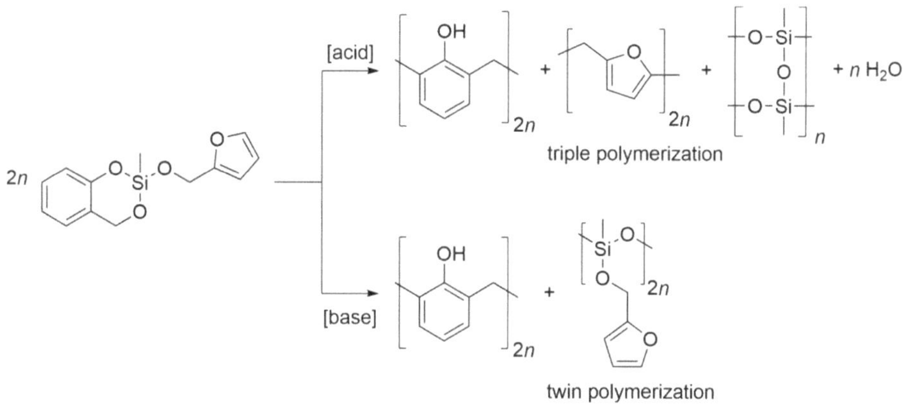

Scheme 4.4: Possible reaction cascades as a function of catalysis among triplet polymerization of 2-furfuryloxy-2-methyl-4*H*-1,3,2-benzoxasiline to the ternary hybrid material phenolic resin-PFA/polymethylsiloxane as well as twin polymerization of the monomer. Reproduced from Ref. [26] with permission from the Royal Society of Chemistry.

furfuryloxy moiety remains intact. This is illustrated in Scheme 4.4. A post-treatment of the resulting hybrid material with an acid in principle gives a ternary hybrid material which is also received when the starting triplet monomer is instantly polymerized by an acid. Each individual polymerization step can be clearly evidenced by the formed oligo(methylsiloxane) molecular species which is shown by solid state ^{29}Si NMR spectroscopy. The nanostructure dimension of the hybrid material can be controlled in the same way.

4.1.5 Template-assisted twin polymerization

Twin polymerization was proven to give nanostructured hybrid materials consisting of organic and inorganic polymers that are closely mixed on a nearly molecular level. The idea behind using templates is the opportunity to further structure the final material at a different length scale and to end up with hierarchically structured materials. There are two methodologies using templates for twin polymerization. Firstly, soft templates such as surfactants are mixed together with the twin monomer and then twin polymerization is catalyzed by acid or base catalysts. This strategy is adopted from template-assisted sol-gel processing to fabricate mesoporous silica materials such as MCM-41 [37]. Secondly, hard templates such as inorganic or organic particles are employed. In this case, the polymerization is started on the surface of the particles and the core/shell geometries are targeted. It is principally possible to use particles only as filler component. However, this procedure is not attractive for the application of twin polymerization because conventional synthetic routes already use this method.

Various soft templates were used to check their effect on the morphology of the resulting hybrid material. Especially, the effect of polyethylene glycol (PEG) on acid-catalyzed twin polymerization of TFOS was studied. Generally, the use of PEG is found suitable in particular in enhancing the surface area of the resulting silica. For this purpose, the synthesized PFA-PEG/silica hybrid materials were burned according to Figure 4.6.

Usually, silica produced from PFA/silica made by TFOS shows surface areas of about 700 to 800 $m^2 \cdot g^{-1}$. Adding low molecular weight PEG (M_n = 400 to 1.000 $g \cdot mol^{-1}$) shows only negligible effects on BET values. Increasing PEG concentrations disturb the formation of solid monolithic hybrid materials due to phase separation, but 1 to 5 wt% of PEG in relation to TFOS is suitable. However, using PEG with M_n = 10.000 $g \cdot mol^{-1}$ (5 wt%) results in silica material with a very large BET surface area up to 1.800 $m^2 \cdot g^{-1}$. However, this procedure delivers waxy and soft black oils of PFA/silica hybrids, independent of the PEG concentration, and the products need 14 days to solidify. Therefore, the addition of soft polymers as structuring agents was unsuitable because the twin polymerization process is affected in an unfortunate way by disturbing the catalysis process. To circumvent this problem, polymerizable furfuryl alcohol derivatives such as the ester of myristic acid were used as soft templates. There is a clear effect on the morphology as observed in TEM pictures [38]. However, the X-ray diffraction patterns of the hybrid materials indicate only amorphous sub-structures, thus, the ordering of the mesostructured silica domains is not uniform. In spite of this effect, the nanostructure is not influenced by the use of soft templates. In conclusion, there are drawbacks in soft-template-assisted twin polymerizations.

In contrast, hard templates such as silica particles, which contain acidic groups at the surface, are well suited as templates for twin polymerization. Also, acid adsorption on silica particles works very effectively for this purpose improving their catalytic activity with regard to TP. The activated particles are dispersed in a nonpolar solvent like toluene, in which the acidic catalyst is not soluble, in contrast to the twin monomer. Then, twin polymerization proceeds solely at the surface which is advantageous for the production of core shell particles. The synthetic procedure is illustrated in Figure 4.7. However, basic catalysis on the particle surface is also possible. It works effectively for the coating of carbon materials as long as the catalyst is completely adsorbed at the particle surface. Sometimes, twin polymerization also occurs in the surrounding of the particles, which is not desired. It is noteworthy that experimental conditions for polymerization must be carefully adjusted for each material composition.

The procedure of surface twin polymerization has been successfully applied to coat silica and carbon particles, respectively, of various sizes and shapes with PFA/SiO_2 (blue shell in Fig. 4.7) or phenolic resin/SiO_2 (red shell in Fig. 4.7) layers [27, 30–33]. The thickness of the nanostructured hybrid material layer can be precisely adjusted by the amount of twin monomer used. This is a great advantage of twin polymerization because the stoichiometry of the surface layer composition is exactly determined by the molecular structure of the twin monomer. A decisive advantage

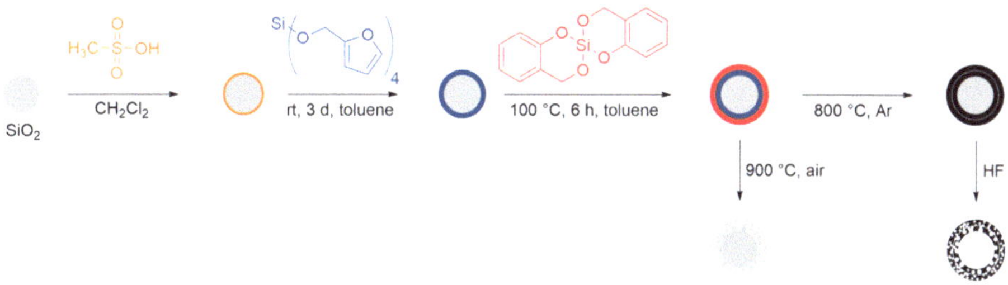

Fig. 4.7: General synthesis protocol for the production of hybrid-material–coated particles by twin polymerization and fabrication of multi-shell composite and hollow particle materials [33].

of surface twin polymerization coatings compared to the sol-gel coating technology is that water as reactant is not required. Organic solvent slurries of dichloromethane, toluene, or di-*iso*-propylnaphtalene have been found suitable for surface twin polymerization to coat solid particles. The comparison of sol-gel coatings with twin polymerization (TP) coatings is demonstrated in Table 4.1 using different criteria for assessment.

Furthermore, the choice of the twin monomer also decides whether microporous or mesoporous carbon shells are finally obtained after transformation of the hybrid material into carbonaceous materials. Hence, PFA/SiO_2 delivers mesoporous carbon

Tab. 4.1: Criteria for producing polymer/SiO_2 nanocomposite coatings by the use of surface twin polymerization of Si-Spiro compared to aqueous sol-gel approach.

criterion	sol-gel coating	TP coating
raw materials	nanostructure material or monomers that react consecutively on the same time scale	twin monomer solvent (optional)
number of components/ reactants	at least two including water	one or more water is not needed
handling	solution	melt/solution
reaction time	days (aging)	minutes to hours
triggering	acid or base	thermal, acid, or base
cleaning steps	drying (slow evaporation of solvent)	thermal induction: none catalysis with acid: filtering (removal of solvent) drying in vacuum
process time	days to weeks	minutes to hours
drying	indispensable	not necessary
contraction	at all times	shrinking $< 5\%$
procedure for coating	dip coating, spraying, spin coating	dip coating, spraying, spin coating

Tab. 4.2: Suitable surface twin polymerization processes used as methods for surface coating of solid substrates.

substrate	monomer	catalyst	final product	reference
silica particle	TFOS	CH_3SO_3H	carbon hollow spheres	[30]
silica particle	Si-Spiro	CH_3SO_3H	carbon hollow spheres	[30]
glass fiber	Si-Spiro	CH_3SO_3H	macroscopic carbon tubes	[30]
graphite	Si-Spiro	DABCO	binary carbon hybrid material	[32]
MWCNT	Si-Spiro	DABCO	binary carbon hybrid material	[38]
carbon black	Si-Spiro/TFOS	DABCO/acid	multi-shell carbon hollow spheres*	[38]
graphene	Si-Spiro	graphene-SO_3H	layered carbon composite	[38]
$LiNi_{0.5}Mn_{1.5}O_4$	Si-Spiro	DABCO	phenolic resin/silica/ $LiNi_{0.5}Mn_{1.5}O_4$	[38]
silica	TTHS**	CH_3SO_3H	S-doped carbon hollow spheres	[39]

* two-step synthetic route
** tetra(thiophene-2-ylmethoxy)silane

whereas a phenolic resin/SiO_2 layer is more suitable for the production of microporous carbon. This feature is discussed in more detail in Chapter 5.1 of this book, because these types of carbon materials are important for energy storage applications.

Twin polymerization of Si-Spiro is a well suited method to coat a variety of solid materials which are different in size, shape, and chemical composition. For instances, carbon materials, glass fibers, flat polymeric substrates, inorganic oxidic particles such as ZnO, TiO_2, SiO_2, as well as metal nanoparticles and others can be straightforwardly functionalized by a hybrid material layer. Table 4.2 shows a compilation of solid substrates which were chemically modified using surface twin polymerization.

4.1.6 Ternary hybrid materials by simultaneous twin polymerization (STP)

Ternary hybrid materials of targeted composition and morphology can be fabricated by simultaneous twin polymerization (STP), which has been introduced in Chapter 2.

It should be emphasized that STP is the most important synthesis process for silicatic hybrid materials made by twin polymerization. Especially, twin polymerization of Si-Spiro with 2,2-dimethyl-4*H*-1,3,2-benzodioxasiline or related twin monomers has been established as an elegant route to produce hybrid materials which consist of three different polymer components: silica, disubstituted polysiloxanes such as poly(dimethylsiloxane), and phenolic resin [17, 28]. The dimethylsiloxane intermediate can undergo chemical reactions with nanoscale silica domains during twin polymerization. Hence, $O_3\,Si\text{–}O\text{–}Si(CH_3)_2\text{–}O$-linkages are formed within the hybrid material. If this happens, this specific type of polymerization process has

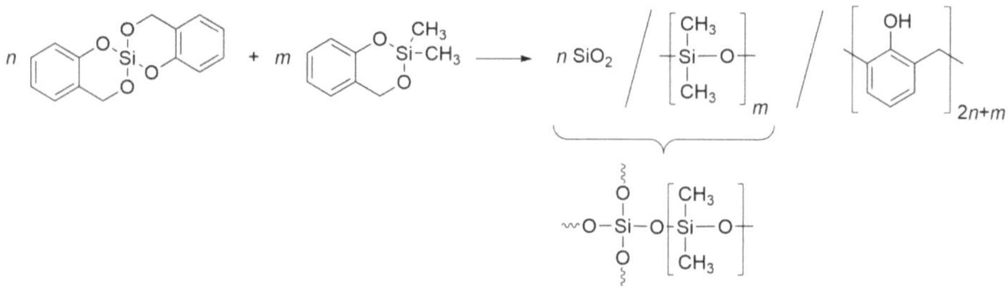

Scheme 4.5: Simultaneous twin copolymerization of Si-Spiro with 2,2-dimethyl-4*H*-1,3,2-benzodioxasiline.

been classified as simultaneous twin copolymerization because two of the three theoretically possible polymer fragments undergo copolymerization according to Scheme 4.5.

An advantage of this type of hybrid material synthesis becomes clear in the following. Reactions of silanol groups with polysiloxane are hard to realize because base catalysis is necessary and incomplete conversion is usually observed. Generally, the final product must be laboriously purified. Classical synthetic reactions of surface silanol groups with polysiloxane $-Si(CH_3)_2-OH$ end groups occur spontaneously, but water release is inevitable [40].

Compared to other methods, the STP of Si-Spiro with 2,2-dimethyl-4*H*-1,3,2-benzodioxasiline can be carried out in melt, because both monomer species are completely miscible. The liquid mixture can be easily handled at T< 80 °C and it does not polymerize spontaneously. Pot life depends on the nature of the catalyst used and can be adjusted precisely. STP can be catalyzed by various catalysts such as lactic acid, trifluoroacetic acid, or even by basic catalysts such as DBU. The choice of the appropriate catalytic process depends on the field of application. Specific twin monomers which contain functional alkaline groups can serve as both co-monomer and catalyst (see Chapter 4.1.7). In the following, the specific STP of Si-Spiro with 2,2-dimethyl-4*H*-1,3,2-benzodioxasiline will be presented as particularly elegant procedure because it demonstrates the realization of covalent linking of silica with PDMS. The covalent linkage of the silica component to the polysiloxane (or oligosiloxane) chains is proven by the solid state ^{29}Si NMR spectrum of the resulting hybrid material which shows the characteristic D signal that is attributed to the structure motif silica–O–$Si(CH_3)_2$– [18].

The suggested molecular structure of the organic polymers of the ternary hybrid material is confirmed by the solid state ^{13}C NMR spectrum of the same sample. The decisive molecular structure elements in relation to the NMR signals are shown in the inset of Figure 4.8.

The nanostructure phase domain size is a function of the initial monomer composition. Increasing Si-Spiro content definitively decreases the dimension of the resulting silicatic domain in the final hybrid material. This fact can be clearly demonstrated by the corresponding HAADF-STEM pictures shown in Figure 4.9.

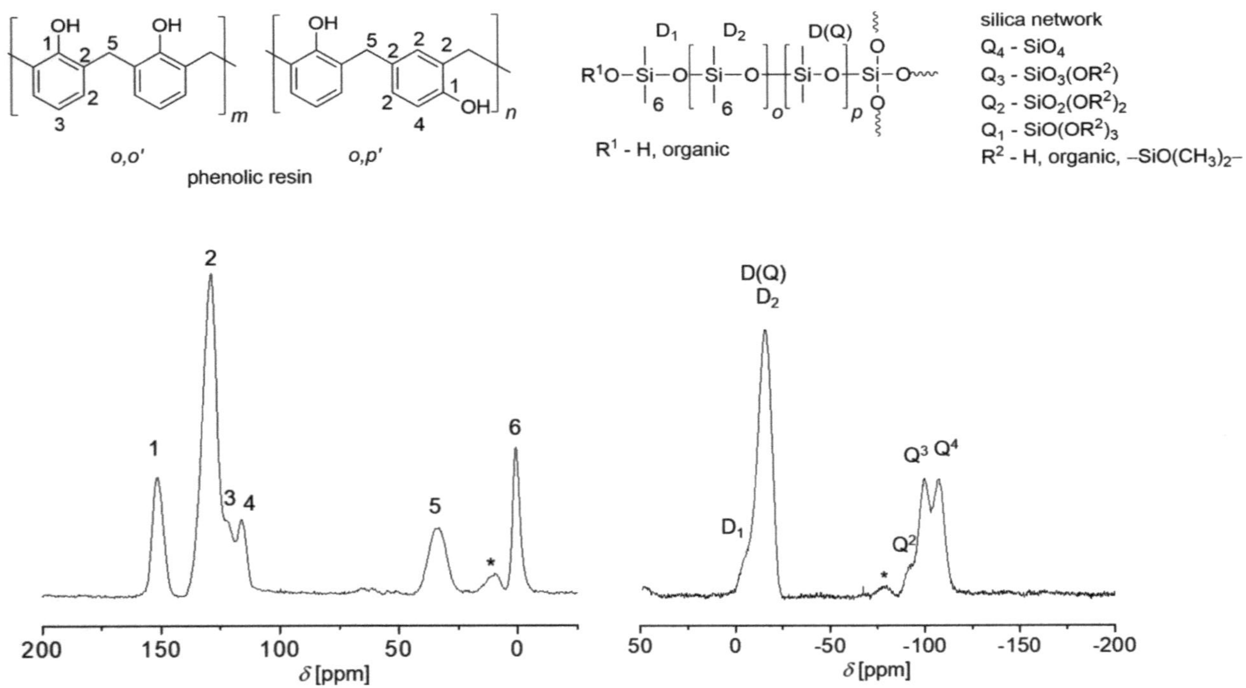

Fig. 4.8: Left: solid state $^{13}C\{^1H\}$-CP-MAS NMR spectrum (12 kHz) and right: solid state $^{29}Si\{^1H\}$-CP-MAS NMR spectrum (5 kHz) of a nanostructured phenolic resin/polysiloxane-silica hybrid material. The mixture was polymerized in the melt at 85 °C for 4 h under trifluoroacetic acid catalysis. Composition of the initial monomer mixture Si-Spiro to 2,2-dimethyl-4*H*-1,3,2-benzodioxasiline was 3:2. *Spinning side band.

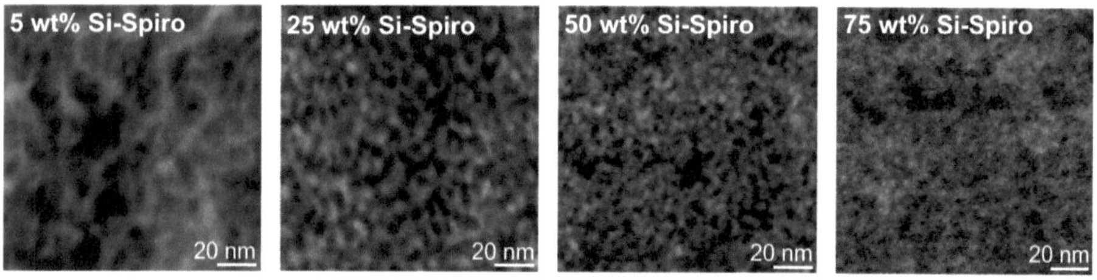

Fig. 4.9: HAADF-STEM pictures of nanostructured phenolic resin/polysiloxane-silica ternary hybrid material (the weight content of Si-Spiro in the initial monomer mixture is indicated inside) [41]. Copyright John Wiley & Sons, Inc. Reproduced with permission.

According to the extended *Carothers* equation (see Chapter 1), a small Si-Spiro content decreases the ability for crosslinking during simultaneous polymerization of the two monomers because the average number of functionality decreases with an increasing amount of 2,2-dimethyl-4*H*-1,3,2-benzodioxasiline. Thus, the nanostructure phase domain size of the Si-rich phase domains (white areas in Fig. 4.9) decreases with increasing Si-Spiro content. This result is completely in accord with the theoretically expected low degree of crosslinking with increasing amount of the polydimethylsiloxane.

Hardness and E-modulus of the hybrid material increase in the same way as the Si-Spiro content (Fig. 4.10). The E-modulus of the phenolic resin/silica sample (100 mol% Si-Spiro) is close to that of a commercial phenolic resin (E-modulus = 7.5 GPa) that is strongly crosslinked (see Fig. 4.10).

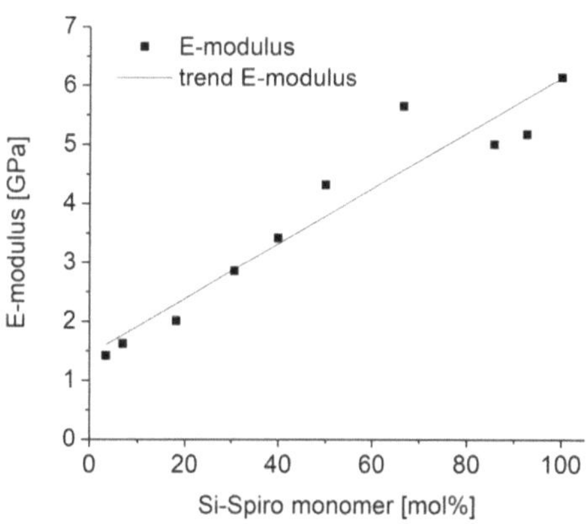

Fig. 4.10: Modulus of elasticity (E-modulus) of the ternary hybrid material made from Si-Spiro and 2,2-dimethyl-4*H*-1,3,2-benzodioxasiline as a function of initial monomer composition. Synthesis was carried out using an acid catalyst (CF_3COOH) in the melt at 85 °C [41]. Copyright John Wiley & Sons, Inc. Reproduced with permission.

In addition to 2,2-dimethyl-4*H*-1,3,2-benzodioxasiline, other 2,2-dialkyl-4*H*-1,3,2-benzodioxasiline derivatives are also useable for STP. However, bulky alkyl substituents such as cyclohexyl or phenyl disturb the reaction of the silica fragments with the dialkyl-substituted silicon moieties and the mechanical properties (E-modulus) of the resulting hybrid material are significantly worsened [18].

The second twin monomer component for an STP can be modified to such an extent that a polymer chain is linked at the silicon atom. The conceptual background for this approach is that the covalently linked polymer chain shall serve as structure-directing agent. This methodology was employed because "free" soft templates did not work as expected for structuring. However, the PEG-functionalized twin monomer is unsuitable for usual twin polymerization because the PEG moiety disturbs the acid catalysis and the polymerization process [24, 25]. A similar effect occurs when Si-Spiro is polymerized in the presence of large amounts of PEG (see above).

Therefore, STP of Si-Spiro with a PEG-functionalized twin monomer has been applied (see Scheme 4.6). This procedure is the method of choice as long as the portion of the PEG-functionalized TM does not exceed about 30 wt% of the monomer composition. The molecular structure of the hybrid material as suggested in Scheme 4.6 is clearly confirmed by solid state ^{13}C and ^{29}Si NMR spectroscopies [24]. The polysiloxane moiety bearing the PEG chain is covalently linked to the silicon dioxide framework. The STP of Si-Spiro with PEG-functionalized TMs delivers hybrid materials with phase nano-domains of about 5–6 nm in size (Fig. 4.11).

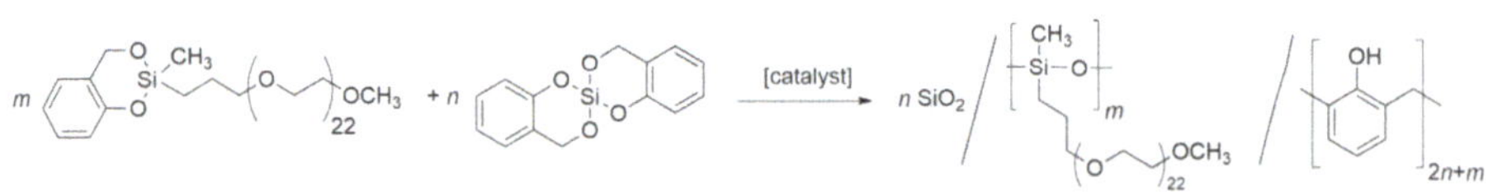

Scheme 4.6: Suggested scenario for STP of PEG-functionalized twin monomer with Si-Spiro.

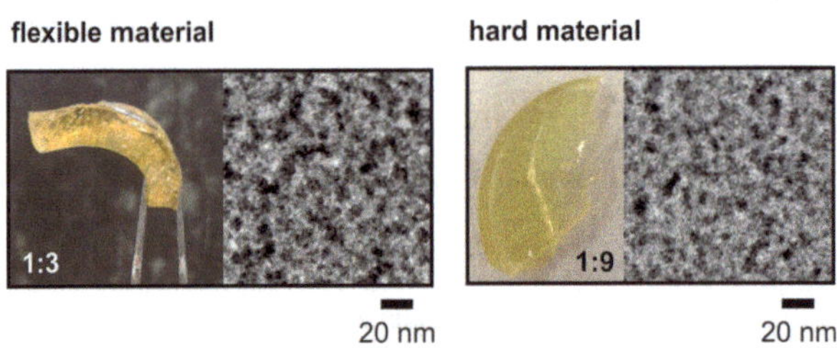

Fig. 4.11: Photographs and TEM images of hybrid materials produced by STP of PEG-functionalized twin monomer with Si-Spiro in given molar ratios [24]. Copyright John Wiley & Sons, Inc. Reproduced with permission.

The integration of a flexible PEG chain within the hybrid material has a clear impact on the hardness and flexibility rather than on the dimension of the nanostructures. Electron microscopic images show only little influence of the PEG-functionalized TM percentage on the dimension of the resulting phase nano-domain. However, the trend is similar to that observed for other combinations: The higher the Si-Spiro content, the finer the nanostructure. Thus, the covalent linking of the PEG chain is beneficial for adjusting the physical properties such as flexibility of the hybrid material which makes this technology suitable for production of membranes and furling films.

4.1.7 Polymer/Silica hybrid foams by simultaneous twin polymerization

In this chapter, the STP of twin monomers with polymerizable organic carbonates is discussed. Organic carbonates containing the same or structurally similar polymerizable moiety as in the twin monomer were used. According to the STP of two twin monomers, the STP of two different monomer species can also be modified in such a way that the second component does not form a polymer product during the reaction. The idea is that a gaseous product such as carbon dioxide is released instead of SiO_2 formation. It is intended that the "by-product" carbon dioxide serves as foaming agent similar to the established polyurethane foam production process. For instance, difurfurylcarbonate (DFC) readily polymerizes by acid catalysis to a PFA resin, water, and carbon dioxide, as theoretically expected (Scheme 4.7). The carbon dioxide formation is attributed to the fact that the furfurylium ion species is highly resonance stabilized and thus acid-induced ester cleavage occurs easily. Afterwards, the resulting semi-ester of the carbonic acid readily decomposes to carbon dioxide and furfuryl alcohol which polymerizes as well as used in the twin polymerization of TFOS [42].

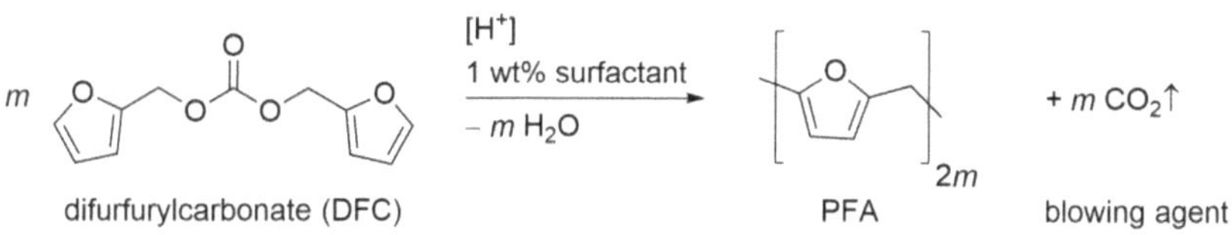

Scheme 4.7: Production of PFA foams by cationic polymerization of DFC with *p*-toluenesulfonic acid as catalyst. Polymerization is carried out in the melt using 1 wt% of the surfactant.

In spite of the chemical reaction proceeding smoothly, gaseous carbon dioxide is only trapped within the crosslinked PFA resin if an appropriate surfactant is used. Otherwise, the carbon dioxide release takes place too rapidly and it is not trapped in the final product. For foam stabilization, a low quantity (approximately 1 wt%) of a polysiloxane derivative with pendant PEG chains (Dabco® DC 193) has been found suitable (see Fig. 4.12 left). Other surfactants, such as pure polysiloxanes or pure PEG, do not work appropriately for this purpose.

The PFA foam formation can be readily combined with the twin polymerization of TFOS. The organic fragments of both monomers, TFOS and DFC, can undergo homopolymerization to produce polyfurfuryl alcohol whereas silica and gaseous carbon dioxide arise simultaneously as coproducts (Scheme 4.8). That scenario is completely in accord with the concept of STP.

The STP of DFC with TFOS works well because the monomer mixture can be handled as a melt at low temperature. DFC is liquid at room temperature and TFOS is completely miscible with it. The polymerization can be triggered by heating up the

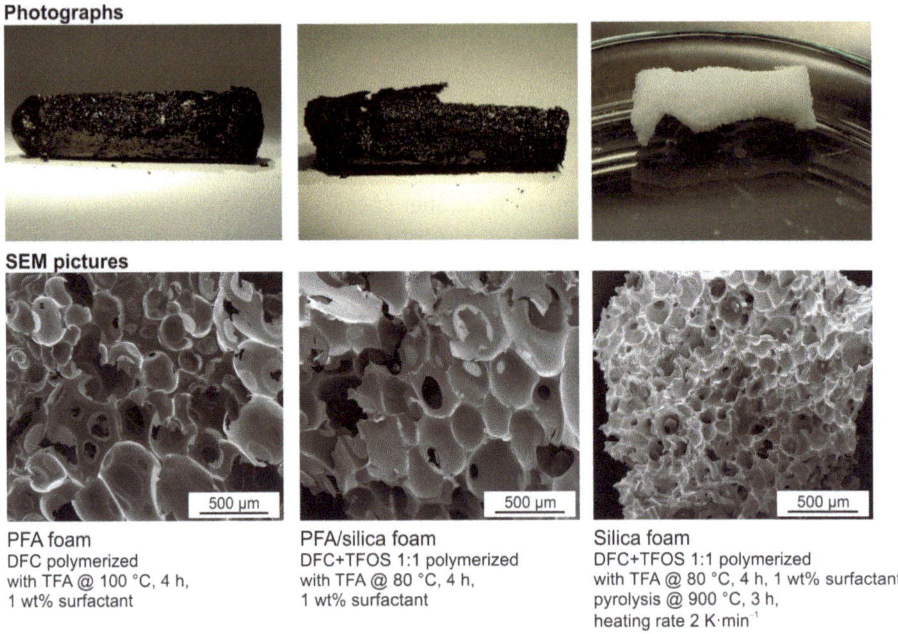

Fig. 4.12: SEM pictures and photographs of PFA foam (left), PFA/silica foam (middle), and silica foam (right) produced from STP of TFOS with DFC with trifluoroacetic acid (TFA) as catalyst. Reproduced from Ref. [43] with permission from the Royal Society of Chemistry.

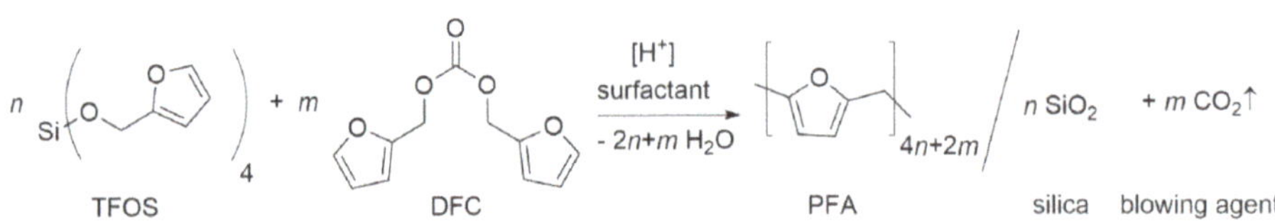

Scheme 4.8: STP of tetrafurfuryloxysilane with difurfurylcarbonate to fabricate a nanostructured PFA/SiO_2 hybrid foam [44].

homogeneous reaction mixture of DFC, TFOS, and the acid catalyst to 50 °C. Volume expansion caused by carbon dioxide release during polymerization depends on many factors such as molecular ratio of the two monomers, surfactant concentration, and acid catalyst used. The foamy hybrid materials can be treated in the same way to produce silica foam or carbon foam as demonstrated for the compact hybrid material monoliths. Burning out the carbon component is suitable to fabricate silica foam. Figure 4.12 shows typical SEM pictures of the foamy hybrid material and the silica foam as well as photographs depicting the materials which were produced by this synthetic concept. The STP of various polymerizable carbonates with twin monomers has been established as a suitable tool to fabricate hierarchically structured carbon and silica foams [43].

4.1.8 Functional twin monomers

Functional twin monomers are defined as special twin monomers which contain an additional functional group (Y) suitable to undergo or cause a chemical reaction with a second component after the twin polymerization. In principle, the functional group Y can be positioned nearly everywhere in the monomer molecule: the organic as well as the inorganic part may be modified. A general chemical formula for simple twin polymerization of C–A–Y monomers is given in Scheme 4.9.

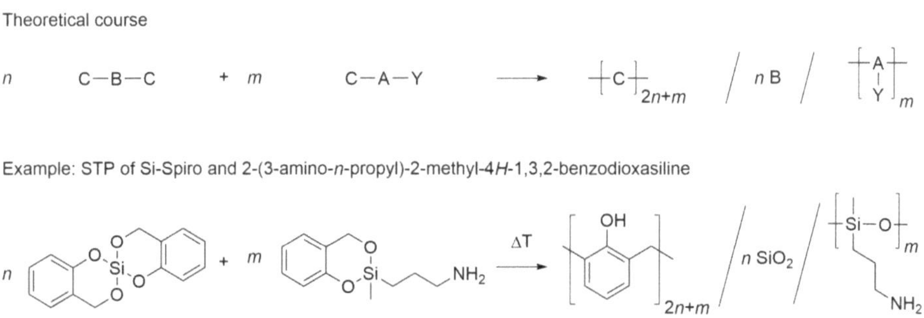

Scheme 4.9: Theoretical course of twin polymerization of a functionalized twin monomer with Y as functional group (above) and an example for the self-catalyzed STP of Si-Spiro with an amino-functionalized twin monomer (below).

Scheme 4.9 describes a concrete twin polymerization of a functional twin monomer for which the functional group Y does not interfere with the twin polymerization process. A very versatile class of functional twin monomers are those which contain various amino groups. They can readily be synthesized from salicyl alcohol by transesterification with various γ-aminopropylmethyldialkoxysilanes according to Scheme 4.2. Table 4.3 shows a compilation of synthetically available amino-functionalized twin monomers and their trigger temperatures T_T necessary to start a homo-TP or an STP in combination with the Si-Spiro twin monomer.

Amino-functionalized twin monomers do not spontaneously undergo twin polymerization at ambient temperatures. Most of them are stable up to at least 140 °C (see Tab. 4.3).

Si-Spiro is more reactive towards bases than the 2,2-dialkyl-4*H*-1,3,2-benzodioxasiline moiety. Therefore, STP of amino-functionalized twin monomers with Si-Spiro in molar ratio 1:1 proceeds readily to hybrid materials upon processing in the melt at significantly lower temperatures necessary for the homo-twin polymerization. As a result of the alkaline effect of the amino group, this type of amino-functionalized twin monomer serves both as catalyst in STP with Si-Spiro and as co-component for the production of ternary hybrid materials consisting of phenolic resin/silica-poly-(aminopropylsiloxane) (Scheme 4.9).

The STP of Si-Spiro with amino-functionalized twin monomers can be triggered by heat in the temperature interval 100–200 °C. The trigger temperature for STP (T_T (STP), see Tab. 4.3) depends on monomer composition and molecular structure of the amino-functionalized twin monomer in the monomer mixture. The catalytic activity of the amino-functionalized twin monomer correlates roughly only with its basicity. This result is in agreement with the effect of base structure on external alkaline-catalyzed twin polymerization of Si-Spiro using various amines (see Chapter 3.3). A significant decrease of basicity, as expected for the $-NH-C_6H_5$ or for the morpholine substituent, is associated with an increase of the necessary trigger temperature (see Tab. 4.3).

Tab. 4.3: Physical properties and chemical reactivity of various amino-functionalized twin monomers. Trigger temperatures (onset temperatures) are determined by DSC measurements. For simultaneous polymerization, molar ratios of 1:1 with Si-Spiro were used.

formula of twin monomer	name	T_T /°C (homo-TP)	T_T/°C (STP)
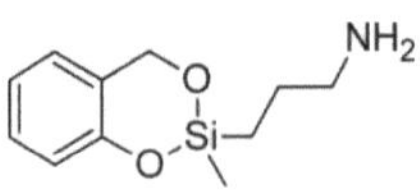	2-(3-amino-*n*-propyl)-2-methyl-*4H*-1,3,2-benzodioxasiline	190	105
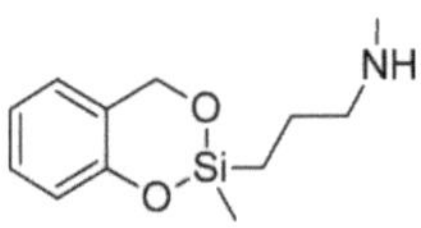	2-(*N*-methyl-3-amino-*n*-propyl)-2-methyl-*4H*-1,3,2-benzodioxasiline	140	100
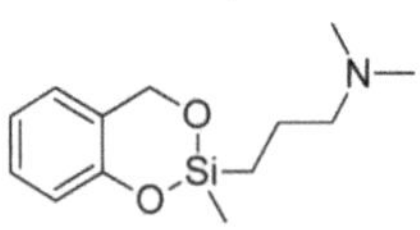	2-(*N,N*-dimethyl-3-amino-*n*-propyl)-2-methyl-*4H*-1,3,2-benzodioxasiline	177	109
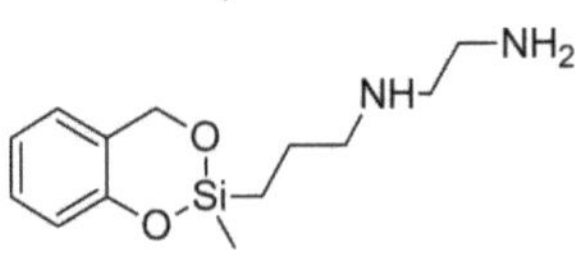	2-(*N*-(2-aminoethyl)-3-amino-*n*-propyl)-2-methyl-*4H*-1,3,2-benzodioxasiline	136	99
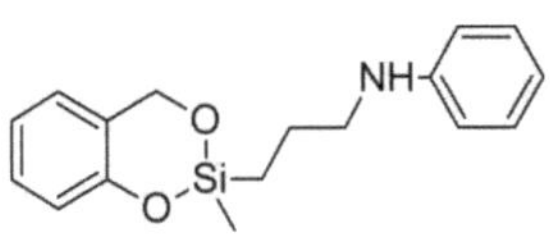	2-(*N*-phenyl-3-amino-*n*-propyl)-2-methyl-*4H*-1,3,2-benzodioxasiline	250	205
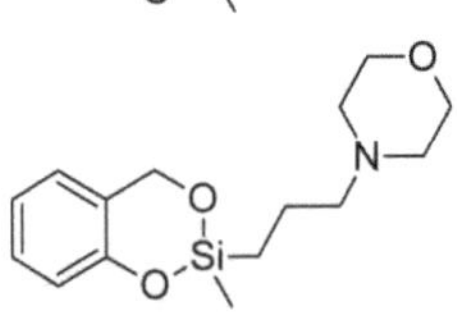	2-(3-morpholino-*n*-propyl)-2-methyl-*4H*-1,3,2-benzodioxasiline	233	197

Furthermore, primary and secondary amino groups can readily react with epoxides, isocyanates, and carboxylic acid derivatives. Thus, they can serve as the bridge to connect twin polymerization with conventional polymer synthesis based on polyepoxides, polyurethanes, polyamides, or polyesters. This important topic is discussed in Chapter 5.2 of this book.

4.1.9 Silicon monomers based on Si–N-containing moieties

In principle, the oxygen atoms of the Si-Spiro monomer can be replaced stepwise by other elements such as sulfur or nitrogen (see also Chapter 3.1). Nitrogen-substitution is promising because silicon–nitride-based hybrid materials are of interest for several applications, e.g., as basic catalysts, or as components in energy storage. However, as discussed in Chapter 3.1, spirocyclic azasilines and diazasilines, which contain N–Et or N–Pr substituents instead of oxygen, were unsuitable for polymerization. Neither a thermal nor a proton-assisted catalytic process enables polymerization of these compounds [45]. Therefore, it was necessary to search for other Si–N-containing twin monomers. Hence, such spiro monomer species were synthesized for twin polymerization (Scheme 4.10) containing NH moieties rather than N–alkyl groups. The synthesis of the NH-analogs of Si-Spiro can be performed by using the corresponding nitrogen compounds of salicyl alcohol such as 2-aminobenzyl alcohol, 2-(aminomethyl)phenol, or 2-aminobenzylamine and their derivatives. In contrast to the synthesis of Si-Spiro, transamidation reactions of siliconalkoxy compounds with amino alcohols or 2-aminobenzylamine towards the Si–N compounds do not work for thermodynamic reasons. Thus, silicon tetrachloride must be used as reactant. Due to the risk of salt formation with the product, the removal of the generated HCl is difficult. For that reason, strong bases such as tertiary aliphatic amines must be used as trapping reagent for HCl. The synthesis of several twin monomer species such as 1,1',4,4'-tetrahydro-2,2'-spirobi[benzo[*d*][1,3,2]oxazasiline] and 2,2'-spirobi[3,4-dihydro-1*H*-1,3,2-benzodiazasiline], which contain nitrogen in form of NH moieties instead of O, was successful.

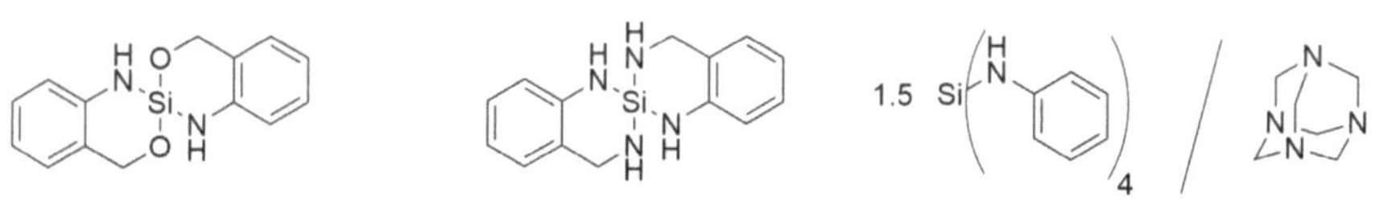

1,1',4,4'-tetrahydro-2,2'-spirobi[benzo[*d*][1,3,2]oxazasiline] — 2,2'-spirobi[3,4-dihydro-1*H*-1,3,2-benzodiazasiline] — tetrakis(phenylamino)-silane — hexamethylene-tetramine

Scheme 4.10: Synthesized nitrogen-containing twin monomers derived from 2-aminobenzyl alcohol and 2-aminobenzylamine as well as the deficient twin monomer derived from aniline combined with hexamethylenetetramine (HMTA) for hybrid materials synthesis.

Additionally, the alternative synthetic concept about the apparent twin polymerization is employed by using a deficient (see Chapter 2.5.3) twin monomer tetrakis(phenylamino)silane in combination with hexamethylenetetramine (HMTA) [46–48].

Fortunately, all twin monomers which contain NH groups instead of oxygen are suitable to be polymerized merely by heating above 100 °C without the need for any additional catalyst. Figure 4.13 shows exemplary DSC measurements of 1,1',4,4'-tetrahydro-2,2'-spirobi[benzo[*d*][1,3,2]oxazasiline], 2,2'-spirobi[3,4-dihydro-1*H*-1,3,2-benzodiazasiline], and tetrakis(phenylamino)silane in combination with HMTA using a constant heating rate. The measured effects of heat flow processes are summarized in Table 4.4.

Tab. 4.4: Thermally induced processes of 1,1',4,4'-tetrahydro-2,2'-spirobi[benzo[*d*][1,3,2]-oxazasiline], 2,2'-spirobi[3,4-dihydro-1*H*-1,3,2-benzodiazasiline], and deficient twin monomer tetrakis(phenylamino)silane in combination with HMTA during DSC measurement in the range of 20–300 °C.

monomer	temperature [°C]	effect	observation
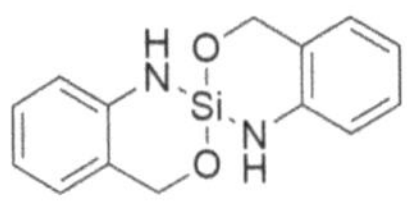	>100	exotherm	polymerization
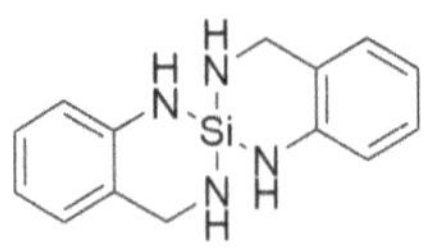	128	endotherm	melting, rearrangement, crosslinking
	232	endotherm	polymerization, gas release
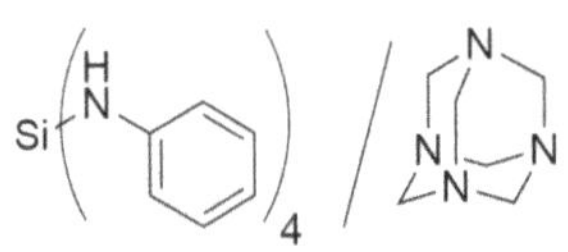	112	endotherm	melting
	120–280	exo- and endotherm	polymerization, rearrangement, gas release, crosslinking, decomposition of HMTA

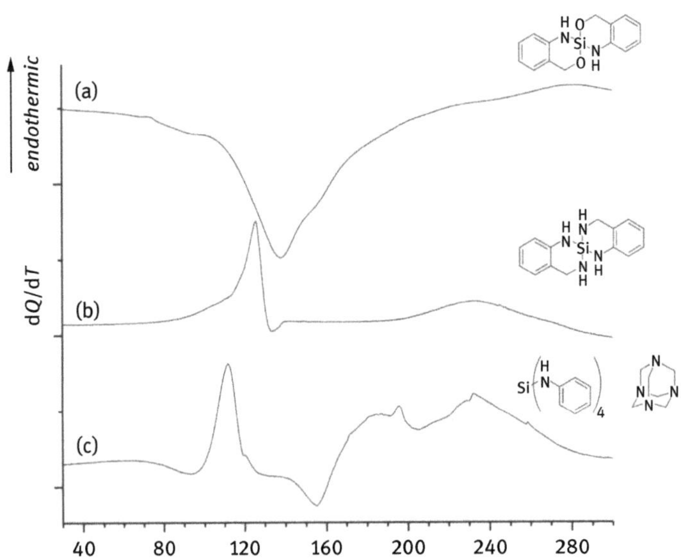

Fig. 4.13: DSC measurements of (a) 1,1',4,4'-tetrahydro-2,2'-spirobi[benzo[*d*][1,3,2]oxazasiline], (b) 2,2'-spirobi[3,4-dihydro-1*H*-1,3,2-benzodiazasiline], and (c) tetrakis(phenylamino)silane in combination with HMTA with constant heating rate of 10 K·min^{-1}, 40 µL aluminum pan with hole in the lid, N_2 flow.

For comparison, Si-Spiro melts at 82–84 °C and the thermally induced polymerization of Si-Spiro occurs as an exothermic reaction at temperatures above 180 °C with silica as inorganic product and phenolic resin as organic phase [16].

In contrast, the thermal behavior of nitrogen-containing twin monomers is difficult to interpret. Due to the complex curve shape of the DSC plots, all occurring processes during melting and polymerization could not be assigned. For example, 2,2'-spirobi[3,4-dihydro-1*H*-1,3,2-benzodiazasiline] shows an asymmetric melting peak with a maximum at around 128 °C (Fig. 4.13). The asymmetry is a result of different superimposed effects like melting, rearrangement, and crosslinking. Furthermore, in this special case, an exothermic polymerization peak could not be observed due to overlying endothermic contribution of releasing ammonia and other organic volatile by-products [46, 47]. As a result, the polymerization reaction of 2,2'-spirobi[3,4-dihydro-1*H*-1,3,2-benzodiazasiline] is an endothermic process under formation of polysilazanes and a polyaniline-formaldehyde polymer.

In contrast, the monomer 1,1',4,4'-tetrahydro-2,2'-spirobi[benzo[*d*][1,3,2]oxazasiline] lacks of the melting peak. The substance polymerizes exothermic at temperatures above 100 °C to SiO_2 and a polyaniline-formaldehyde resin in the form of a powder in a solid state polymerization [47, 48].

Furthermore, tetrakis(phenylamino)silane in combination with hexamethylenetetramine (HMTA) shows also a complex sequence of endo- and exothermic processes due to the combination of two different substances which melt, rearrange, or decompose at different temperatures [46].

Scheme 4.11 shows the polymerization reactions of the three investigated twin monomers: 1,1',4,4'-tetrahydro-2,2'-spirobi[benzo[*d*][1,3,2]oxazasiline], 2,2'-spirobi-[3,4-dihydro-1*H*-1,3,2-benzodiazasiline], and a deficient twin monomer tetrakis(phenylamino)silane in combination with HMTA.

The organic and inorganic molecular structures were investigated by solid state $^{13}C\{^1H\}$-CP-MAS and $^{29}Si\{^1H\}$-CP-MAS NMR spectroscopies (see Fig. 4.14). The monomer 1,1',4,4'-tetrahydro-2,2'-spirobi[benzo[*d*][1,3,2]oxazasiline] provides a polyaniline-formaldehyde resin/silica hybrid material. The monomer 2,2'-spirobi[3,4-dihydro-1*H*-1,3,2-benzodiazasiline] as well as the deficient twin monomer tetrakis(phenylamino)-silane in combination with HMTA form polyaniline-formaldehyde resin/polysilazane hybrid materials. There are some differences in the substitution pattern of the formed polyaniline-formaldehyde polymers of the three different hybrid materials as a function of the monomer composition observed as proven by solid state $^{13}C\{^1H\}$-CP-MAS NMR spectroscopy (Fig. 4.14, left).

Thus, the different hybrid materials obtained at reaction temperatures below 300 °C show different solubility properties due to different crosslinking patterns of the organic polymers (see Scheme 4.11 and Fig. 4.14).

The hybrid materials obtained by polymerization of 2,2'-spirobi[3,4-dihydro-1*H*-1,3,2-benzodiazasiline] is solid and soluble in organic solvents. The organic polyaniline-formaldehyde polymer is connected over *N*,2- or *N*,4-linkages (see Fig. 4.14

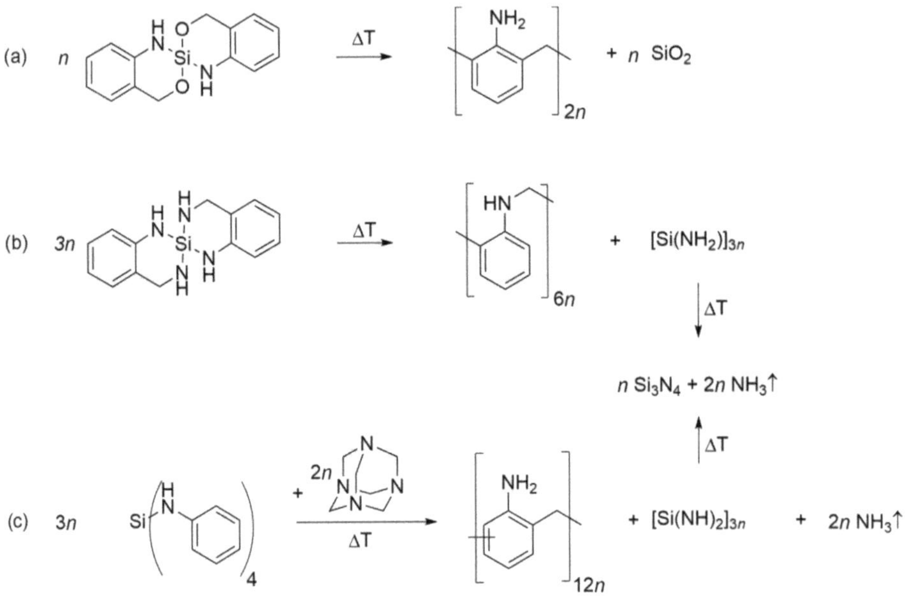

Scheme 4.11: Schemes of polymerization reactions of (a) 1,1',4,4'-tetrahydro-2,2'-spirobi[benzo[*d*]-[1,3,2]oxazasiline], (b) 2,2'-spirobi[3,4-dihydro-1*H*-1,3,2-benzodiazasiline], and (c) a deficient twin monomer tetrakis(phenylamino)silane in combination with HMTA.

left). Also, the inorganic polysilazane network is still soluble, e.g., in dichloromethane or tetrahydrofuran (Fig. 4.14 right), when produced at reaction temperatures below 280 °C. In contrast, a hybrid material prepared from 1,1',4,4'-tetrahydro-2,2'-spirobi[benzo[*d*][1,3,2]oxazasiline] forms nearly insoluble polyaniline-formaldehyde resin/silica hybrid materials after annealing at 280 °C. The organic polymer is crosslinked *via* 2,6- or 2,4-linkages at the aromatic ring. Silica is formed as inorganic network. Products from polymerization of tetrakis(phenylamino)silane in combination with HMTA are partly soluble, depending on reaction temperature and the ratio of monomer:HMTA. The organic polymer is mainly crosslinked by 2,6- and 2,4-linkages with a minor amount of *N*,2- or *N*,4-linkages. Additionally, a polysilazane network is formed [46–48]. Finally, Si_3N_4 hybrid materials can be obtained from 2,2'-spirobi[3,4-dihydro-1*H*-1,3,2-benzodiazasiline] or tetrakis(phenylamino)silane in combination with HMTA at temperatures above 1100 °C [46, 47].

However, the mechanism of thermally induced polymerization of 2,2'-spirobi[3,4-dihydro-1*H*-1,3,2-benzodiazasiline] is not yet fully understood. In any case, redox processes take place because large amounts of *o*-toluidine are formed as by-product during polymerization. This indicates a disproportionation reaction rather than common electrophilic substitution reaction of polyaniline-formaldehyde polymer formation. A similar redox process involving a *β*-hydride elimination is also observed for thermal polymerization of 3,3'-diphenyl-3,3',4,4'-tetrahydro-2,2'-spiro[benzo[*e*][1,3,2]-oxagermine to Ge@C materials [45].

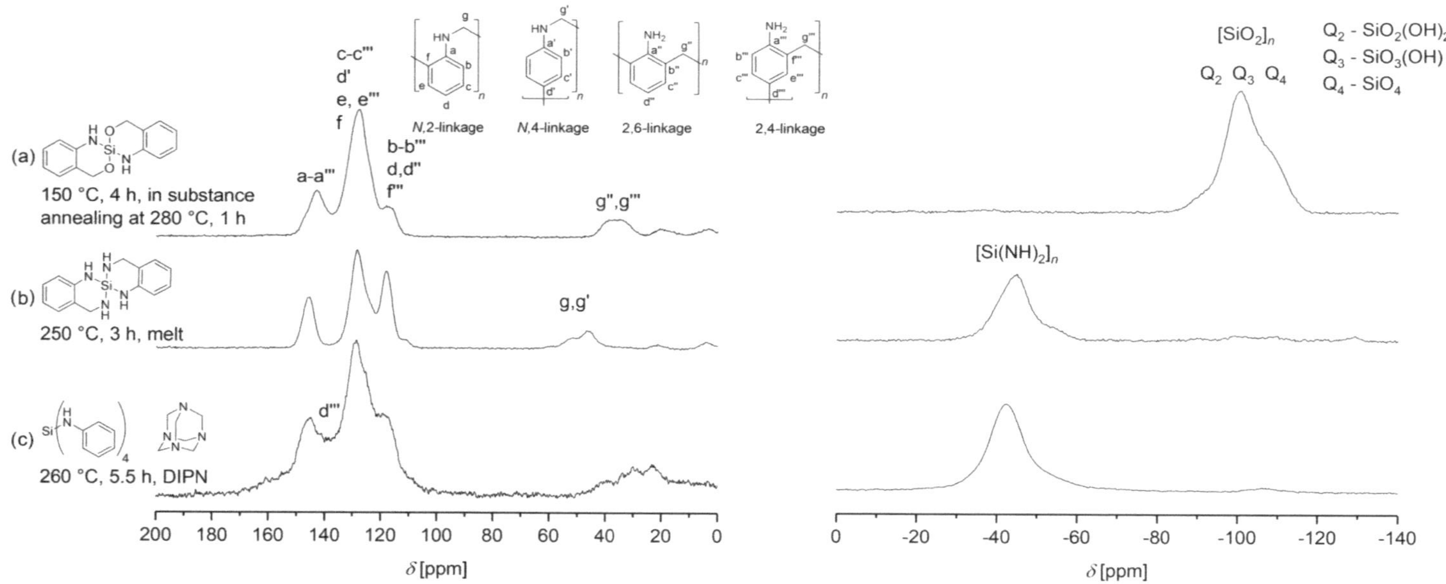

Fig. 4.14: $^{13}C\{^1H\}$-CP-MAS NMR spectra (left) and $^{29}Si\{^1H\}$-CP-MAS NMR spectra (right) of products from polymerization of (a) 1,1',4,4'-tetrahydro-2,2'-spirobi[benzo[*d*][1,3,2]oxazasiline, (b) 2,2'-spirobi[3,4-dihydro-1*H*-1,3,2-benzodiazasiline], and (c) tetrakis(phenylamino)silane in combination with HMTA with assignment of signals to the corresponding structural units. The rotational frequences were 15 kHz for ^{13}C NMR spectra and 12 kHz for ^{29}Si NMR spectra.

2,2'-Spirobi[3,4-dihydro-1*H*-1,3,2-benzodiazasiline] is found to be the most promising monomer for producing silicon-nitride-related hybrid materials [46, 47]. The composition of the resulting hybrid material strongly depends on the reaction temperature used. The formation of a silicon nitride component is only observed at higher temperatures. Carbon/Si_3N_4-like composite materials can be produced if the hybrid material is treated at 1100 °C under inert atmosphere [46, 47]. However, silica instead of any Si–N-containing hybrid material is produced as the main product if oxygen is present in any bonded form and in the surroundings during the polymerization. It does not make a difference whether oxygen is part of the monomer, the solvent, or the glass wall of the reaction vessel. Therefore, all twin polymerizations of 2,2'-spirobi[3,4-dihydro-1*H*-1,3,2-benzodiazasiline] at temperatures below 300 °C yielding a polyaniline-formaldehyde polymer/polysilazane hybrid material must be carried out under completely oxygen-free conditions such as in inert atmosphere and in Teflon® containers. The same conditions are applicable for tetrakis(phenylamino)silane in combination with hexamethylenetetramine. Both processes for the synthesis of polyaniline-formaldehyde polymer/polysilazane hybrid materials take place under volatilization of by-products such as ammonia, or different aromatic nitrogen compounds [46, 47]. Therefore, the synthetic concept works, but the yield of the desired products, sometimes, is rather low with respect to the reactants.

In summary, twin polymerization of *N*-analogs of Si-Spiro is possible but its scope is rather limited in comparison to its oxygen counterparts.

4.1.10 Polyamide 6/Silica composite production

In this subchapter, a special synthetic approach for the production of various polyamide 6/SiO_2 composites is presented. This process is not a genuine twin polymerization. However, the overall process is related to the simultaneous twin polymerization for the following reasons. The employed silicon monomers contain polymerizable fragments for generation of the organic as well as inorganic polymer for the targeted composite materials. For this purpose, silicon-containing lactam monomers are available by substitution reaction of ε-caprolactam with various chlorosilanes as demonstrated in Scheme 4.12 [47, 49]. The chemical reaction of 1,1',1'',1'''-silanetetrayltetrakis(azepan-2-one) (Si(ε-CL)$_4$) yielding polyamide 6/SiO_2 composites is theoretically possible if water is used as additional reactant. Moreover, the stoichiometry for composite synthesis must be fulfilled (Scheme 4.13). Thus, the specific types of monomers presented in Scheme 4.12 lack molecular entities that are necessary for the desired target composite material. Therefore, these types of monomers may be named as deficient twin monomers.

Despite the fact that polyamide 6 produced from ε-caprolactam works well with water as co-component in industry, the monomers presented in Scheme 4.12 do not polymerize to the target composite material when free water is used as reactant (Scheme 4.13). However, the process works well if ε-aminocaproic acid (ε-ACA) is used as reactant instead of water. ε-ACA serves as the source for water and reacts directly

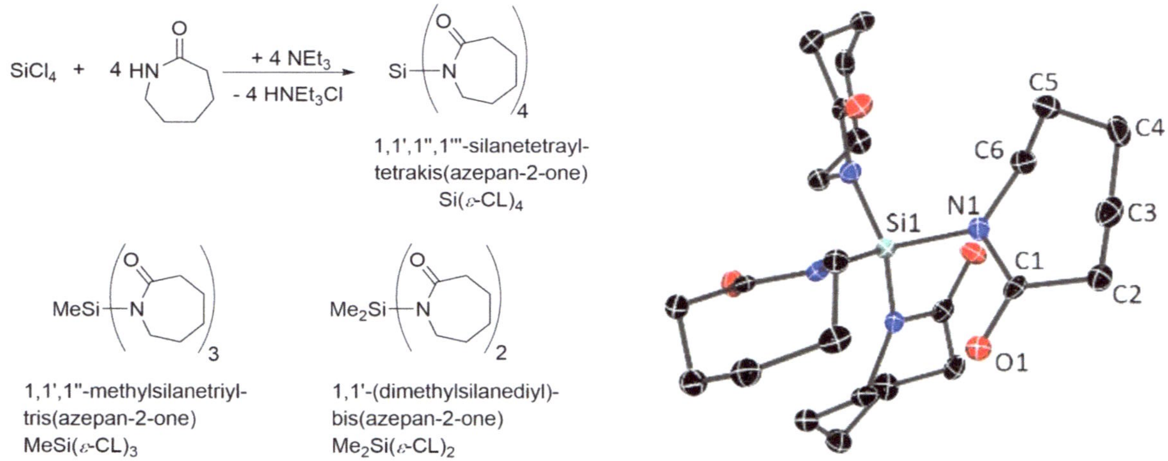

Scheme 4.12: Si–N lactam monomer species as used for polyamide 6/silica composite synthesis (left) and molecular structure (right) (ORTEP) of Si(ε-CL)$_4$ [49]. Published by The Royal Society of Chemistry.

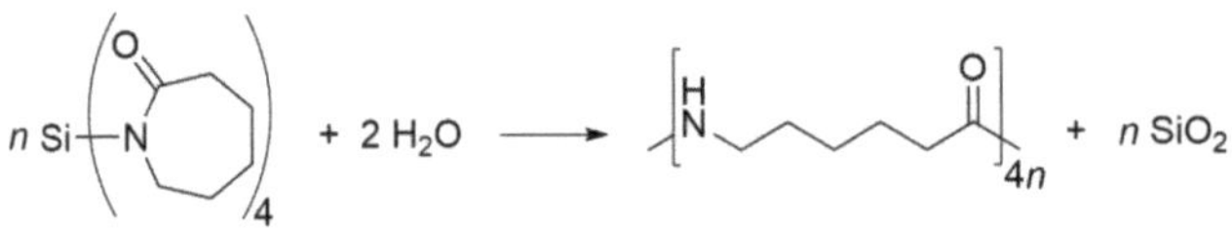

Scheme 4.13: Hypothetical reaction of Si(ε-CL)$_4$ with water to a polyamide 6/SiO_2 composite material which does not work in experiment.

with the Si(ε-CL)$_4$ monomer. If the reaction of Si(ε-CL)$_4$ with ε-ACA is carried out in the presence of ε-caprolactam, the mixture of monomers polymerizes simultaneously, as shown in Scheme 4.14. The kinetic and thermodynamic couplings of three independent reactions are crucial for the success of this procedure. Therefore, the relationship with simultaneous twin polymerization (STP) is obvious. According to the partial reaction 2 of Scheme 4.14, the amount of water required for the formation of polyamide 6 and SiO_2 from Si(ε-CL)$_4$ (see Scheme 4.13) originates from the condensation and/or cyclization of ε-aminocaproic acid. Additionally, ε-aminocaproic acid acts as catalyst for the hydrolytic polymerization of ε-caprolactam (ε-CL) as indicated in the partial reaction 3 of Scheme 4.14. Furthermore, ε-caprolactam is necessary for homogenization of the reaction melt and adjustment of the filler content, which can be tuned independently up to 5 wt%. The best results can be obtained using conditions similar to technical PA6 production. Therefore, composite synthesis was performed in a laboratory autoclave at a starting pressure of 8 bar and 230 °C for 210 min [47, 49].

The reactant ratio of ε-aminocaproic acid regarding Si(ε-CL)$_4$ is of great importance. The best results were obtained when the molar ratio of ε-ACA to the Si-monomer is 4:1. If the amount of the ε-ACA is too high, the molecular weight of the resulting polyamide 6 shares will be low. Smaller amounts of ε-ACA lead to low yields and also a slow increase in viscosity during polymerization that induces strong agglomeration of the formed silica particles [47, 49].

The molecular structure of the produced polyamide 6/silica composite materials was examined by solid state NMR (Fig. 4.15) and FTIR spectroscopies (Fig. 4.16). Solid state ^{13}C NMR spectra prove the formation of polyamide 6 with a high degree in α-crystallinity. In the solid state ^{29}Si NMR spectra, Q_3 and Q_4 signals at –100 and –110 ppm, respectively, indicate a highly condensed silica network. The investigations of the composite materials by FTIR spectroscopy confirm the NMR spectroscopy results. Typical bands for polyamide 6 like the amide I band at 1636 cm^{-1} and an amide II band at 1536 cm^{-1} were observed. Moreover, crystallization in α-modification was proven by the amide V band at 690 cm^{-1} and an amide VI band at 580 cm^{-1} that are shifted to 712 cm^{-1} and 625 cm^{-1}, respectively, for γ-modification. Higher filler amounts lead to an increase in intensity of the Si–O stretching vibration at ≈ 1070 cm^{-1} [47, 49].

Despite the short reaction time of 210 min of the polymerization process, relatively high molecular weights up to 22900 g·mol^{-1} are achievable, which are comparable with reference experiments using ε-caprolactam and ε-ACA in a two-component system. The reduction of the amount of the amino acid in three-component systems

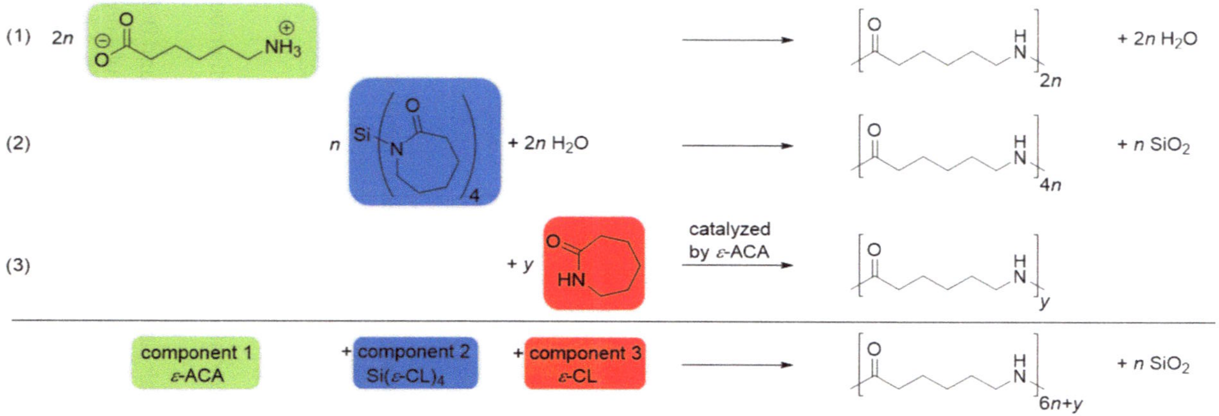

Scheme 4.14: Formation of PA6/SiO_2 composites by a coupled reaction procedure consisting of three partial reactions. The release of water by condensation of ε-ACA facilitates the formation of PA6 and silicic acid followed by condensation of the latter to form SiO_2.

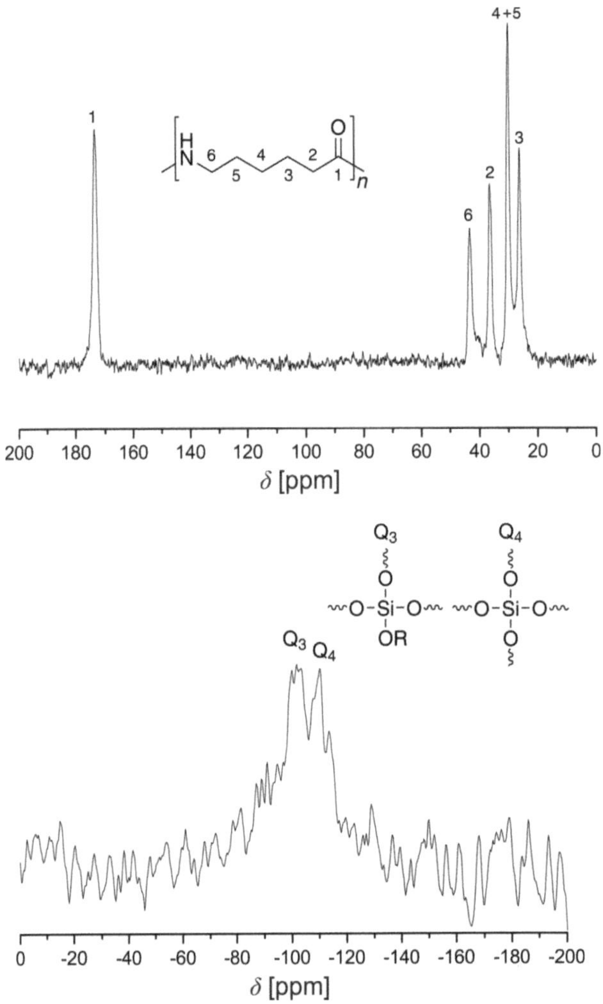

Fig. 4.15: Solid state $^{13}C\{^1H\}$- (top) and $^{29}Si\{^1H\}$-CP-MAS NMR spectrum (below) of a polyamide 6/SiO_2 composite material with 2 wt% filler content; R = alkyl, H. Adapted from [49]. Published by The Royal Society of Chemistry.

leads to a decrease of the molecular mass. Monomodal molecular weight distributions with polydispersities in the range of 2.5 to 3.5 can be observed [47, 49].

Thermal properties of the composite materials have been studied by TGA and DSC measurements (Fig. 4.17). DSC analysis shows the typical melting behavior of polyamide 6 with the melting point at approximately 211 °C and 221 °C, indicating melting of the *γ*- and *α*-modifications, respectively, with no significant differences between the samples. TGA results show a small increase in thermal stability for 1 wt% filler

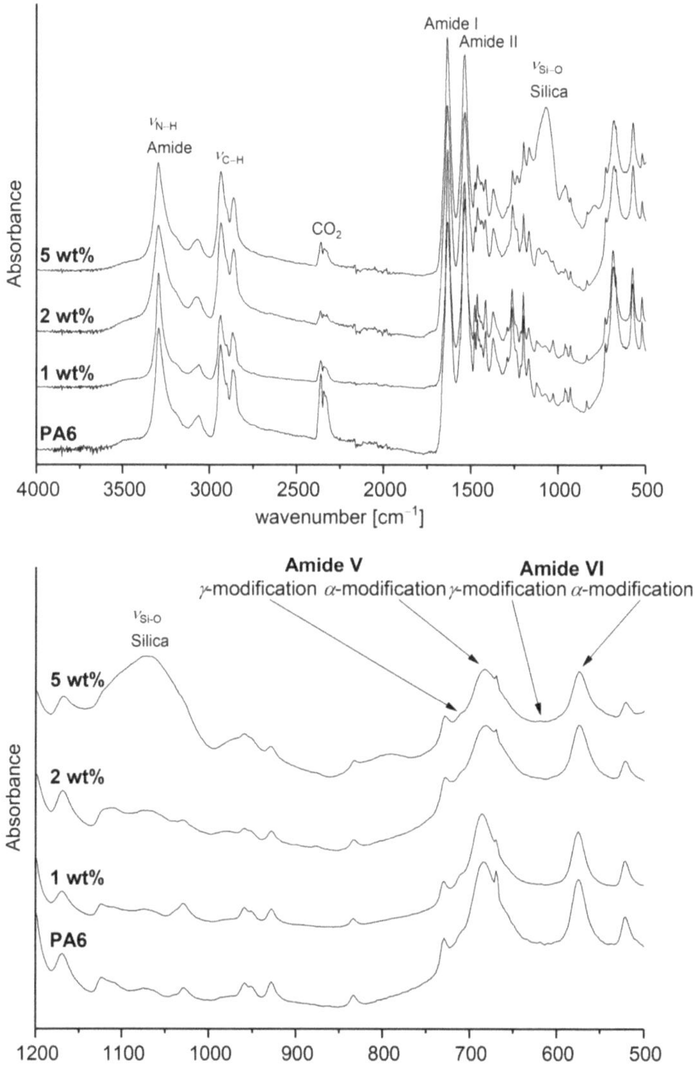

Fig. 4.16: FTIR spectra of polyamide 6/SiO_2 composites with different filler contents compared to PA6. Adapted from [49]. Published by The Royal Society of Chemistry.

content. Higher amounts of silica lead to a decrease in the decomposition temperature, probably due to agglomeration of the filler particles [47, 49].
TEM images of thin sections of a sample with 2 wt% silica show primary particles with a size of 35–60 nm, which form larger agglomerates of several micrometers during processing in melt (Fig. 4.18).

The size of the agglomerates increases with higher filler contents. Furthermore, the molar ratios of reactants have an influence on the agglomeration behavior.

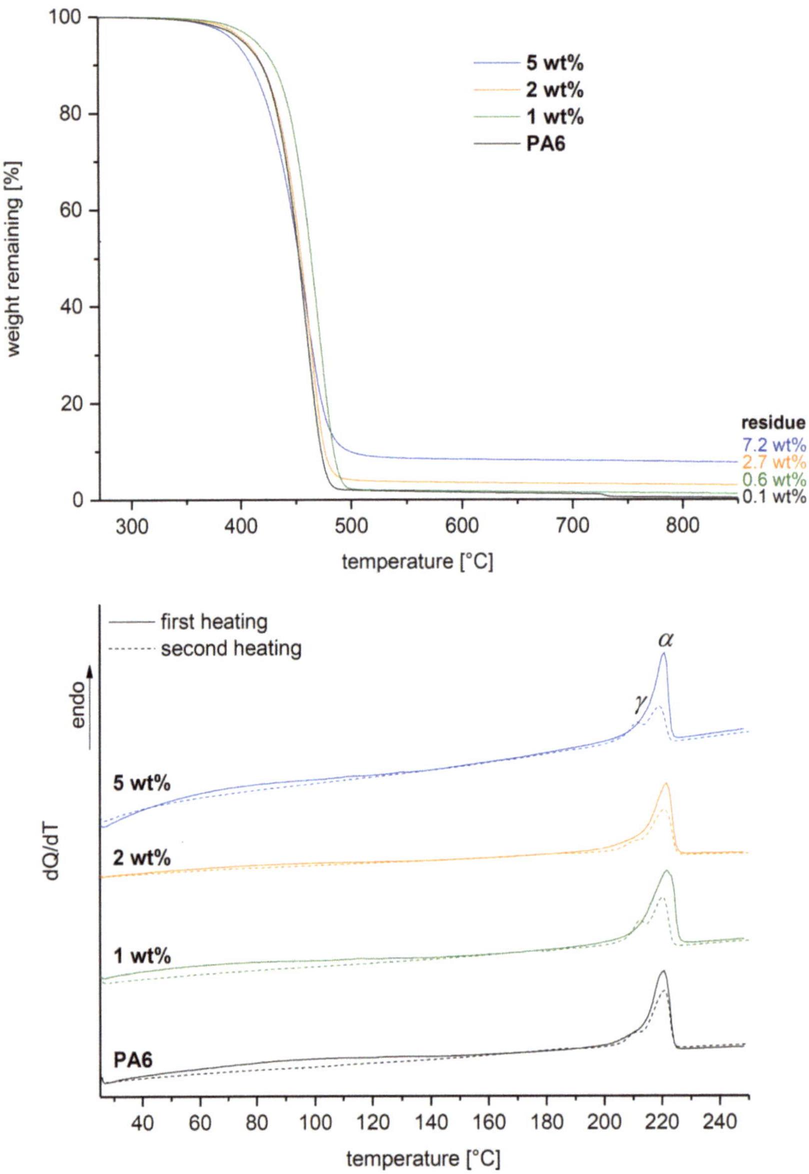

Fig. 4.17: TGA (top) and DSC curves (below) of polyamide 6/SiO_2 composites with different filler contents compared to PA6. Adapted from [49]. Published by The Royal Society of Chemistry.

Experiments with constant silica content show that an increasing amount of ε-ACA causes a decrease in particle size [47, 49].

The agglomeration of the formed silica particle fractions within the polyamide 6 composites is disadvantageous. The agglomeration is due to the high reaction

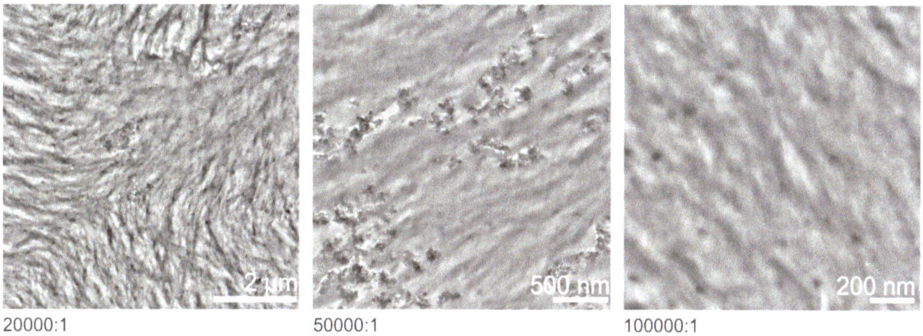

Fig. 4.18: TEM images of a monolithic sample with 2 wt% silica shown at different magnifications. Samples were prepared by ultra-thin sections [49]. Published by The Royal Society of Chemistry.

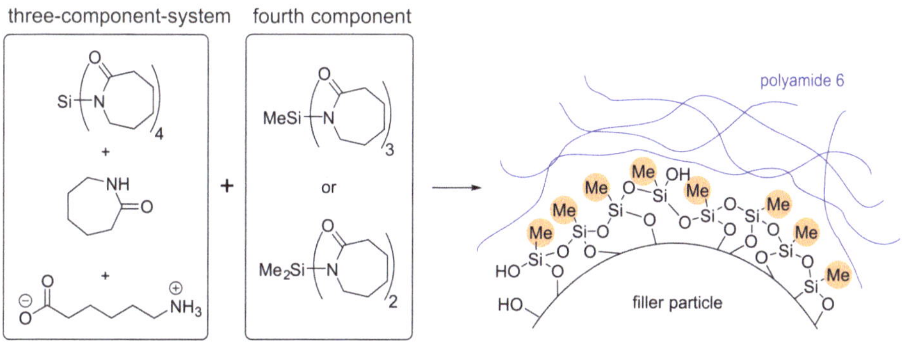

Fig. 4.19: Preparation of polyamide 6/SiO_2-polymethylsiloxane composites using a four-component system.

temperature above the melting point of PA6, which decreases the overall viscosity. Hence, it is driven by thermodynamics. One possibility to inhibit agglomeration of silica is the modification of the filler surface to increase hydrophobicity (decrease the surface energy of silica) and to enhance compatibility with the polymer matrix. For this reason, the methyl-substituted Si-monomers 1,1',1''-methylsilanetriyltris(azepan-2-one) and 1,1'-(dimethylsilanediyl)bis(azepan-2-one) with lactam moieties were added to the three-component system as shown in Figure 4.19 [50].

The incorporation of methyl-substituted silicon units into the silica network can be clearly proven by solid state NMR spectroscopy. ^{13}C NMR spectra show weak signals for methyl groups around 0 ppm along with the expected ^{13}C NMR signals for α-crystalline PA6 [50].

The ^{29}Si NMR spectra confirm the formation of the siloxane as well as silicatic hybrid structures. Q_3 and Q_4 signals are observed, resulting from a highly condensed silica network. Composites with polymethylsiloxane units deriving from $MeSi(\varepsilon\text{-CL})_3$ (Scheme 4.12) show mainly T_3 signals. D_2 signals for composites with PDMS units from $Me_2Si(\varepsilon\text{-CL})_2$ (Scheme 4.12) are found at –16 ppm. The observed chemical shift to lower field and broadening of the signal compared to pure PDMS at –22 ppm can be explained by the presence of short $(Me_2SiO)_n$ chain segments that are bonded to the silica surface. Thus, the formation of surface-grafted polydimethylsiloxane-silica particles is very likely [50].

Another approach to improve homogeneity of the composite materials is the formation of covalent bonds between the PA6 matrix and the formed silica particles. For this reason, 3-aminopropyltriethoxysilane (APTES) was used as fourth reaction component in order to achieve a surface functionalization of the silica filler *via* the triethoxysilane moiety as established in literature. Furthermore, the amino functionality of APTES can form amide bonds with carbonic acid moieties from PA6. Additionally, electrostatic interactions between carboxylate and ammonium units as well as hydrogen bonding can cause a better compatibility between the two phases (see Fig. 4.20) [50].

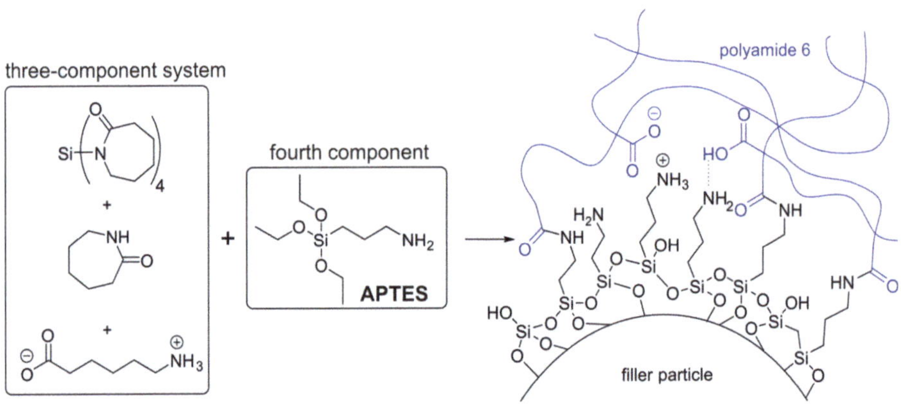

Fig. 4.20: Preparation of polyamide 6/SiO_2-poly(3-aminopropyl)siloxane composites by a four-component system.

Compared to the PA6 composite obtained from the corresponding three-component system (Fig. 4.20), no significant changes in FTIR and solid state ^{13}C NMR spectra, attributed to the addition of APTES, were detected for these four-component systems due to the low content of that compound. As expected, solid state ^{29}Si NMR spectra show the additional T_3 signal for the aminosilane moiety of APTES along with Q_3 and Q_4 signals, indicating the formation of a condensed silica network. The formation of covalent bonds between the organic and inorganic phase could be proven indirectly by extraction experiments using 2,2,2-trifluoroethanol and subsequent investigation of the residues by FTIR spectroscopy and gravimetric analysis [50].

Thermal properties have been studied by TGA and DSC measurements. All composites with APTES as additive show an enhanced thermal stability compared to the three-component system. For a system with 2 wt% filler content and a molar reactant ratio of 4:1 [here, ratio of Si(ε-CL)$_4$ to APTES], the temperature of the highest mass loss (T_D) increased to 487 °C compared to 455 °C if Si(ε-CL)$_4$ is used alone. DSC measurements indicate a higher proportion of the γ-modification in the composite if APTES is added [50]. SEM images show a homogeneous distribution of the *in situ* formed silica filler particles for APTES proportions of 0.5 mol% with regard to the overall amount of used silane species including Si(ε-CL)$_4$. Lower amounts of APTES are unsuitable, as formation of silica particle agglomerates with sizes comparable to the three-component system is obtained [50].

4.1.11 Summary

Siliceous twin monomers possess an enormous potential for the production of a variety of polymeric organic/inorganic nanostructured hybrid materials. The basic structures of 2,2'-spirobi[4*H*-1,3,2-benzodioxasiline] and tetrafurfuryloxysilane are the most frequently used twin monomer frames for the design of twin monomer libraries. The advantage of this modular concept is that commercially available alkoxysilane compounds can be used as starting materials in combination with benzyl alcohol derivatives as second component to achieve various twin monomers. Especially, the STP of two or more different twin monomers or other common monomers, which contain the same or closely related molecular structure of the polymerizable fragment, has proven to be a particularly successful synthetic strategy. With the help of this concept, nanostructured hybrid material foams and ternary hybrid materials of desired molecular composition and morphology control are readily available. These ternary hybrid materials are useful as precursor for micro- and mesoporous silica and carbon materials.

The robust twin polymerization process of Si-Spiro is also structure-directing for other purposes such as the production of microporous carbon composites (Chapter 4.2), mixed oxides (Chapter 4.3), and metal particles generation within nanostructured carbon matrices (Chapter 4.4). The use of amino-functionalized twin monomers is particularly advantageous because twin polymerization can be applied to improve polyurethane structures and that of related polymer materials which is of importance for various types of applications. This feature will be demonstrated in Chapter 5.2.

The use of twin monomers which contain Si–N bonds is possible for the production of silicon-nitride-related hybrid materials. The appropriate twin monomers are available from the corresponding chlorosilane compounds and *o*-aminobenzylamine or aniline. Their synthesis requires more effort due to the hydrolytic sensitivity of these monomer classes. Similar types of monomers based on lactams are suitable for the fabrication of polyamide 6/silica composite materials.

Thomas Ebert

4.2 Carbon materials

Carbon naturally occurs in its crystalline forms as diamond or as graphite. Besides, amorphous carbons also exist which consist of variable ratios of sp^2- and sp^3-hybridized carbon atoms. Furthermore, carbon nanotubes (CNT), fullerenes, and graphene can be synthesized. These different forms show different properties which lead to their use in very different applications in many fields [51–59]. Coal is formed in nature by the carbonization of plant matter under elevated temperature and pressure. The technological production of carbon from biomass is one of the oldest chemical processes. In its course, the carbon precursors are heated to higher temperature in oxygen-free atmosphere, which leads to the removal of small molecules like water from the precursors and thus increases the carbon content of the material. Cellulose, which has been used as carbon precursor for ages, has in many cases been replaced by synthetic organic polymers, which induce a better processability of the precursors before the pyrolysis step. Good carbon precursors show a dimensional stability and a high carbon yield upon pyrolysis. Among others, the use of pitch [60], or organic polymers like phenolic resin [35], polyfurfuryl alcohol [35], and polyacrylonitrile [61] have been successfully demonstrated. Typically, the annealing takes place under inert atmosphere at temperatures of around 800 °C. Depending on the temperature and time of the pyrolysis step, the carbon content of the final materials can be adjusted. Longer carbonization times and higher temperatures lead to higher carbon contents. Furthermore, many approaches focus on a more sustainable hydrothermal production of carbon materials from biomass. This process is typically carried out in a closed reaction vessel in water at comparatively low temperatures of less than 200 °C under self-generated pressure [62, 63].

4.2.1 Synthesis of porous carbon materials

Carbon materials show a chemical inertness, mechanical stability, and high conductivity which makes them ideal candidates for many applications in various fields. Furthermore, they are easily available and can be synthesized with tunable properties, e.g., high surface area and adjustable porosity. Carbon materials are used in different energy storage and conversion techniques, as catalyst supports, as adsorbents, and in filtration or separation processes.

For the synthesis of porous carbon materials, different strategies can be applied. Besides physical or chemical activation strategies, especially templating methods are of great importance. In this process, a sacrificial component is intermixed with a carbon precursor and, after a pyrolysis step, is removed. In a typical example, a silica

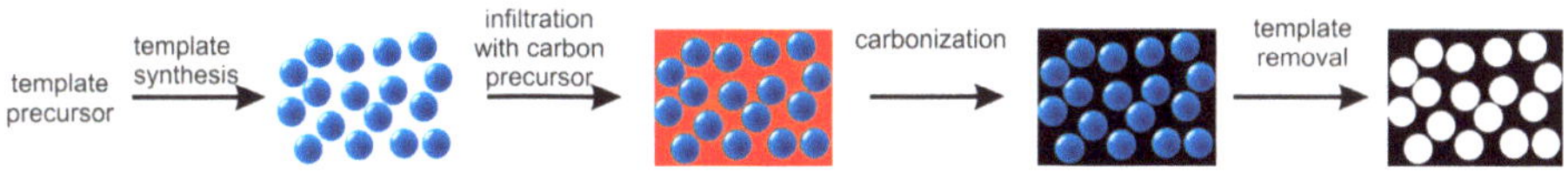

Fig. 4.21: Schematic depiction of the synthesis of well-defined porous carbon materials by infiltrating a previously synthesized template with a carbon precursor and subsequent carbonization and template removal [27]. Reproduced and modified with permission from T. Ebert.

template is synthesized in a first step and infiltrated with a carbon precursor. After a subsequent pyrolysis step, the silica template is removed leading to pores which are replica of the silica templates (see Fig. 4.21).

The synthesis of ordered mesoporous silica molecular sieves using molecular precursors opened up a synthetic path for the synthesis of numerous well-defined silica particles [37, 64]. This shows the potential of this templating approach for the production of porous carbon materials with tailor-made morphology and porosity.

An important milestone was the synthesis of highly ordered mesoporous carbon material CMK-1 (carbon material of Korea-1) which were obtained by carbonizing sucrose in the pores of a silica molecular sieve [65]. After finding the optimal hydrothermal carbon replication conditions for mesoporous silica, a replica of an ordered SBA-15 (Santa Barbara Amorphous-15) material has also been successfully produced [66].

4.2.2 Carbon from hybrid materials synthesized by twin polymerization

As shown in the previous chapters, twin polymerization offers the chance to synthesize organic/inorganic hybrid materials with phase domains in the range of a few nanometers. In many cases, the well-known carbon precursors phenolic resin and polyfurfuryl alcohol are obtained as organic polymers. Therefore, the hybrid materials can easily be transformed into nanoporous carbon materials. The synthesis can be performed in three steps: (I) synthesis of the hybrid material, (II) carbonization, and (III) the removal of the inorganic phase domains (see Fig. 4.22).

During the pyrolysis process, the interpenetrating inorganic networks can stabilize the morphology and minimize the shrinkage [68]. The twin monomer 2,2'-spirobi[*4H*-1,3,2-benzodioxasiline] (Si-Spiro) has been shown to be the most versatile and widely used. Its polymerization yields a hybrid material of phenolic resin and silica with silica phase domains in the order of 2 nm as shown in the HAADF-STEM images in Figure 4.22. The size of the inorganic phase domains does not change significantly during the carbonization step as seen in Figure 4.22. Consequently, the conversion of the hybrid material to porous carbon according to the scheme above mainly yields microporous carbon materials. As Si-Spiro can be polymerized in different ways,

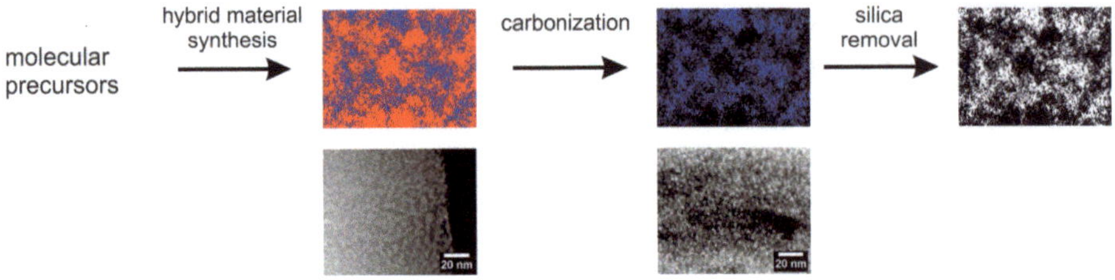

Fig. 4.22: Schematic depiction of the conversion of hybrid materials into porous carbon by carbonization and removal of the inorganic phase domains and corresponding HAADF-STEM micrographs of the phenolic resin/SiO_2 hybrid material and of the C/SiO_2 composite material [27, 67]. Reproduced and modified with permission of T. Ebert.

carbon materials with different morphologies can be synthesized. The polymerization in melt yields a monolithic hybrid material, which can be converted into a porous carbon monolith. Using precipitation polymerization, spherical hybrid material particles can be synthesized which can be converted into porous carbon particles. Furthermore, the twin polymerization on different surfaces can lead to porous carbon composites and, by using the substrates as templates, can lead to additionally structured porous carbon materials (see Chapters 4.2.3 and 4.2.4). Examples of the different accessible carbon materials are shown in Figure 4.23.

Additionally, the size of the phase domains in the hybrid material can be adjusted using the simultaneous twin polymerization (STP) of Si-Spiro together with other twin monomers (see Chapter 2 and Chapter 4.1). This already indicates the high potential for the carbon material synthesis. The simultaneous twin polymerization of Si-Spiro and 2,2-dimethyl-4*H*-1,3,2-benzodioxasiline leads to a hybrid material composed of phenolic resin as organic part along with polydimethylsiloxane and silica as inorganic components which form copolymers (see Fig. 4.24 (a)) [17]. By variation of the ratio of the two monomers, the size of the inorganic phase domains in the hybrid material can be controlled. It has been shown that an increasing content

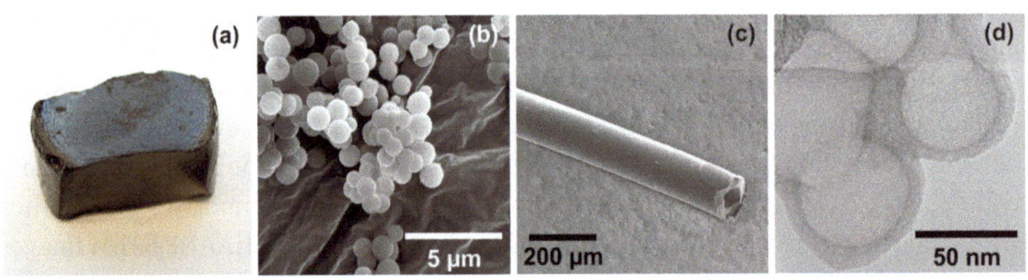

Fig. 4.23: (a) Photographic image of a carbon monolith [15], (b) SEM image of carbon particles, (c) carbon tubes [31], and (d) TEM image of hollow carbon spheres [30,31]; modified. (a) and (d) Copyright John Wiley & Sons, Inc. Reproduced with permission. (c) Reproduced with kind permission of F. Böttger-Hiller [31].

of 2,2-dimethyl-4*H*-1,3,2-benzodioxasiline leads to larger inorganic phase domains. After carbonization and removal of the inorganic part, porous carbon materials reflecting the size of the inorganic phase domains are obtained. Thus, an increasing content of 2,2-dimethyl-4*H*-1,3,2-benzodioxasiline in the STP leads to larger pore sizes, lower micropore content, and thus lower specific surface areas of the final carbon materials [28, 41]. Figure 4.24 (b) shows the nitrogen sorption isotherms of the porous carbons obtained from the hybrid materials synthesized by STP of various ratios of 2,2-dimethyl-4*H*-1,3,2-benzodioxasiline and Si-Spiro. All isotherms show a strong rise at low values of relative pressure, which is associated with the presence of micropores. The stronger the rise at the beginning, the higher the amount of micropores that can be found in the system. Only the isotherm of the carbon from the hybrid material obtained by simultaneous twin polymerization with the highest content of 2,2-dimethyl-4*H*-1,3,2-benzodioxasiline shows a pronounced hysteresis. It should be noted that with an increasing fraction of Si-Spiro during the polymerization, the sorption isotherms of the final porous carbon materials look gradually more like a typical type I isotherm according to IUPAC [69] and thus the carbon can increasingly be described as a microporous solid. This is reflected in the calculated pore size distribution shown in Figure 4.24 (c) and clearly shows the strong influence

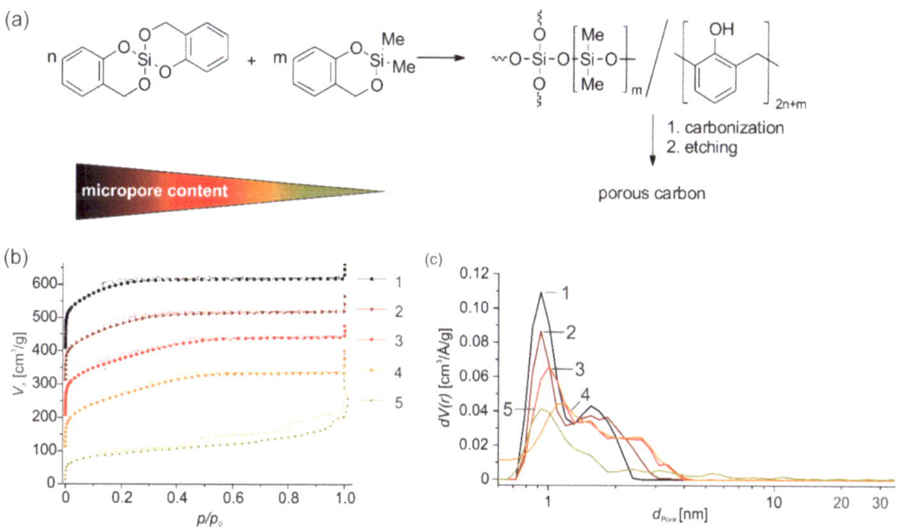

Fig. 4.24: Schematic description of the simultaneous twin polymerization of 2,2-dimethyl-4*H*-1,3,2-benzodioxasiline and 2,2'-spirobi[4*H*-1,3,2-benzodioxasiline] and the influence of the monomer ratio on the porosity of the final (a) carbon material, (b) nitrogen sorption isotherms (Y-offset 100 $cm^3 \cdot g^{-1}$), and (c) pore size distribution (QSDFT method, adsorption branch, slit pores) of the porous carbon obtained from the hybrid materials synthesized by STP with 10 mol% (5), 25 mol% (4), 50 mol% (3), 75 mol% (2), and 90 mol% (1) 2,2'-spirobi[4*H*-1,3,2-benzodioxasiline] after carbonization and removal of the silica using hydrofluoric acid; modified and reproduced with permission of T. Mark [28].

of the size of the phase domains in the hybrid material on the porosity of the resulting porous carbons.

It has been shown that a variation in the substitution pattern of the salicyl alcohol fragment of twin monomers is another way to influence the pore size distribution and specific surface area of the porous carbon materials which are obtained by carbonization of the hybrid materials and subsequent etching of the silica using hydrofluoric acid. The introduction of substituents, which facilitate the thermally induced twin polymerization of the respective twin monomers, leads to porous carbons with higher specific surface areas. Nevertheless, the hybrid material obtained by the polymerization of unsubstituted Si-Spiro can be converted to porous carbon with the highest specific surface area. This is attributed to a possible stronger crosslinking due to unsubstituted *ortho*- and *para*-positions of the salicyl alcohol [70].

When tetrafurfuryloxysilane is used as twin monomer, the resulting polyfurfuryl alcohol/silica hybrid materials can be converted into porous carbon in the same ways. However, the monomer is not as widely studied since its polymerization only occurs when catalyzed by acids and water is released during the polymerization process. Compared to the carbon from the phenolic resin/silica hybrid materials, a higher amount of mesopores is found in the tetrafurfuryloxysilane-based carbon materials. Besides the variation of the pore size distribution and specific surface area, it is also possible to introduce heteroatoms in the porous carbon materials. For this purpose, the polymerization of twin monomers based on the heteroaromatic organic monomer fragments thiophene [31, 39] and pyrrole [71] has been described (see Scheme 4.15).

Tetra(thiophene-2-ylmethoxy)silane has been used in twin polymerization for the synthesis of a poly(2-thiophenemethanol)/SiO_2 hybrid material which can be converted into porous carbon. After pyrolysis at 800 °C for 3 h and etching of the silica using hydrofluoric acid, the final porous carbon shows a specific surface area of about 790 $m^2 \cdot g^{-1}$ and a sulfur content of 15.5%. If the temperature of the carbonization process is increased to 1100 °C, the sulfur content is reduced to only 2–3% [39].

In a similar way, pyrrole-based twin monomers can be converted into nitrogen-doped porous carbon materials. For this purpose, tetra[(*N*-methylpyrrol)methoxy]-silane and tetra(pyrrolmethoxy)silane have been synthesized. These twin monomers were polymerized under different conditions; such as in a STP with Si-Spiro. The obtained hybrid materials were carbonized at 800 °C for 15 min and the silica was removed using 5 M sodium hydroxide solution. The carbon materials from the hybrid materials synthesized from the pyrrole-based monomers showed a nitrogen content of merely around 8–9%.

4.2.3 Porous carbon composites by twin polymerization on surfaces

As shown in the previous chapter, Si-Spiro can be polymerized under different conditions. Besides a thermally induced polymerization, the use of acid or base catalysts

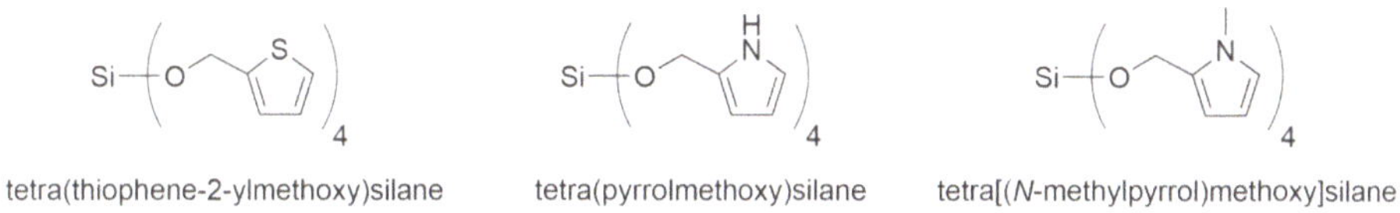

Scheme 4.15: Molecular structure of twin monomers which can be polymerized to hybrid materials that can be converted into heteroatom-doped carbon materials.

has been described [15, 16, 32]. This versatile polymerization behavior makes it an ideal candidate for the polymerization on different surfaces. The hybrid-material-coated substrate particles can be converted into porous carbon according to the method described in Chapter 4.2.2. The stability of the substrate in this process is decisive for the final material obtained. If the substrate particles are stable during the pyrolysis and etching steps, carbon composite materials can be obtained (see Fig. 4.25). It is crucial for the purity of the final material that the twin polymerization mainly occurs on the substrate surface. If the polymerization also occurs in solution, there will additionally be porous carbon spheres besides porous carbon composites in the final product.

This procedure has first been described for the DABCO-catalyzed surface twin polymerization of Si-Spiro on graphite particles. Using this approach by a simple slurry polymerization, graphite can be coated with a phenolic resin/silica hybrid material. Surprisingly, almost no spherical particles originating from a precipitation polymerization were found in the final material. Therefore, the variation of the ratio of substrate and monomer directly influences the thickness of the hybrid material layer (see Fig. 4.26). In all cases, the size of the phase domains of the hybrid materials is similar; and furthermore, the thickness of the hybrid material coating remains unaltered during the pyrolysis and etching process [32]. Therefore, the pore sizes and thus the shape of the sorption isotherms are expected to be similar. It has also been demonstrated that the silica can be removed from these C/SiO_2 coatings using a 5 M sodium hydroxide solution. In this way, porous carbon/graphite composites

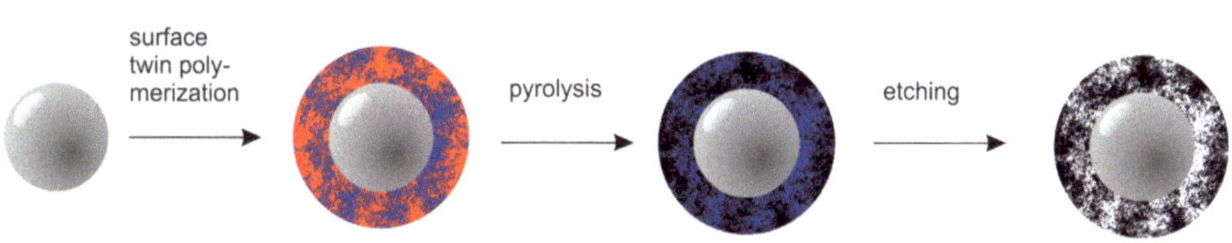

Fig. 4.25: Schematic description of the synthesis of porous carbon composites using surface twin polymerization and subsequent pyrolysis of the hybrid material and removal of the inorganic component [27]. Reproduced and modified with permission of T. Ebert.

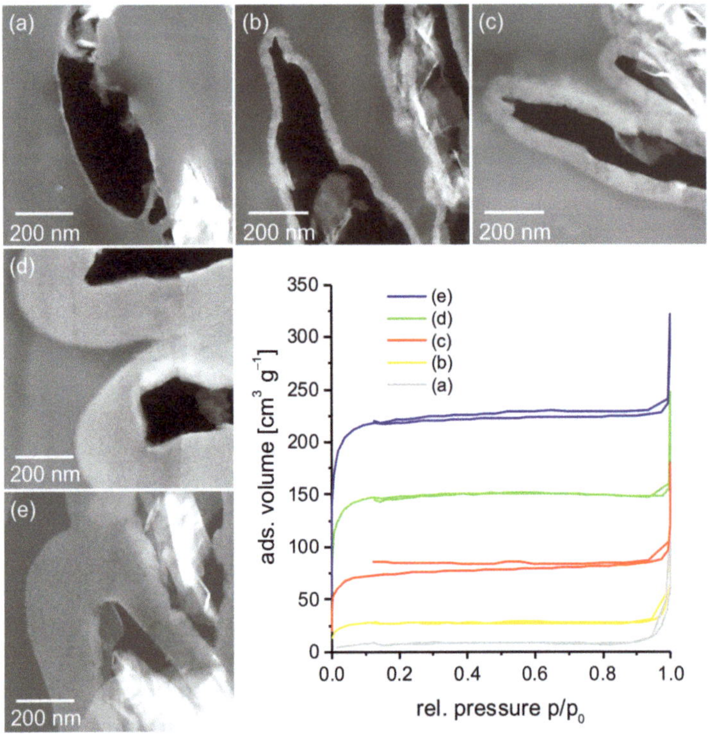

Fig. 4.26: HAADF-STEM images of the hybrid materials obtained by DABCO-catalyzed twin polymerization of (a) 10 wt%, (b) 30 wt%, (c) 50 wt%, (d) 70 wt%, (e) 91 wt% Si-Spiro on graphite particles and nitrogen sorption isotherms of the resulting carbon/graphite composites. Reproduced and modified from Ref. [32] with permission of The Royal Society of Chemistry.

are accessible which show a specific surface area depending on the thickness of the coating. The nitrogen sorption isotherms prove that all materials are microporous. The specific BET surface areas range between 25 $m^2 \cdot g^{-1}$ and 900 $m^2 \cdot g^{-1}$, depending on the amount of monomer used for the coating.

In a similar way, it is possible to coat carbon black [33] and multi-walled carbon nanotubes (MWCNT). Due to the low accessible MWCNT content of the dispersions and the requirement of a minimum monomer concentration, the variation of the coating thickness was only possible in a small range. The coating of commercial graphene was also realized in a similar way. Owing to acidic surface groups of the graphene substrate, no additional catalyst was required. Furthermore, graphene can also be coated by monomers that can only be polymerized using acid catalysts. Consequently, besides the coating of graphene with Si-Spiro, tetrafurfuryloxysilane was polymerized on its surface. These hybrid materials can be converted into porous carbon/carbon/silica composites by annealing in inert atmosphere. After removal of the silica, a range of different porous carbon/carbon composites are accessible. The specific surface area of these composites can be adjusted by the ratio of substrate:monomer [27]. Therefore, surface twin polymerization on different stable substrates offers a new synthetic route towards tailor-made carbon composite materials.

4.2.4 Structured porous carbon materials by twin polymerization on hard templates

In the next paragraphs, it will be shown that it is also possible to combine the twin polymerization with the templating effect of hard templates for the synthesis of structured porous carbon materials. It has also been reported that structured porous carbon materials can be obtained using chemical foaming in combination with twin polymerization. This was achieved by the polymerization of carbonates under the release of CO_2 with a simultaneously occurring twin polymerization [43]. The resulting hybrid material foam can be converted into porous carbon by pyrolysis followed by an etching step. While this elegant approach shows that the morphology and porosity of the final carbon materials can be influenced, a precise control of the shape or the synthesis of well-defined particles is not possible. The hard templating approach offers an additional way of precisely structuring porous carbon materials. The porous carbon tube shown in Figure 4.23 (c) was prepared in the same way. The general strategy is equal to the synthesis of porous carbon composites; only the substrate is removed either during carbonization or during etching of the silica (Fig. 4.27). It is also possible that the substrate from a porous carbon composite is removed in an additional step, but this has not been described so far.

As in the previous chapter, it is important to note that the polymerization mainly occurs on the surface of the template particles. Therefore, different strategies have been developed to modify the template particles in a way that they act as catalyst for

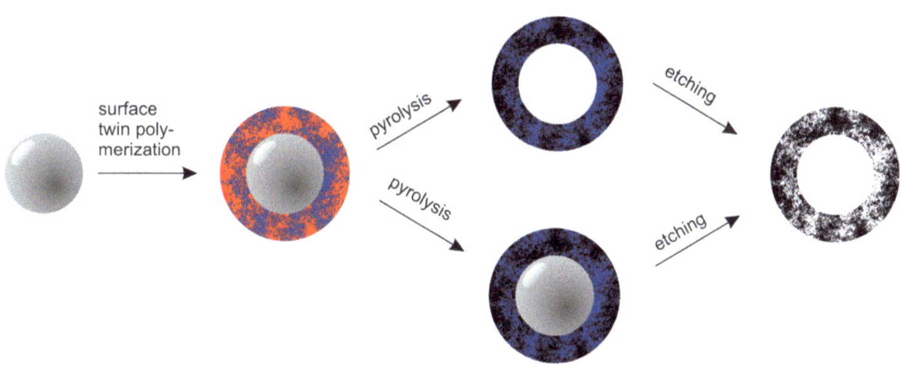

Fig. 4.27: Schematic description of the synthesis of porous hollow carbon spheres by surface twin polymerization and subsequent pyrolysis of the hybrid material followed by removal of the inorganic component [27]. Reproduced and modified with permission of T. Ebert.

the twin polymerization. The cationic polymerization has been shown to be a good option for the surface modification of silica surfaces in general [72]. Consequently, one approach uses the adsorption of methanesulfonic acid on silica substrates. These modified silica surfaces can act as catalyst for the surface twin polymerization of different twin monomers. Using this strategy, various commercially available silica particles, such as Aerosil and LiChrospher particles, were coated with hybrid materials in a slurry polymerization [30, 33, 39, 73]. After pyrolysis, the silica of the coating as well as the silica template were removed in one reaction step to yield porous hollow carbon spheres. These hollow carbon spheres showed high specific surface areas of 690–1370 $m^2 \cdot g^{-1}$ and large total pore volumes of 0.49–2.33 $cm^3 \cdot g^{-1}$. The shape of the porous hollow carbon spheres directly depends on the substrate used. When Aerosil particles with 7 nm average diameter were used, the final porous carbon materials showed a foam-like morphology; whereas when 40 nm Aerosil particle substrates were utilized, individual hollow spheres were observed. When the micrometer sized LiChrospher 100 particles were used as substrates, smaller carbon particles were found besides the hollow carbon spheres, resulting from the precipitation polymerization of the respective monomers (see Fig. 4.28). The obtained porous carbon materials were mixed with sulfur and investigated as cathode material in lithium-sulfur batteries (see Chapter 5.1) [30, 31, 74]. It has also been shown that by using a sulfur-containing twin monomer, porous hollow carbon spheres with a sulfur content of 2.3–16.5 % can be synthesized [39]. As mentioned earlier in this chapter, the sulfur content of these materials largely depends on the carbonization temperature. When the samples

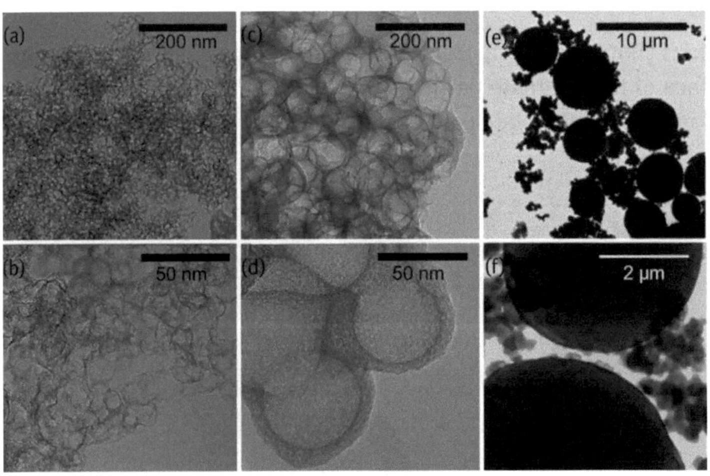

Fig. 4.28: TEM images of hollow carbon spheres obtained by surface twin polymerization of Si-Spiro, subsequent annealing, and removal of the silica obtained from the polymerization process and the silica substrate. Aerosil 300 (a,b), Aerosil OX50 (c,d), and LiChrospher 100 (e,f) particles have been used as templates for the hollow sphere synthesis [30]. Copyright John Wiley & Sons, Inc. Reproduced with permission.

are carbonized at 800 °C, a sulfur content of 11.5–16.5 % is found. Whereas a carbonization at 1100 °C, as expected, leads to a much lower sulfur content of only 2.3 %.

To synthesize more uniform hollow carbon spheres, the substrates need a very narrow size distribution. This was realized after using Stöber particles as substrates instead of commercially available silica [33]. In another approach, monodisperse polystyrene particles have been used. The surfaces of these particles could successfully be functionalized by covalently bonded sulfonic acid groups in presence of sulfuric acid as reactant. Therefore, the particles themselves act as catalyst for the twin polymerization which reduces the precipitation polymerization as side reaction. Figure 4.29 shows the SEM images of polystyrene particles coated by surface twin polymerization of (b) 40 wt% and (c) 50 wt% Si-Spiro. After the carbonization, it was found that the lower amount of hybrid material coating is not sufficient to withstand the forces caused by the volume shrinkage during the depolymerization of polystyrene when heating the hybrid material above the ceiling temperature of polystyrene. Therefore, shrinkage of the particle shell is observed which leads to a wrinkled morphology. For the particles with a higher coating thickness, this change in morphology is not observed which leads to the conclusion that the thickness of the hybrid material shell is high enough to withstand the forces during the depolymerization of polystyrene in the carbonization processes. Figure 4.29 (d) and (e) show the porous carbon particles obtained after carbonization and HF etching of the silica. The removal of the SiO_2 shows no influence on the particle morphology.

Fig. 4.29: (a) SEM images of uncoated sulfonated polystyrene template particles, (b) particles coated with phenolic resin/silica hybrid material from the polymerization of 40 wt%, and (c) 50 wt% Si-Spiro compared to the porous hollow carbon spheres thereof (d,e) obtained by carbonization and etching of the silica with hydrofluoric acid [27]. Reproduced and modified with permission of T. Ebert.

Besides this difference in morphology, the nitrogen sorption isotherms of the hollow carbon spheres are almost identical. The wrinkled hollow carbon spheres show a slightly higher specific surface area and micropore content (1389 $m^2 \cdot g^{-1}$ BET surface area, 29.7% micropores) compared with the spherical hollow carbon spheres (1262 $m^2 \cdot g^{-1}$ BET surface area, 23.6 % micropores), whereas the pore volume of both samples show the same value of 1.16 $cm^3 \cdot g^{-1}$. Both porous carbons were investigated as components of cathodes in lithium-sulfur batteries (see Chapter 5.1).

4.2.5 Strategies for the synthesis of hierarchically porous carbon materials by consecutive twin polymerizations on hard templates

The high potential of carbon materials for different applications has motivated the search for synthesis routes towards tailor-made materials. Therefore, the combination of micro- and mesoporous carbons for the synthesis of hierarchically structured materials has also been studied. Mesopores contribute to higher pore volumes of the final material, whereas micropores lead to carbon materials with high specific surface areas. To achieve this goal, two different synthetic approaches were investigated. First, mesoporous carbon black particles have been used as substrate for the surface twin polymerization of Si-Spiro. The hybrid material coating was then converted into porous carbon to yield mesoporous carbon black particles with a microporous carbon coating (see Fig. 4.30). It has been found that using this method, the mesopores of the carbon black particles are partly filled with hybrid

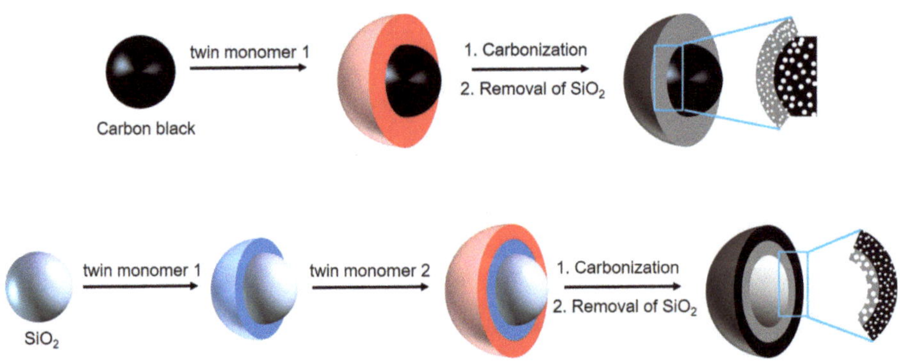

Fig. 4.30: Schematic depiction of the formation of hierarchically structured carbon/carbon nanocomposites with adjustable porosity by surface twin polymerization on a porous carbon substrate or subsequent twin polymerization of different twin monomers onto a silica template. In both cases, the porous carbons are obtained by carbonization of the hybrid materials and HF etching of the silica phase domains. The carbon black substrate is not altered/inert throughout the process. Reprinted from [33] with permission from Elsevier.

material during the polymerization process. Therefore, the mesopores are not fully accessible for later application. In a second approach, silica particles were subsequently coated with two different twin monomers. These hybrid material coatings were then converted into porous carbon. The monomers were chosen with respect to the porosity of the carbon material accessible after the carbonization and etching of the silica. Tetrafurfuryloxysilane was polymerized as a first layer and an additional layer was obtained by the subsequent polymerization of Si-Spiro. Therefore, hollow carbon spheres with a mesoporous and microporous shell are accessible (see Fig. 4.30).

It becomes obvious that the concept of twin polymerization offers a broad range of possibilities for the synthesis of carbon materials. Besides the adjustment of the specific surface area, the pore volumes, and diameters, it is also possible to introduce heteroatoms into the carbon. Additionally structured porous carbon materials are accessible using external templates and also combining different strategies to obtain tailor-made carbon materials in a modular approach.

Philipp Kitschke, Michael Mehring

4.3 Inorganic oxides from twin polymerization

Twin polymerization provides simple synthesis protocols to obtain a whole variety of porous inorganic oxides and mixtures of porous inorganic oxides such as B_2O_3 [13, 75], GeO_2 [76, 77], SnO_2 [78], TiO_2 [79, 80], and WO_3 [81], as well as In_2O_3/SiO_2 [82], SnO_2/SiO_2 [78], and ZrO_2/SiO_2 [83], respectively. In general, three conceptual protocols have been developed for the synthesis of (i) inorganic oxides, (ii) mixtures of inorganic oxides, and (iii) mixed oxides that exhibit large BET surface areas using the process of twin polymerization. These synthesis concepts are discriminated by the type of twin polymerization, e.g., twin polymerization of single types of precursors, simultaneous twin polymerization of at least two different types of precursors, or twin polymerization of precursors with auxiliary additives, which was particularly applied to synthesize nanocomposites or rather hybrid materials on demand to finally obtain nanostructured porous inorganic oxides. Notably, the oxide materials are termed as porous oxides hereafter, because they exhibit large BET surface areas and their nitrogen physisorption isotherms are characteristic for highly porous materials (type I and type IV isotherms). However, the morphology of these oxide materials may differ. They can either be monoliths consisting of continuous oxide phases possessing a porous structure, porous materials consisting of crystallites exhibiting crystallite sizes down to the nanometer range, or an oxide material combining the latter features. Details about these synthesis concepts are outlined in the following section.

4.3.1 Synthesis of inorganic oxides and mixed inorganic oxides

Porous inorganic oxides can be synthesized starting from single source precursors that are suitable for twin polymerization. These precursors are based on metal and metalloid alkoxides providing organic/inorganic nanocomposites or rather hybrid materials consisting of an organic polymer and the respective metal oxide or metalloid oxide after twin polymerization. Oxidation of the as-obtained nanocomposites or rather hybrid materials produce porous inorganic oxides (Scheme 4.16) [13, 15, 70, 75–78, 82].

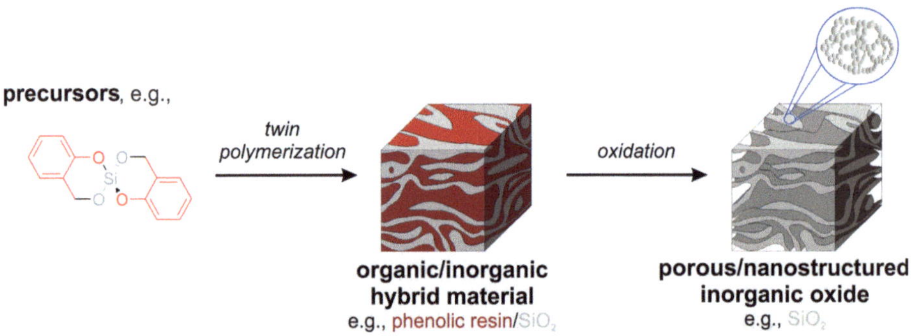

Scheme 4.16: Synthesis of porous inorganic oxides using the concept of twin polymerization. Please note that the exemplarily illustrated SiO_2 monolith consists of nanostructured SiO_2 phase domains forming a continuous and microporous phase. Inorganic oxides obtained from different precursors, e.g., $[Sn(OCH_2C_4H_3S)_3(\mu_2\text{-}OCH_2C_4H_3S)(HOCH_2C_4H_3S)]_2$, may give materials consisting of single inorganic oxide particles exhibiting sizes down to the nanometer range.

Various inorganic oxides, which in particular are B_2O_3 [13, 75], In_2O_3 [82], SiO_2 [13, 15, 70, 75], GeO_2 [76, 77], SnO_2 [78], TiO_2 [79, 80], ZrO_2 [83], HfO_2 [83], and WO_3 [81], were accessible using this general synthesis protocol. A summary of the porous inorganic oxides, the specific synthesis conditions that were applied, and their BET surface areas is presented in Table 4.5.

Mixtures of porous inorganic oxides (e.g., SnO_2/SiO_2 monoliths) and porous mixed oxides (e.g., $(In_2O_3)_{\approx 0.9}(SnO_2)_{\approx 0.1}$) are accessible using the general concept of twin polymerization. Two synthesis protocols that differ conceptually have been developed so far [78, 84]. First, simultaneous twin polymerization of at least two precursors being suitable for twin polymerization provides hybrid materials that can be converted into porous mixtures of inorganic oxides and porous mixed oxides by oxidation procedures (Scheme 4.17). Notably, according to the condition that the precursors consist of the same organic molecular moiety, it is assumed that the organic part combines to a single polymer, e.g., a phenolic resin. The two inorganic components usually do not mix to give a single compound, but rather form separate metal oxide phase domains

Tab. 4.5: Summary of porous inorganic oxides synthesis using the concept of twin polymerization. Step I refers to the twin polymerization and steps II and III describe the post-treatment process.

inorganic oxides	precursor	preparation	BET surface area[#]	reference
B_2O_3	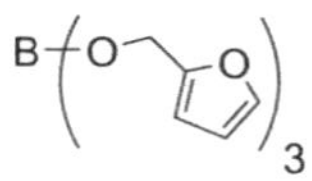	I. acid-catalyzed twin polymerization II. oxidation in air flow (200 L·h^{-1}) at 900 °C for 3 hours with a heating rate of 2.3 K·min^{-1}	20 m^2·g^{-1}	[75]
In_2O_3	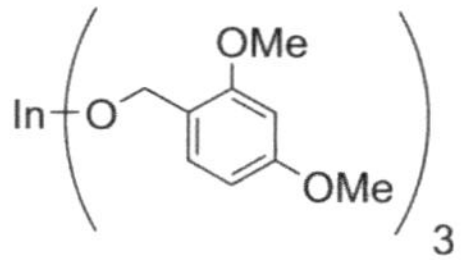	I. acid-catalyzed twin polymerization II. oxidation in air flow (200 L·h^{-1}) at 600 °C for 4 h with a heating rate of 2.3 K·min^{-1}	22 m^2·g^{-1}	[82]
SiO_2	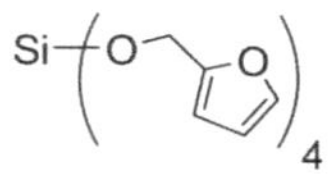	I. acid-catalyzed twin polymerization II. oxidation in air flow (200 L·h^{-1}) at 900 °C for 1 h with a heating rate of 2.3 K·min^{-1}	518 m^2·g^{-1}	[13, 75]
SiO_2	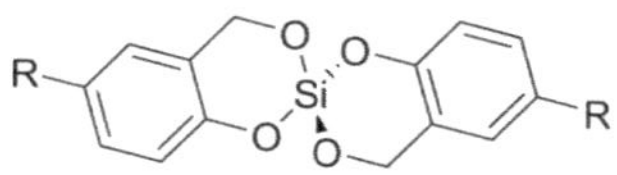 R = H, Me, Br, CN	I. acid-catalyzed/thermally induced twin polymerization II. oxidation in air flow (200 L·h^{-1}) at 900 °C for 3 h with a heating rate of 10 K·min^{-1}	889 m^2·g^{-1}	[15, 70]
GeO_2	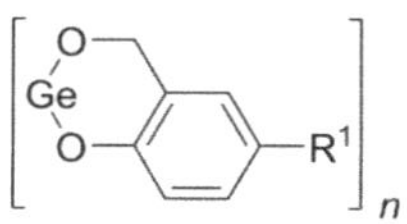 R^1 = H, Me, Br	I. thermally induced twin polymerization II. oxidation in air flow (200 L·h^{-1}) at 800 °C for 3 h with a heating rate of 10 K·min^{-1}	20 m^2·g^{-1}	[77]

(continued)

Tab. 4.5 (continued)

inorganic oxides	precursor	preparation	BET surface area#	reference
GeO_2	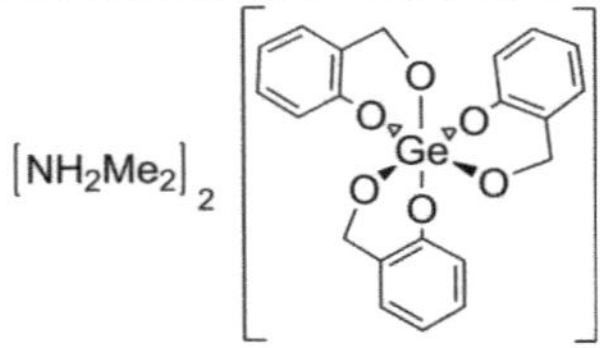	I. thermally induced twin polymerization II. oxidation in air flow (200 L·h^{-1}) at 800 °C for 3 h with a heating rate of 10 K·min^{-1}	27 $m^2 \cdot g^{-1}$	[76]
SnO_2	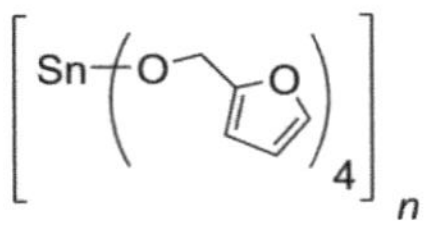	I. acid-catalyzed twin polymerization II. oxidation in air flow (200 L·h^{-1}) at 500 °C for 10 h with a heating rate of 10 K·min^{-1}	87 $m^2 \cdot g^{-1}$	[78]
SnO_2	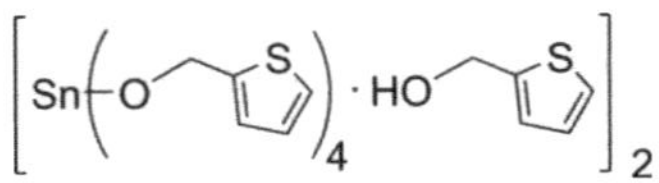	I. acid-catalyzed twin polymerization II. oxidation in air flow (200 L·h^{-1}) at 600 °C for 10 h with a heating rate of 10 K·min^{-1}	178 $m^2 \cdot g^{-1}$	[78]
SnO_2	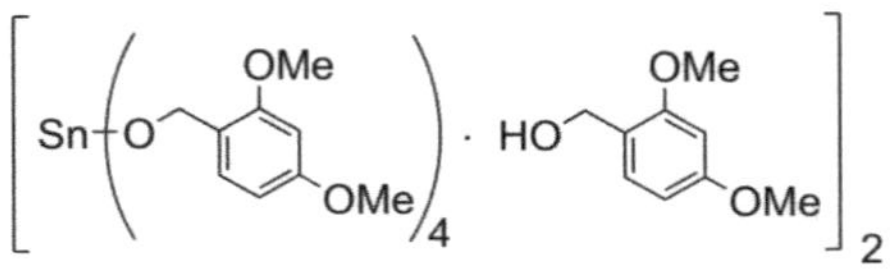	I. acid-catalyzed twin polymerization II. oxidation in air flow (200 L·h^{-1}) at 450 °C for 5 h with a heating rate of 10 K·min^{-1}	137 $m^2 \cdot g^{-1}$	[78]
SnO_2	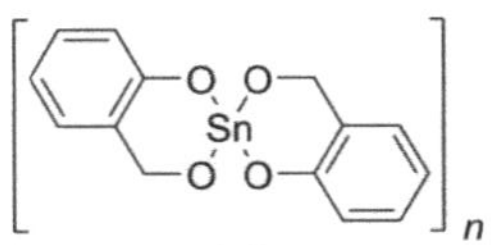 $n > 3$	I. acid-catalyzed twin polymerization II. oxidation in air flow (200 L·h^{-1}) at 600 °C for 1 h with a heating rate of 10 K·min^{-1}	143 $m^2 \cdot g^{-1}$	[78]

TiO_2	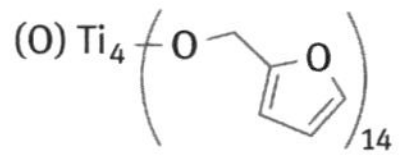	I. acid-catalyzed twin polymerization II. stirring of the hybrid material in water at 100 °C for 12 h III. oxidation in air flow (200 $L \cdot h^{-1}$) at 400 °C for 5 h with a heating rate of 2.1 $K \cdot min^{-1}$	109 $m^2 \cdot g^{-1}$	[79]
TiO_2	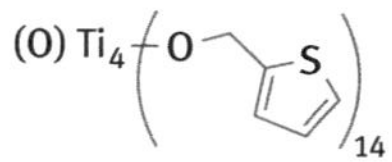		177 $m^2 \cdot g^{-1}$	[79]
ZrO_2	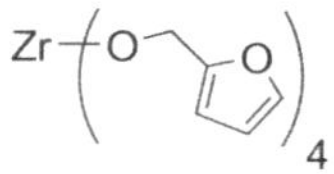	I. acid-catalyzed twin polymerization II. carbonization in argon flow (10 $L \cdot h^{-1}$) at 800 °C for 30 min with a heating rate of 50 $K \cdot min^{-1}$ III. oxidation in air flow (200 $L \cdot h^{-1}$) at 450 °C for 48 h with a heating rate of 50 $K \cdot min^{-1}$	100 $m^2 \cdot g^{-1}$	[83]
ZrO_2	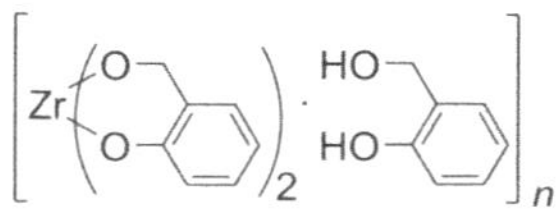		100 $m^2 \cdot g^{-1}$	[83]
HfO_2	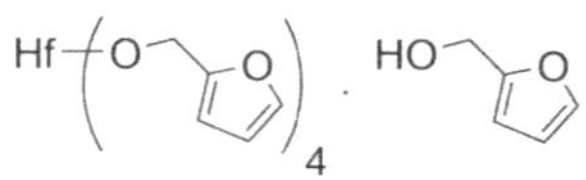	I. acid-catalyzed twin polymerization II. carbonization in argon flow (10 $L \cdot h^{-1}$) at 800 °C for 30 min with a heating rate of 50 $K \cdot min^{-1}$ III. oxidation in air flow (200 $L \cdot h^{-1}$) at 450 °C for 48 h with a heating rate of 50 $K \cdot min^{-1}$	100 $m^2 \cdot g^{-1}$	[83]
HfO_2	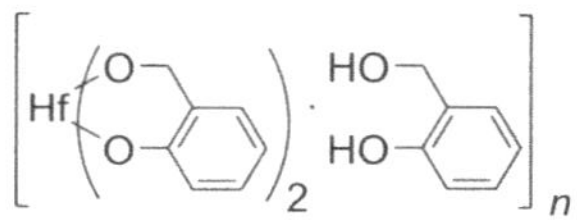		100 $m^2 \cdot g^{-1}$	[83]
WO_3	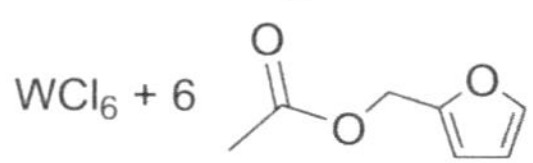	I. *in situ* twin polymerization II. oxidation in air flow (400 $L \cdot h^{-1}$) at 400 °C for 3 h with a heating rate of 2 $K \cdot min^{-1}$	56 $m^2 \cdot g^{-1}$	[81]
WO_3	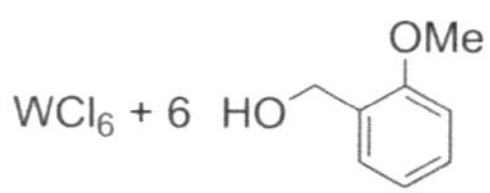		56 $m^2 \cdot g^{-1}$	[81]

Maximum value is given as published in the literature.

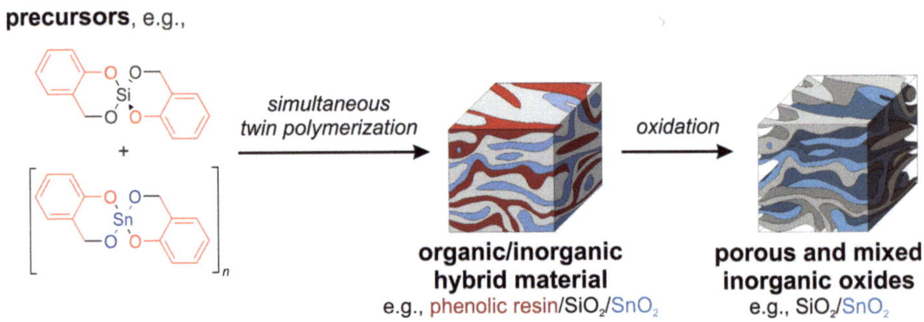

Scheme 4.17: Synthesis of porous and mixed inorganic oxides using the concept of simultaneous twin polymerization.

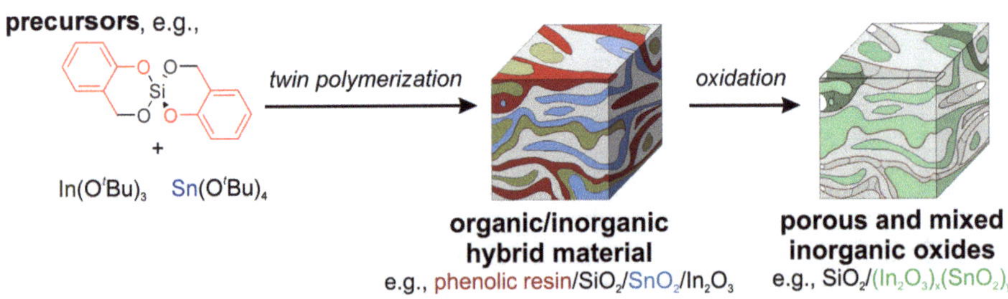

Scheme 4.18: Synthesis of porous and mixed inorganic oxides using the concept of twin polymerization with additives.

(Scheme 4.17). Following this synthesis concept, porous oxide mixtures consisting of either B_2O_3/SiO_2 [75], In_2O_3/SiO_2 [82], SnO_2/SiO_2 [78], ZrO_2/SiO_2 [83], or HfO_2/SiO_2 [83] and the porous mixed oxide $(In_2O_3)_{\approx 0.9}(SnO_2)_{\approx 0.1}$ [82] were synthesized. It is assumed that SiO_2 is well suited to provide a highly porous matrix, whereas other metal oxides do not.

Another approach to obtain mixtures of porous inorganic oxides and porous mixed oxides is given by twin polymerization of at least one single type of precursors being suitable for twin polymerization in the presence of appropriate additives such as metal alkoxides, metalloid alkoxides, and metal carboxylates [82, 84]. Here, precursors undergoing twin polymerization are typically derivatives of Si-Spiro, which easily polymerize and trap the auxiliary component during its polymerization process. Thus, hybrid materials are obtained, whose specific composition can be tuned by the choice of the twin monomers and additives as well as by the adjustment of the precursors' ratios. Oxidation of the organic/inorganic hybrid materials eventually provides porous inorganic oxide materials (Scheme 4.18). Generally, porous mixtures of In_2O_3/SiO_2 and $(In_2O_3)_{\approx 0.9}(SnO_2)_{\approx 0.1}/SiO_2$ materials were synthesized using the concept of twin polymerization with additives [82].

Tab. 4.6: Overview of porous and mixed inorganic oxides synthesized using the concept of simultaneous twin polymerization or twin polymerization with additives.

inorganic oxides	precursors	preparation	BET surface area#	reference
B_2O_3/SiO_2	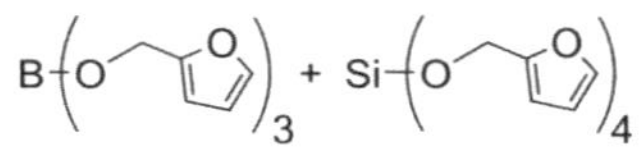	I. thermally induced simultaneous twin polymerization II. oxidation in air flow (200 L·h^{-1}) at 900 °C for 3 h with a heating rate of 2.3 K·min^{-1}	647 m^2·g^{-1}	[75]
$C/B_2O_3/SiO_2$	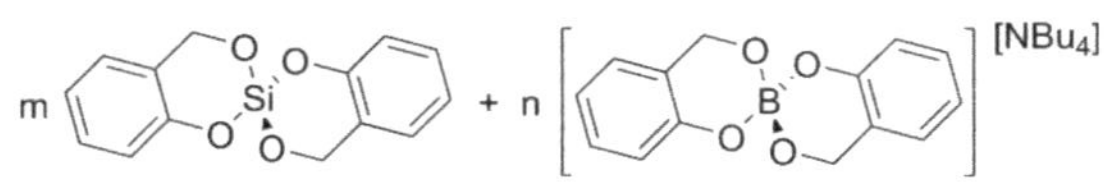	I. thermally induced simultaneous twin polymerization II. oxidation in air flow at 900 °C for 3 h with a heating rate of 2.3 K·min^{-1}	not given	[85]
In_2O_3/SiO_2	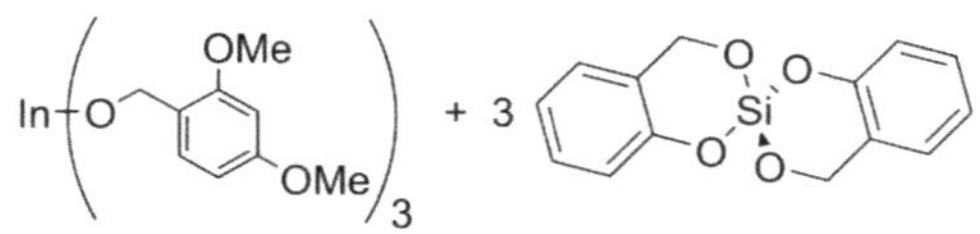	I. thermally induced simultaneous twin polymerization II. oxidation in air flow (200 L·h^{-1}) at 700 °C for 4 h with a heating rate of 10 K·min^{-1}	312 m^2·g^{-1}	[82]
In_2O_3/SiO_2	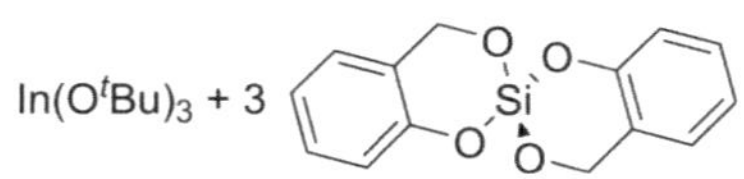	I. thermally induced twin polymerization with additives II. oxidation in air flow (200 L·h^{-1}) at 700 °C for 4 h with a heating rate of 10 K·min^{-1}	334 m^2·g^{-1}	[82]
$(In_2O_3)_{\approx 0.9}(SnO_2)_{\approx 0.1}$	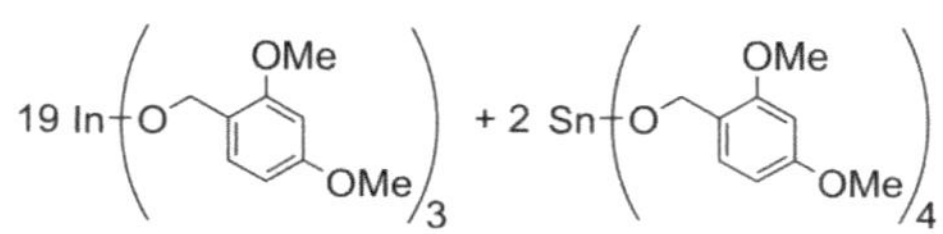	I. acid-catalyzed simultaneous twin polymerization II. oxidation in air flow (200 L·h^{-1}) at 600 °C for 4 h with a heating rate of 10 K·min^{-1}	19 m^2·g^{-1}	[82]

(continued)

Tab. 4.6 (continued)

inorganic oxides	precursors	preparation	BET surface area#	reference
$(In_2O_3)_{\approx 0.9}(SnO_2)_{\approx 0.1}$/ SiO_2	10 $In(O^tBu)_3$ + $Sn(O^tBu)_4$ + 33 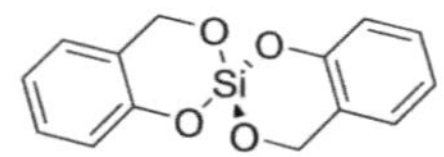	I. thermally induced twin polymerization with additives II. oxidation in air flow (200 $L \cdot h^{-1}$) at 700 °C for 4 h with a heating rate of 10 $K \cdot min^{-1}$	248 $m^2 \cdot g^{-1}$	[82]
SnO_2/SiO_2	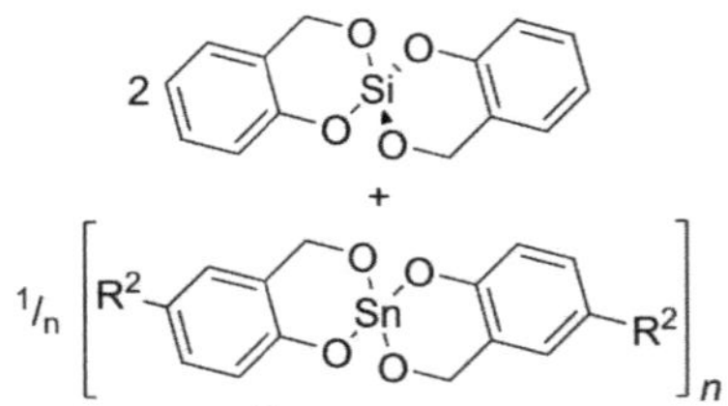 R^2 = H, Me, OMe	I. acid-catalyzed simultaneous twin polymerization II. oxidation in air flow (200 $L \cdot h^{-1}$) at 700 °C for 3 h with a heating rate of 10 $K \cdot min^{-1}$	378 $m^2 \cdot g^{-1}$	[78]
ZrO_2/SiO_2	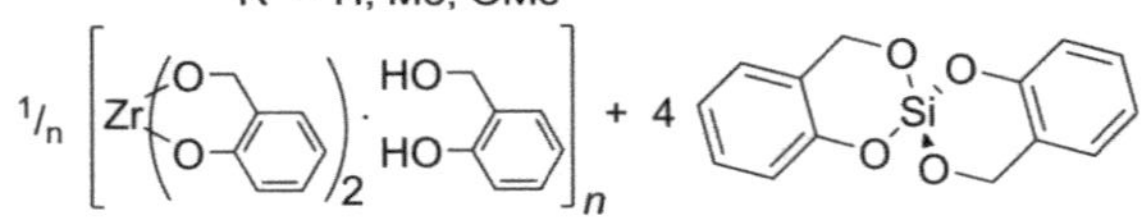	I. acid-catalyzed simultaneous twin polymerization II. carbonization in argon flow (10 $L \cdot h^{-1}$) at 800 °C for 30 min with a heating rate of 50 $K \cdot min^{-1}$ III. oxidation in air flow (200 $L \cdot h^{-1}$) at 450 °C for 48 h with a heating rate of 50 $K \cdot min^{-1}$	684 $m^2 \cdot g^{-1}$	[83]
HfO_2/SiO_2	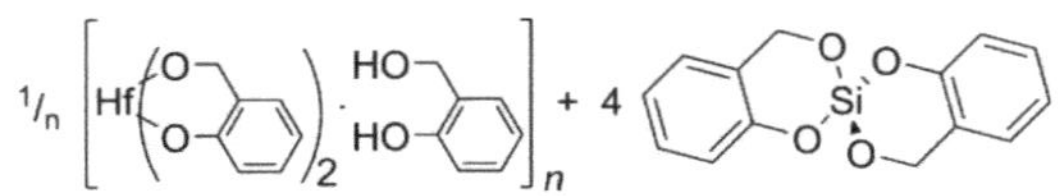		550 $m^2 \cdot g^{-1}$	[83]

Maximum value is given as published in the literature.

All mixtures of porous inorganic oxides and porous mixed oxides were synthesized using the latter synthesis concepts based on twin polymerization. The specific conditions that were applied to obtain them and their BET surface areas are summarized in Table 4.6.

Notably, the BET surface areas of the porous inorganic oxides, the mixtures of inorganic oxides, and the mixed oxides differ significantly ranging from a few $m^2{\cdot}g^{-1}$ up to hundreds of $m^2{\cdot}g^{-1}$. The differences in their BET surface areas indicate that each inorganic oxide material possesses specific properties determining its porosity. Several factors were identified influencing the structure formation of the inorganic oxides. First, the synthesis conditions that are applied to convert the hybrid materials into the porous inorganic oxides affect the porosity of the resulting oxides. Generally speaking, higher oxidation temperatures result in inorganic oxides exhibiting smaller BET surface areas. This effect was best shown by the comparison of the BET surface areas of mixed inorganic oxides consisting of ZrO_2/SiO_2 that were obtained from the same hybrid material. Its oxidation at 450 °C for 48 h after calcination of the material at 800 °C for 30 min gave porous ZrO_2/SiO_2 materials exhibiting a BET surface area of 684 $m^2{\cdot}g^{-1}$ (see Tab. 4.6), whereas further treatment of the material at 1000 °C for another 48 h gave ZrO_2/SiO_2 material possessing a rather low BET surface area of only 20 $m^2{\cdot}g^{-1}$ [83]. The enormous reduction of the surface area was attributed to crystallization processes and the collapse of the porous oxide matrix due to rearrangement processes being only feasible at elevated temperatures [83]. Second, the synthesis conditions that are applied to obtain the hybrid material on demand also affect the porosity of the final inorganic oxide. For instance, studies on the porosity of In_2O_3 materials that were obtained starting from indium(III) (2,4-dimethoxyphenyl)methanolate revealed the influence of the catalyst concentration on the porosity of the resulting oxide material. Smaller BET surface areas were determined for the In_2O_3 materials that were obtained from the hybrid materials synthesized using larger compound to catalyst ratios (M:I), but applying otherwise exactly the same synthesis conditions. Here, an M:I of 1:0.05 gave an In_2O_3 material exhibiting a BET surface area of 48 $m^2{\cdot}g^{-1}$, whereas an M:I of 1:0.1 gave an In_2O_3 material exhibiting a BET surface area of 70 $m^2{\cdot}g^{-1}$ [82]. It is assumed that an enhanced reaction rate resulting from the presence of larger amount of catalyst gives smaller phase domain sizes of the single polymeric components in the hybrid materials. Consequently, oxidation of hybrid materials possessing smaller phase domain sizes of the single organic and inorganic components provides porous oxides exhibiting larger BET surface areas. Notably, studies on the porosity of porous mixed inorganic oxides (e.g., B_2O_3/SiO_2) revealed that the acidity of the Brønsted-Lowry acid used for the synthesis of the hybrid materials by a proton-catalyzed twin polymerization process also influences the porosity of the final oxides [75]. However, the mutual dependencies of the nature of the precursors, the nature of the catalysts, and the specific synthesis conditions on the porosity of the final inorganic oxides are complex and not fully understood yet. In addition, if twin polymerization of the precursors is induced by different processes,

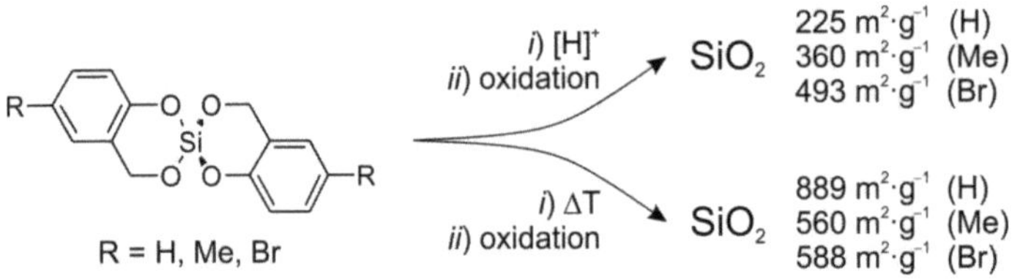

Scheme 4.19: BET surface areas of porous SiO_2 obtained starting from spirocyclic silicon salicyl alcoholates (R = H, Me, Br) either by proton-catalyzed (top) or thermally induced (bottom) twin polymerization. The data were taken from Ref. [70].

the distinguishable natures of the polymerization processes will also affect the porosity of the finally obtained porous inorganic oxides. For instance, twin polymerization of a series of spirocyclic silicon salicyl alcoholates that were either polymerized by addition of a Brønsted-Lowry acid or thermally induced gave hybrid materials that are powdery or monolithic, respectively. Oxidation of these hybrid materials always gave mesoporous SiO_2 materials. Notably, larger BET surface areas were always determined for the SiO_2 materials that were obtained starting from a thermally induced twin polymerization process. For example, a BET surface area of 889 $m^2 \cdot g^{-1}$ was detected for the mesoporous SiO_2 material obtained from thermally induced twin polymerization of Si-Spiro, whereas the SiO_2 material obtained from its proton-catalyzed twin polymerization exhibits a BET surface area of 225 $m^2 \cdot g^{-1}$ (Scheme 4.19) [70].

The significant difference in their porosity was attributed to the different temperature at which the polymerization processes were performed. Larger amounts of residual Si–OH groups were assumed to be present due to $^{29}Si\{^1H\}$-CP-MAS NMR spectroscopy data within the hybrid materials that were obtained by proton-catalyzed twin polymerization of the spirocyclic silicon salicyl alcoholates in comparison to the hybrid materials as obtained by their thermally induced twin polymerization [70]. Thus, the polymerization process at elevated temperatures gave hybrid materials possessing a more completed SiO_2-network. The latter may be comprehensible considering the fact that condensation reactions of Si–OH groups to give the SiO_2 material as a part of the twin polymerization process possess rather high reaction barriers (≈ 50–150 $kJ \cdot mol^{-1}$) [86]. Another aspect further contributing to the observation is that the larger BET surface areas determined for the SiO_2 material as obtained by thermally induced twin polymerization was deduced by comparison of the BET surface areas of the respectively obtained porous carbon materials. Thus, it was assumed that the process of proton-catalyzed twin polymerization results in smaller phase domain sizes of the single components within the hybrid materials in the first place. However, the smaller phase domain sizes and the incomplete formation of the SiO_2 network facilitate sintering processes during the oxidation procedure to give porous SiO_2 possessing larger pores in comparison to the porous SiO_2 material obtained by thermally induced twin polymerization [70]. Third, the studies on the spirocyclic silicon salicyl alcoholates revealed that the features of the mesoporous SiO_2 materials are also affected by the substituents at the aromatic moieties of these precursors. Two aspects of the substituents were determined to affect the porosity of the SiO_2 materials. (i) Precursors possessing substituents that facilitate the initial process of twin polymerization and, thus, enhancing the reaction rates gave SiO_2 materials exhibiting smaller BET surface areas due to the formation of smaller phase domain sizes of the single components within the nanostructured hybrid materials. As mentioned earlier, smaller phase domain sizes of the single components within the as-obtained hybrid materials facilitate sintering processes during the oxidation process, at least under the applied harsh oxidation condition (at 900 °C in an air flux of 200 $L{\cdot}h^{-1}$ for 3 h), resulting in larger pore sizes and, thus, in smaller BET surface areas within the final SiO_2 materials [70]. (ii) The largest BET surface areas were determined for the SiO_2 material that was obtained starting from thermally induced twin polymerization of Si-Spiro (Scheme 4.19). This precursor does not possess any substituent at its aromatic moieties. Thus, its twin polymerization gives the hybrid materials exhibiting the less ordered structure of the single polymeric components due to an extended formation of crosslinkages by bridging methylene groups during the phenolic resin formation process. Consequently, smaller phase domain sizes of the single components within the nanostructured hybrid materials are formed during the polymerization processes. In the case of its thermally induced twin polymerization, the polymeric SiO_2 network was almost completely formed and only small amounts of Si–OH groups were left. Thus, the SiO_2 network underwent only minor sintering processes during the oxidation procedure. Consequently, the small phase domain sizes of the single components within the nanostructured hybrid materials were preserved within the mesoporous SiO_2 materials that were obtained by thermally induced twin polymerization of Si-Spiro leading to their large BET surface areas (up to 889 $m^2{\cdot}g^{-1}$) [70]. Notably, the smallest BET surface area (225 $m^2{\cdot}g^{-1}$) of the SiO_2 materials as obtained starting from spirocyclic silicon salicyl alcoholates was detected for the SiO_2 material that was synthesized starting from Si-Spiro by proton-catalyzed twin polymerization (Scheme 4.19) [70]. Thus, the formation of very small phase domain sizes of the single components within the hybrid materials is only one side of the coin. The complete formation of the inorganic oxide component during the polymerization process represents the second side being a prerequisite to obtain highly porous inorganic oxide materials. Fourth, the molecular nature of the precursors affects the properties of the final inorganic oxide materials as well, because the course of the twin polymerization process and the specific type of its triggering process are related to their molecular structures. For instance, porous SiO_2 materials were obtained using the concept of twin polymerization starting from either tetrafurfuryloxysilane (TFOS) or spirocyclic silicon salicyl alcoholates [13, 14, 70]. The largest BET surface areas that were determined for the SiO_2 materials obtained starting from TFOS were about 518 $m^2{\cdot}g^{-1}$ [13]. Contrastingly, the SiO_2 materials that were obtained starting from spirocyclic silicon salicyl alcoholates such as Si-Spiro exhibited BET surface areas up to 889 $m^2{\cdot}g^{-1}$ (Scheme 4.20) [70].

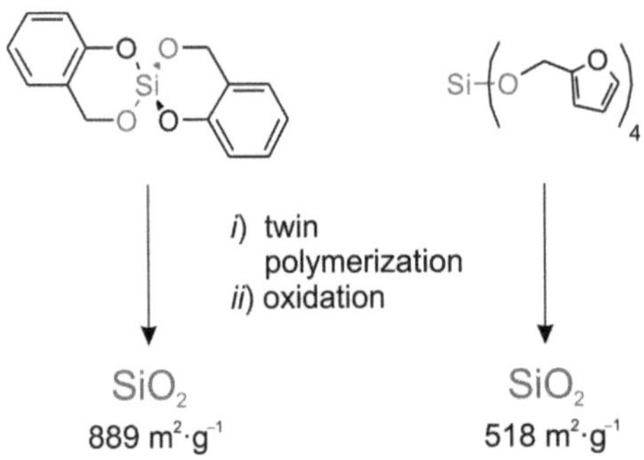

Scheme 4.20: BET surface areas of porous SiO_2 that were obtained using the concept of twin polymerization starting from Si-Spiro (left side) and TFOS (right side). The data were taken from Refs. [13, 70].

Notably, the oxidation procedures of both hybrid materials were carried out at similar conditions (at 900 °C in an air flux of 200 $L\cdot h^{-1}$ for either 1 h starting from TFOS or 3 h starting from spirocyclic silicon salicyl alcoholates). Thus, the differences of their porosities must originate in the structure of the as-obtained hybrid materials, e.g., the phase domain sizes of their single components being either polyfurfuryl alcohol and SiO_2 starting from TFOS or phenolic resin and SiO_2 starting from the spirocyclic silicon salicyl alcoholates. Notably, twin polymerization of TFOS can only be catalyzed upon addition of an acid such as a Brønsted-Lowry acid. Moreover, water is formed as volatile by-product during the polymerization process resulting in larger phase domain sizes of the single components within the as-obtained hybrid material. Contrastingly, thermally induced twin polymerization of the spirocyclic silicon salicyl alcoholates provided the hybrid materials that were eventually converted into the highly porous SiO_2 materials. Here, the polymerization process was performed at elevated temperature resulting in a highly condensed polymeric SiO_2 network within the as-obtained phenolic resin/SiO_2 hybrid materials as compared to the as-formed SiO_2 network within the polyfurfuryl alcohol/SiO_2 material. In addition, no volatile by-products are formed during the polymerization process of the spirocyclic silicon salicyl alcoholates. Thus, smaller phase domain sizes of the single components can be formed during the polymerization process due to the lack of segregation processes (see Chapter 3.1). Both aspects contribute to the observation that the SiO_2 materials obtained starting from the spirocyclic silicon salicyl alcoholates are of higher porosity than the SiO_2 materials as obtained starting from TFOS. Notably, simultaneous twin polymerization of TFOS with Si-Spiro gave nanostructured hybrid materials consisting of phenolic resin, polyfurfuryl alcohol, and SiO_2 whose oxidation (at 900 °C in an air flux of 200 $L\cdot h^{-1}$ for 1 h) provided microporous SiO_2 monoliths exhibiting BET surface areas up to 923 $m^2\cdot g^{-1}$ depending on the molar ratios of the precursors and the catalyzing Brønsted-Lowry acid [75]. Thus, the latter indicates that beneath the concentration of the triggering reagent, the ratio of the precursors also significantly influences the porosity of the resulting inorganic oxide material. Fifth, the chemical nature of the precursors determines the chemical nature of the inorganic oxides. This is trivial to begin with. However, the ability of the inorganic oxides to form a porous material is related to their structural features on a molecular level. For instance, B_2O_3 forms a three-dimensional network of crossing zigzag chains of corner-connected planar BO_3–units that are not expected to form highly porous structures [87]. On the other hand, SiO_2 consists of corner-connected tetrahedral SiO_4-units that are able to form highly porous structures [87]. As a consequence, it is expected that materials consisting of B_2O_3 exhibit significantly smaller BET surface areas than SiO_2 materials which were both obtained using the concept of twin polymerization starting from either boron alkoxides or silicon alkoxides, respectively. B_2O_3 materials that were obtained starting from tris(furan-2-ylmethyl) borate [$B(OFur)_3$ with $OFur = OCH_2C_4H_3O$] provided BET surface areas only up to 20 $m^2 \cdot g^{-1}$, whereas SiO_2 material that was obtained by a similar synthesis protocol (acid-catalyzed twin polymerization of the precursors and oxidation of the resulting hybrid materials at 900 °C in an air flux of 200 $L \cdot h^{-1}$ for 1 h) starting from tetrafurfuryloxysilane (TFOS) exhibited BET surface areas up to 518 $m^2 \cdot g^{-1}$ (Tab. 4.5) [75]. Thus, the porosity of the inorganic oxides obtained using the concept of twin polymerization is also determined by the chemical nature of the inorganic oxide itself, which is defined by the chemical nature of the precursor in the first place. Notably, simultaneous twin polymerization of $B(OFur)_3$ with TFOS gave hybrid materials consisting of polyfurfuryl alcohol, B_2O_3, and SiO_2 whose oxidation (at 900 °C in an air flux of 200 $L \cdot h^{-1}$ for 1 h) provided microporous B_2O_3/SiO_2 monoliths exhibiting BET surface areas up to 647 $m^2 \cdot g^{-1}$ (Tab. 4.6) [75]. In addition, simultaneous twin polymerization of Si-Spiro and the boron compound [$B(O_2C_7H_6)_2$][NBu_4] gave a hybrid material which is transferred to $C/B_2O_3/SiO_2$ composites by either oxidation or carbonization. The hybrid material might be promising with regard to application as flame retardant material [88].

In summary, the properties of the porous inorganic oxides, mixtures of porous inorganic oxides, and porous mixed oxides that are accessible using the different synthetic concepts of twin polymerization are determined by several factors. These are in particular (i) the oxidation conditions; (ii) the synthesis conditions, e.g., type of twin polymerization and/or the triggering reagent to precursor ratio as well as performance of the polymerization process in solution or in melt; (iii) the nature of the precursors, e.g., tetrafurfuryloxysilane (TFOS) *versus* Si-Spiro [$Si(O_2C_7H_6)_2$]; (iv) the chemical nature of the inorganic oxides; and (v) the ratio of the precursors on the condition of the usage of a multi-precursor synthesis protocol. Here, the porosity of the resulting oxide materials can be tuned by the content of SiO_2 that can be adjusted by the type and ratio of the silicon alkoxide precursor. The BET surface areas being presented in this chapter are values that were reported in the given references, respectively. These values should be considered as rough estimates for the porosity of the inorganic oxides. Variation of the BET surface areas is attributed to the complexity of the polymerization and oxidation process and we did observe that even slight changes of the reaction conditions might influence the final porosity.

4.3.2 Potential applications

Metal oxides and metalloid oxides are important classes of materials regarding their usage in many technological and industrial applications, e.g., in catalysis as supporting material, as catalyst promoter, or even as catalytically active material themselves, as sensors, and as semiconducting materials being photocatalysts for water splitting and water purification or in microelectronics industry [89–94]. Most of the inorganic oxides possess preferable mechanical and chemical properties such as thermal stability, hardness, and chemical resistance that may be combined with physical features of the materials being interesting in the field of optics, magnetism, or electrics [90, 92, 95–97]. Notably, many of these applications depend on phenomena, interactions, and/or reactions that are performed on the surface of the inorganic oxide. Thus, if these materials possess high surface to volume ratios, their performances in such surface sensitive applications will be enhanced. Therefore, there has been a vast interest in the development of synthesis protocols providing highly porous metal oxide and metalloid oxide materials within the last decades [96–99]. As outlined above, the synthesis concepts exploiting twin polymerization processes provide a simple and effective alternative to known processes. Up to now, a couple of studies were performed targeting on specific applications of materials that were synthesized using twin polymerization. For instance, porous $(In_2O_3)_{\approx 0.9}(SnO_2)_{\approx 0.1}$ and porous $(In_2O_3)_{\approx 0.9}(SnO_2)_{\approx 0.1}/SiO_2$ materials were synthesized aiming at porous transparent electrode materials [82]. Lang and coworkers reported on twin polymerization of Si-Spiro in the presence of a silver carboxylate providing porous materials consisting of silver nanoparticles being incorporated in a porous SiO_2-matrix [100]. Notably, such silica-supported metal-nanoparticle-containing materials represent a class of materials possessing promising features regarding their potential in applications such as oxidation, hydrogenation, and electrooxidation [101–106]. Moreover, the reduction of nitric oxides in the presence of methane to give N_2 and CO may be performed using porous ZrO_2 and ZrO_2/SiO_2 materials [83]. Mesoporous SnO_2/SiO_2 materials were tested as sensors for CO detection exhibiting good sensitivities at low concentrations of the CO gas being in the ppm range [107].

Please note that the synthetic concepts that were developed for the synthesis of porous inorganic oxide materials can be altered to give porous metal-oxide-based carbon materials exhibiting metal nanoparticles within its porous matrix. For instance, simultaneous twin polymerization of Si-Spiro with a tin-containing precursor in the presence of a nickel carboxylate provided a hybrid material whose reduction by thermal annealing under a hydrogen-containing argon atmosphere gave a porous silica-based carbon material exhibiting Ni_3Sn_2 nanoparticles (Scheme 4.21). Following this approach, several microporous hybrid materials (e.g., $Sn/C/SiO_2$, $Sn_2Co/C/SiO_2$, $Sn_2Fe/C/SiO_2$, $Ni_3Sn_2/C/SiO_2$, and $Sn_5Cu_6/C/SiO_2$) were synthesized representing a promising class of materials for prospective applications (e.g., as anode material for Li-ion batteries or in heterogeneous catalysis) [84, 108, 109].

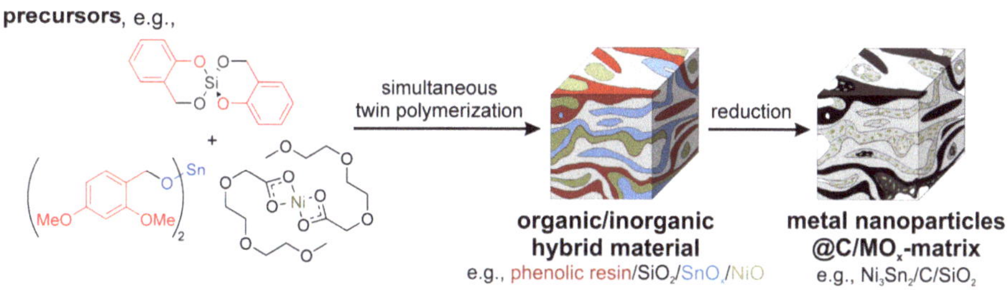

Scheme 4.21: Synthesis of porous metal-oxide-based carbon material exhibiting metal nanoparticles using the concept of simultaneous twin polymerization in the presence of additives.

Despite some promising first studies, the potential of metal oxides as obtained from twin polymerization still needs to be exploited in the future.

Heinrich Lang

4.4 Twin polymerization – a unique and efficient methodology for decorating porous carbon and silica materials with nanoparticles

4.4.1 Introduction

Metal nanoparticles (M-NPs) are well known for their special and unique properties, based on, for example, their size and shape. Their optical, electronic, and catalytic attributes make them of significant interest for diverse technological applications in the area of both science and industry, including catalysis, optoelectronics, and biochemistry [110–124], where the controlled and reproducible synthesis of defined and stable M-NPs with a small size distribution is of great importance. This led to the accomplishment of diverse synthetic methodology protocols for the formation and stabilization of M-NPs, using either the top-down or bottom-up process. The top-down technology is based on attrition or milling, involving mechanical thermal cycles. Generally, this procedure produces NPs with a broad size particle distribution and the particles are of different morphology. In addition, the as-prepared NPs may contain some impurities due to the milling process. This procedure is often used to obtain nanocomposites or nano-grained bulk materials. The bottom-up synthetic methodology is a chemical approach and implies the assembly of NPs from main-group or transition metal compounds or atoms, using, for example, pyrolysis, inert gas condensation, solvothermal, or sol-gel processes.

Within these synthesis protocols, the chemical preparation of M-NPs by reduction of different metal salts is an apparently very straightforward process, simply requiring the mixing of the respective reagents (metal salt, reducing reagent, stabilizer) of which the reaction conditions affect, i.e., the morphology and size of the particles [125–130]. In this respect, the organic phase synthesis after Brust must be mentioned [131, 132], involving a two phase process, as well as the single phase water-based reduction of a gold or silver salt by citrate, firstly introduced by Turkevich and coworkers [133, 134] and refined by Frens [135, 136], producing spherical particles over a tunable range of sizes.

Small M-NPs are only kinetically stable and combine *via* agglomeration to thermodynamically more favored larger particles, based on the high surface energy as well as large surface area of the respective M-NPs. To avoid metal particle agglomeration, the corresponding M-NPs need to be stabilized with strongly coordinating protective ligands, providing electrostatic and/or steric protection. As stabilizing component, for example, polymers and surfactants are applied [110–124].

The most favored synthesis method is the treatment of transition metal salts with a reducing agent in presence of a stabilizing component in an appropriate solvent. Therefore, a variety of reductive reagents were explored of which $Na[BH_4]$ is preferred [137–139]. Polysaccharides are a green alternative for environmentally friendly reducing agents [140–142]. Another methodology to form M-NPs involves ultra-violet irradiation of a metal source in presence of a stabilizing component [143–147]. In this respect, the sensitiveness of the respective system can be tuned by adding a chromophoric species producing radicals, which are essential for the reduction process. Reducing agent-free methods were also established; commonly they involve metal carbonyls, diluted in ionic liquids, which were irradiated with ultra-violet light to form the appropriate NPs [111, 148–150]. Recently, we did report on a further synthetic methodology for the generation and stabilization of M-NPs by using a *single-source* approach without the addition of any reducing agent and any stabilizing component. Thus, our group succeeded in the preparation and stabilization of, for example, copper [151–153], silver [106, 154–160], gold [161–165], rhodium [166], palladium [167, 168], platinum [169], and ruthenium NPs [167, 168, 170] through the thermal decomposition of mainly metal-organic carboxylate complexes of the general type $[M(O_2CCH_2(OCH_2CH_2)_nOMe)_m]$ and $[L_xM(O_2CCH_2(OCH_2CH_2)_nOMe)_m]$ (M = Cu, Ag, Au, Pd, Pt, Ru, Rh; $n = 2, 3, 4, 5, 7.33$; $m = 1, 2$; L = phosphine, isonitrile; $x = 1, 2$) at low temperature. When metal-organic carboxylates such as $[M(O_2CCH_2(OCH_2CH_2)_2OMe)_2]$ (M = Mn, Fe, Co, Ni) were used, the corresponding metal oxides Mn_2O_3, Fe_2O_3, Co_2O_3, or NiO were formed [171–173]. Recently, this method was extended to a new advanced procedure for using a *single-source* precursor, which combines light sensitive functionalities, a reducing agent, and the metal source in one molecule, simplifying the preparation and optimization and hence the efficiency of the whole process [143], allowing to avoid the use of hazardous chemicals.

Twin polymerization was recently introduced by Spange and co-workers, allowing the non-aqueous preparation of organic/inorganic hybrid materials composed of interpenetrating phase nano-domains within one step [13–15, 36, 70, 174]. This convenient procedure benefits from the strong mechanistic coupling between the formation of inorganic and organic networks. It involves, for example, a silicon orthoester bearing polymerizable organic groups such as 2-furfurylic or salicylic alcoholates. By reacting these compounds under acid catalysis (e.g., $MeSO_3H$) or by thermally induced polymerization, reactive sites on the inorganic and organic part of the monomer are formed. Thus, the formation of the phenolic resin proceeds faster and hence phase separation of inorganic SiO_2 is not observed, leading to nanostructured hybrid materials. Both networks can be separated from each other. Oxidizing the hybrid material resulted in the formation of mesoporous silica, while carbonization under inert gas atmosphere and removal of the SiO_2 lattice by addition of HF or NaOH yielded microporous carbon. This concept was recently extended to main-group and transition metal-containing twin monomers [76–81, 83, 175].

Within this chapter, straightforward approaches for the decoration of porous carbon and SiO_2 materials with metal-NPs or tin alloys, as well as the encapsulation of M-NPs within a porous carbon shell by using the twin polymerization process and mainly metal carboxylates and alkoxides as the metal or metal oxide nanoparticle source will be discussed.

4.4.2 Metal and metal oxide nanoparticles by thermal decomposition of *single-source* metal carboxylate precursors

A typical nanoparticle preparation protocol for the generation and stabilization of metal nanoparticles (M-NPs) without the addition of any reducing agent and any stabilizing component (for more details see references [106, 143, 151–173, 176]) is the usage of the *single-source* methodology, starting from metal carboxylates of the general type $[M(O_2CCH_2(OCH_2CH_2)_nOMe)_m]$ and $[L_xM(O_2CCH_2(OCH_2CH_2)_nOMe)_m]$ (M = Cu, Ag, Au, Pd, Pt, Ru, Rh; n = 2, 3, 4, 5, 7.33; m = 1, 2; L = phosphine, isonitrile; x = 1, 2). The M-NPs' formation proceeds by heating the appropriate metal precursor in various organic solvents to temperatures in the range of 110–330 °C, depending on the appropriate coordination complex applied (Fig. 4.31) [106, 143, 151–173, 176]. After precipitation of the NPs, they were separated by centrifugation, washed once with ethanol and dried or re-dispersed in n-hexane for analysis. In contrast to the NPs' dispersions, the dried particles are stable *ad infinitum* under inert conditions. In general, a virtually quantitative conversion of the corresponding precursor complexes to metal(0) is observed. The NPs were characterized by, for example, UV-Vis spectroscopy, thermogravimetry, dynamic light scattering, electron transmission microscopy, etc.

The mechanism of the thermally induced decomposition of the metal carboxylates was investigated by thermogravimetry-coupled mass spectrometry (TG-MS) [106, 151–160, 176, 177]. It was found that, for example, $[AgO_2CCH_2(OCH_2CH_2)_2OMe]$ initially releases CO_2, forming $[AgCH_2(OCH_2CH_2)_2OMe]$, while in the final step this metal-organic complex decomposes *via* Ag–C, C–C, and C–O bond cleavages [106, 154–160, 177]. The decomposition of $[L_xM(O_2CCH_2(OCH_2CH_2)_nOMe)_m]$ occurs in a similar way; after decarboxylation, however, most likely elimination of *x* L follows before M–C, C–C, and C–O bond breaking arises, as could be shown by TG-MS studies on the example of $[Au(O_2CCH_2(OCH_2CH_2)_2OMe)L]$ (L = PPh_3, C≡N-*t*-Bu) [161–165].

When the metal-organic complexes $[M(O_2CCH_2(OCH_2CH_2)_2OMe)_2]$ (M = Mn, Fe, Co, Ni) are used, the respective metal-oxide-NPs, such as Mn_2O_3, Fe_2O_3, Co_2O_3, or NiO, are formed [171–173]. Exemplarily, the synthesis of $[Co(O_2CCH_2(OCH_2CH_2)_2OMe)_2]$ and its thermal decomposition behavior, which was studied by thermogravimetry and TG-MS, are outlined in Figure 4.32. Metal-organic $[Co(O_2CCH_2(OCH_2CH_2)_2OMe)_2]$ was prepared by an anion exchange reaction of cobalt(II) acetate tetrahydrate with the corresponding carboxylic acid. High-resolution ESI-TOF mass spectrometry confirmed the formation of this coordination complex. For a detailed decomposition mechanism, see Refs. [171–173].

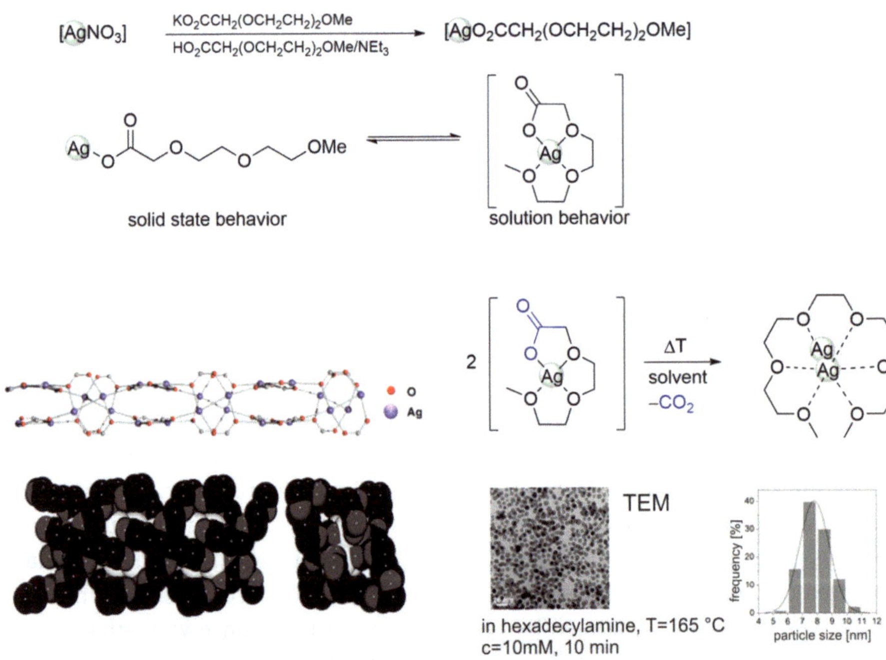

Fig. 4.31: Synthesis of M-NPs shown on the example of $[AgO_2CCH_2(OCH_2CH_2)_2OMe]$ [106, 154–160].

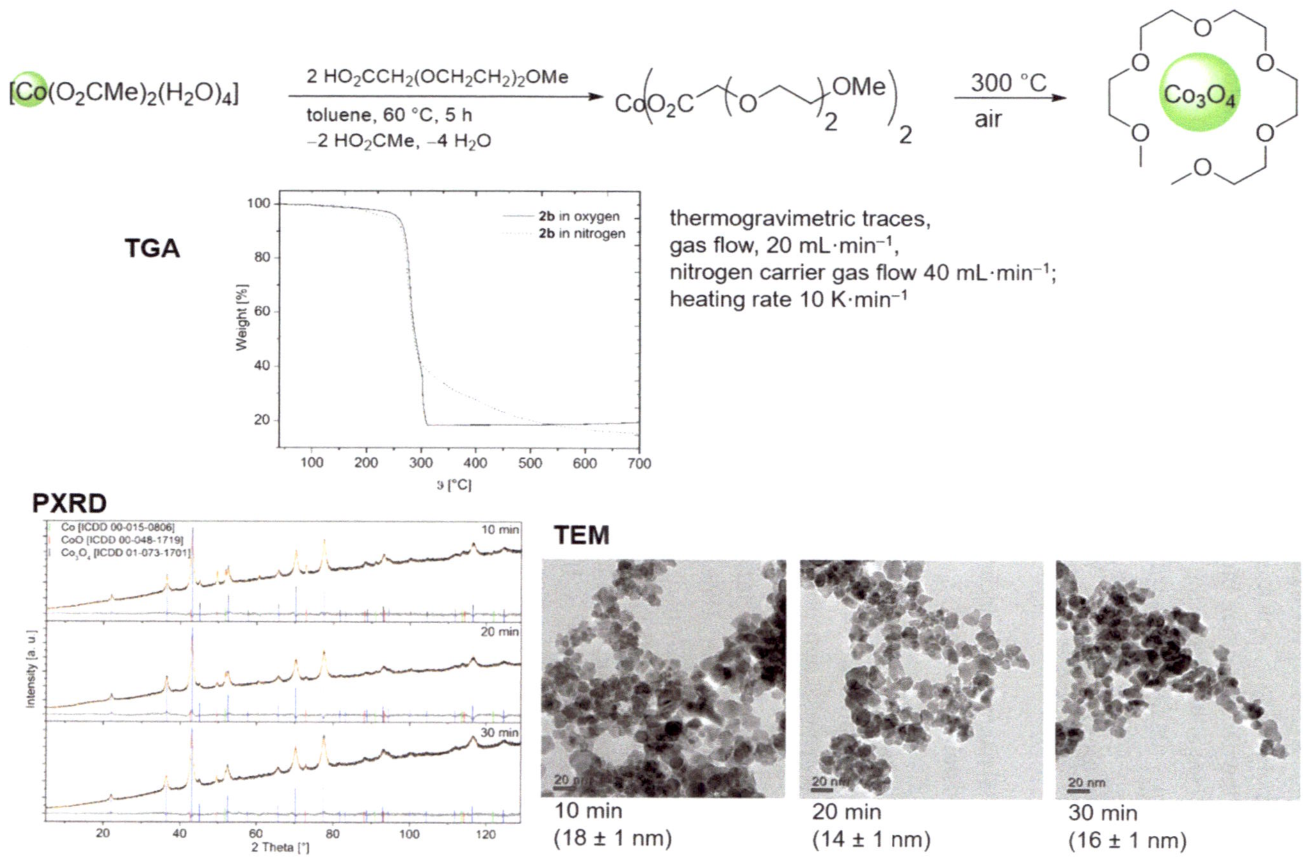

Fig. 4.32: Synthesis of $[Co(O_2CCH_2(OCH_2CH_2)_2OMe)_2]$ and its thermal decomposition behavior affording Co_3O_4-(CoO)-NPs [171–173].

Cobalt-oxide-NPs were prepared by solid state thermal decomposition of [$Co(O_2C\text{-}CH_2(OCH_2CH_2)_2OMe)_2$] at 300 °C in air (heating rate 10 $°C\cdot min^{-1}$) and applying decomposition times of 10, 20, and 30 min (Fig. 4.32) [171–173]. PXRD measurements confirmed the formation of cubic Co_3O_4 (89–94 %) along with small amounts of CoO (4–8 %) and Co (2–3 %). Increasing the decomposition time from 10 over 20 to 30 min afforded mainly Co_3O_4, which is ascribed to the gradual oxidation of Co and CoO to Co_3O_4, as the decomposition increases. The Rietveld method was also used to determine the crystallite size of the Co_3O_4-NPs showing particle sizes of 18±1 nm (after 10 min), 14±1 nm (after 20 min), and 16±1 nm (after 30 min) (Fig. 4.32) [171–173]. TEM studies were carried out, confirming that the Co_3O_4-NPs are strongly agglomerated and are of various shapes and sizes (Fig. 4.32).

4.4.3 Metal nanoparticle-decorated porous carbon and silicon dioxide materials by twin polymerization

Within this chapter, the incorporation of metal and partly metal-oxide-NPs within porous carbon and silica networks using the twin polymerization and simultaneous twin polymerization process is reported. Therefore, two methodologies were developed:

(*i*) The incorporation of, for example, metal carboxylates of the general type [$M(O_2CCH_2(OCH_2CH_2)_nOMe)_m$] and [$L_xM(O_2CCH_2(OCH_2CH_2)_nOMe)_m$] (see earlier) during the twin polymerization process (or other metal-organic or organometallic compounds, like metal carbonyls) [111], and

(*ii*) the polymerization of twin monomers such as the organometallic compound $Si(OCH_2Fc)_4$ (Fc = $Fe(\eta^5\text{–}C_5H_4)(\eta^5\text{–}C_5H_5)$) and their simultaneous twin polymerization with, for example, 2,2'-spirobi[4*H*-1,3,2-benzodioxasiline] [100].

The appropriate M- and M_xO_y-NP-decorated hybrid materials are of great interest, due to their manifold applications in the field of catalysis, magnetism, etc. [13, 15, 36, 70, 76–81, 106, 151, 152, 154–165, 174–176, 178, 179].

In the following, the preparation of M-NPs-supported carbon and silica materials will be discussed on the example of a silver(I) carboxylate coordination complex. In this respect, an efficient approach for the preparation of highly dispersed Ag-NPs of size <5 nm in a porous carbon or SiO_2 matrix through twin polymerization is possible as outlined in Scheme 4.22 [100]. The obtained microporous carbon (IUPAC-type I isotherm [180]) and mesoporous silica materials (IUPAC-type IV isotherm [180]) possess BET surface areas of 1034 $m^2.g^{-1}$ (carbon) or 666 $m^2.g^{-1}$ (SiO_2).

In order to obtain Ag-NP-enriched porous carbon or silica materials, silver(I) carboxylates, for example, [$Ag(PPh_3)(CO_2CH_2\text{-}2\text{-}c\text{-}C_4H_3S$] (**1**) were introduced during the twin polymerization process (Scheme 4.22) [100]. The PPh_3 ligand in **1** is advantageous for increasing the solubility of this coordination complex. For a typical

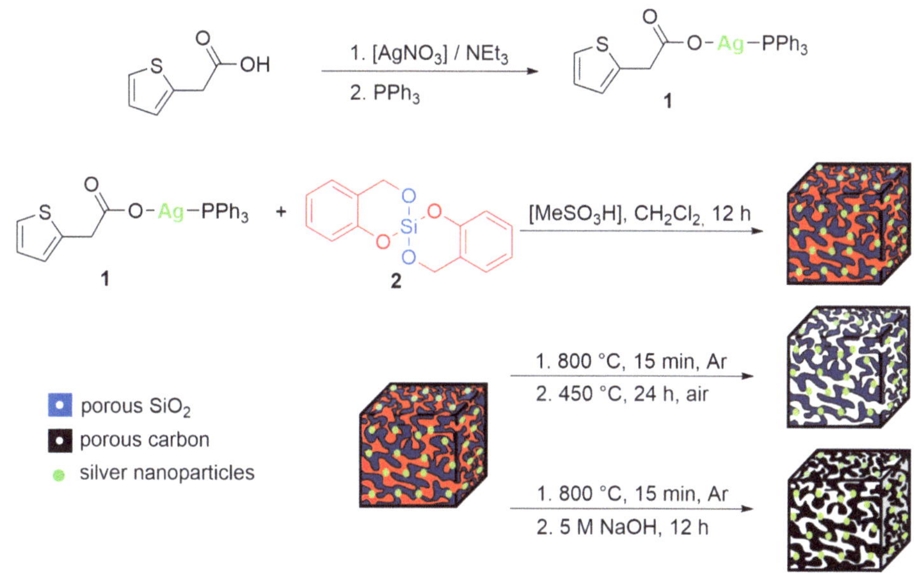

Scheme 4.22: Preparation of silver-decorated porous carbon and silicon dioxide composites. Reproduced from [100] by permission of John Wiley & Sons, Inc.

polymerization experiment (for more details see Ref. [100]), the appropriate amount of 2,2'-spirobi[4*H*-1,3,2-benzodioxasiline] (Si-Spiro, **2**) and 15 mol% of **1** in dichloromethane were warmed to 40 °C in the presence of methanesulfonic acid as catalyst, whereby the hybrid material precipitated. This hybrid material was then either treated with a refluxing NaOH solution to remove the silica template (formation of the respective carbon@Ag-NPs-decorated material) or oxidized in air to get the porous inorganic material (silica@Ag-NP) (Scheme 4.22).

The formation of Ag-NPs during the twin polymerization process was proven by PXRD studies of the corresponding C and SiO_2 composite. Broad reflexes of Ag [111] indicated the formation of silver particles of rather small size [100].

In addition, HAADF-STEM (high-angle annular dark-field scanning transmission electron microscopy) studies were carried out, indicating that the composite material is decorated with Ag-NPs as can be seen from the images depicted in Figure 4.33. As evidenced by PXRD and HAADF-STEM studies, the C and SiO_2 materials show dispersed silver-NPs with the majority being smaller than 5 nm in diameter [100].

The inner surface of the carbon and SiO_2 materials was investigated by nitrogen adsorption-desorption isotherms at 77 K, whereby a IUPAC-type I [180] isotherm with an inner surface of 1034 $m^2 \cdot g^{-1}$ was found for the carbon material, while for the silica component a IUPAC-type IV isotherm [180] with the characteristic hysteresis, caused by capillary condensation in mesoporous regions of the materials, is characteristic.

These studies show that the twin polymerization concept makes it possible to generate carbon or silica-supported M-NPs in a straightforward manner with the benefit of a

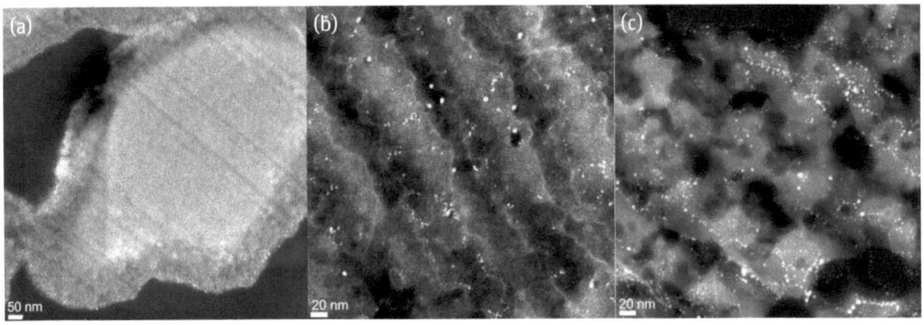

Fig. 4.33: High-angle annular dark-field STEM images of the composite material obtained from **1** and **2** (**A**, 50 nm scale bar), the derived carbon component (**B**, 20 nm scale bar), and the silicon dioxide material (**C**, 20 nm scale bar). Reproduced from [100] by permission of John Wiley & Sons, Inc.

convenient "*one-pot*" procedure, involving [Ag(PPh_3)(CO_2CH_2-2-*c*-C_4H_3S)] (**1**) and the twin monomer Si-Spiro, **2** as precursor molecules. The extension of this procedure to other silver(I)-based metal-organic complexes [106, 154–160], as well as metal carboxylate coordination complexes of the structural type [Au($O_2CCH_2(OCH_2CH_2)_2OMe$)L] (L = PPh_3, P(*n*-Bu)$_3$, C≡NR; R = single bonded organic group) [161–165], [Cu($O_2CCH_2(OCH_2CH_2)_2OMe$)L_x] (L = PPh_3, P(*n*-Bu)$_3$; x = 2, 3) [151–153, 176], [Cu($O_2CCH_2(OCH_2CH_2)_2OMe$)$_2$] [151–153, 176], [Pd($O_2CCH_2(OCH_2CH_2)_nOMe$)$_2L_2$] [167, 168], [Pt($O_2CCH_2(OCH_2CH_2)_nOMe$)$_2L_2$] (n = 1, 2; L = PPh_3, P(*n*-Bu)$_3$) [169], [Ru($O_2CCH_2(OCH_2CH_2)_nOMe$)$_2$(CO)$_2L_2$] (n = 1, 2; L = P(*n*-Bu)$_3$) [170], and [Rh_2($O_2CCH_2(OCH_2CH_2)_2OCH_3$)$_4$]$_n$ as well as [Rh(μ-$O_2CCH_2(OCH_2CH_2)_2OCH_3$)$_2$($PPh_3$)]$_2$ [166] is possible. The corresponding hybrid materials featuring the respective M-NPs could be applied in catalysis, including *C,C* cross-coupling (Ru, Pd), reduction (Ag, Au), and hydrogenation (Pd, Pt, Ru) processes.

Based on these studies, [Co($O_2CCH_2(OCH_2CH_2)_2OMe$)$_2$] (**3**) was also used in the preparation of cobalt(II,III) oxide-supported porous materials, wherein Si-Spiro **2** was used as the twin monomer and **3** as the cobalt source (Fig. 4.34) [171–173]. In boiling dichloromethane, the respective phenolic resin/SiO_2 hybrid material, containing **3**, was obtained. After heating this hybrid material to 800 °C, the as-prepared system was either treated with a 5 M NaOH solution to remove the silicon dioxide template obtaining the porous carbon matrix, or heated to 450 °C in air to achieve the appropriate silica component. Both materials are decorated with Co_3O_4-NPs [171–173].

The appropriate porous nature of these materials was investigated by N_2 adsorption-desorption isotherms, signifying a IUPAC type IV isotherm with a small hysteresis for C. Quenched solid density functional theory (for C) and non-linear density functional theory calculations (for silica) gave a surface area of 1040 $cm^2 \cdot g^{-1}$ for carbon and 336 $cm^2 \cdot g^{-1}$ for silica, whereas the presence of micro- and mesopores was confirmed for both carbon (53 % below 2 nm) and silica (5 % below 2 nm).

PXRD measurements confirmed the presence of solely Co_3O_4-NPs inside the carbon and silica matrix (Fig. 4.34). The Rietveld refinement method proved an

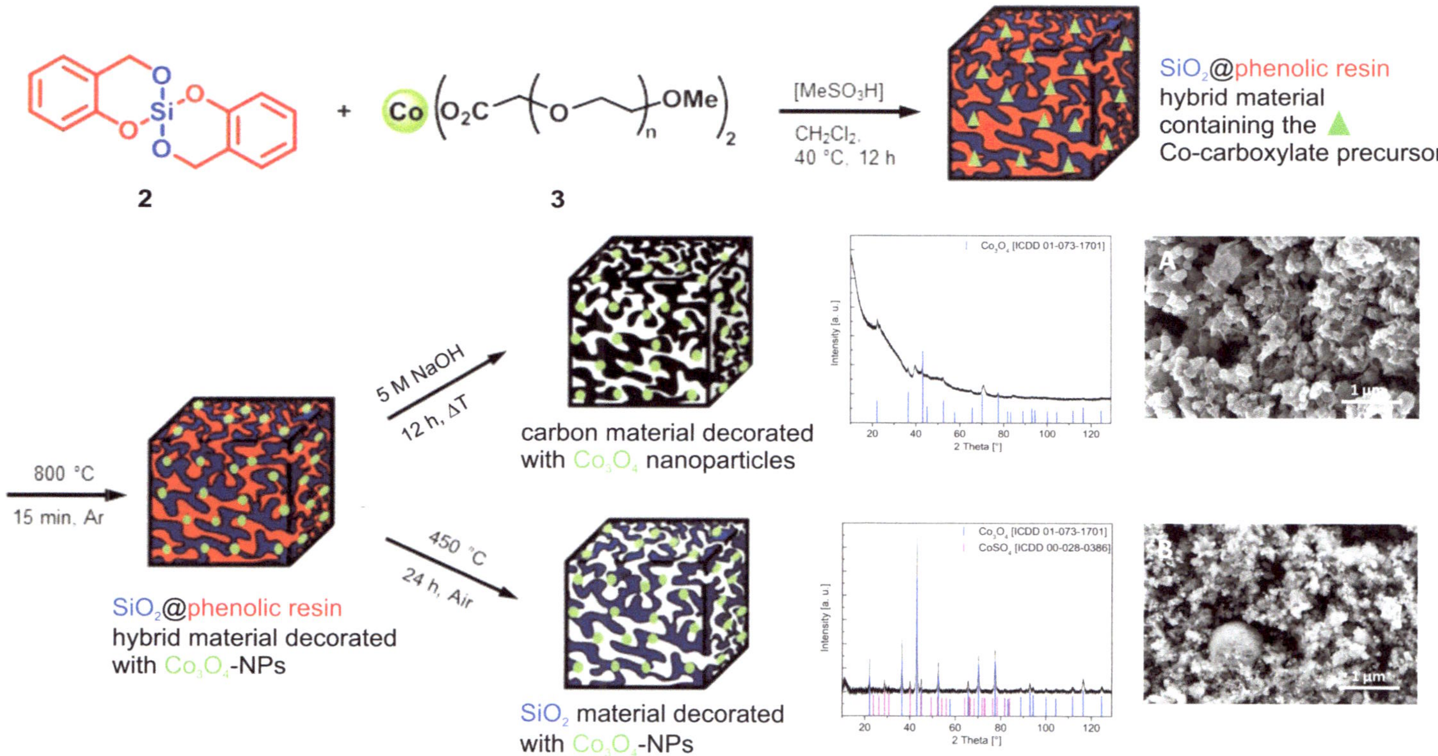

Fig. 4.34: Synthesis protocol for the preparation of Co_3O_4-NP-decorated silica and carbon materials by the twin polymerization of Si-Spiro **2** with **3**. PXRD and SEM images are shown for both materials ((**a**), SEM image for carbon; (**b**), for silicate) [171].

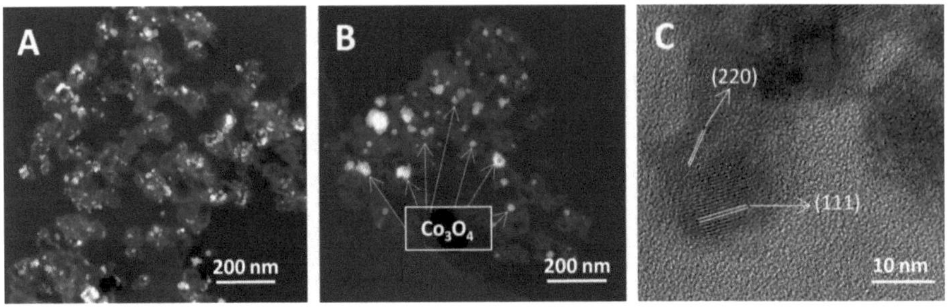

Fig. 4.35: HAADF-STEM images of the derived silicon dioxide component with Co_3O_4-NPs (**A** and **B**) obtained by the twin polymerization of Si-Spiro **2** in presence of **3**, high resolution TEM image of Co_3O_4 indicating the respective (111) and (220) orientations of crystalline NPs (**C**). Reproduced from [171] by permission of Springer Science + Business Media.

average particle size of 40±1 nm for crystallites embedded in silica, whereas significantly smaller ones were determined in the carbon matrix (22±1.0 nm).

The morphology of the porous supports and the NP distribution within both structures was proven by HAADF-STEM (Figs. 4.34 and 4.35). Additionally, EDX spectroscopy was conducted in order to determine the cobalt content. The SEM studies evidenced the porous structures of both materials, confirming the large surface areas measured by N_2 adsorption-desorption isotherms (*vide supra*). In addition, HAADF-STEM was used to distinguish between supports and the embedded NPs. Figure 4.35 presents the respective images for the silica component, where well-dispersed NPs with a majority in the range of 10–12 nm in diameter were found. While the silica component revealed a cobalt content of 10 %, the carbon material exhibited only 3.5 % of Co. It is estimated that the wet-etching process of SiO_2 by NaOH not only removed the silica template but also part of the Co_3O_4-NPs. Moreover, the HAADF-STEM images revealed a grainy structure for the silica matrix (Fig. 4.35).

Since cobalt oxides including CoO and Co_3O_4 are considered as potential electrode materials for lithium ion batteries, the electrochemical reactivity of the derived Co_3O_4-based carbon and silica nanostructured composite materials were investigated towards lithium [171–173]. Within the electrochemical studies, the cyclic voltammogram of the silica component showed redox events associated with the reduction of Co_3O_4 to CoO and CoO to Co, while no redox potentials occurred for the Co_3O_4-NPs embedded in the carbon matrix [171–173].

Another possibility for the formation of metal and/or metal-oxide-NP-decorated porous silica materials exists in the polymerization of the twin monomer $Si(OCH_2Fc)_4$ (**4**) (Fc = $Fe(\eta^5\text{-}C_5H_4)(\eta^5\text{-}C_5H_5)$), which is accessible by the reaction of $SiCl_4$ with four equivalents of $FcCH_2OH$ (Scheme 4.23) [181].

The thermal solid state behavior of $Si(OCH_2Fc)_4$ (**4**) was studied by differential scanning calorimetry. It was found that **4** melts at 128 °C (endothermic process), followed by a strong exothermic process at 210 °C. The latter event can be ascribed to

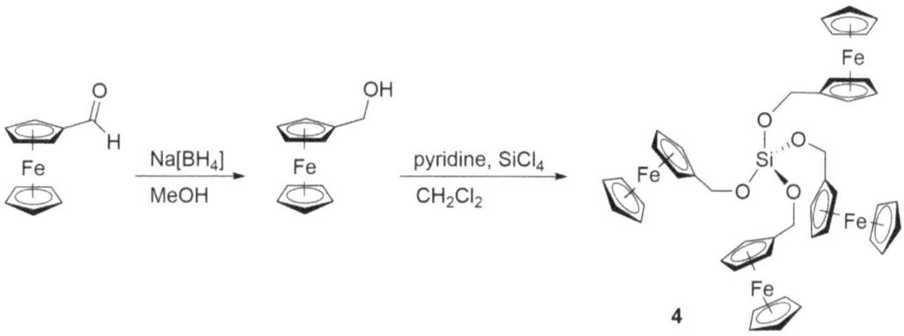

4

Scheme 4.23: Synthesis of $Si(OCH_2Fc)_4$ (**4**) (Fc = Fe(η^5-C_5H_4)(η^5-C_5H_5)) [181].

a thermally induced condensation reaction, since during the heating process of **4**, evaporation of water occurred at 200–230 °C as verified by TG-MS coupling experiments [181].

Upon heating a suspension of **4** in di-*iso*-propylnaphthalene to 230 °C, homopolymerization occurred and a hybrid material was formed showing similarities with known systems derived from, for example, Si-Spiro, **2** including the nano-patterning of the formed silica (*vide supra*), e.g., see [13–15, 36, 70, 174]. While reaching the condensation temperature of **4** (210 °C), water evaporated and a brown solid precipitated. The thus obtained material was characterized by solid state $^{13}C\{^1H\}$-CP-MAS NMR spectroscopy. The signals for the Fc cyclopentadienyl groups C_5H_5 and C_5H_4 were observed at 70 and 90 ppm, respectively [181]. In contrast to **4**, no resonance signal for the **C**H_2OSi groups (62 ppm) could be detected, indicating the complete conversion of **4**. The signal at 29 ppm can be assigned to the carbon atom of the formed Fc**C**H_2Fc bridging motif, which is in agreement with the chemical shift of the respective unit in, for example, diferrocenyl methane [182].

However, solid state NMR studies on the appropriate incorporated silica material failed, due to the small amount of Si (3.2 %) within $Si(OCH_2Fc)_4$ [182] and hence REM, EDX, and TEM studies were carried out to determine the incorporated Si content. It was found that silicon is homogeneously distributed within the respective hybrid material as evidenced by the EDX pattern and is present in the form of a SiO_2 matrix. This corroborates with the TEM images for this material (Fig. 4.36) showing a homogeneous texturing with particles exhibiting a higher contrast than the surrounding matrix, attributed to SiO_2 clusters formed during twin polymerization.

Electron microscopy studies carried out with this material exposed the absence of the distinctive micro-structuring characteristics for other twin polymers. Based on this, the respective material can be characterized as a borderline system, when compared to the twin monomer Si-Spiro (**2**) [13–15, 36, 70, 174, 181].

In addition, the simultaneous twin polymerization (STP) of **4** with Si-Spiro (**2**) in dichloromethane gave a hybrid material from which porous carbon or silica materials containing iron-oxide-NPs were obtained [181]. The oxidation state of the incorporated

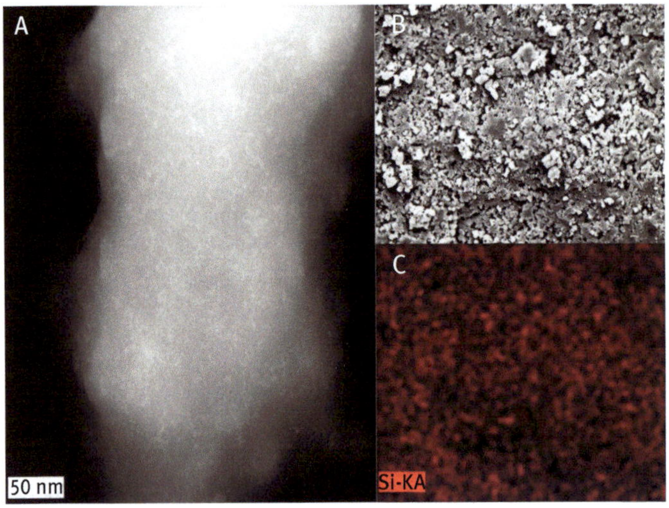

Fig. 4.36: HAADF-STEM image of the obtained hybrid material from **4** by thermally induced polymerization (**A**), REM image (**B**), and EDX analysis (**C**) showing the homogeneous distribution of silicon within the sample. Reproduced from [181] by permission of John Wiley & Sons, Inc.

particles was examined by Mössbauer experiments, confirming that solely Fe(III) was incorporated within the porous carbon and the silica networks. The as-prepared hybrid materials were investigated by $^{13}C\{^1H\}$-CP-MAS and $^{29}Si\{^1H\}$-CP-MAS NMR spectroscopies. All expected resonance signals for the phenolic resin were observed, as is typical for a twin polymer derived from Si-Spiro (**2**) [13–15, 36, 70, 174, 181]. The $^{29}Si\{^1H\}$-CP-MAS NMR spectrum showed, as expected, Q_3 and Q_4 signals, which are characteristic for the SiO_2 network. In addition, the incorporation of Fc functionalities during simultaneous twin polymerization was set by AAS measurements and an iron content of 3.8 % was found. This value was also supported by EDX experiments (3.6 %) and a homogeneous distribution over the total sample was achieved [181].

Carbonization of the hybrid material obtained from the acid-catalyzed simultaneous twin polymerization of Si-Spiro (**2**) and **4** followed by oxidation of the obtained material in air producing the inorganic silica matrix, or, when refluxed with a 5 M NaOH solution produced the porous C material [181]. The porous character of the appropriate networks was examined by N_2 adsorption-desorption isotherms. For the carbon material, a IUPAC-type I isotherm was found, indicating microporosity of the material, while the silica component possesses a IUPAC-type IV isotherm [180] with a small hysteresis, which is mostly caused by capillary condensation within mesoporous structures. The measured surface areas (C, 858 $m^2 \cdot g^{-1}$; SiO_2, 555 $m^2 \cdot g^{-1}$) and pore volumes (C, 0.43 $cm^3 \cdot g^{-1}$; SiO_2, 0.66 $cm^3 \cdot g^{-1}$) are somewhat smaller than with the related pure carbon or silica components obtained from the twin monomer Si-Spiro (**2**). This effect is most probably attributed to the incorporation of iron-oxide-NPs into the network, since they were mostly situated within the porous structure and therefore, are decreasing the surface area [181].

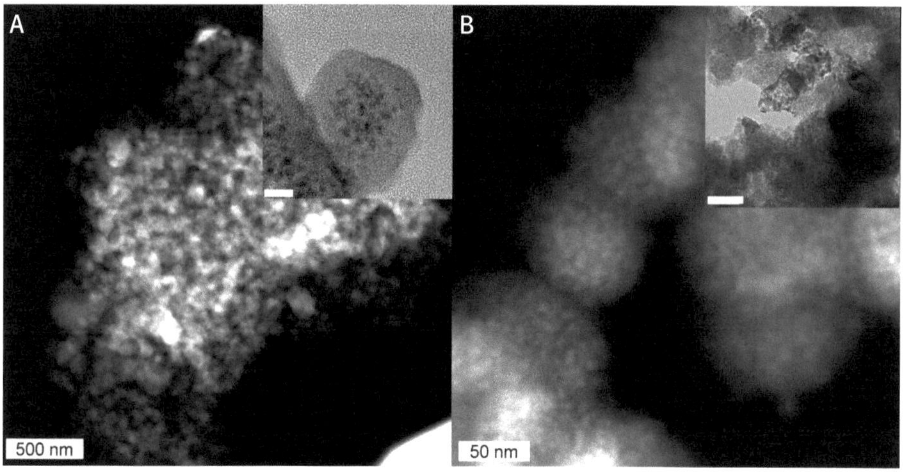

Fig. 4.37: HAADF-STEM images of the carbon (**A**) and the silica material (**B**). Bright field TEM images of the specific materials are displayed as insets (scale bar inset A: 20 nm, scale bar inset B: 100 nm). Reproduced from [181] by permission of John Wiley & Sons, Inc.

AAS experiments revealed a Fe content of 7.1 % for the carbon material, while the respective silica system shows 15.0 % of iron. These values were confirmed by EDX measurements. The iron distribution within both networks was homogeneous, since the EDX analysis showed no agglomeration. The incorporated NPs were determined by TEM measurements as shown in Figure 4.37. The bright field TEM images show a homogeneous distribution, the HAADF-STEM images reveal some separated areas with a higher z-contrast, probably denoting iron oxide-rich areas. The particle sizes of the observable structures are ≤10 nm in both materials [181].

Mössbauer experiments were carried out to determine the oxidation state of the particles (Fig. 4.38). The NPs within the matrix are rather small as already seen from the TEM images and have a non-negligible surface area, which causes a line broadening (different iron sites on the surface that contribute to the overall spectrum). Comparing the obtained isomer shifts of the carbon sample (0.32 $mm\cdot s^{-1}$) with literature results, it fits to either Fe_3C ($Fe_3C = 0.29$ $mm\cdot s^{-1}$) or Fe_2O_3 (0.37 $mm\cdot s^{-1}$) [181, 183–189]. With the obtained parameters (quadrupole splitting 0.71 $mm\cdot s^{-1}$), however, it was not possible to differentiate between Fe_2O_3 and Fe_3O_4, respectively. Nevertheless, due to the preparation under an oxidizing atmosphere and at elevated temperature, the incorporation of iron oxides rather than Fe_3C within the sample is more likely [181]. In addition, the magnetic properties of the materials were determined by SQUID experiments. The silica material showed a superparamagnetic behavior, while the carbon system confirmed a paramagnetic one, however, not following the Brillouin function. Recently, superparamagnetic iron-carbide-NPs were obtained by carbonization of ferrocene [183–189] or $FeCl_3$ [190–197], indicating that the silica material

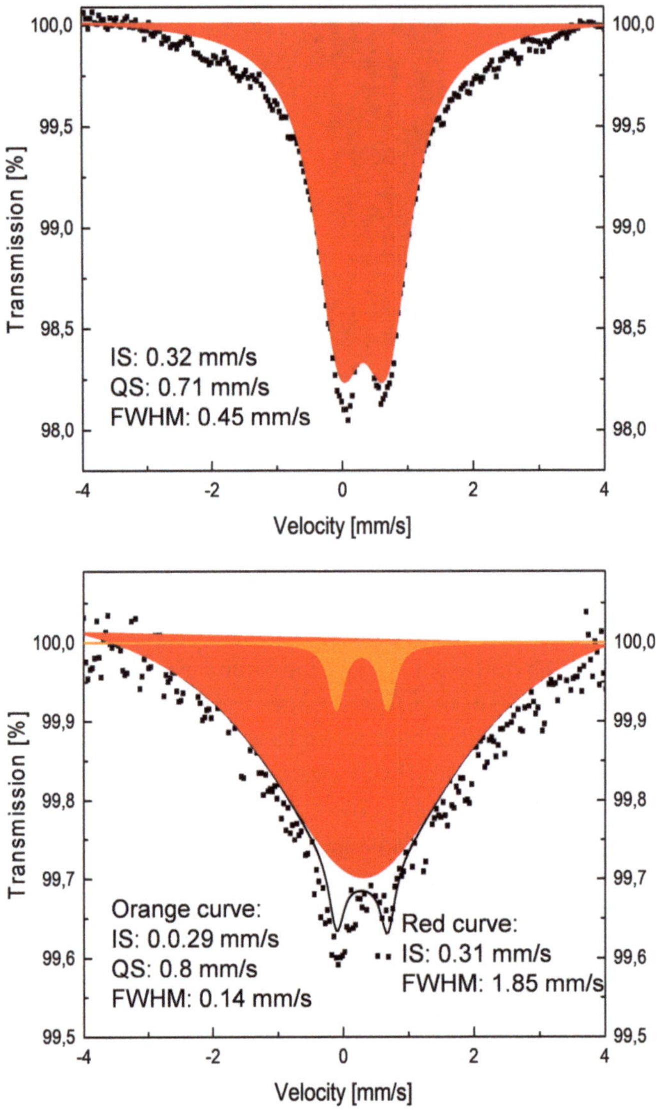

Fig. 4.38: Results of the Mössbauer experiments for the carbon (top) and silica sample (bottom) and the fitting of the experimental data (IS: Isomer Shift; QS: Quadrupole Splitting: FWHM: Full Width at Half Maximum). The very wide line widths of the Mössbauer spectra are due to the very small size of the NPs. Reproduced from [181] by permission of John Wiley & Sons, Inc.

contains Fe_3C. Based on the Mössbauer results (Fig. 4.38), no explanation was given for the differences in magnetism. The different possible iron species (*vide supra*) could not be distinguished and the formation of superparamagnetic NPs is possible in all cases. Furthermore, the thus prepared particles were not crystalline, which also influences their magnetic behavior [181].

The concept of twin polymerization was also applied in the synthesis of Sn/C/SiO_2 composites by using two different synthetic methodologies implying two different monomer types (Scheme 4.24) [109]. For additional information, see Chapter 5.1.

In the first approach, simultaneous twin polymerization was initiated by applying Si-Spiro, **2** with the tin alkoxides **5a**–**c** (tin(IV) species, in melt) and **6a**,**b** (tin(II) species; in melt and solution, respectively) as shown in Scheme 4.24 (monomer type I and type II approach). The reduced number of organic ligands in **6a**,**b**, when compared with **5a**–**c**, decreases the C content in the as-formed hybrid materials. These materials are composed of tin oxides, silica, and a phenolic resin (Scheme 4.24) [109].

Treatment of the as-prepared hybrid materials in a reducing atmosphere of Ar/H_2 (95/5) at a temperature of 600 °C gave the corresponding Sn/C/SiO_2 materials **7a-red**–**7d-red** (Scheme 4.24) (Brunauer-Emmett-Teller surface areas: 40–260 $m^2 \cdot g^{-1}$), containing Sn-NPs in the range of 28–48 nm [109]. The formation of metallic Sn-NPs was confirmed by electron diffraction studies as well as PXRD measurements. It was found that for **7a-red**–**7c-red**, the distribution of the Sn-NPs is homogeneous over large areas within the C/SiO_2 matrix (Fig. www).

In the second approach, twin polymerization was carried out by treatment of Si-Spiro (**2**) in the presence of additives, such as $Sn(OR)_4$ (**8a**, R = *t*-Bu; **8b**, R = *n*-Bu; **8c**, R = Ac) and Sn(O-*n*-Bu)$_2$ (**9a**) or $Sn(OAc)_2$ (**9b**) (Scheme 4.25) [109]. The polymerization process is initiated by the Lewis acidic character of the respective tin species affording the corresponding hybrid materials. These materials gave the respective Sn/C/SiO_2 networks under reducing conditions at 600 °C (Scheme 4.25). Within these materials, the Sn-NPs are embedded in the C/SiO_2 matrix. Nevertheless, a homogeneous distribution of the Sn-NPs is only observed when **8a** was used as tin additive.

Characterization of the as-prepared materials has been carried out using, for example, solid state ^{119}Sn NMR spectroscopy, showing that hexacoordinated tin species were incorporated in the polymer matrix [109]. In addition, HAADF-STEM studies were performed on the respective hybrid materials indicating the formation of characteristic morphologies of twin polymerization with 2 nm sized phase domains (Fig. 4.40).

In summary, both approaches did allow to control the resulting particle size of the embedded Sn-NPs in the range of 10–200 nm. In addition, they allowed to carry out the twin polymerization process in the melt and in solution of which, however, the polymerization in the melt enabled the formation of denser polymer matrices [109]. Hence, the leaching of Sn-NPs and sintering them during the reduction process for the formation of the respective composite materials is prevented. In contrast, the materials obtained from the polymerization in solution show larger surfaces, possessing an increased leaching of the particles. Using the tin(II) and tin(IV) additives $Sn(OR)_n$ (R = *t*-Bu, *n*-Bu, Ac; $n = 2, 4$) (*vide supra*) (**8**, **9**) gave hybrid materials with a high degree of particle leaching and a large particle distribution, which can be explained by the porous character of the respective materials.

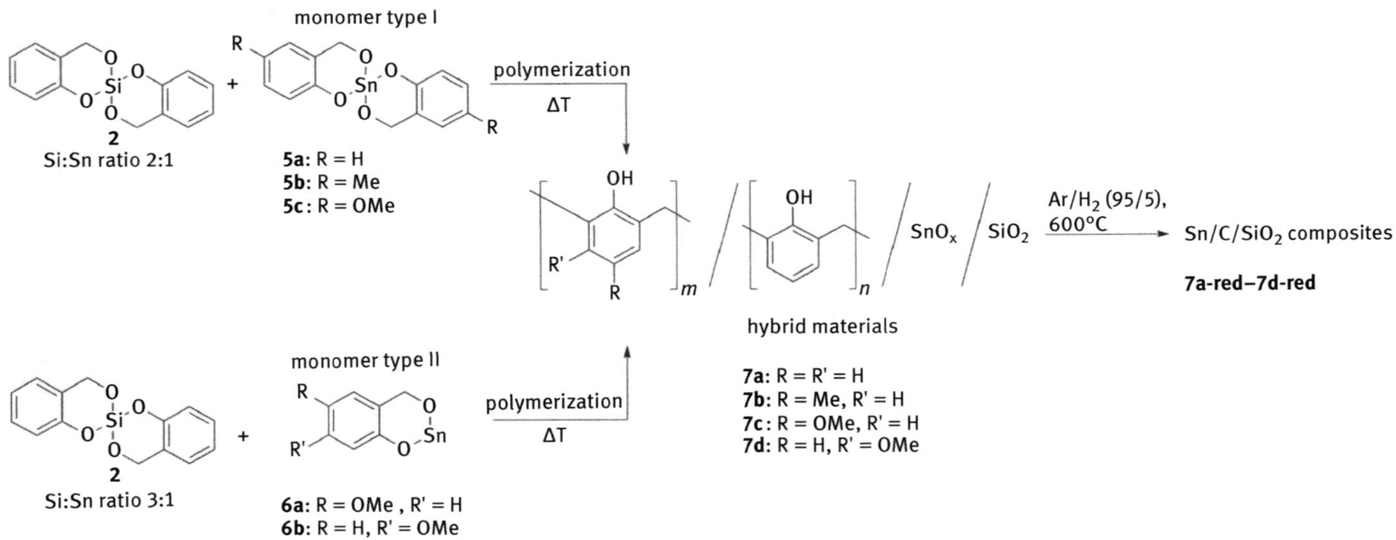

Scheme 4.24: Simultaneous twin polymerization (STP) of tin alkoxides **5a–c** and **6a,b** and the treatment of the respective hybrid materials under reducing conditions. Please note that STP of **7a–d** was performed in the melt [109].

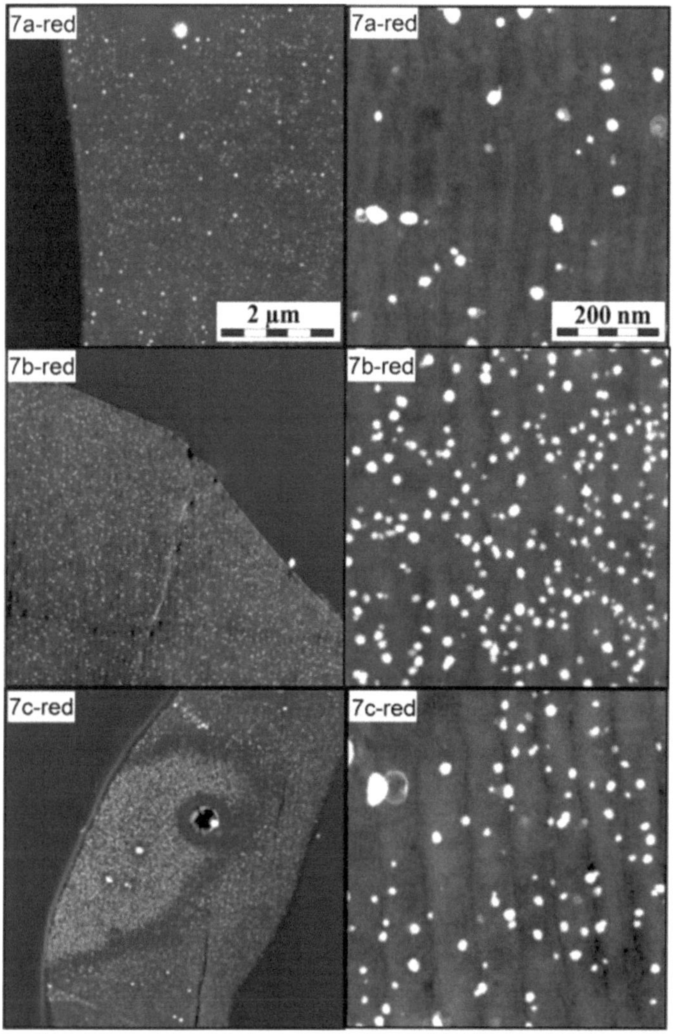

Fig. 4.39: HAADF-STEM images of the reduced hybrid materials **7a-red–7c-red**. Reproduced from [109] by permission of John Wiley and Sons, Inc.

4.4.4 Tin alloys embedded in carbon and silica materials by twin polymerization

A possibility to generate tin alloys, e.g., of the compositions Sn_2Co, Sn_2Ni_3, Sn_4Ni_3, and Sn_5Cu_6, embedded in a carbon or silica matrix is provided by the use of simultaneous twin polymerization in the presence of additives [84]. In this respect, the twin monomer Si-Spiro (**2**) was polymerized in the presence of tin-2,4-dimethoxyphenylmethanolate (**9c**) mixed with diverse metal additives, including metal carboxylates such as $[M(O_2CR)_2]$ ($R = CH_2(OCH_2CH_2)_2OMe$: **16a**, M = Co; **17a**, M = Ni; **18a**, M = Cu; R = $(CH_2)_9Me$: **16b**,

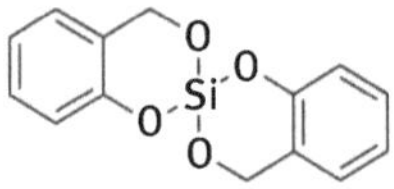

2

+ "tin additive" 1. polymerization 2. reduction → **10-red–15-red** Sn/C/SiO_2 composite materials

Sn(O-*t*-Bu)$_4$ (**8a**) in melt
Si:Sn ratio 2:1 hybrid material **10**
1:1 hybrid material **11**
1:2 hybrid material **12**

Sn(O-*t*-Bu)$_4$ (**8a**) in solution
Si:Sn ratio 2:1 hybrid material **10s**

Sn(O-*n*-Bu)$_2$ (**9a**) in solution
Si:Sn ratio 2:1 hybrid material **13**

Sn(OAc)$_2$ (**9b**) in solution
Si:Sn ratio 2:1 hybrid material **14**

Sn(OAc)$_4$ (**8c**) in solution
Si:Sn ratio 2:1 hybrid material **15**

Scheme 4.25: Polymerization of Si-Spiro (**2**) with different tin additives (**8**, **9**) to produce SnO_x/polymer/SiO_2 hybrid materials **10–15** and their conversion into Sn/C/SiO_2 composites **10-red–15-red** applying an atmosphere of Ar/H_2 (95/5) at 600 °C [109].

M = Co; **17b**, M = Ni; **18b**, M = Cu) and [Fe(O_2CR)$_3$] (**19a**, R = $CH_2(OCH_2CH_2)_2OMe$; **19b**, R = $(CH_2)_9Me$) (Scheme 4.26) (for metal-oxide-NP formation using these coordination complexes, see earlier). These metal carboxylates were chosen as metal source in the polymerization process due to their high solubility in toluene [171–173, 198].

The samples (Scheme 4.26) were polymerized in toluene at a temperature of 105 °C yielding amorphous solids after appropriate work-up [84]. The as-prepared hybrid materials were characterized by solid state ^{13}C NMR spectroscopy. Accordingly, the spectra contain resonance signals which can be assigned to a phenolic resin with an *o,o'*- and *o,p'*-substitution pattern. Additional signals were observed between 55–95 ppm which are attributed to the polymerized tin(II) alkoxide **9c** and the ethoxy functionalities of the metal carboxylates [84].

Treatment of the hybrid materials **20–24** (Scheme 4.26) under reducing conditions (Ar/H_2, 95/5) at 600–780 °C produced black solid nanocomposites **20-red–24-red** [84]. These materials consist of different tin alloys (Sn_2Co, Sn_2Ni_3, Sn_4Ni_3, Sn_5Cu_6), which are embedded in a C/SiO_2 matrix with particle sizes in the nanometer range (average particle size 20–80 nm) and with a homogeneous distribution. BET analyses of the final composites show surface areas between 10 and 327 $m^2 \cdot g^{-1}$, where less amounts of the metal additives correlate with the higher values. The powder X-ray diffraction and sealed-area electron diffraction studies on the corresponding nanocomposites confirmed the formation of the respective alloys. Only low amounts of elemental tin were detected, which can be explained by leaching of tin, due to the high temperature during the reduction process (see earlier), which is above the melting point of tin (232 °C). When the reduction was carried out below 600 °C, only amorphous composites were formed.

It was further found that variation of the Sn/M ratio allowed the formation of alloys of different compositions, e.g., Sn/Ni gave either Sn_2Ni_3 or Sn_4Ni_3 in the according nanocomposite **23-red** (**23a-red**, Sn_2Ni_3; **23b-red**, Sn_4Ni_3). HAADF-STEM images

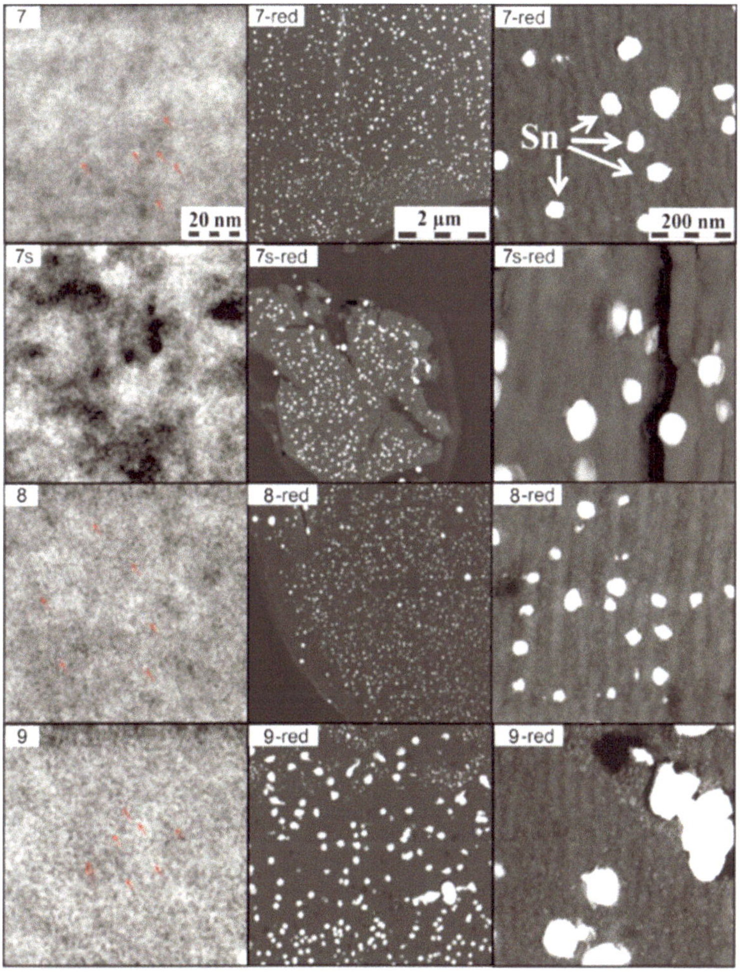

Fig. 4.40: HAADF-STEM images of the as-prepared hybrid materials (left) and the respective Sn/C/SiO_2 materials (right). Reproduced from [109] by permission of John Wiley & Sons, Inc.

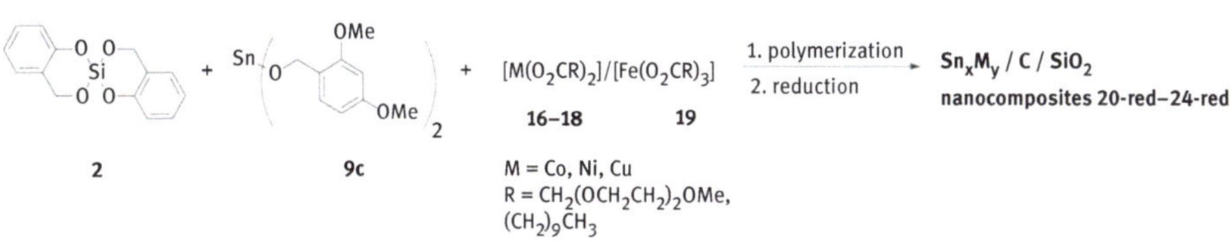

Scheme 4.26: Simultaneous twin polymerization of Si-Spiro (**2**) with **9c** in the presence of metal additives **16–19**, affording after reduction (Ar/H_2, 95/5) the respective nanocomposites **20-red–24-red** [84].

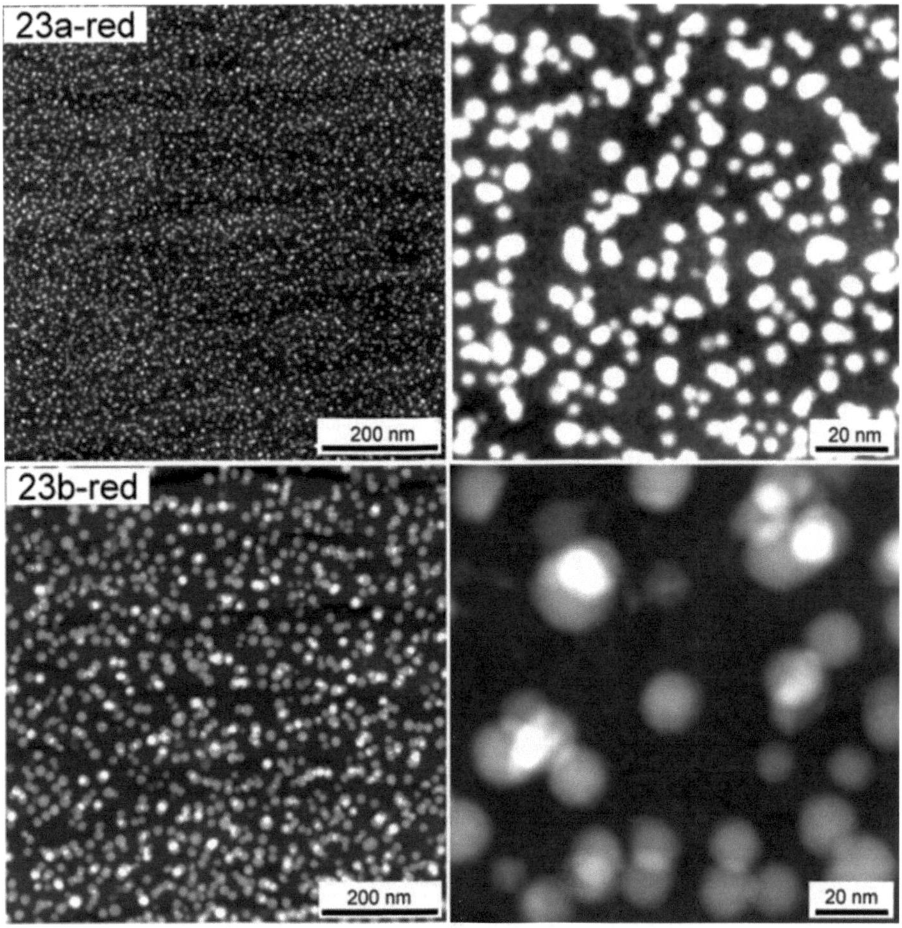

Fig. 4.41: HAADF-STEM images of the nanocomposites $Sn_2Ni_3/C/SiO_2$ (**23a-red**) (top) and $Sn_4Ni_3/C/SiO_2$ (**23b-red**) (bottom). Reproduced from [84] with permission of the Royal Society of Chemistry.

of these nanocomposites are depicted in Figure 4.41, showing a homogeneous distribution of the nanoparticular tin alloys in the C/SiO_2 matrix [84]. The same observations were made for all other alloy/C/SiO_2 materials.

Within the C/SiO_2 matrix, the smallest particles with a narrow size distribution of 3–10 nm were observed for the nanocomposite composed of Si_2Ni_3 [84]. As outlined earlier, in most material residues some amounts of elemental Sn were detected, with exception of composite $Sn_5Cu_6/C/SiO_2$, where solely the pure alloy is present. The authors point out that this novel approach offers access to different porous alloy/C/SiO_2 composites, which may be used for the development of novel nanocomposites as anode materials for lithium ion batteries, as exemplarily demonstrated for Sn_2Ni_3 (see Chapter 5.1) [84].

4.4.5 Metal-loaded porous hollow carbon spheres by twin polymerization

Within this chapter, the generation of silver and gold nanoparticles as well as iron-oxide-encapsulated hollow carbon spheres (HCS) by twin polymerization is reported [199].

Recently, the encapsulation of reactive species, such as M-NPs, gained great attraction [30, 199–205]. These materials benefit from a variety of advantages provided by the capsule itself, including the prevention of phase coalescence of M-NPs, enabling mass transport between the hollow core and the surrounding environment. The preparation of functional species mainly involves a mesoporous silica template, coated with Me_3Si octadecane [206]. In this respect, HCS with a mesoporous hull ($S_{BET} = 1620\ m^2.g^{-1}$) could be isolated after carbonization and removal of the template with HF. The respective HCS were then infiltrated by incipient wetness methods with platinum or ruthenium salts, affording, after reduction, $Pt_{50}Ru_{50}$ clusters. These materials were applied as catalysts in methanol or formic acid fuel cells, showing an improved catalytic activity compared to commonly used carbon black supports [207–209]. In addition, the incorporation and encapsulation of, for example, $CoFe_2O_4$, Fe_3O_4, Sn, and SnO_2-NPs in porous carbon were possible [210–214]. Magnetically separable HCS ($CoFe_2O_4$ and Fe_3O_4) or novel anode materials (Sn and SnO_2) for high capacity Li ion batteries were also accessible [74]. Moreover, HCS can be functionalized with NPs by using the appropriate NPs themselves as templates [30, 199]. For example, Au or Fe_3O_4-NPs could be encapsulated in a SiO_2 material followed by grafting a mesoporous silica layer onto these particles [141, 215]. Then the metal-containing silica template was coated with a carbon source such as polystyrene and carbonization of the thus obtained material was carried out. Removal of the silica template with HF led to mesoporous HCS featuring the corresponding NPs. Another example for direct encapsulation of M-NPs was achieved by reacting a solution of glucose and [$AgNO_3$] under hydrothermal conditions, whereby carbon-coated Ag-NPs were formed [216–218].

Towards the preparation of metal-loaded porous HCS (Fig. 4.42), silica spheres with different diameters (Aerosil® AS90, d ≈ 20 nm; Aerosil® OX50, d ≈ 40 nm; Stöber particles, d ≈ 200 nm) were, for example, coated with [$AgO_2CCH_2(OCH_2CH_2)_2OMe$] (**25**) and [$AuO_2CCH_2(OCH_2CH_2)_2OMe(PPh_3)$] (**26**), respectively [199]. Thermal treatment of these templates gave the respective M-NP-functionalized systems as depicted in Figure 4.42. They were characterized by PXRD and TEM.

Afterwards, the precursor-loaded templates (Fig. 4.42) were heated under an atmosphere of argon for 1 h at 250 °C, whereupon a color change of the AS90 and OX50 spheres from colorless to yellow (Ag) or red (Au) was observed, attributed to the plasmon resonance of the as-formed M-NPs (Fig. 4.43) [199].

The NP-loaded silica templates were covered with the twin monomer Si-Spiro (**2**); polymerization was catalyzed by $MeSO_3H$ and the thus formed twin-polymer-coated templates were carbonized at 800 °C under argon. The inorganic lattice of the twin

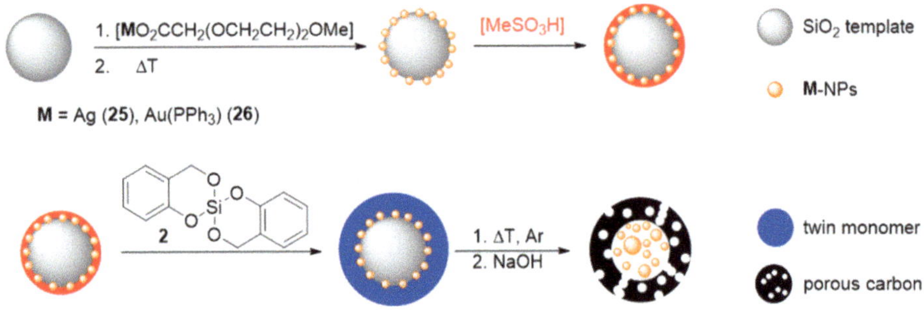

Fig. 4.42: Preparation of M-NPs-filled HCS by twin polymerization. Reproduced from [199] with some changes and by permission of Christian Schliebe.

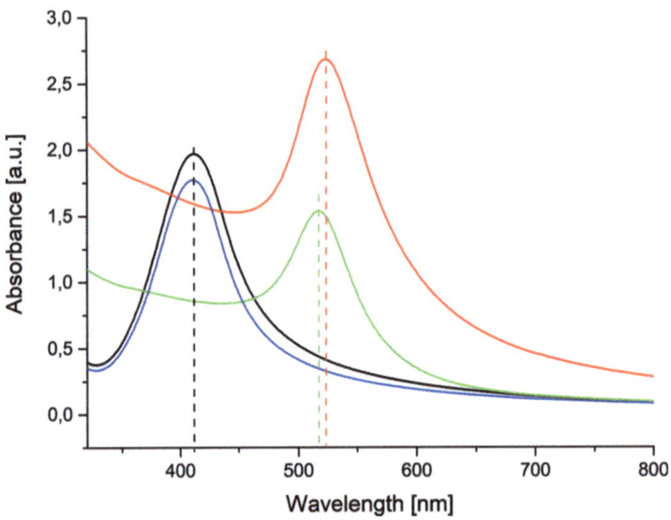

Fig. 4.43: UV-Vis spectra showing the plasmon resonance of the deposited NPs suspended in dichloromethane (Ag on AS90 (black), Ag on OX50 (blue), Au on AS90 (red), and Au on OX50 (green)). Reproduced from [199] by permission of Christian Schliebe.

monomer and the spherical template were removed by refluxing with a 5 M NaOH solution, giving the respective M-NPs-loaded HCS [199]. The metal loading was confirmed by AAS studies. However, this was only successful for the gold-containing HCS and the silver-containing material derived from the Stöber particle template (AS90: Ag (0.96 %), Au (7.1 %); OX50: Ag (0.92 %), Au (7.6 %); Stöber particle: Ag (3.8 %), Au (4.0 %)). For the Ag-containing HCS, prepared with the AS90 and OX50 template, the metal loading was determined from hybrid-material-coated compounds before carbonization, since the concentration in the carbon materials was too low

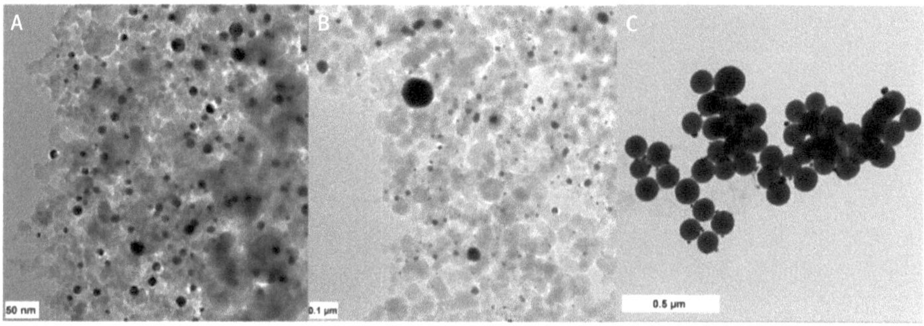

Fig. 4.44: TEM images of the used templates with the deposited NPs (**A**, Au on AS90; **B**, Au on OX50, and **C**, Ag on Stöber particles). Reproduced from [199] by permission of Christian Schliebe.

for successful AAS analysis. TEM studies confirmed the formation of Au- and Ag-NPs on all substrates with M-NP sizes between 10–20 nm for supports AS90 and OX50 (Fig. 4.44). The Stöber particles contain larger particles with 40–45 nm in diameter as evidenced by PXRD (substrate AS90: Ag (14±1 nm), Au (19±1 nm); OX50: Ag (17±1 nm), Au (34±3 nm); Stöber particle: Ag (22±1 nm), Au (44±4 nm)) [199].

The crystallite sizes were calculated using the Scherrer equation for CSP and by TEM measurements for the CAS90 and COX50 materials (CAS90: Au, 15±4 nm; COX50: Ag, 24±11 nm; Au, 21±9 nm; CSP: Ag, 58±10 nm; Au, 48±6 nm). The crystallite sizes have been increased for both metals in comparison to the M-NPs-functionalized templates, which is caused by the thermal treatment during the carbonization process [199].

In addition, the thickness of the carbon hull was determined by TEM studies (Fig. 4.45). The carbon materials derived from the AS90 templates exhibit a foam-like structure with thin walls (thickness 3–4 nm) and the incorporated M-NPs seem not to be encapsulated, rather situated in the cavities between the HCS. A possible explanation is given in Figure 4.46. Although the Ag-NPs within CAS90 could not be confirmed by TEM measurements, AAS studies proved the existence of silver in this material. In comparison, the COX50 carbon spheres have a much thicker capsule wall with ca. 32 nm, which can be explained by the decreased template surface area, while keeping the used amount of monomer equal within the different polymerization procedures [199]. In contrast to the AS90 substrate, the respective M-NPs are encapsulated within the cavities (Fig. 4.45). Increasing the template diameter to 200 nm (Stöber particles), however, did not improve the encapsulation process. Due to their uniform spherical shapes and size distribution, homogeneously formed carbon spheres were obtained. The wall thickness increased somewhat to 35 nm.

To confirm the porous character of COX50, CAS90, and CSP, N_2 adsorption-desorption isotherms were established. Surface areas reaching from 1165 to 1250 $m^2 \cdot g^{-1}$ were obtained (Fig. 4.47). For the Stöber particles, mainly IUPAC-type I isotherms were determined, while the smaller template sizes (AS90, 20 nm → OX50, 40 nm) exhibit

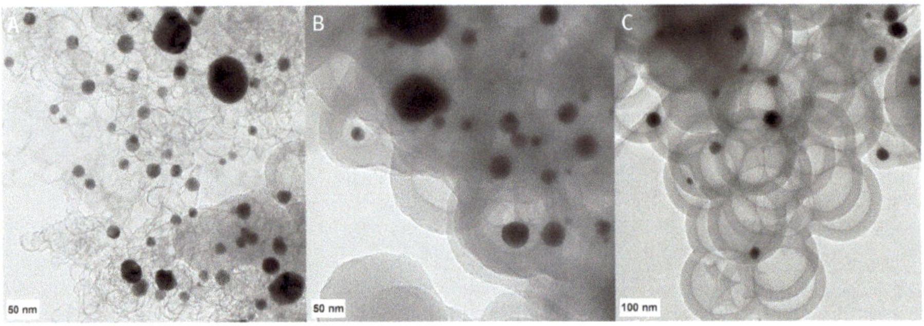

Fig. 4.45: Selected TEM images of the obtained carbon materials (**A**, Au on AS90; **B**, Ag on OX50; **C**, Au on Stöber particle). Reproduced from [199] by permission of Christian Schliebe.

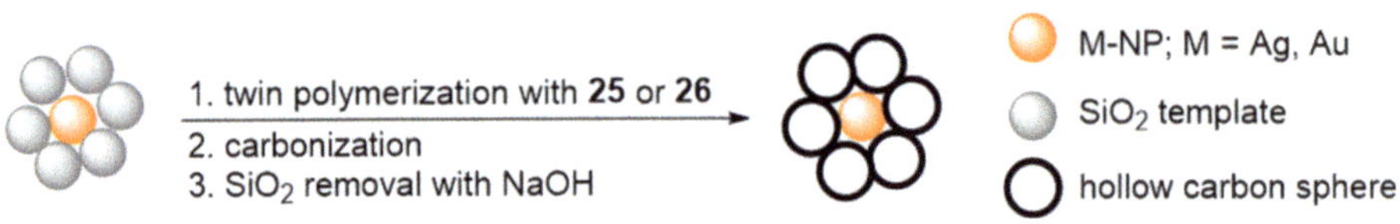

Fig. 4.46: Idealized M-NP (M = Ag, Au) stabilization on the AS90 template and resulting encapsulation within the formed cavities between the carbon capsules. Reproduced from [199] by permission of Christian Schliebe.

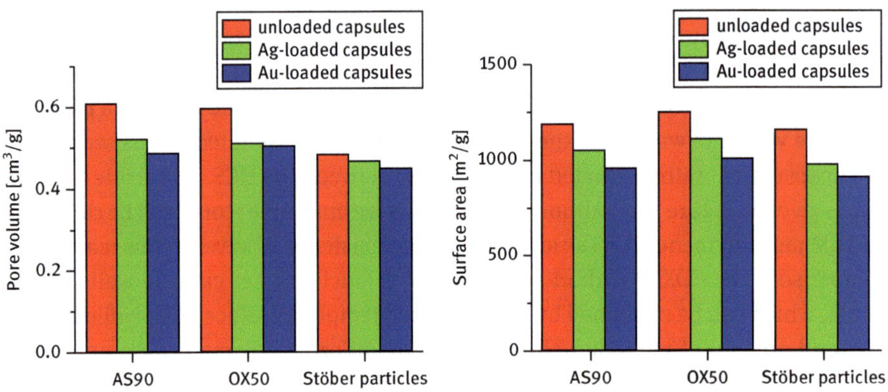

Fig. 4.47: Comparison of the obtained pore volumes and calculated surface areas of the loaded CAS90, COX50, and CSP materials (derived from nitrogen adsorption-desorption isotherms). Reproduced from [199] by permission of Christian Schliebe.

a slightly macroporous behavior (IUPAC-type II isotherm) [180]. Considering the pore size distribution, mainly microporous carbons were obtained. It could be shown that encapsulation of Ag- or Au-NPs only slightly influences the porous nature of the final HCS (Tab. 4.7). The incorporation of Au-NPs generally leads to overall smaller surface areas than compared with Ag, while the pore volumes somewhat decrease.

Tab. 4.7: Surface areas and specific pore volumes obtained from the N_2 adsorption-desorption isotherms by using the DFT method.

substrates	unloaded		Ag		Au	
	S_v [$m^2 \cdot g^{-1}$]	V_p [$cm^3 \cdot g^{-1}$]	S_v [$m^2 \cdot g^{-1}$]	V_p [$cm^3 \cdot g^{-1}$]	S_v [$m^2 \cdot g^{-1}$]	V_p [$cm^3 \cdot g^{-1}$]
CAS90	1189	0.608	1049	0.521	958	0.486
COX50	1250	0.598	1108	0.510	1001	0.505
CSP	1164	0.485	972	0.466	911	0.446

To investigate the influence of the different carbon shells, the reduction of methylene blue and 4-nitrophenol as an efficient benchmark system was applied.

The accessibility of the encapsulated Ag-NPs within the HCS was exemplarily investigated by the reduction of methylene blue with Na[BH_4] to give the respective colorless form of methylene blue, which was monitored by UV-Vis spectroscopy (Fig. 4.48).

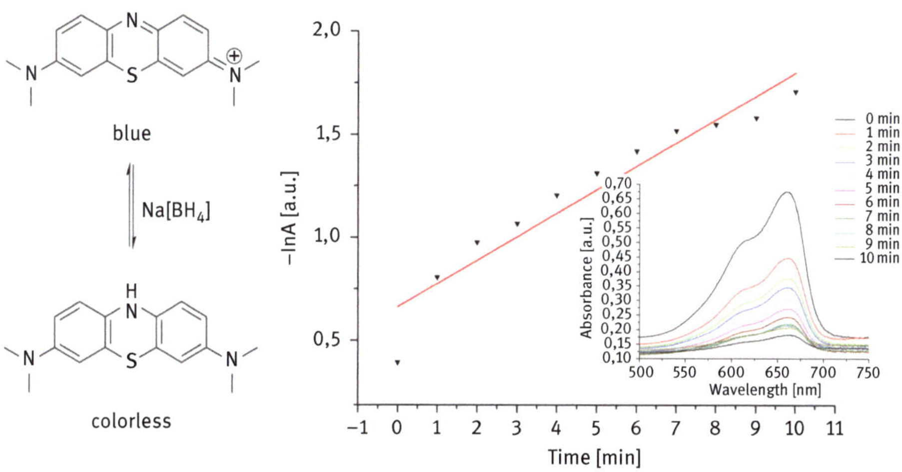

Fig. 4.48: Left: Redox reaction of methylene blue and formation of its colorless form (counter ions have been omitted for clarity). Right: Logarithmic plot of the maximum absorbance (▾, catalyst: Ag on CSP) and the linear fitting to give the *k* values from the slope (red line). Inset: UV-Vis spectra of methylene blue during the reduction with sodium borohydride and CSP. Reproduced from [199] with some changes and by permission of Christian Schliebe.

The silver-containing CAS90 (Aerosil® AS90, d ≈ 20 nm) showed the lowest activity for reducing methylene blue, while the highest activity was observed for COX50. The CSP exhibited a lower *k* value, since the crystallite size of the incorporated M-NPs (before encapsulation 22±1 nm, after encapsulation 58±10 nm) had increased. The reactions were followed by UV-Vis spectroscopy (Fig. 4.48). At the beginning of the reaction, an intense absorption at 650 nm was observed, which decreases during the reduction progress.

The experiment shows that for all prepared Ag-containing HCS, a first order kinetic is characteristic. The natural logarithm of the absorption maxima was plotted over the reduction time (Fig. 4.48). The *k* values for the specific material were determined from the slope. The *k* values for the smallest and largest template (CAS90, $k = 0.096\ s^{-1}$; CSP, $k = 0.114\ s^{-1}$) are similar, while the midsized material COX50 ($k = 0.217\ s^{-1}$) shows a higher value. The obtained results are in accordance with comparable systems described for Ag-NPs supported on silica spheres [216–218]. The metal content (see above) of the legacy hybrid materials, from where CAS90 and COX50 were obtained, was nearly the same and therefore, the difference in the rate constant *k* must be caused by the surrounding carbon sphere. To verify this hypothesis, the template only modified with the Ag-NPs was studied. Regarding the obtained *k* values for AS90 ($k = 0.018\ s^{-1}$) and OX50 ($k = 0.361\ s^{-1}$), the same trend was observed. It is assumed that the smallest template agglomerates around the NPs and so the accessibility is diminished as compared to COX50. This fact also has to be considered during the coating process. Hence, the NPs will be surrounded by empty HCS, which complicate the approachability to the M-NPs [199].

The accessibility of incorporated Au-NPs could also be shown by the reduction of 4-nitrophenol to 4-aminophenol in the presence of $Na[BH_4]$ (Fig. 4.49) [199]. The experiments were performed in a similar manner as discussed for the reduction of methylene blue. The reduction of 4-nitrophenol catalyzed by the gold-containing CSP material was not observed, which can be ascribed to the large crystallite size of the Au-NPs. However, smaller particles of the other materials facilitated the catalytic

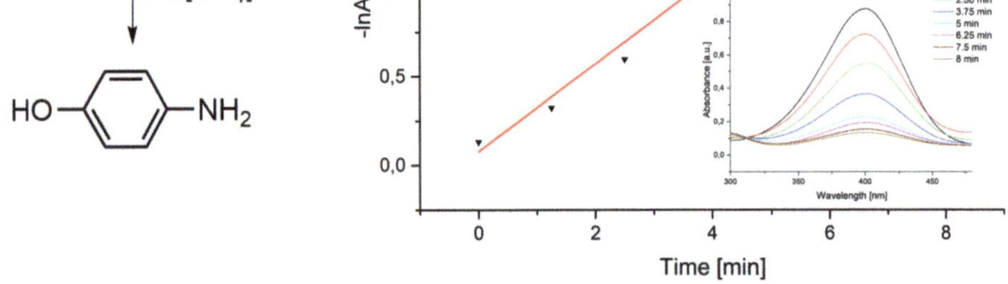

Fig. 4.49: Left: Reduction of 4-nitrophenol with $Na[BH_4]$ to give 4-aminophenol. Right: Logarithmic plot of the absorbance maximum (▾, catalyst: Au on COX50) and the linear fitting to give the *k* values from the slope (red line). Inset: UV-Vis spectra showing the reduction of the 4-nitrophenol. Reproduced from [199] with some changes and by permission of Christian Schliebe.

reduction process. This observation is supported by the literature, where the catalytic influence on the reduction reaction between growing and fully grown NPs was investigated, showing excellent activity with small NPs (ca. 5 nm) [199]. Since the Au-NPs in CAS90 and COX50 are smaller than in the CSP material (see above), a successful reduction of 4-nitrophenol to 4-aminophenol is observed (Fig. 4.49). As demonstrated for facetted Au-NPs, the concentration of $Na[BH_4]$ and 4-nitrophenol has an influence on the reaction kinetic as well [199]. Therefore, the apparent *k* value has to be considered (k_{app}). The obtained k_{app} values for CAS90 ($k_{app} = 0.114\ s^{-1}$) and COX50 ($k_{app} = 0.250\ s^{-1}$) exhibit the same trend as observed for the methylene blue experiment. Based on this, it is assumed that the same structural parameters, like cavity entrapped NPs in CAS90, cause this difference. The obtained results are in agreement with, for example, bimetallic NPs or Au particles deposited on ceria [199, 216–218].

Considering the obtained results, the incorporated NPs can be accessed by the used chemicals and can act as reduction catalyst towards 4-nitrophenol. It was also shown that the template diameter plays a crucial part within the encapsulation process and the best results could be reached with COX50 [199].

The preparation of iron oxide porous carbon capsules is possible by the reaction sequence shown in Figure 4.50.

The coating abilities of **4** and Si-Spiro (**2**) upon spherical silica templates were investigated [181]. Therefore, plain silica particles (200 nm in diameter) were functionalized with methanesulfonic acid. After addition of the monomer mixture during 12 h, the coated templates precipitated. The thus prepared material was carbonized in an atmosphere of argon at 800 °C. Afterward, the incorporated SiO_2 was removed by refluxing the carbonized material in a 5 M NaOH solution (Fig. 4.50).

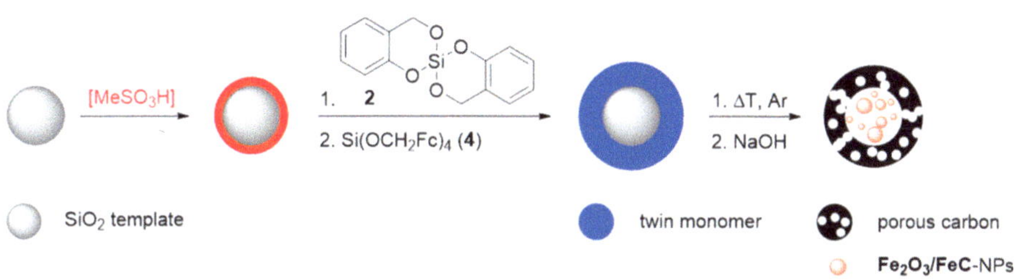

Fig. 4.50: Preparation of Fe_2O_3-filled porous carbon capsules (Fc = Fe(η^5-C_5H_4)(η^5-C_5H_5)) [181].

The obtained carbon capsules were characterized by N_2 adsorption-desorption isotherms. The observed isotherm looks like a type I isotherm according to IUPAC [180] (Fig. 4.51), which is caused by the microporous carbon obtained from the twin polymer of Si-Spiro (**2**). However, considering the desorption part of the observed isotherm, a small hysteresis typical for IUPAC-type IV isotherms [180] could be noticed. This is probably caused by cavities formed between connecting capsules.

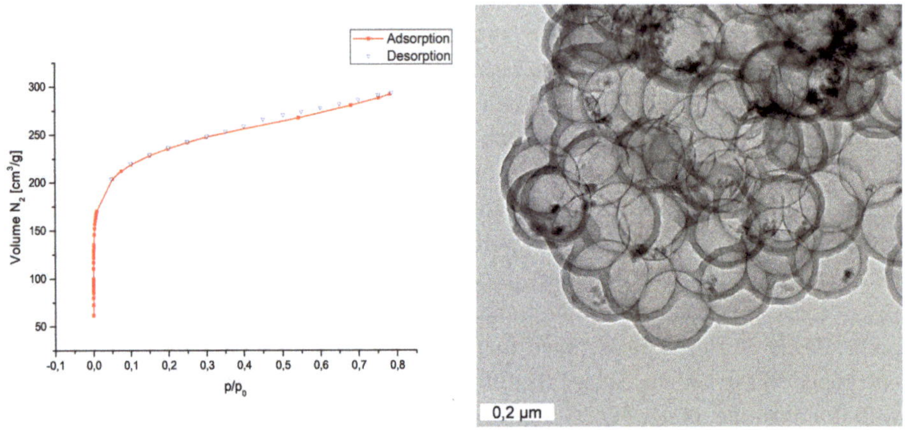

Fig. 4.51: Left: N_2 adsorption-desorption isotherms of the obtained carbon capsules. Right: bright field TEM images showing the prepared porous carbon capsules with the incorporated NPs. Reproduced from [181] by permission of John Wiley & Sons, Inc.

The calculated surface area (915 $m^2 \cdot g^{-1}$) and the obtained pore volume (0.42 $cm^3 \cdot g^{-1}$) are smaller than the values for the reference material (surface area 1180 $m^2 \cdot g^{-1}$, pore volume 0.48 $cm^3 \cdot g^{-1}$) derived by using solely Si-Spiro (**2**) as twin monomer. TEM studies reveal the formation of spherical carbon capsules with incorporated NPs. This is most probably ascribed to the finite carbon hull thickness, since within the bulk material the incorporated NPs are even better separated. A Fe content of 5 % was determined by AAS.

Mössbauer and SQUID experiments showed that the incorporated NPs are of the same nature as those obtained by simultaneous twin polymerization of Si-Spiro (**2**) with **4** (see earlier) and therefore, also exhibiting a paramagnetic behavior (Fig. 4.52).

4.4.6 Summary

Within this article, the twin polymerization as well as simultaneous twin polymerization of main-group- and transition-metal-containing twin monomers, including 2,2'-spirobi[4*H*-1,3,2-benzodioxasiline] and $Si(OCH_2Fc)_4$ ($Fc = Fe(\eta^5\text{-}C_5H_4)(\eta^5\text{-}C_5H_5)$) and additives such as tin(II) and tin(IV) alkoxides as well as metal carboxylates of the type $[M(O_2CCH_2(OCH_2CH_2)_2OMe)_m]$ and $[M(O_2CCH_2(OCH_2CH_2)_2OMe)_mL_x]$ (M = Cu, Ag, Au, Pd, Pt, Ru, Rh; $m = 1, 2$; L = phosphine, isonitrile; $x = 1, 2$), $[M(O_2CR)_2]$ and $[Fe(O_2CR)_3]$ ($R = CH_2(OCH_2CH_2)_2OMe$, $(CH_2)_9Me$; M = Co, Ni, Cu) as metal source were used for the preparation of hybrid materials. In this respect, nanomaterials of porous carbon and silica, decorated with metal or metal-oxide-NPs or tin alloys were obtained. In addition, the encapsulation of metal or metal-oxide-NPs within porous carbon shells by using the twin polymerization approach is reported. The chemical, analytical, and physical properties of these materials are discussed.

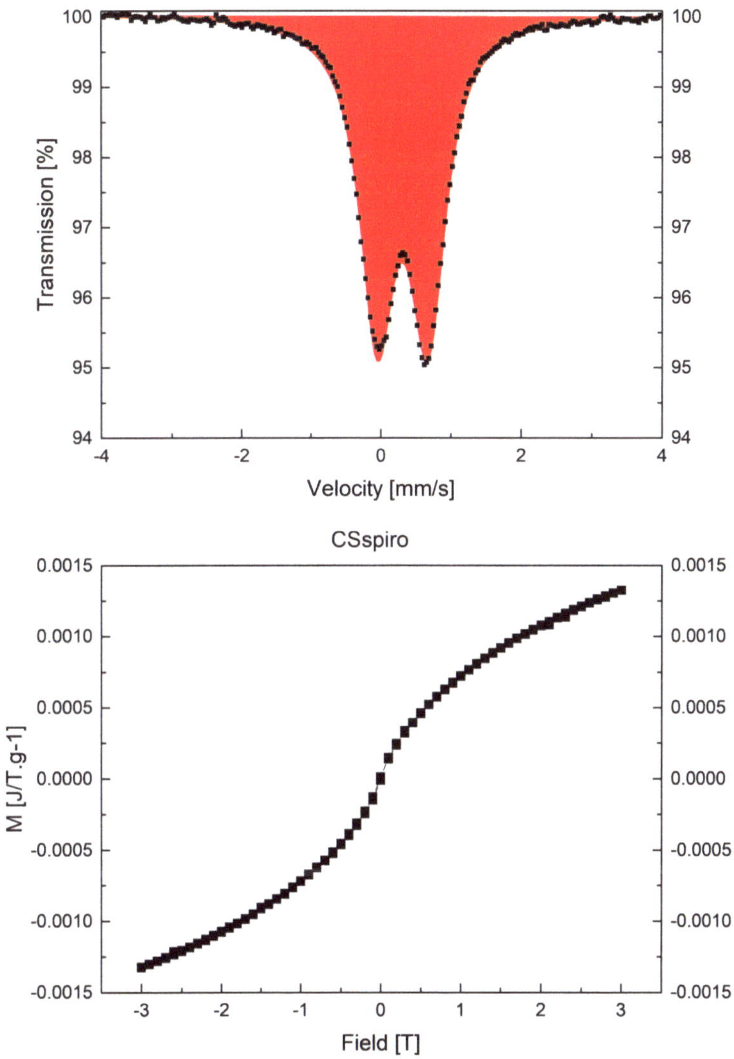

Fig. 4.52: Results of the Mössbauer experiments with the fitting of the experimental data (IS = 0.31 mm·s^{-1}, QS = 0.72 mm·s^{-1}, FWHM = 0.30 mm·s^{-1}) for the iron-containing carbon capsules (top) and the respective SQUID experiments (bottom). Reproduced from [181] by permission of John Wiley & Sons, Inc.

References

[1] Sanchez C, Julián B, Belleville P, Popall M. Applications of hybrid organic–inorganic nanocomposites. J Mater Chem 2005;15:3559. doi:10.1039/b509097k.

[2] Faulkner R, DiVerdi J, Yang Y, Kobayashi T, Maciel G. The Surface of Nanoparticle Silicon as Studied by Solid-State NMR. Materials 2012;6:18–46. doi:10.3390/ma6010018.

[3] Kickelbick G. Hybrid materials synthesis, characterization, and applications. Weinheim: Wiley-VCH; 2007.
[4] Gómez-Romero P, Sanchez C, editors. Functional hybrid materials. Weinheim: Wiley-VCH; 2004.
[5] Rózga-Wijas K, Chojnowski J, Ścibiorek M, Fortuniak W. Polysiloxane–silica hybrids from novel precursors by the sol–gel process. J Mater Chem 2005;15:2383. doi:10.1039/b501833c.
[6] Breiner JM, Mark JE, Beaucage G. Preparations, Structures, and Properties of Polysiloxane-Silica Composites Prepared from a Variety of Hydrolyzable Precursors. MRS Proceedings 1998;520. doi:10.1557/PROC-520-275.
[7] Brinker CJ, Scherer GW. Sol-gel science: the physics and chemistry of sol-gel processing. Boston: Academic Press; 1990.
[8] Wen J, Wilkes GL. Organic/Inorganic Hybrid Network Materials by the Sol-Gel Approach. Chem Mater 1996;8:1667–81. doi:10.1021/cm9601143.
[9] Schubert U, Huesing N, Lorenz A. Hybrid Inorganic-Organic Materials by Sol-Gel Processing of Organofunctional Metal Alkoxides. Chem Mater 1995;7:2010–27. doi:10.1021/cm00059a007.
[10] Schubert U, Hüsing N. Synthesis of inorganic materials. 2., revised and updated ed. Weinheim: Wiley-VCH; 2005.
[11] Staudinger von H, Hahn W. Versuche zur Darstellung von polymeren Kieselsäureestern. Die Makromolekulare Chemie 1953;11:24–50. doi:10.1002/macp.1953.020110103.
[12] Löschner T. Synthese und Zwillingspolymerisation neuer Silicium-Monomere. Diploma Thesis. Technische Universität Chemnitz, 2008.
[13] Grund S, Kempe P, Baumann G, Seifert A, Spange S. Nanocomposites Prepared by Twin Polymerization of a Single-Source Monomer. Angew Chem Int Ed 2007;46:628–32. doi:10.1002/anie.200504327.
[14] Spange S, Grund S. Nanostructured Organic-Inorganic Composite Materials by Twin Polymerization of Hybrid Monomers. Adv Mater 2009;21:2111–6. doi:10.1002/adma.200802797.
[15] Spange S, Kempe P, Seifert A, Auer AA, Ecorchard P, Lang H, et al. Nanocomposites with Structure Domains of 0.5 to 3 nm by Polymerization of Silicon Spiro Compounds. Angew Chem Int Ed 2009;48:8254–8. doi:10.1002/anie.200901113.
[16] Kempe P, Löschner T, Auer AA, Seifert A, Cox G, Spange S. Thermally Induced Twin Polymerization of 4*H*-1,3,2-Benzodioxasilines. Chem Eur J 2014;20:8040–53. doi:10.1002/chem.201400038.
[17] Löschner T, Mehner A, Grund S, Seifert A, Pohlers A, Lange A, et al. A Modular Approach for the Synthesis of Nanostructured Hybrid Materials with Tailored Properties: The Simultaneous Twin Polymerization. Angew Chem Int Ed 2012;51:3258–61. doi:10.1002/anie.201108011.
[18] Weißhuhn J, Mark T, Martin M, Müller P, Seifert A, Spange S. Ternary organic-inorganic nanostructured hybrid materials by simultaneous twin polymerization. Polym Chem 2016;7:5060–8. doi:10.1039/C6PY00903D.
[19] Plueddemann EP. Silane coupling agents. 2. ed. New York: Plenum Press; 1991.
[20] Brook MA, Chen Y, Guo K, Zhang Z, Brennan JD. Sugar-modified silanes: precursors for silica monoliths. J Mater Chem 2004;14:1469–79. doi:10.1039/B401278J.
[21] Niu S, Yan H, Chen Z, Li S, Xu P, Zhi X. Unanticipated bright blue fluorescence produced from novel hyperbranched polysiloxanes carrying unconjugated carbon-carbon double bonds and hydroxyl groups. Polym Chem 2016;7:3747–55. doi:10.1039/C6PY00654J.
[22] Koltzenburg S, Maskos M, Nuyken O. Polymere: Synthese, Eigenschaften und Anwendungen. Berlin Heidelberg: Springer Spektrum; 2014.
[23] Spange S, Mark T, Weißhuhn J, Friebe N. Unpublished Results 2016.

[24] Oschatz S, Lange A, Csihony S, Cox G, Gronwald O, Seifert A, et al. Functional twin monomers and twin macro monomers as components for the synthesis of hierarchically nanostructured hybrid materials. J Polym Sci Part A: Polym Chem 2016;54:2312–20. doi:10.1002/pola.28102.

[25] Göring M, Seifert A, Schreiter K, Müller P, Spange S. A non-aqueous procedure to synthesize amino group bearing nanostructured organic–inorganic hybrid materials. Chem Commun 2014;50:9753–6. doi:10.1039/C4CC03640A.

[26] Ebert T, Wollbrink A, Seifert A, John R, Spange S. Multiple polymerization – formation of hybrid materials consisting of two or more polymers from one monomer. Polym Chem 2016;7:6826–33. doi:10.1039/C6PY01619G.

[27] Ebert T. Mehrlingspolymerisation in Substanz und an Oberflächen zur Synthese nanostrukturierter und poröser Materialien. Dissertation. Technische Universität Chemnitz, 2016.

[28] Löschner T. Synthese nanostrukturierter, organisch-anorganischer Hybridmaterialien über Zwillingspolymerisation. Dissertation. Technische Universität Chemnitz, 2013.

[29] Príncipe M, Suárez H, Jimenez GH, Martínez R, Spange S. Composites prepared from silica gel and furfuryl alcohol with *p*-toluenesulfonic acid as the catalyst. Polym Bull 2007;58:619–26. doi:10.1007/s00289-006-0701-5.

[30] Böttger-Hiller F, Kempe P, Cox G, Panchenko A, Janssen N, Petzold A, et al. Twin Polymerization at Spherical Hard Templates: An Approach to Size-Adjustable Carbon Hollow Spheres with Micro- or Mesoporous Shells. Angew Chem Int Ed 2013;52:6088–91. doi:10.1002/anie.201209849.

[31] Böttger-Hiller F. Nanostrukturierter Kohlenstoff durch Zwillingspolymerisation an Hart-Templaten. Dissertation. Technische Universität Chemnitz, 2012.

[32] Ebert T, Cox G, Sheremet E, Gordan O, Zahn DRT, Simon F, et al. Carbon/carbon nanocomposites fabricated by base catalyzed twin polymerization of a Si-spiro compound on graphite sheets. Chem Commun 2012;48:9867–9. doi:10.1039/c2cc34775j.

[33] Windberg T, Ebert T, Uhlig D, Schulze S, Spange S. Hierarchically structured carbon/carbon nanocomposites with adjustable porosity fabricated by twin polymerization. Microporous Mesoporous Mater 2017;246:62–71. doi:10.1016/j.micromeso.2017.03.012.

[34] Trick KA, Saliba TE. Mechanisms of the pyrolysis of phenolic resin in a carbon/phenolic composite. Carbon 1995;33:1509–15. doi:10.1016/0008-6223(95)00092-R.

[35] Fitzer E, Schaefer W, Yamada S. The formation of glasslike carbon by pyrolysis of polyfurfuryl alcohol and phenolic resin. Carbon 1969;7:643–8. doi:10.1016/0008-6223(69)90518-1.

[36] Ebert T, Seifert A, Spange S. Twin Polymerization – a New Principle for Hybrid Material Synthesis. Macromol Rapid Commun 2015;36:1623–39. doi:10.1002/marc.201500182.

[37] Kresge CT, Leonowicz ME, Roth WJ, Vartuli JC, Beck JS. Ordered mesoporous molecular sieves synthesized by a liquid-crystal template mechanism. Nature 1992;359:710–2. doi:10.1038/359710a0.

[38] Ebert T. Herstellung neuer Siliziummonomere für die Synthese neuer SiO_x/C-Hybridmaterialien. Diploma Thesis. Technische Universität Chemnitz, 2010.

[39] Böttger-Hiller F, Mehner A, Anders S, Kroll L, Cox G, Simon F, et al. Sulphur-doped porous carbon from a thiophene-based twin monomer. Chem Commun 2012;48:10568–70. doi:10.1039/C2CC35112A.

[40] Prause S, Spange S, Barthel H. Surface Polarity of Dimethylsiloxane-Grafted Silica Particles. Macromol Chem Phys 2005;206:364–71. doi:10.1002/macp.200400399.

[41] Löschner T, Seifert A, Lungwitz R, Cox G, Lange A, Hähnle H-J, et al. Simultaneous Twin Polymerization – Controlling the Nanostructure Formation of Hybrid Materials. Macromol Rapid Commun, vol. 32, Freiburg: 2011, p. F60–1.

[42] Wöckel L, Seifert A, Mende C, Roth-Panke I, Kroll L, Spange S. Resin and carbon foam production by cationic step-growth polymerization of organic carbonates. Polym Chem 2017;8:404–13. doi:10.1039/C6PY01572G.

[43] Wöckel L, Windberg T, John R, Seifert A, Spange S. Hierarchically structured carbon and silica by chemical foaming. Polym Chem 2018;9:1385–96. doi:10.1039/C7PY01888F.
[44] Spange S, Wöckel L, Windberg T, John R, Seifert A. Hierarchically Structured Resin-, Hybrid Material-, Carbon- and Silica-Foams. Macromol Rapid Commun, Freiburg: 2018.
[45] Dannenberg F, Thiele G, Dornsiepen E, Dehnen S, Mehring M. Synthesis, structure and thermolysis of oxazagermines and oxazasilines. New J Chem 2017;41:4990–7. doi:10.1039/C7NJ00894E.
[46] Kaßner L, Knoblauch A, Seifert A, Grützner R-E, Cox G, Lange A, et al. Nanostructured Aniline Formaldehyde Resin/Polysilazane Hybrid Materials by Twin Polymerization. Macromol Chem Phys 2016;217:2462–72. doi:10.1002/macp.201600152.
[47] Kaßner L. Stickstoffhaltige Monomere zur Herstellung von Hybridmaterialien. Dissertation. Technische Universität Chemnitz, 2015.
[48] Windberg T. Herstellung und Charakterisierung substituierter 2-Aminobenzylalkohole als neue Precursoren für Silicium-Spiroverbindungen. Bachelor Thesis. Technische Universität Chemnitz, 2011.
[49] Kaßner L, Nagel K, Grützner R-E, Korb M, Rüffer T, Lang H, et al. Polyamide 6/silica hybrid materials by a coupled polymerization reaction. Polym Chem 2015;6:6297–304. doi:10.1039/C5PY00815H.
[50] Nagel K, Kaßner L, Seifert A, Grützner R-E, Cox G, Spange S. Ternary composites by an *in situ* hydrolytic polymerization process. RSC Adv 2018;8:14713–21. doi:10.1039/C8RA02402B.
[51] Antonietti M, Müllen K, editors. Chemical Synthesis and Applications of Graphene and Carbon Materials. Weinheim, Germany: Wiley-VCH Verlag GmbH & Co. KGaA; 2017. doi:10.1002/9783527648160.
[52] Schlögl R. Carbon in Catalysis. Advances in Catalysis, vol. 56, Elsevier; 2013, pp. 103–85. doi:10.1016/B978-0-12-420173-6.00002-4.
[53] Mauter MS, Elimelech M. Environmental Applications of Carbon-Based Nanomaterials. Environ Sci Tech 2008;42:5843–59. doi:10.1021/es8006904.
[54] Sengupta R, Bhattacharya M, Bandyopadhyay S, Bhowmick AK. A review on the mechanical and electrical properties of graphite and modified graphite reinforced polymer composites. Prog Polym Sci 2011;36:638–70. doi:10.1016/j.progpolymsci.2010.11.003.
[55] Bakry R, Vallant RM, Najam-ul-Haq M, Rainer M, Szabo Z, Huck CW, et al. Medicinal applications of fullerenes. Int J Nanomedicine 2007;2:639–49.
[56] Casiraghi C, Robertson J, Ferrari AC. Diamond-like carbon for data and beer storage. Mater Today 2007;10:44–53. doi:10.1016/S1369-7021(06)71791-6.
[57] Zhu Y, Murali S, Cai W, Li X, Suk JW, Potts JR, et al. Graphene and Graphene Oxide: Synthesis, Properties, and Applications. Adv Mater 2010;22:3906–24. doi:10.1002/adma.201001068.
[58] Raccichini R, Varzi A, Passerini S, Scrosati B. The role of graphene for electrochemical energy storage. Nature Mater 2015;14:271–9. doi:10.1038/nmat4170.
[59] De Volder MFL, Tawfick SH, Baughman RH, Hart AJ. Carbon Nanotubes: Present and Future Commercial Applications. Science 2013;339:535–9. doi:10.1126/science.1222453.
[60] Ōtani S. On the carbon fiber from the molten pyrolysis products. Carbon 1965;3:31–8. doi:10.1016/0008-6223(65)90024-2.
[61] Shindo A. Report of the Government Industrial Research Institute, Osaka, Japan 1961.
[62] Titirici M-M, Antonietti M. Chemistry and materials options of sustainable carbon materials made by hydrothermal carbonization. Chem Soc Rev 2009;39:103–16. doi:10.1039/B819318P.

[63] Berl E, Schmidt A. Über die Entstehung der Kohlen. II. Die Inkohlung von Cellulose und Lignin in neutralem Medium. Justus Liebigs Ann Chem 1932;493:97–123. doi:10.1002/jlac.19324930106.

[64] Beck JS, Vartuli JC, Roth WJ, Leonowicz ME, Kresge CT, Schmitt KD, et al. A new family of mesoporous molecular sieves prepared with liquid crystal templates. J Am Chem Soc 1992;114:10834–43. doi:10.1021/ja00053a020.

[65] Ryoo R, Joo SH, Jun S. Synthesis of Highly Ordered Carbon Molecular Sieves via Template-Mediated Structural Transformation. J Phys Chem B 1999;103:7743–6. doi:10.1021/jp991673a.

[66] Titirici M-M, Thomas A, Antonietti M. Aminated hydrophilic ordered mesoporous carbons. J Mater Chem 2007;17:3412–8. doi:10.1039/B703569A.

[67] Wollbrink A, Ebert T, Koch J, Tegenkamp C, Spange S, Caro J. Private Communication 2017.

[68] Liu R, Shi Y, Wan Y, Meng Y, Zhang F, Gu D, et al. Triconstituent Co-assembly to Ordered Mesostructured Polymer–Silica and Carbon–Silica Nanocomposites and Large-Pore Mesoporous Carbons with High Surface Areas. J Am Chem Soc 2006;128:11652–62. doi:10.1021/ja0633518.

[69] Sing KSW. Reporting physisorption data for gas/solid systems with special reference to the determination of surface area and porosity (Recommendations 1984). Pure Appl Chem 1985;57:603–19. doi:10.1351/pac198557040603.

[70] Kitschke P, Auer AA, Löschner T, Seifert A, Spange S, Rüffer T, et al. Microporous Carbon and Mesoporous Silica by Use of Twin Polymerization: An Integrated Experimental and Theoretical Approach to Precursor Reactivity. ChemPlusChem 2014;79:1009–23. doi:10.1002/cplu.201402029.

[71] Schliebe C, Noll J, Scharf S, Gemming T, Seifert A, Spange S, et al. Nitrogen-containing porous carbon materials by twin polymerization. Colloid Polym Sci 2018;296:413–26. doi:10.1007/s00396-017-4254-y.

[72] Spange S. Silica surface modification by cationic polymerization and carbenium intermediates. Prog Polym Sci 2000;25:781–849. doi:10.1016/S0079-6700(00)00014-9.

[73] Böttger-Hiller F, Kempe P, Baumann G, Hietschold M, Schäfer P, Zahn DRT, et al. The Controlled Synthesis of Carbon Tubes and Rods by Template-Assisted Twin Polymerization. Adv Mater Sci Eng 2013;2013:1–8. doi:10.1155/2013/872019.

[74] Brückner J, Thieme S, Böttger-Hiller F, Bauer I, Grossmann HT, Strubel P, et al. Carbon-Based Anodes for Lithium Sulfur Full Cells with High Cycle Stability. Adv Funct Mater 2014;24:1284–9. doi:10.1002/adfm.201302169.

[75] Grund S. Synthese anorganisch-organischer Polyfurfurylalkohol-Nanokomposite durch die Zwillingspolymerisation. Dissertation. Technische Universität Chemnitz, 2010.

[76] Kitschke P, Walter M, Rüffer T, Lang H, Kovalenko MV, Mehring M. From molecular germanates to microporous Ge@C via twin polymerization. Dalton Trans 2016;45:5741–51. doi:10.1039/C6DT00049E.

[77] Kitschke P, Walter M, Rüffer T, Seifert A, Speck F, Seyller T, et al. Porous Ge@C materials via twin polymerization of germanium(II) salicyl alcoholates for Li-ion batteries. J Mater Chem A 2016;4:2705–19. doi:10.1039/C5TA09891B.

[78] Leonhardt C, Brumm S, Seifert A, Cox G, Lange A, Rüffer T, et al. Tin Oxide Nanoparticles and SnO_2/SiO_2 Hybrid Materials by Twin Polymerization Using Tin(IV) Alkoxides. ChemPlusChem 2013;78:1400–12. doi:10.1002/cplu.201200242.

[79] Mehner A, Pohlers A, Hoyer W, Cox G, Spange S. Cationic Twin Polymerization Versus Simultaneous Polymerization of Titanium Compounds to Fabricate Nanostructured Organic Polymer/TiO_2 Hybrid Materials. Macromol Chem Phys 2013;214:1000–10. doi:10.1002/macp.201200629.

[80] Mehner A, Rüffer T, Lang H, Pohlers A, Hoyer W, Spange S. Synthesis of Nanosized TiO_2 by Cationic Polymerization of (μ4-oxido)-hexakis(μ-furfuryloxo)-octakis(furfuryloxo)-tetra-titanium. Adv Mater 2008;20:4113–7. doi:10.1002/adma.200801376.
[81] Böttger-Hiller F, Lungwitz R, Seifert A, Hietschold M, Schlesinger M, Mehring M, et al. Nanoscale Tungsten Trioxide Synthesized by *In Situ* Twin Polymerization. Angew Chem Int Ed 2009;48:8878–81. doi:10.1002/anie.200903636.
[82] Böhme S. Heterogene Metalloxide durch Zwillingspolymerisation. Diploma Thesis. Technische Universität Chemnitz, 2012.
[83] Schliebe C, Gemming T, Noll J, Mertens L, Mehring M, Seifert A, et al. Zirconium and Hafnium Twin Monomers for Mixed Oxides. ChemPlusChem 2015;80:559–67. doi:10.1002/cplu.201402338.
[84] Leonhardt C, Seifert A, Csihony S, Sommer H, Mehring M. Nanocomposites by the use of simultaneous twin polymerization: tin alloys in a carbon/silica matrix. RSC Adv 2016;6:3091–8. doi:10.1039/C5RA18574B.
[85] Weißhuhn J, Seifert A, Dzhagan V, Palaniyappan S, Zahn DRT, Wagner G, et al. B_2O_3/SiO_2/Phenolic Resin Hybrid Materials Produced by Simultaneous Twin Polymerization of Spiromonomers. Macromol Chem Phys 2018;219:1700487. doi:10.1002/macp.201700487.
[86] Auer AA, Richter A, Berezkin AV, Guseva DV, Spange S. Theoretical Study of Twin Polymerization – From Chemical Reactivity to Structure Formation. Macromol Theory Simul 2012;21:615–28. doi:10.1002/mats.201200036.
[87] Holleman AF, Wiberg E, Wiberg N. Lehrbuch der Anorganischen Chemie. 102nd ed. Berlin: de Gruyter; 2007.
[88] Weißhuhn J, Seifert A, Dzhagan V, Palaniyappan S, Zahn DRT, Wagner G, et al. B_2O_3/SiO_2/Phenolic Resin Hybrid Materials Produced by Simultaneous Twin Polymerization of Spiromonomers. Macromol Chem Phys 2018;219:1700487. doi:10.1002/macp.201700487.
[89] Fierro JLG. Metal Oxides: Chemistry and Applications. Boca Raton: CRC press; 2005.
[90] Rodríguez J, Fernández-García M. Synthesis, Properties, and Applications of Oxide Nanomaterials. John Wiley & Sons, Ltd; 2007.
[91] Jones AC, Hitchman ML. Chemical Vapour Deposition: Precursors, Processes and Applications. Cambridge: Royal Society of Chemistry; 2008.
[92] Fernandez-Garcia M, Martinez-Arias A, Hanson JC, Rodriguez JA. Nanostructured Oxides in Chemistry: Characterization and Properties. Chem Rev 2004;104:4063–104. doi:10.1021/cr030032f.
[93] Kudo A, Miseki Y. Heterogeneous photocatalyst materials for water splitting. Chem Soc Rev 2009;38:253–78. doi:10.1039/b800489g.
[94] Knapp CE, Carmalt CJ. Solution based CVD of main group materials. Chem Soc Rev 2016;45:1036–64. doi:10.1039/C5CS00651A.
[95] Jolivet J-P, Henry M, Livage J, Bescher E. Metal Oxide Chemistry and Synthesis: From Solution to Solid State. John Wiley & Sons, Ltd; 2000.
[96] Boyjoo Y, Wang M, Pareek VK, Liu J, Jaroniec M. Synthesis and applications of porous non-silica metal oxide submicrospheres. Chem Soc Rev 2016;45:6013–47. doi:10.1039/C6CS00060F.
[97] Joo SH, Zhao D. Environmental dynamics of metal oxide nanoparticles in heterogeneous systems: A review. J Hazard Mater 2017;322, Part A:29–47. doi:10.1016/j.jhazmat.2016.02.068.
[98] Sanchez C, Rozes L, Ribot F, Laberty-Robert C, Grosso D, Sassoye C, et al. "Chimie douce": A land of opportunities for the designed construction of functional inorganic and hybrid organic-inorganic nanomaterials. C R Chim 2010;13:3–39. doi:10.1016/j.crci.2009.06.001.

[99] Sanchez C, Soler-Illia G, Ribot F, Lalot T, Mayer CR, Cabuil V. Designed hybrid organic-inorganic nanocomposites from functional nanobuilding blocks. Chem Mater 2001;13:3061–83. doi:10.1021/cm011061e.
[100] Schliebe C, Gemming T, Mertens L, Mehring M, Lang H. Twin Polymerization: A Unique and Efficient Tool for Supporting Silver Nanoparticles on Highly Porous Carbon and Silica. Eur J Inorg Chem 2014;3161–3. doi:10.1002/ejic.201402299.
[101] Chen D, Qu Z, Shen S, Li X, Shi Y, Wang Y, et al. Comparative studies of silver based catalysts supported on different supports for the oxidation of formaldehyde. Catal Today 2011;175:338–45. doi:10.1016/j.cattod.2011.03.059.
[102] Chen L, Ma D, Bao X. Hydrogen Treatment-Induced Surface Reconstruction: Formation of Superoxide Species on Activated Carbon over Ag/Activated Carbon Catalysts for Selective Oxidation of CO in H_2-Rich Gases. J Phys Chem C 2007;111:2229–34. doi:10.1021/jp0665628.
[103] Chen L, Ma D, Li X, Bao X. Silver catalysts supported over activated carbons for the selective oxidation of CO in excess hydrogen: effects of different treatments on the supports. Catal Lett 2006;111:133–9. doi:10.1007/s10562-006-0139-2.
[104] Chen Y, Wang C, Liu H, Qiu J, Bao X. Ag/SiO_2: a novel catalyst with high activity and selectivity for hydrogenation of chloronitrobenzenes. Chem Commun 2005:5298–300. doi:10.1039/B509595F.
[105] Salaev MA, Krejker AA, Magaev OV, Malkov VS, Knyazev AS, Borisova ES, et al. Ethylene glycol oxidation over supported catalyst in tubular reactor. Chem Eng J 2011;172:399–409. doi:10.1016/j.cej.2011.05.079.
[106] Steffan M, Jakob A, Claus P, Lang H. Silica supported silver nanoparticles from a silver(I) carboxylate: Highly active catalyst for regioselective hydrogenation. Catal Commun 2009;10:437–41. doi:10.1016/j.catcom.2008.10.003.
[107] Leonhardt C, Selter S, Paul A, Tiemann M, Mehring M. Nanoscaled Tin Oxide for Gas Sensing Synthesized via Twin Polymerization. GDCh-Wissenschaftsforum Chemie, Dresden (Germany): 2015.
[108] Leonhardt C. Zinnalkoxide als Präkursoren für zinnhaltige Nanokomposite. Dissertation. Technische Universität Chemnitz, 2016.
[109] Leonhardt C, Brumm S, Seifert A, Lange A, Csihony S, Mehring M. Tin Nanoparticles in Carbon/Silica Hybrid Materials by the Use of Twin Polymerization. ChemPlusChem 2014;79:1440–7. doi:10.1002/cplu.201402137.
[110] Schmid G, editor. Nanoparticles: from theory to application. 2nd ed. Weinheim: Wiley-VCH; 2010.
[111] Vollmer C, Janiak C. Naked metal nanoparticles from metal carbonyls in ionic liquids: Easy synthesis and stabilization. Coord Chem Rev 2011;255:2039–57. doi:10.1016/j.ccr.2011.03.005.
[112] Pelaz B, Jaber S, de Aberasturi DJ, Wulf V, Aida T, de la Fuente JM, et al. The State of Nanoparticle-Based Nanoscience and Biotechnology: Progress, Promises, and Challenges. ACS Nano 2012;6:8468–83. doi:10.1021/nn303929a.
[113] Mahmoodi NO, Ghavidast A, Amirmahani N. A comparative study on the nanoparticles for improved drug delivery systems. J Photochem Photobiol B 2016;162:681–93. doi:10.1016/j.jphotobiol.2016.07.037.
[114] Wais U, Jackson AW, He T, Zhang H. Nanoformulation and encapsulation approaches for poorly water-soluble drug nanoparticles. Nanoscale 2016;8:1746–69. doi:10.1039/c5nr07161e.
[115] Ahmed S, Ahmad M, Swami BL, Ikram S. A review on plants extract mediated synthesis of silver nanoparticles for antimicrobial applications: A green expertise. J Adv Res 2016;7:17–28. doi:10.1016/j.jare.2015.02.007.

[116] Astruc D, editor. Nanoparticles and catalysis. Weinheim: Wiley-VCH; 2008.
[117] Tao FF. Excavation of Precious-Metal-Based Alloy Nanoparticles for Efficient Catalysis. Angew Chem Int Ed 2016;55:15212–4. doi:10.1002/anie.201606324.
[118] Nguyen-Phan T-D, Baber AE, Rodriguez JA, Senanayake SD. Au and Pt nanoparticle supported catalysts tailored for H_2 production: From models to powder catalysts. Appl Catal A Gen 2016;518:18–47. doi:10.1016/j.apcata.2015.12.012.
[119] Lu A-H, Salabas EL, Schüth F. Magnetic nanoparticles: synthesis, protection, functionalization, and application. Angew Chem Int Ed 2007;46:1222–44. doi:10.1002/anie.200602866.
[120] Niu M, Pham-Huy C, He H. Core-shell nanoparticles coated with molecularly imprinted polymers: a review. Microchim Acta 2016;183:2677–95. doi:10.1007/s00604-016-1930-4.
[121] Aliofkhazraei M, editor. Comprehensive guide for nanocoatings technology. New York: Nova Publishers; 2015.
[122] Zhou K, Li Y. Catalysis Based on Nanocrystals with Well-Defined Facets. Angew Chem Int Ed 2012;51:602–13. doi:10.1002/anie.201102619.
[123] Maurer T, Adam P-M, Lévêque G. Coupling between plasmonic films and nanostructures: from basics to applications. Nanophotonics 2015;4:363–82. doi:10.1515/nanoph-2014-0015.
[124] Wei S, Wang Q, Zhu J, Sun L, Lin H, Guo Z. Multifunctional composite core-shell nanoparticles. Nanoscale 2011;3:4474–502. doi:10.1039/c1nr11000d.
[125] Jun Y, Choi J, Cheon J. Shape Control of Semiconductor and Metal Oxide Nanocrystals through Nonhydrolytic Colloidal Routes. Angew Chem Int Ed 2006;45:3414–39. doi:10.1002/anie.200503821.
[126] Romo-Herrera JM, Alvarez-Puebla RA, Liz-Marzán LM. Controlled assembly of plasmonic colloidal nanoparticle clusters. Nanoscale 2011;3:1304–15. doi:10.1039/c0nr00804d.
[127] Elechiguerra JL, Reyes-Gasga J, Yacaman MJ. The role of twinning in shape evolution of anisotropic noble metal nanostructures. J Mater Chem 2006;16:3906–19. doi:10.1039/B607128G.
[128] Tao AR, Habas S, Yang P. Shape Control of Colloidal Metal Nanocrystals. Small 2008;4:310–25. doi:10.1002/smll.200701295.
[129] Xia Y, Xiong Y, Lim B, Skrabalak SE. Shape-Controlled Synthesis of Metal Nanocrystals: Simple Chemistry Meets Complex Physics? Angew Chem Int Ed 2009;48:60–103. doi:10.1002/anie.200802248.
[130] Zhang H-X, Wang H, Re Y-S, Cai W-B. Palladium nanocrystals bound by {110} or {100} facets: from one pot synthesis to electrochemistry. Chem Commun 2012;48:8362–4. doi:10.1039/C2CC33941B.
[131] Brust M, Walker M, Bethell D, Schiffrin DJ, Whyman R. Synthesis of thiol-derivatised gold nanoparticles in a two-phase Liquid–Liquid system. J Chem Soc, Chem Commun 1994:801–2. doi:10.1039/C39940000801.
[132] Kanaras AG, Kamounah FS, Schaumburg K, Kiely CJ, Brust M. Thioalkylated tetraethylene glycol: a new ligand for water soluble monolayer protected gold clusters. Chem Commun (Camb) 2002:2294–5.
[133] Turkevich J, Stevenson PC, Hillier J. A study of the nucleation and growth processes in the synthesis of colloidal gold. Discuss Faraday Soc 1951;11:55–75. doi:10.1039/DF9511100055.
[134] Kimling J, Maier M, Okenve B, Kotaidis V, Ballot H, Plech A. Turkevich Method for Gold Nanoparticle Synthesis Revisited. J Phys Chem B 2006;110:15700–7. doi:10.1021/jp061667w.
[135] Frens G. Particle size and sol stability in metal colloids. Kolloid-Z u Z Polymere 1972;250:736–41. doi:10.1007/BF01498565.
[136] Frens G. Controlled Nucleation for the Regulation of the Particle Size in Monodisperse Gold Suspensions. Nature 1973;241:20–2. doi:10.1038/10.1038/physci241020a0.
[137] Liu Q, Zhou D, Yamamoto Y, Ichino R, Okido M. Preparation of Cu nanoparticles with $NaBH_4$ by aqueous reduction method. Transactions of Nonferrous Metals Society of China 2012;22:117–23. doi:10.1016/S1003-6326(11)61149-7.

[138] Song KC, Lee SM, Park TS, Lee BS. Preparation of colloidal silver nanoparticles by chemical reduction method. Korean J Chem Eng 2009;26:153–5.
[139] Samal AK, Sreeprasad TS, Pradeep T. Investigation of the role of $NaBH_4$ in the chemical synthesis of gold nanorods. J Nanopart Res 2010;12:1777–86. doi:10.1007/s11051-009-9733-8.
[140] Liu Z, Jiao Y, Wang Y, Zhou C, Zhang Z. Polysaccharides-based nanoparticles as drug delivery systems. Adv Drug Deliv Rev 2008;60:1650–62. doi:10.1016/j.addr.2008.09.001.
[141] Sharma VK, Yngard RA, Lin Y. Silver nanoparticles: green synthesis and their antimicrobial activities. Adv Colloid Interface Sci 2009;145:83–96. doi:10.1016/j.cis.2008.09.002.
[142] Hernández-Sánchez H, Gutiérrez-López GF, editors. Food nanoscience and nanotechnology. Cham: Springer; 2015.
[143] Schliebe C, Jiang K, Schulze S, Hietschold M, Cai W-B, Lang H. A convenient light initiated synthesis of silver and gold nanoparticles using a single source precursor. Chem Commun 2013;49:3991–3. doi:10.1039/C3CC41445K.
[144] Aherne D, Ledwith DM, Gara M, Kelly JM. Optical Properties and Growth Aspects of Silver Nanoprisms Produced by a Highly Reproducible and Rapid Synthesis at Room Temperature. Adv Funct Mater 2008;18:2005–16. doi:10.1002/adfm.200800233.
[145] Fujita S. Organic Chemistry of Photography. Berlin, Heidelberg: Springer Berlin Heidelberg; 2004.
[146] Stamplecoskie KG, Scaiano JC. Light emitting diode irradiation can control the morphology and optical properties of silver nanoparticles. J Am Chem Soc 2010;132:1825–7. doi:10.1021/ja910010b.
[147] Sudeep PK, Kamat PV. Photosensitized Growth of Silver Nanoparticles under Visible Light Irradiation: A Mechanistic Investigation. Chem Mater 2005;17:5404–10. doi:10.1021/cm0512777.
[148] Janiak C. Ionic Liquids for the Synthesis and Stabilization of Metal Nanoparticles. Z Naturforsch B 2013;68:1059–89. doi:10.5560/znb.2013–3140.
[149] He Z, Alexandridis P. Nanoparticles in ionic liquids: interactions and organization. Phys Chem Chem Phys 2015;17:18238–61. doi:10.1039/C5CP01620G.
[150] Prechtl M, Dupont J, editors. Nanocatalysis in Ionic liquids. Weinheim: Wiley-VCH; 2016.
[151] Adner D, Korb M, Schulze S, Hietschold M, Lang H. A straightforward approach to oxide-free copper nanoparticles by thermal decomposition of a copper(I) precursor. Chem Commun 2013;49:6855. doi:10.1039/c3cc42914h.
[152] Adner D, Möckel S, Korb M, Buschbeck R, Rüffer T, Schulze S, et al. Copper(II) and triphenylphosphine copper(I) ethylene glycol carboxylates: synthesis, characterisation and copper nanoparticle generation. Dalton Trans 2013;42:15599–609. doi:10.1039/C3DT51913A.
[153] Blaudeck T, Adner D, Hermann S, et al. Wafer-level decoration of carbon nanotubes in field-effect transistor geometry with preformed gold nanoparticles using a microfluidic approach. Microelectronic Eng 2015;137:135–40.
[154] Adner D, Noll J, Schulze S, Hietschold M, Lang H. Aspherical silver nanoparticles by thermal decomposition of a single-source-precursor. Inorganica Chim Acta 2016;446:19–23. doi:10.1016/j.ica.2016.02.059.
[155] Rodriguez RD, Blaudeck T, Kalbacova J, Sheremet E, Schulze S, Adner D, et al. Metal nanoparticles reveal the organization of single-walled carbon nanotubes in bundles. RSC Adv 2016;6:15753–8. doi:10.1039/C5RA28181D.
[156] Struppert T, Jakob A, Heft A, Grünler B, Lang H. The use of silver(I)-2-[2-(2-methoxyethoxy) ethoxy]acetate as precursor in the deposition of thin silver layers on float glass by the atmospheric pressure combustion chemical vapor deposition process. Thin Solid Films 2010;518:5741–4. doi:10.1016/j.tsf.2010.05.082.

[157] Jahn SF, Jakob A, Blaudeck T, Schmidt P, Lang H, Baumann RR. Inkjet printing of conductive patterns with an aqueous solution of [$AgO_2C(CH_2OCH_2)_3H$] without any additional stabilizing ligands. Thin Solid Films 2010;518:3218–22. doi:10.1016/j.tsf.2010.01.030.
[158] Jahn SF, Blaudeck T, Baumann RR, Jakob A, Ecorchard P, Rüffer T, et al. Inkjet Printing of Conductive Silver Patterns by Using the First Aqueous Particle-Free MOD Ink without Additional Stabilizing Ligands. Chem Mater 2010;22:3067–71. doi:10.1021/cm9036428.
[159] Oestreicher A, Röhrich T, Wilden J, Lerch M, Jakob A, Lang H. An innovative method for joining materials at low temperature using silver (nano)particles derived from [$AgO_2C(CH_2OCH_2)_3H$]. Appl Surf Sci 2013;265:239–44. doi:10.1016/j.apsusc.2012.10.186.
[160] Gilles S, Tuchscherer A, Lang H, Simon U. Dip-pen-based direct writing of conducting silver dots. J Colloid Interface Sci 2013;406:256–62. doi:10.1016/j.jcis.2013.05.047.
[161] Kirillova A, Schliebe C, Stoychev G, Jakob A, Lang H, Synytska A. Hybrid Hairy Janus Particles Decorated with Metallic Nanoparticles for Catalytic Applications. ACS Appl Mater Interfaces 2015;7:21218–25. doi:10.1021/acsami.5b05224.
[162] Tuchscherer A, Schaarschmidt D, Schulze S, Hietschold M, Lang H. Gold nanoparticles generated by thermolysis of "all-in-one" gold(I) carboxylate complexes. Dalton Trans 2012;41:2738–46. doi:10.1039/C2DT11748G.
[163] Schoner C, Tuchscherer A, Blaudeck T, Jahn SF, Baumann RR, Lang H. Particle-free gold metal-organic decomposition ink for inkjet printing of gold structures. Thin Solid Films 2013;531:147–51. doi:10.1016/j.tsf.2013.01.027.
[164] Tuchscherer A, Schaarschmidt D, Schulze S, Hietschold M, Lang H. Simple and efficient: Gold nanoparticles from triphenylphosphane gold(I) carboxylates without addition of any further stabilizing and reducing agent. Inorg Chem Commun 2011;14:676–8. doi:10.1016/j.inoche.2011.02.003.
[165] Tuchscherer A, Schaarschmidt D, Schulze S, Hietschold M, Lang H. Simple and Efficient: Ethylene Glycol Isonitrile Gold(I) Chlorides for the Formation and Stabilization of Gold Nanoparticles. Eur J Inorg Chem 2011;4421–8. doi:10.1002/ejic.201100324.
[166] Tuchscherer A, Packheiser R, Rüffer T, Schletter H, Hietschold M, Lang H. Rhodium Nanoparticles from Dirhodium(II) Ethylene Glycol Tetracarboxylates. Eur J Inorg Chem 2012;2251–8. doi:10.1002/ejic.201101424.
[167] Jeschke J. Metall-Carboxylate: Synthese, Charakterisierung und Anwendung. Master Thesis. Technische Universität Chemnitz, 2013.
[168] Wiese A. Synthese, Charakterisierung und Anwendung von Palladium-Nanopartikeln ausgehend von einem Phosphan-stabilisierten Palladium-Carboxylat. Bachelor Thesis. Technische Universität Chemnitz, 2016.
[169] Künzel-Tenner A. Synthese und Charakterisierung von Platin-Nanopartikeln ausgehend von Single-Source Precursoren. Bachelor Thesis. Technische Universität Chemnitz, 2016.
[170] Böhme N. Synthese, Charakterisierung und Katalyseeigenschaften von Rutheniumnano-partikel. Bachelor Thesis. Technische Universität Chemnitz, 2016.
[171] Assim K, Schulze S, Pügner M, Uhlemann M, Gemming T, Giebeler L, et al. Co(II) ethylene glycol carboxylates for Co_3O_4 nanoparticle and nanocomposite formation. J Mater Sci 2017;52:6697–711. doi:10.1007/s10853-017-0904-1.
[172] Assim K, Melzer M, Korb M, Rüffer T, Jakob A, Noll J, et al. Bis(β-diketonato)- and allyl-(β-diketonato)-palladium(II) complexes: synthesis, characterization and MOCVD application. RSC Adv 2016;6:102557–69. doi:10.1039/C6RA22887A.
[173] Assim K, Jeschke J, Jakob A, Dhakal D, Melzer M, Georgi C, et al. Manganese half-sandwich complexes as metal-organic chemical vapor deposition precursors for manganese-based thin films. Thin Solid Films 2016;619:265–72. doi:10.1016/j.tsf.2016.10.052.

[174] Kempe P, Löschner T, Adner D, Spange S. Selective ring opening of 4*H*-1,3,2-benzodioxasiline twin monomers. New J Chem 2011;35:2735–9. doi:10.1039/C1NJ20654K.

[175] Mehner A, Rüffer T, Lang H, Schlesinger M, Mehring M, Spange S. Chirality and reversible solid state phase transition of a tetranuclear titanium oxido cluster. New J Chem 2013;37:1290–3. doi:10.1039/C3NJ40775F.

[176] Adner D, Wolf FM, Möckel S, Perelaer J, Schubert US, Lang H. Copper(II) ethylene glycol carboxylates as precursors for inkjet printing of conductive copper patterns. Thin Solid Films 2014;565:143–8. doi:10.1016/j.tsf.2014.06.054.

[177] Tamuliene J, Noll J, Frenzel P, Rüffer T, Jakob A, Walfort B, et al. Synthesis of $[\{AgO_2CCH_2OMe(PPh_3)\}_n]$ and theoretical study of its use in focused electron beam induced deposition. Beilstein J Nanotech 2017;8:2615–24. doi:10.3762/bjnano.8.262.

[178] Chandra S, Dietrich S, Lang H, Bahadur D. Dendrimer–Doxorubicin conjugate for enhanced therapeutic effects for cancer. J Mater Chem 2011;21:5729–37. doi:10.1039/C0JM04198J.

[179] Dietrich S, Chandra S, Georgi C, Thomas S, Makarov D, Schulze S, et al. Design, characterization and magnetic properties of Fe_3O_4-nanoparticle arrays coated with PEGylated-dendrimers. Mater Chem Phys 2012;132:292–9. doi:10.1016/j.matchemphys.2011.11.015.

[180] Donohue M, Aranovich G. Classification of Gibbs adsorption isotherms. Adv Colloid Interface Sci 1998;76–77:137–52. doi:10.1016/S0001-8686(98)00044-X.

[181] Schliebe C, Pfaff U, Gemming T, Lochenie C, Weber B, Lang H. $Si(OCH_2Fc)_4$: Synthesis, Electrochemical Behavior, and Twin Polymerization. Eur J Inorg Chem 2015;3850–60. doi:10.1002/ejic.201500464.

[182] Suzaki Y, Yoshigoe Y, Osakada K. Cyclic and linear poly(ferrocenylene alkylene)s synthesized from addition-condensation polymerization of ferrocene with aldehydes. J Polym Sci Part A: Polym Chem 2013;51:3627–35. doi:10.1002/pola.26758.

[183] Davydov V, Rakhmanina A, Kireev I, Alieva I, Zhironkina O, Strelkova O, et al. Solid state synthesis of carbon-encapsulated iron carbide nanoparticles and their interaction with living cells. J Mater Chem B 2014;2:4250–61. doi:10.1039/C3TB21599G.

[184] Hu Y, Jensen JO, Zhang W, Cleemann LN, Xing W, Bjerrum NJ, et al. Hollow spheres of iron carbide nanoparticles encased in graphitic layers as oxygen reduction catalysts. Angew Chem Int Ed 2014;53:3675–9. doi:10.1002/anie.201400358.

[185] Schnitzler MC, Oliveira MM, Ugarte D, Zarbin AJG. One-step route to iron oxide-filled carbon nanotubes and bucky-onions based on the pyrolysis of organometallic precursors. Chem Phys Lett 2003;381:541–8. doi:10.1016/j.cplett.2003.10.037.

[186] Abd-El-Aziz AS. Organometallic Polymers of the Transition Metals. Macromol Rapid Commun 2002;23:995–1031. doi:10.1002/marc.200290003.

[187] Hudson RDA. Ferrocene polymers: current architectures, syntheses and utility. J Organomet Chem 2001;637:47–69. doi:10.1016/S0022-328X(01)01142-1.

[188] Ni Y, Rulkens R, Manners I. Transition Metal-Based Polymers with Controlled Architectures: Well-Defined Poly(ferrocenylsilane) Homopolymers and Multiblock Copolymers via the Living Anionic Ring-Opening Polymerization of Silicon-Bridged [1]Ferrocenophanes. J Am Chem Soc 1996;118:4102–14. doi:10.1021/ja953805t.

[189] Manners I. Poly(ferrocenylsilanes): novel organometallic plastics. Chem Commun 1999:857–65. doi:10.1039/A810043H.

[190] Fan N, Ma X, Ju Z, Li J. Formation, characterization and magnetic properties of carbon-encapsulated iron carbide nanoparticles. Mater Res Bull 2008;43:1549–54. doi:10.1016/j.materresbull.2007.06.017.

[191] Wu A, Liu D, Tong L, Yu L, Yang H. Magnetic properties of nanocrystalline Fe/Fe_3C composites. CrystEngComm 2011;13:876–82. doi:10.1039/C0CE00328J.

[192] Gao J, Wang X, Pan X, Ren X, Han Y, Yang X, et al. Facile synthesis of nanocrystalline Fe/Fe_3C induced by bromide. J Mater Sci: Mater Electron 2016;27:64–9. doi:10.1007/s10854-015-3717-5.

[193] Dong X, Chen H, Zhao W, Li X, Shi J. Synthesis and Magnetic Properties of Mesostructured γ-Fe_2O_3/Carbon Composites by a Co-casting Method. Chem Mater 2007;19:3484–90. doi:10.1021/cm0709065.

[194] Fuertes AB, Tartaj P. A Facile Route for the Preparation of Superparamagnetic Porous Carbons. Chem Mater 2006;18:1675–9. doi:10.1021/cm052695e.

[195] Xu X, Wan Y, Sha Y, Deng W, Xue G, Zhou D. Nanoporous iron oxide@carbon composites with low carbon content as high-performance anodes for lithium-ion batteries. RSC Adv 2015;5:89092–8. doi:10.1039/C5RA16460E.

[196] Hassan MF, Rahman MM, Guo ZP, Chen ZX, Liu HK. Solvent-assisted molten salt process: A new route to synthesise α-Fe_2O_3/C nanocomposite and its electrochemical performance in lithium-ion batteries. Electrochim Acta 2010;55:5006–13. doi:10.1016/j.electacta.2010.04.006.

[197] Zhang J, He D, Su H, Chen X, Pan M, Mu S. Porous polyaniline-derived FeN_xC/C catalysts with high activity and stability towards oxygen reduction reaction using ferric chloride both as an oxidant and iron source. J Mater Chem A 2013;2:1242–6. doi:10.1039/C3TA14065B.

[198] Yang Z, Shen J, Archer LA. An in situ method of creating metal oxide–carbon composites and their application as anode materials for lithium-ion batteries. J Mater Chem 2011;21:11092–7. doi:10.1039/C1JM10902B.

[199] Schliebe C. Zwillingspolymerisation in Gegenwart von Übergangsmetallen. Dissertation. Technische Universität Chemnitz, 2015.

[200] Kim M, Yoon SB, Sohn K, Kim JY, Shin C-H, Hyeon T, et al. Synthesis and characterization of spherical carbon and polymer capsules with hollow macroporous core and mesoporous shell structures. Microporous Mesoporous Mater 2003;63:1–9. doi:10.1016/S1387-1811(03)00410-4.

[201] Zhang Y, Xu S, Luo Y, Pan S, Ding H, Li G. Synthesis of mesoporous carbon capsules encapsulated with magnetite nanoparticles and their application in wastewater treatment. J Mater Chem 2011;21:3664–71. doi:10.1039/C0JM03727C.

[202] Sun, Li. Ag@C Core/Shell Structured Nanoparticles: Controlled Synthesis, Characterization, and Assembly. Langmuir 2005;21:6019–24. doi:10.1021/la050193+.

[203] Coville NJ, Mhlanga SD, Nxumalo EN, Shaikjee A. A review of shaped carbon nanomaterials. S Afr J Sci 2011;107. doi:10.4102/sajs.v107i3/4.418.

[204] Hu J, Chen M, Fang X, Wu L. Fabrication and application of inorganic hollow spheres. Chem Soc Rev 2011;40:5472–91. doi:10.1039/c1cs15103g.

[205] Liu J, Wickramaratne NP, Qiao SZ, Jaroniec M. Molecular-based design and emerging applications of nanoporous carbon spheres. Nature Mater 2015;14:763–74. doi:10.1038/nmat4317.

[206] Büchel G, Unger KK, Matsumoto A, Tsutsumi K. A Novel Pathway for Synthesis of Submicrometer-Size Solid Core/Mesoporous Shell Silica Spheres. Adv Mater 1998;10:1036–8. doi:10.1002/(SICI)1521-4095(199809)10:13<1036::AID-ADMA1036>3.0.CO;2-Z.

[207] Chai GS, Yoon SB, Kim JH, Yu J-S. Spherical carbon capsules with hollow macroporous core and mesoporous shell structures as a highly efficient catalyst support in the direct methanol fuel cell. Chem Commun 2004:2766–7. doi:10.1039/B412747C.

[208] Hu FP, Wang Z, Li Y, Li C, Zhang X, Shen PK. Improved performance of Pd electrocatalyst supported on ultrahigh surface area hollow carbon spheres for direct alcohol fuel cells. J Power Sources 2008;177:61–6. doi:10.1016/j.jpowsour.2007.11.024.

[209] Fang B, Kim M, Yu J-S. Hollow core/mesoporous shell carbon as a highly efficient catalyst support in direct formic acid fuel cell. Appl Catal B: En 2008;84:100–5. doi:10.1016/j.apcatb.2008.03.005.
[210] Lou XW, Deng D, Lee JY, Archer LA. Preparation of SnO_2/Carbon Composite Hollow Spheres and Their Lithium Storage Properties. Chem Mater 2008;20:6562–6. doi:10.1021/cm801607e.
[211] Valle-Vigón P, Fuertes AB. Magnetically separable carbon capsules loaded with laccase and their application to dye degradation. RSC Adv 2011;1:1756–62. doi:10.1039/C1RA00333J.
[212] Valdés-Solís T, Valle-Vigón P, Sevilla M, Fuertes AB. Encapsulation of nanosized catalysts in the hollow core of a mesoporous carbon capsule. J Catal 2007;251:239–43. doi:10.1016/j.jcat.2007.07.006.
[213] Lou XW, Li CM, Archer LA. Designed Synthesis of Coaxial SnO_2@carbon Hollow Nanospheres for Highly Reversible Lithium Storage. Adv Mater 2009;21:2536–9. doi:10.1002/adma.200803439.
[214] Zhang W-M, Hu J-S, Guo Y-G, Zheng S-F, Zhong L-S, Song W-G, et al. Tin-Nanoparticles Encapsulated in Elastic Hollow Carbon Spheres for High-Performance Anode Material in Lithium-Ion Batteries. Adv Mater 2008;20:1160–5. doi:10.1002/adma.200701364.
[215] Iravani S, Korbekandi H, Mirmohammadi SV, Zolfaghari B. Synthesis of silver nanoparticles: chemical, physical and biological methods. Res Pharm Sci 2014;9:385–406.
[216] González-Castillo J, Rodriguez E, Jimenez-Villar E, Rodríguez D, Salomon-García I, de Sá GF, et al. Synthesis of Ag@Silica Nanoparticles by Assisted Laser Ablation. Nanoscale Res Lett 2015;10:399. doi:10.1186/s11671-015-1105-y.
[217] Muthuchamy N, Gopalan A, Lee K-P. A new facile strategy for higher loading of silver nanoparticles onto silica for efficient catalytic reduction of 4-nitrophenol. RSC Adv 2015;5:76170–81. doi:10.1039/C5RA11892A.
[218] Choi H, Lee J-P, Ko S-J, Jung J-W, Park H, Yoo S, et al. Multipositional Silica-Coated Silver Nanoparticles for High-Performance Polymer Solar Cells. Nano Lett 2013;13:2204–8. doi:10.1021/nl400730z.

5 Applications

Felix Roschke and Michael Mehring

5.1 Porous materials from twin polymerization for energy storage

5.1.1 Introduction

High consumption of fossil energy sources associated with a rising CO_2 emission and the effect of global climate change have led to a growing interest in renewable energy sources and energy storage devices. Development of the latter plays a key role in future electric mobility concepts (e.g., in vehicles and bicycles), storage of electricity from renewable sources (e.g., wind and solar energy), in portable electronics (e.g., laptops, mobile phones, and in implantable medical devices), and in flexible electronic devices (e.g., glasses and smart textiles) [1–3]. So far, lithium-ion batteries are the most promising candidates to overcome the technical challenges in energy storage due to the lowest value of the reduction potential of all elements for lithium (–3.04 V), the low weight and ionic radii of its cation, and the resulting high gravimetric and volumetric capacity and power density of lithium-based batteries [1, 3, 4]. Thus, the interest in the advancement of electrode materials for lithium-ion batteries was steadily increasing within the last decades since *Sony* launched the first commercially available lithium-ion battery in 1990 [5]. Currently, carbon-based anodes are used most commonly as a result of their high cycling stability and low price, but compared to other main group IV elements, the theoretical gravimetric capacity of carbon with a value of 372 $mAh \cdot g^{-1}$ (for LiC_6) is quite low. Hence, the usage of the heavier analogs, e.g., silicon (4200 $mAh \cdot g^{1}$ for $Li_{22}Si_5$), germanium (1600 $mAh \cdot g^{-1}$ for $Li_{17}Ge_4$),[1] and tin (990 $mAh \cdot g^{-1}$ for $Li_{17}Sn_4$)[1] is highly promising [5, 6, 8], but their industrial application is hampered by poor cycling stability caused by large volume changes up to 400 %, 246 % and 257 % for silicon, germanium, and tin, respectively, during lithiation/delithiation [3, 5, 9–11]. In addition to the use of group IV elements, the use of either lithium metal oxides, e.g., Li_xMO_2 (M = Co, Ni, Mn, Mo) with theoretical capacities of 274 $mAh \cdot g^{-1}$ ($LiCoO_2$), 275 $mAh \cdot g^{-1}$ ($LiNiO_2$), 285 $mAh \cdot g^{-1}$ ($LiMnO_2$), 209 $mAh \cdot g^{-1}$ ($LiMoO_2$), 240 $mAh \cdot g^{-1}$ ($LiNi_{0.5}Mn_{0.5}O_2$), and 275 $mAh \cdot g^{-1}$ ($LiCo_{0.33}Ni_{0.33}Mn_{0.33}O_2$), or sulfur-based materials with a theoretical capacity of 1675 $mAh \cdot g^{-1}$ as cathodes in lithium-ion batteries is intensively studied [4, 12–17].

1 The groups of Nazar and Reichmann studied the composition of lithium-intercalated metal phases in detail. They postulate on the basis of crystallographic studies that the ratio of lithium and germanium, and lithium and tin, respectively, are in accordance with the compositions of $Li_{17}Ge_4$ and $Li_{17}Sn_4$ (Li:M = 4.25:1; M = Ge, Sn).[6, 7] The more commonly used formulas of $Li_{22}Ge_5$ and $Li_{22}Sn_5$ (Li:M = 4.4:1; M = Ge, Sn) are noted in the majority of publications.

https://doi.org/10.1515/9783110499360-005

To date, the materials used in commercially available cells are those with working capacities of 145 mAh · g^{-1} ($LiCoO_2$), 170 mAh · g^{-1} ($LiCo_{0.33}Ni_{0.33}Mn_{0.33}O_2$), and 200 mAh · g^{-1} ($LiNi_{0.8}Co_{0.15}Al_{0.05}O_2$) [4]. Notably, cathode materials based on oxide materials intercalate lithium ions with a negligible structural change [16, 18]. However, nanostructuring in combination with high surface areas and incorporation of active species in a porous carbon matrix decreases the effect of volume expansion in electrodes, which leads to loss of contact by destruction of the material and ensures short diffusion paths. Thus, nanostructuring of the active material and its embedding into a porous carbon matrix leads to (i) short distances for lithium-ion transport in electrodes, (ii) high contact areas by high surface areas, (iii) increased electrolyte access by ordered pore structure and, thus, to an improved cycling stability and rate capability [1, 5, 19]. This concept was successfully applied to different electrode materials. For instance, single-wall and multi-wall carbon nanotubes showed increased capacities up to 1000 mAh · g^{-1} and 400 mAh · g^{-1}, respectively, which is significantly higher compared to pure graphite (372 mAh · g^{-1} for LiC_6), but their cycling stability is still poor [18]. Contrastingly, pure silicon is the most promising candidate by its highest gravimetric capacity, but high volume expansion up to 400 % during lithiation/delithiation is still a challenge with regard to practical applications. A huge drop of capacity during five cycles from 2900 mAh · g^{-1} to under 500 mAh · g^{-1} using micron-scale material is quite common. Nanostructuring was shown to increase the cycling stability significantly, e.g., nanomaterials with capacities of 1300 mAh · g^{-1} after 22 cycles and silicon nanowires with a capacity of 2000 mAh · g^{-1} after 80 cycles were reported [18]. Moreover, silicon films of nanoscale thickness are shown to exhibit a capacity of 3600 mAh · g^{-1} after 200 cycles [18]. Notably, a transfer of silicon electrode materials to commercially applicable lithium-ion batteries has not yet been implemented and the automotive industry is searching for practicable and less challenging solutions. Therefore, materials such as tin and tin alloys, although showing lower performance, are of interest as a result of smaller volume changes. For example, surface-stabilized tin nanoparticles with average particle sizes of 10 and 20 nm are accessible with capacities of 1000 mAh · g^{-1} and 940 mAh · g^{-1} in the first cycle, respectively. After 60 cycles, a negligible capacity fading was obtained with final capacities of 940 mAh · g^{-1} and 818 mAh · g^{-1}, respectively [18, 20]. Furthermore, $Sn@SiO_2/C$ nanoparticles were encapsulated in hollow carbon nanofibers, which showed a very low fading with a reversible capacity of 731 mAh · g^{-1} after 200 cycles and composite nanomaterials consisting of $Si/Sn@SiO_2/C$ exhibited a very high reversible capacity of 1073 mAh · g^{-1} after 50 cycles [21, 22]. In addition, tin alloys were synthesized and tested, e.g., Sn_2Co-containing carbon composite retained of a capacity of 900 mAh · g^{-1} after 100 cycles [23].

Within the following chapter, the synthesis of porous carbon materials following the concept of twin polymerization is described with the focus on materials for energy storage. The process of twin polymerization combines both nanostructuring of materials and embedding of active species in a porous carbon matrix, which is important for potential application as electrode materials. Selected anode materials as obtained by twin polymerization and used for rechargeable lithium-ion batteries on the basis of germanium (Ge@C), tin ($Sn@SiO_2/C$), and tin alloys ($Sn_xM_y@SiO_2/C$; M = Co, Ni, Sb) as well as cathode materials consisting of sulfur-loaded hollow carbon spheres are presented within this chapter [11, 24–28]. Notably, several carbon materials synthesized by twin polymerization, which hold potential as electrode materials in lithium-ion batteries, are described in patents so far [29–36]. In general, the synthesis protocol starts with polymerization of a suitable monomer or a monomer mixture either with or without additives and carbonization of the as-obtained hybrid materials under argon or argon/hydrogen atmosphere results in porous carbon materials, following a similar concept as presented in Chapter 4.2. In the first part, the syntheses and electrochemical measurements of the sulfur-loaded hollow carbon spheres, which are potential cathode materials, are described in more detail. Subsequently, the syntheses and electrochemical tests of anode materials based on either germanium or tin are discussed.

5.1.2 Syntheses and electrochemical tests of porous materials

5.1.2.1 Cathode materials

The incorporation of sulfur in carbon materials and their usage as electrode materials is promising with regard to the high theoretical capacity [17]. Therefore, carbon materials, especially hollow carbon spheres, were synthesized using the concept of twin polymerization, from which materials with promising properties, e.g., high surface areas, are accessible. Additionally, the incorporation of sulfur is achieved either during the synthesis or by post-functionalization strategies. The synthesis of hollow carbon spheres using the concept of twin polymerization starts by adsorption of an acid as promotor for a subsequent polymerization, e.g., methanesulfonic acid at the surface of spherical SiO_2 particles. After addition of a twin monomer, either tetrafurfuryloxysilane or Si-Spiro, the polymerization occurs instantaneously at the surface of the SiO_2 particles and a nanostructured polymer/SiO_2 hybrid material is formed. Carbonization of the hybrid material in argon atmosphere at 800 °C or 1100 °C and subsequent treatment with aqueous hydrofluoric acid provides porous hollow carbon spheres (Scheme 5.1). Different pore textures depending on the precursors were obtained. The use of the spirocyclic compound Si-Spiro resulted in microporous hollow carbon spheres with surface areas in the range of 790 $m^2 \cdot g^{-1}$ to 1370 $m^2 \cdot g^{-1}$. Contrastingly, mesoporous materials with surface areas ranging from 690 $m^2 \cdot g^{-1}$ up to 1450 $m^2 \cdot g^{-1}$ were accessible starting from tetrafurfuryloxysilane. Subsequently, molten sulfur is infiltrated at 140 °C in the case of hollow carbon spheres based on Si-Spiro or at 155 °C in the case of hollow carbon spheres based on tetrafurfuryloxysilane after removal of silica [24, 25]. Notably, it has been shown that sulfur is exclusively located in pores and at the surface of the hollow carbon spheres, but was not detected in the inner space of the hollow spheres [24]. The general synthetic pathway for the synthesis of sulfur-loaded hollow carbon spheres using the twin polymerization process is depicted in Scheme 5.1. It should be emphasized that the particle size of the final hollow carbon spheres depends on the diameter of the applied SiO_2 templates and, thus, can be tuned on demand. Particles with diameters of 7 nm (Aerosil 300), 40 nm (Aerosil OX 50), or 5–10 µm (LiChrospher 100) have been tested so far.

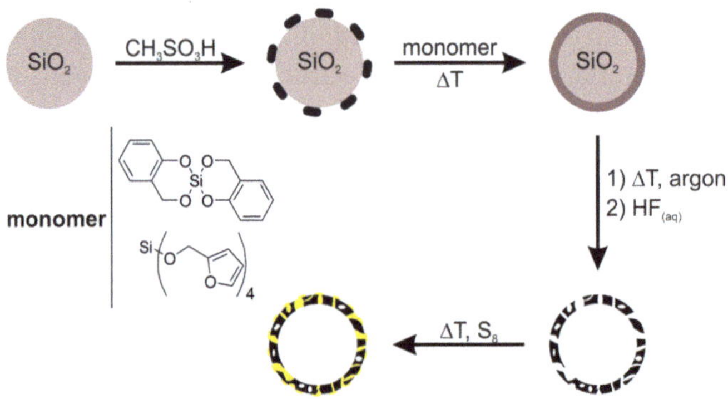

Scheme 5.1: Synthesis of sulfur-loaded porous hollow carbon spheres by the use of surface twin polymerization at silica template particles starting either from Si-Spiro or tetrafurfuryloxysilane. The figure was created according to the style in [24].

An alternative option to synthesize sulfur-loaded hollow carbon spheres using the twin polymerization process is the polymerization of a sulfur-containing thiophene-based twin monomer (Scheme 5.2). Therefore, silica templates were acidified with methanesulfonic acid and the subsequent polymerization of tetra(thiophene-2-ylmethoxy)-silane at the surface yielded polymer/silica hybrid materials. The latter are converted to sulfur-doped hollow carbon spheres by carbonization of the as-obtained hybrid material in argon at 800 °C. Removal of silica by aqueous hydrofluoric acid gives microporous hollow carbon spheres with micropore contents of 21 to 30 % and surface areas of 680 $m^2 \cdot g^{-1}$ to 870 $m^2 \cdot g^{-1}$. The sulfur contents of the composites range from 11.5 to 16.5 % [37].

The concept of surface-mediated twin polymerization using various templates is suitable to produce diverse hollow carbon spheres. Depending on synthetic conditions, the physical properties of these hollow carbon spheres are adjustable. For instance, the inner diameter of the as-prepared hollow carbon spheres correlates with the diameter of the template that is used. Moreover, the monomer:template ratio affects the pore size distribution and thickness of the carbon shell and

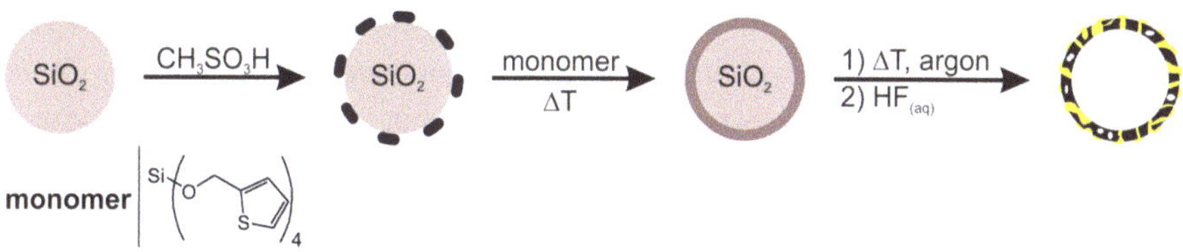

Scheme 5.2: Synthesis of sulfur-doped porous hollow carbon spheres by the use of surface twin polymerization on silica particles starting from tetra(thiophene-2-ylmethoxy)silane and further carbonization of the as-obtained hybrid material with removal of silica by aqueous hydrofluoric acid. The figure was created according to the style in [37].

furthermore, pore textures are selectable by the monomer, e.g., Si-Spiro provided microporous carbon materials and in the case of tetrafurfuryloxysilane, mesoporous carbon materials are obtained. Notably, the micropore content decreases and the total pore volume increases with increasing template ratio. It is noteworthy that the electric conductivity of the resulting hollow carbon sphere depends on the carbonization temperature. An increasing graphitic content is accessible by addition of $FeCl_3$ in a further carbonization process, but unfortunately destroys the spherical morphology. A summary of adjustable properties of hollow carbon spheres *via* the concept of surface twin polymerization on silica templates is given in Figure 5.1.

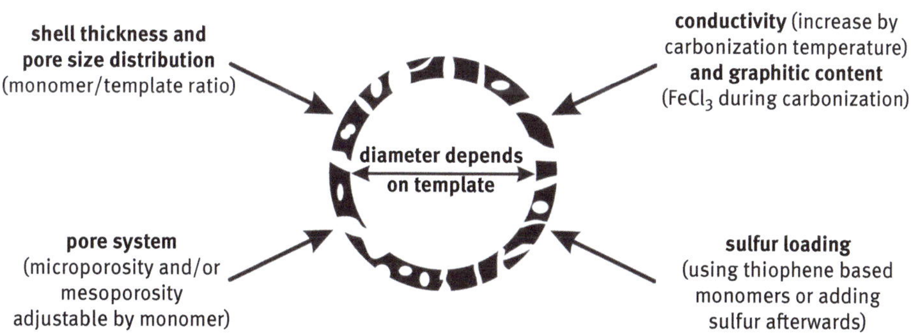

Fig. 5.1: Adjustable properties of hollow carbon spheres synthesized by surface twin polymerization and further carbonization of the as-obtained hybrid material. The figure was created according to the style in [24].

Electrochemical investigations were carried out for sulfur-loaded hollow carbon spheres that were synthesized from Si-Spiro and Aerosil 300 (diameter 7 nm) with 67 wt% twin monomer. Subsequently, molten sulfur was added at 140 °C. The temperature range between 120 °C (melting temperature of S_8) and approximately 159 °C (floor temperature of S_8) is necessary in order to achieve a molecularly well-defined sulfur loading. The cathode material consisted of a coated aluminum foil with an active layer composed of approximately 75.3 wt% of the as-prepared sulfur-containing hollow carbon spheres, 20.7 wt% conductive carbon black, 3.1 wt% synthetic graphite, and 0.9 wt% polyvinyl alcohol. Thus, a half-cell setup with a lithium-foil anode and an electrolyte (1 M solution of $LiN(SO_2CF_3)_2$ in dioxolane/dimethoxyethane) was used. It has been shown that the tested cathode material exhibits a high cycling stability with a capacity of 850 $mAh \cdot g^{-1}_{sulfur}$. After 500 cycles, the final capacity did still amount to 580 $mAh \cdot g^{-1}_{sulfur}$. The cycling performance of sulfur-loaded hollow carbon spheres synthesized from Si-Spiro and Aerosil 300 in a half-cell setup is shown in Figure 5.2. The decrease in the specific capacity was attributed to the volume expansion and compression during lithiation/delithiation that could not fully be reduced to zero. Notably, the coulombic efficiency faded during the measurement and was strongly reduced as of cycle 400. Sulfur-loaded carbon composites synthesized from 100 wt% Si-Spiro without the use of any template were tested as cathode material for comparison. A capacity of 650 $mAh \cdot g^{-1}_{sulfur}$ and a value of 390 $mAh \cdot g^{-1}_{sulfur}$ after 250 cycles were determined, which is significantly lower than observed for materials prepared with the "template approach" [24]. Electrochemical studies have shown that the sulfur-containing carbon materials synthesized using the concept of twin polymerization on the basis of Si-Spiro

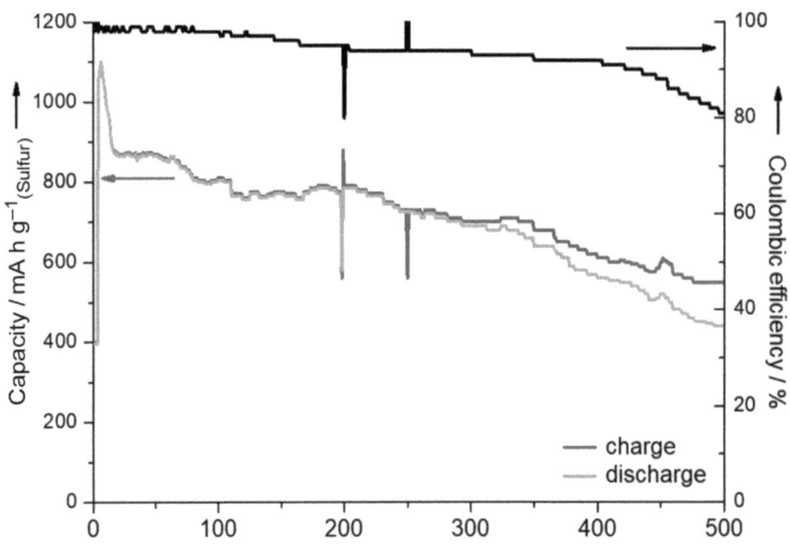

Fig. 5.2: Cycling performance of a lithium-sulfur battery using a cathode composed of sulfur-loaded hollow carbon spheres in a half-cell setup. The hollow carbon spheres were synthesized by surface twin polymerization of 67 wt% monomer (Si-Spiro) and Aerosil 300 (diameter 7 nm). Notably, the anode has been changed due to a rapid wear (e.g., dendrite formation) after 200 cycles, and thus observed amplitudes are most probably preparation artifacts. ©2013 This figure is adapted from Ref. [24] with permission of John Wiley & Sons, Inc.

can compete with published state of the art materials. For instance, final capacities with values of 750 mAh · g^{-1} after 150 cycles for sulfur-containing hollow carbon spheres and 850 mAh · g^{-1} after 200 cycles for sulfur-containing bimodal or ordered mesoporous carbons with reversible capacities up to 1350 mAh · g^{-1} were reported [14, 16, 38].

In addition to hollow carbon spheres as obtained from Si-Spiro, carbon-hollow sphere-based cathodes were synthesized using tetrafurfuryloxysilane. These materials have been tested by J. Brückner *et al.* in half-cell and full-cell setups [25]. They used 33 % monomer and 67 % template by weight and infiltrated sulfur at 155 °C. However, this sulfur-containing material was used to prepare the cathode material. Preparation of the cathode was achieved by mixing 80 wt% of sulfur-containing hollow carbon spheres and 20 wt% of a mixture of polytetrafluoroethylene (PTFE) and multi-walled carbon nanotubes (ratio either 1:1 or 1:3) as conductive additive, yielding cathode materials consisting of approximately 53 wt% sulfur. An electrochemical setup with lithium foil anodes in half-cell measurements, silicon-carbon anodes, or hard-carbon-containing anodes in full-cell measurements, and a specific electrolyte solution (1 M solution of $LiN(SO_2CF_3)_2$ and 0.25 M solution of $LiNO_3$ in dioxolane/dimethoxyethane) was chosen (Fig. 5.3). Cyclization in a half-cell setup at 167 mA · g^{-1}_{sulfur} resulted in reversible cycling over 135 cycles, while cyclization at 836 mA · g^{-1}_{sulfur} (after three cycles at 167 mA · g^{-1}_{sulfur}) caused a cell failure after 47 cycles. This was attributed to dendrite formation, which is enhanced at higher current densities. Furthermore, the use of lithium anodes resulted in a rapid electrolyte degradation by the use of low electrolyte amount (12 µl · mg^{-1}_{sulfur}). However, the coulombic efficiency was stable with values >99 % during cycling.

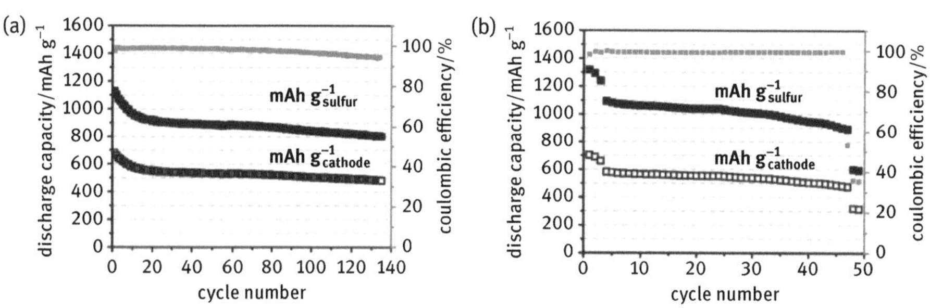

Fig. 5.3: Discharge capacity (black) and coulombic efficiency (grey) of a cathode material as received by the concept of surface twin polymerization using tetrafurfuryloxysilane as monomer (monomer:template ratio 33 %:67 % by weight). The electrochemical measurements were performed in half-cell setups with a silicon-carbon-based anode at (a) 167 mA · g^{-1}_{sulfur} and (b) 836 mA · g^{-1}_{sulfur} (after three cycles at 167 mA · g^{-1}_{sulfur}). This figure is adapted from Ref. [25] with permission of John Wiley & Sons, Inc.

Besides the half-cell setup experiments, the material was tested in a full-cell setup with a pre-lithiated silicon-carbon anode at 836 mA · g^{-1}_{sulfur} (after three cycles at

167 mA · $g^{-1}{}_{sulfur}$). Notably, this setup is stable for more than 1300 reversible cycles with a discharge capacity of 765 mA · $g^{-1}{}_{sulfur}$ in the fifth cycle. Moreover, the coulombic efficiency is approximately 99.8 % in the 1390th cycle and a slow fading of about 0.08 % per cycle was observed. The enhanced cycling stability was attributed to a stable solid electrolyte interface and to the fact that no electrolyte consumption was observed during cycling. Especially, a low amount of electrolyte (12 µl · $mg^{-1}{}_{sulfur}$) and a lithium excess of only 60 % were used. The electrochemical tests are depicted in Figure 5.4.

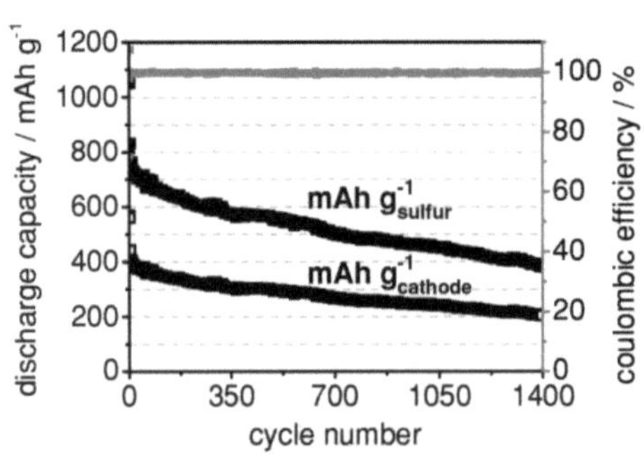

Fig. 5.4: Discharge capacity (black) and coulombic efficiency (grey) of a cathode material as received by the concept of surface twin polymerization using tetrafurfuryloxysilane as monomer at an Aerosil 300 template (monomer:template ratio is 33 %:67 % by weight). The electrochemical measurements were performed in a full-cell setup with a silicon-carbon based anode at 836 mA · $g^{-1}{}_{sulfur}$ (after three cycles at 167 mA · $g^{-1}{}_{sulfur}$). This figure is adapted from Ref. [25] with permission of John Wiley & Sons, Inc.

For comparison, full carbon anodes consisting of commercially available hard carbon were tested as alternative material to silicon-carbon based anodes. Notably, the corresponding sulfur-containing cathode material synthesized by the concept of twin polymerization, was not changed. The carbon material for anode preparation shows broad reflections in the PXRD at either 25.1° (carbon fiber) or 23.1° (hard carbon), which is indicative for non-graphitic carbon. Additionally, in this setup an increased access to sulfur in the cathode material was observed by increasing the amount of carbon nanotubes. Therefore, the PTFE/multi-walled carbon nanotube ratio was changed from 1:1 to 1:3, which gave a better wettability of the cathode, while the sulfur loading was kept constant. In a half-cell setup, the sulfur utilization was about 99.2 % with a capacity of 1658 mAh · $g^{-1}{}_{sulfur}$ in the first cycle. However, a cycling stability over 168 reversible cycles was obtained with a final capacity of approximately 500 mAh · $g^{-1}{}_{sulfur}$ before fading (using a discharge current of 334 mA · $g^{-1}{}_{sulfur}$ and an electrolyte amount of 13 µl · $mg^{-1}{}_{sulfur}$). Contrastingly, an improved cycling stability in the full-cell setup with a hard carbon-based anode was observed. Thus, reversible cycling for more than 550 cycles without cell failure and a capacity fade of only

0.08 % per cycle was determined. Compared to the half-cell setup, high sulfur utilization of 1471 mAh · $g^{-1}{}_{sulfur}$ in the second cycle was reproduced in the full-cell setup and resulted in a high capacity of 753 mAh · $g^{-1}{}_{sulfur}$ with a high coulombic efficiency of 99.1 % in the 550th cycle. Remarkably, the lithium excess in the electrochemical test using full carbon anodes was only 10 %. Both half-cell and full-cell measurements are depicted in Figure 5.5.

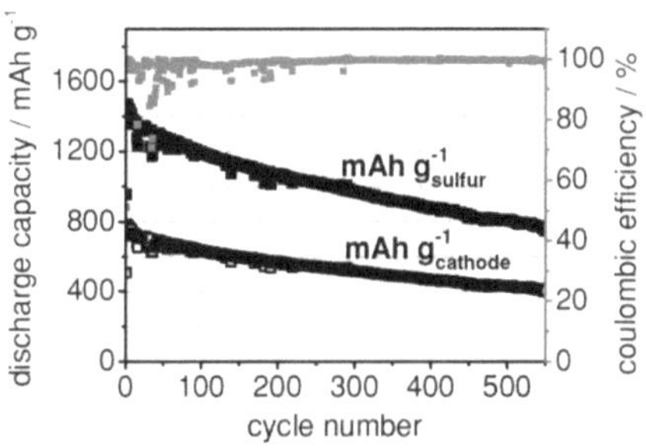

Fig. 5.5: Discharge capacity (black) and coulombic efficiency (grey) of a cathode material received *via* the concept of surface twin polymerization using tetrafurfuryloxysilane as monomer (monomer:template ratio 33 %:67 % by weight). The electrochemical measurements were performed with a hard carbon anode in a full-cell setup at 334 mA · $g^{-1}{}_{sulfur}$ using a PTFE/multi-walled carbon nanotube ratio of 1:3. This figure is adapted from Ref. [25] with permission of John Wiley & Sons, Inc.

Hollow carbon spheres that were synthesized by surface twin polymerization on hard templates, e.g., by usage of silica templates as given above, are also accessible by usage of organic-polymer-based soft templates [28]. Thus, polystyrene microparticles (PS) were synthesized and readily sulfonated to obtain surface-coated microparticles ($PSSO_3H$). Subsequent surface twin polymerization of Si-Spiro on these particles ($PSSO_3H$) proceeds effectively with 100 % coating yield. Notably, the as-obtained acidic particles $PSSO_3H$ serve as both catalyst and template in the twin polymerization process. The complete synthetic pathway starting with sulfonation of polystyrene and further conversion to hollow carbon spheres by usage of twin polymerization is given in Scheme 5.3.

Carbonization of the phenolic resin/SiO_2 hybrid material layer into carbon in an argon atmosphere proceeds as expected, whereupon the polystyrene core collapses during the carbonization process and is completely dismantled. Thus, the outer carbon layer undergoes shrinking and finally a pleated, very robust carbon structure was obtained. That structure is even robust against ball milling, which was used before sulfur loading. Notably, the carbon materials are microporous with BET surface areas up to 1389 $m^2 \cdot g^{-1}$ in case of $PSSO_3H$-Spiro40_C ($PSSO_3H$ particles:Si-Spiro = 60:40). Final sulfur loading was carried out at 155 °C for 5 hours to receive an infiltration into the pores and a homogeneous distribution of sulfur on the carbon surface. The homogeneous distribution of sulfur was proven by SEM/EDX images, which are depicted in Figure 5.6.

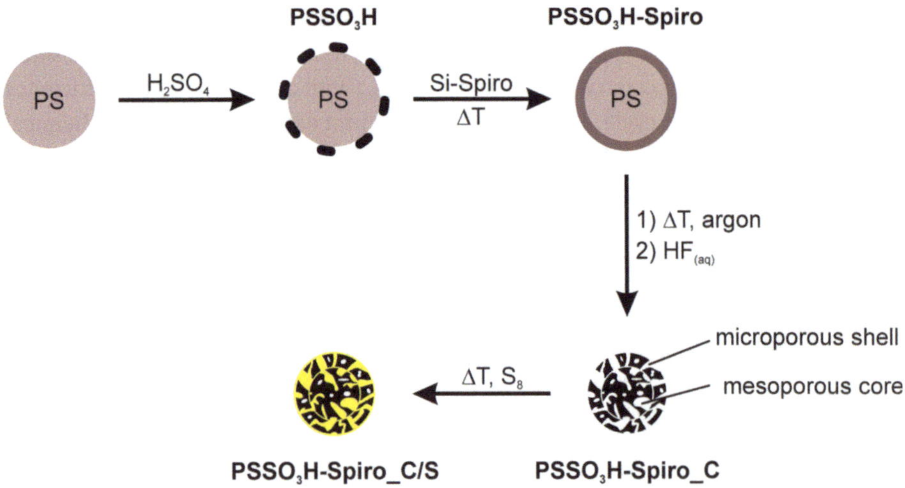

Scheme 5.3: Synthetic pathway for the production of sulfur-loaded hollow carbon spheres starting from polystyrene particles and subsequent surface twin polymerization of Si-Spiro. The as-obtained carbon spheres are surface folded with 200 nm in diameter. The figure was created according to the style in [28].

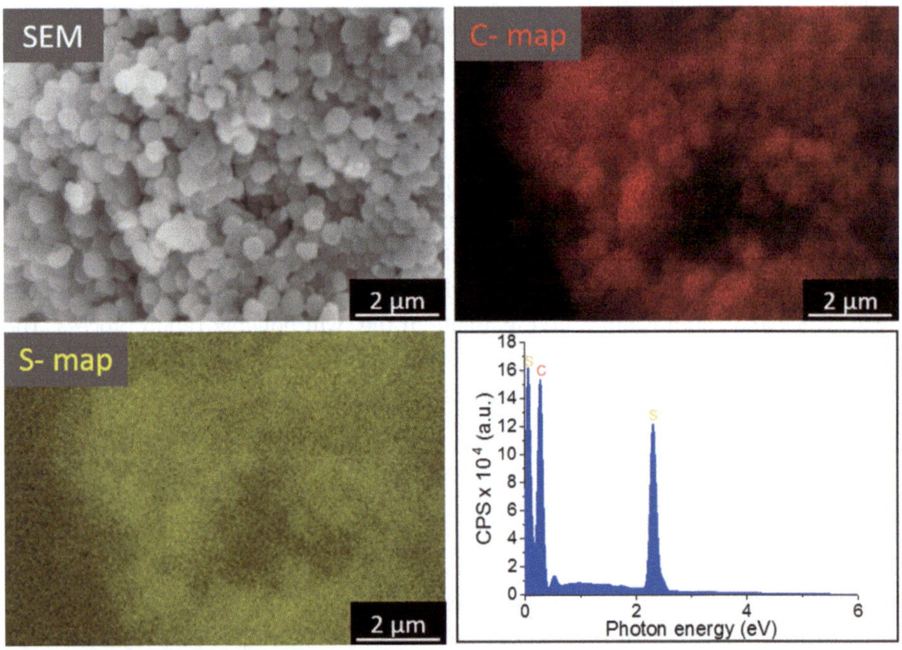

Fig. 5.6: Scanning electron micrographs of carbon/sulfur composite material ($PSSO_3H$-Spiro50_C/S) as well as elemental maps and EDX spectrum. Notably, the material $PSSO_3H$-Spiro50 was synthesized by mass ratio $PSSO_3H$ particles:Si-Spiro = 50:50. This figure was created according to the data from Ref. [28].

The pyrolysis process of the Si-Spiro-coated $PSSO_3H$ microparticles, associated with shrinking of the carbon layer, induces a defined narrowing of the carbon micropores in the carbon wall. The diameter of the micropores is below 0.5 nm. Thus, sulfur is unable to crystallize within the pores due to the confined geometry, while S_8 is adsorbed as proven by X-ray diffraction experiments. Therefore, the S_8-loaded hollow carbon spheres were tested for their usage as cathode materials in lithium-ion batteries. Selected electrochemical measurements are depicted in Figure 5.7. It was shown that the as-obtained materials $PSSO_3H$-Spiro40_C/S and $PSSO_3H$-Spiro50_C/S exhibit a high cycling stability over 100 cycles at low C-rates (charge: 0.2 C and discharge: 0.1 C), while a cycling stability of more than 400 cycles was obtained by an increased C-rate (charge: 2 C and discharge: 1 C). In addition, the highly porous bimodal nanostructure was reported to enhance the lithium-sulfur reactivity significantly. Furthermore, a kinetic inhibition to diffusion and the sorption properties of the carbon might increase the capture of as-formed polysulfides, which leads to the very slow fading.

In conclusion, surface twin polymerization is a promising approach in the synthesis of porous hollow carbon spheres with BET surface areas up to 1450 $m^2 \cdot g^{-1}$, which are suitable as cathode materials in lithium-ion batteries, especially in lithium-sulfur cells. It was shown that a variety of properties, e.g., inner diameter, porosity, shell thickness, and quantity of sulfur loading are adjustable by the use of different templates or monomers and template:monomer ratios. Several of the electrochemical tests were carried out using hollow carbon spheres synthesized either from Si-Spiro or tetrafurfuryloxysilane. The as-obtained hollow carbon spheres

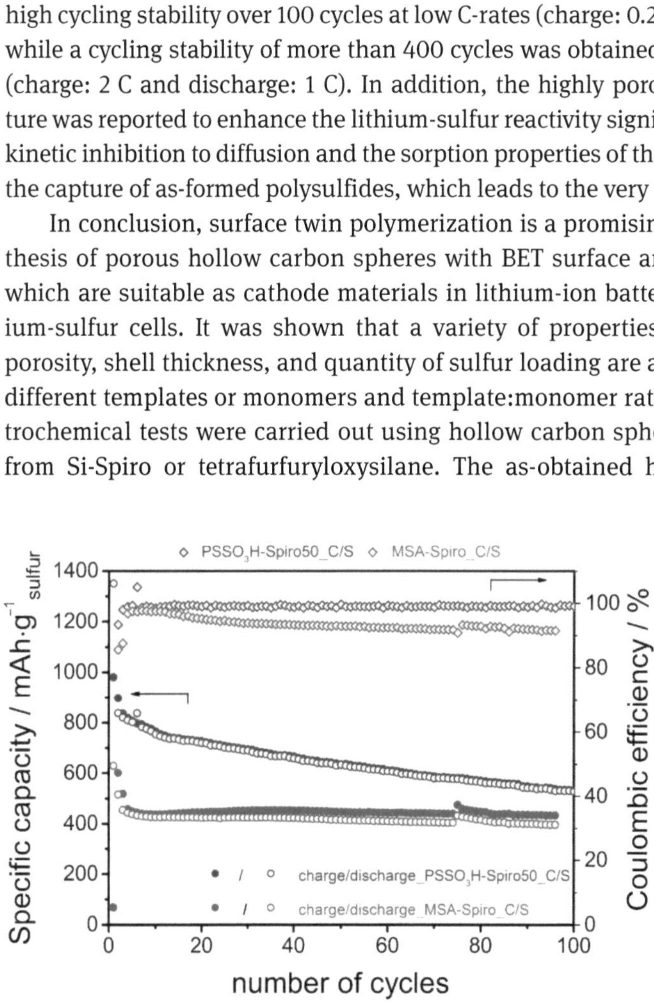

Fig. 5.7: Electrochemical measurements of the carbon/sulfur hybrid material $PSSO_3H$-Spiro50_C/S and MSA-Spiro_C/S, which were synthesized from precipitation of twin polymerization of Si-Spiro by addition of catalytic amounts of methanesulfonic acid without the usage of a template and further conversion to the sulfur-containing carbon material. In case of the capacity a galvanostatic charge/discharge experiment was used. This figure was created according to the generated data from Ref. [28].

were tested in half-cell setups (lithium anode) and full-cell setups (silicon-carbon or full carbon anode). Thus, highly promising cycling stabilities in full-cell setups with reversible cycling for more than 550 charge/discharge cycles in the case of hard carbon anodes or 1390 charge/discharge cycles in the case of silicon-carbon-based anodes with high coulombic efficiencies (>99 %) were obtained. Furthermore, high capacities of up to 1658 $mAh \cdot g^{-1}{}_{sulfur}$ (99.2 %) in the first cycle were achieved, which are close to the theoretical capacity of sulfur [17]. Notably, a significantly lower lithium excess in the presented electrochemical measurements compared with literature values was used [25]. The carbon materials as synthesized using soft templates showed very promising cycling stabilities for more than 400 cycles, while the porous bimodal nanostructure and the increased capture of polysulfides might be the reason for these excellent properties. Thus, the as-obtained hollow carbon spheres are very promising candidates as cathodes in future lithium-ion batteries. However, the development of sustainable materials must focus on the further reduction of the amount of lithium and its effective use to save resources in commercially applied lithium-ion batteries.

5.1.2.2 Anode materials

As mentioned above, nanostructuring within electrode materials and embedding in a carbon matrix are promising strategies to improve cycling stability. The carbon matrix acts as a buffer matrix and reduces the sintering process of the active material whereas nanostructuring provides short diffusion paths. Twin polymerization was shown to combine both features and, thus, germanium- and tin-containing precursors were used to synthesize hybrid materials that were converted to either Ge@C or $Sn@SiO_2/C$ materials with crystallite sizes on the nanometer scale. Tin-containing nanostructured hybrid materials are accessible by simultaneous twin polymerization (STP) of Si-Spiro and tin(II) precursors, e.g., tin(II)-2,4-dimethoxyphenylmethanolate, tin(II)-5-methoxy-2-(oxidomethyl)phenolate, or tin(II)-4-methoxy-2-(oxidomethyl)-phenolate [27, 39, 40]. Notably, twin polymerization of tin(II)-2,4-dimethoxyphenylmethanolate without addition of a second component resulted in a hybrid material consisting of a phenolic resin and tin oxide, in which phase separation with agglomerates on the micrometer-scale was obtained. In STP of tin- and silicon-containing mixtures the Lewis-acidic character of the tin compounds decreases the trigger temperature of the polymerization to 70–83 °C. Reduction of the hybrid material at elevated temperatures with argon:hydrogen gas (95:5) resulted in crystalline tin-containing nanocomposites ($Sn@SiO_2/C$ materials). The synthetic concept is given in Scheme 5.4.

For instance, simultaneous twin polymerization of Si-Spiro and tin(II)-2,4-dimethoxyphenylmethanolate in a 3:1 molar ratio and subsequent carbonization/reduction resulted in a porous $Sn@SiO_2/C$ material that possesses a BET surface area of 313 $m^2 \cdot g^{-1}$ and a micropore content of 25%. Moreover, an average crystallite size of the tin particles of approximately 35 nm was calculated using the Scherrer equation. The

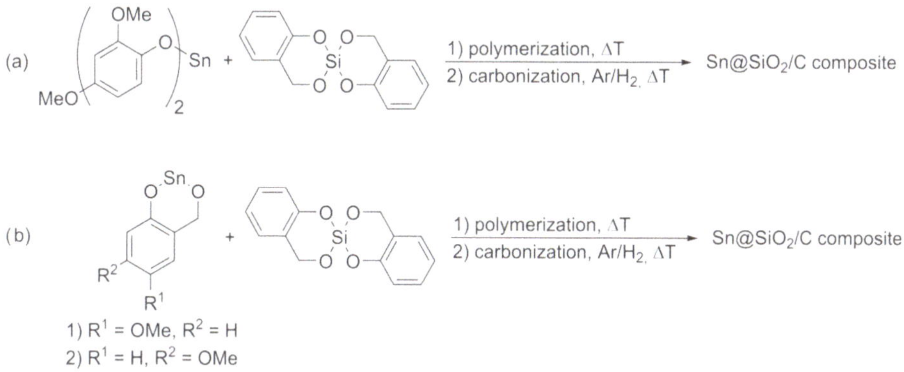

Scheme 5.4: Synthetic routes for tin/silica composite materials following the concept of twin polymerization. Reduction of the as-obtained hybrid materials in an argon/hydrogen atmosphere (Ar/H_2: 95/5, reduction temperature 600–780 °C) gave porous composites [27, 39, 40].

carbon content was 47% by weight and a tin content of 29% was determined using X-ray fluorescence spectroscopy. TEM images indicated a homogeneous distribution of tin nanoparticles in the material (see Fig. 5.8). The as-obtained tin nanocomposite was tested in a half-cell setup with a lithium cathode and a current of 30 $mA \cdot g^{-1}$. A copper sheet was coated with a mixture consisting 87 wt% nanocomposite, 6 wt% carbon black, and 7 wt% binding material. In addition, 1 M solution of $LiPF_6$ in ethylene carbonate and ethyl methyl carbonate (1:1 by volume) was used as electrolyte. These are standard conditions for the electrochemical measurements of the tin-containing materials. In the first cycle, a promising capacity of 545 $mAh \cdot g^{-1}$ was obtained, which decreased further during 17 cycles to 189 $mAh \cdot g^{-1}$. Thus, a reversible

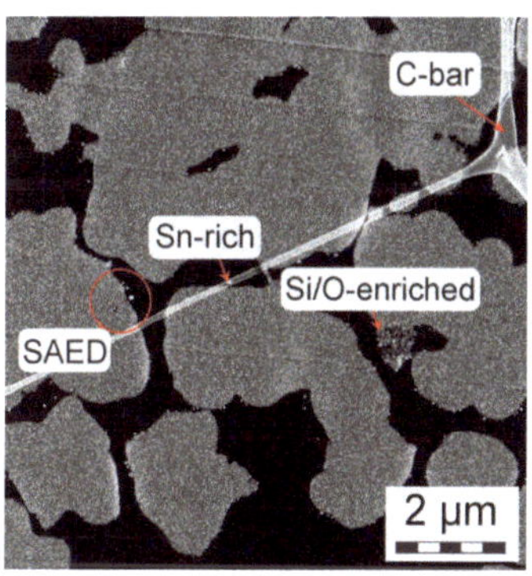

Fig. 5.8: HAADF-STEM image of $Sn@SiO_2/C$ material synthesized by STP of Si-Spiro and tin(II)-2,4-dimethoxyphenylmethanolate in a 3:1 molar ratio. The figure was created according to the generated data from [27].

capacity of 35 % during the first 17 cycles was determined, which indicates a failure of the electrode. The electrochemical tests are depicted in Figure 5.9. A much better cycling stability was expected, because the TEM images showed a homogeneous distribution of tin nanoparticles. Probably, this is indicative for a release of tin from the composite during cycling, which leads to a decreased cycling stability.

Therefore, the addition of appropriate metal reactants to form tin alloys seemed to be promising; the additives act as an additional buffer matrix and should increase the melting temperature. Thus, addition of different metal additives during the nanocomposite synthesis was tested. Therefore, an additive-assisted STP of Si-Spiro and tin(II)-2,4-dimethoxyphenylmethanolate with addition of either cobalt(II) carboxylate or nickel(II) carboxylate, and an additive-assisted TP of Si-Spiro with $Sn(OtBu)_4$ and $Sb(OtBu)_3$ were used to obtain novel organic/inorganic hybrid materials. Carbonization of the hybrid materials under reducing conditions (when $Ar:H_2$ is 95:5) resulted in tin alloy formation with the particles embedded in a carbonized SiO_2/C material [39]. The general concept for the syntheses of tin alloy nanocomposite materials following the twin-polymerization-based approach is given in Scheme 5.5. For further information, see Chapter 4.4.

Several composites were synthesized and first, the Sn_2Co alloy is described in the following. The cobalt-containing composite material was obtained by addition of cobalt(II) carboxylate in the additive-assisted STP. Formation of the alloy as the major phase was confirmed by powder X-ray diffraction, but some additional elemental tin was determined. The composite material exhibits a BET surface area of

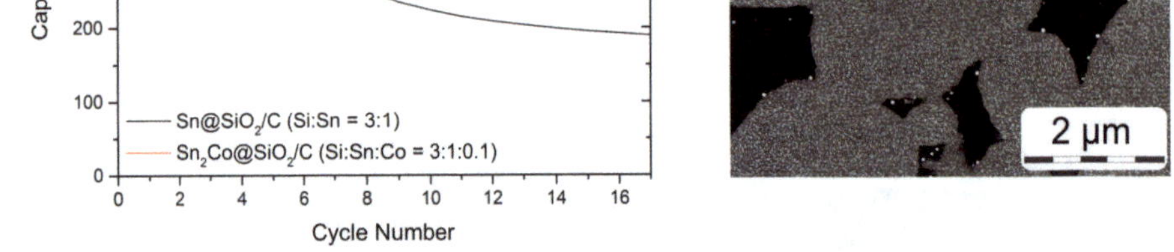

Fig. 5.9: (a) Discharge capacity of $Sn@SiO_2/C$ (black) and $Sn_2Co@SiO_2/C$ (red) composite materials and (b) HAADF-STEM image of $Sn_2Co@SiO_2/C$ composite. The materials were synthesized by STP of Si-Spiro and tin(II)-2,4-dimethoxyphenylmethanolate in a 3:1 molar ratio and in the case of the cobalt-containing material with addition of cobalt(II) carboxylate (Sn:Co = 10:1). The electrochemical measurements were performed in half-cell setup at 30 $mA \cdot g^{-1}$. The graph in (a) was created according to the generated data and the image in (b) was adapted from [27].

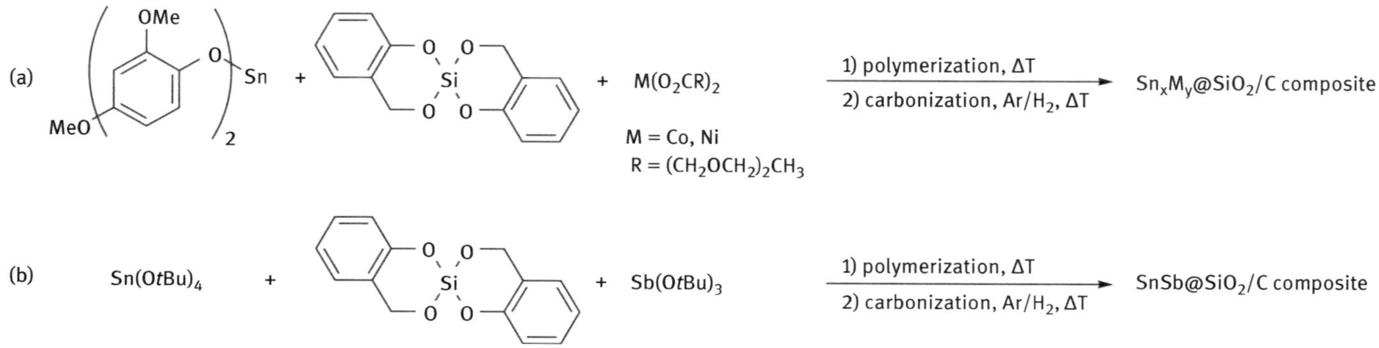

Scheme 5.5: Synthetic routes for tin alloy composite materials following the concept of STP. Reduction in an argon/hydrogen atmosphere (Ar/H_2: 95/5, reduction temperature 600–780 °C) gave porous composites [27, 39]. The figure was created according to the style in Ref. [27].

324 $m^2 \cdot g^{-1}$, an average crystallite size of 24 nm, and a tin and carbon content of 24 wt% and 47 wt%, respectively. Electrochemical measurements were performed using the above-mentioned setup, where the tin-alloy-containing composite was incorporated as anode material. Cycling of the as-obtained composite material in a half-cell indicated an increased cycling stability compared to the tin composite (for comparison see Fig. 5.9a). A reversible capacity of 64 % after 17 cycles was obtained (313 $mAh \cdot g^{-1}$). While the capacity of the first cycle (486 $mAh \cdot g^{-1}$) was lower compared to the cobalt free composite (545 $mAh \cdot g^{-1}$), a decreased fading during the first 17 cycles was given. Cyclic voltammetry showed that typical peaks of lithiation/delithiation processes of tin were still present.

Another type of composites that are based on tin/nickel-alloy-containing materials were synthesized using different Si:Sn:Ni ratios of 3:1:0.75, 1:1:0.3, or 1:1:0.75. In each case, mixtures of different Sn/Ni alloys were formed, most presumably Sn_2Ni_4 and Sn_4Ni_3. In addition, some crystalline tin was also found in the materials according to PXRD analysis. The materials were porous in the case of higher nickel contents (123 $m^2 \cdot g^{-1}$ for 3:1:0.75 and 215 $m^2 \cdot g^{-1}$ for 1:1:0.75). TEM images of the materials (Fig. 5.10) reveal nanoparticles with sizes of approximately 10 nm. Electrochemical measurements of the nickel-containing composites resulted in an improved cycling stability compared to the Sn@SiO_2/C material. As a result, materials synthesized by mixtures of the precursors either with a ratio of 3:1:0.75 or 1:1:0.3 exhibit capacities after the 17th cycle of 339 $mAh \cdot g^{-1}$. However, they showed a different manner during cycling. The composite material with the higher silicon content (ratio 3:1:0.75) had the highest capacity in the first cycle of around 509 $mAh \cdot g^{-1}$ with a reversible capacity of 67 %. In contrast, the composite material with the ratio of 1:1:0.3 exhibited a capacity of 340 $mAh \cdot g^{-1}$ in the first cycle. This indicates a restricted accessibility of the Sn/Ni particles. With regard to further optimization, we conclude that the best cycling and highest capacity in the first cycling were achieved in the case of the composite exhibiting the lowest tin content (14 wt%) and highest silicon content. This is attributed to an intercalation of lithium ions in the C/SiO_2 matrix and a low accessibility of the Sn/Ni particles.

While the Sn/Co- and Sn/Ni-containing composite materials were synthesized using the STP, an antimony-containing composite material was synthesized using the additive-assisted TP of Si-Spiro with $Sn(OtBu)_4$ and $Sb(OtBu)_3$ (see Scheme 5.5). The as-obtained composite material that was obtained after carbonization under reducing conditions, exhibited phase pure SnSb as confirmed by powder X-ray diffraction analysis and electron microscopy. A uniform distribution of the SnSb particles within the SiO_2/C matrix exhibiting particle sizes ranging from 14 to 50 nm and a BET surface area of 272 $m^2 \cdot g^{-1}$ was obtained. It has been demonstrated that these materials showed the best cycling stability of the Sn/M-containing composite materials (with M = Co, Ni, Sb) in combination with a high capacity in the first cycle (567 $mAh \cdot g^{-1}$). Electrochemical testing of the material revealed a reversible capacity of 70 % (397 $mAh \cdot g^{-1}$) after the 12th cycle (Fig. 5.11). Moreover, cyclic voltammetry

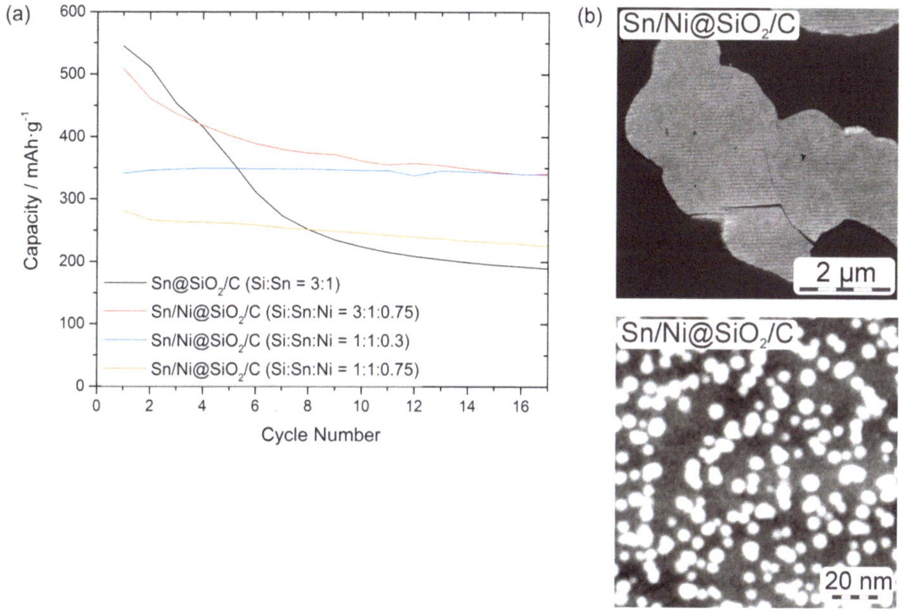

Fig. 5.10: (a) Discharge capacity of $Sn@SiO_2/C$ (black) and $Sn/Ni@SiO_2/C$ (red, blue, and orange) composite materials and (b) HAADF-STEM images of $Sn/Ni@SiO_2/C$ composite. The materials were synthesized by STP of Si-Spiro and tin(II)-2,4-dimethoxyphenylmethanolate in a 3:1 molar ratio and in the case of the nickel-containing material with addition of nickel(II) carboxylate (different ratios). The electrochemical measurements were performed in half-cell setup at 30 $mA \cdot g^{-1}$. The graph in (a) was created according to the generated data from Ref. [27]. The figure in (b) (top) is adapted from Ref. [27] and the figure in (b) (bottom) is reproduced from Ref. [39] with permission of The Royal Society of Chemistry.

measurements indicated typical lithiation/delithiation of the SnSb alloy and intercalation of lithium into the carbon matrix. However, a decrease of activity is also obtained for this nanocomposite and, although promising results were obtained, further improvement on cycling stability is necessary [27].

In conclusion, the embedding of tin nanoparticles or nanoparticulate tin alloys in a carbon matrix was successful following the concept of twin polymerization. The distribution of the particles within the carbon matrix is homogeneous, which is a prerequisite for good cycling stability. The most promising cycling stability was so far obtained for the SnSb-containing composite material but its performance still has to be improved with regard to practical applications. It might be assumed that a low accessibility for lithium ions is related to hindered diffusion in the SiO_2 matrix. It is concluded that the amount of active sites within the composite material must be increased, e.g., by partial removal of silica [27].

An alternative to the usage of tin-containing composite materials is represented by germanium-containing anode materials. Although being expensive, the abundance in the earth crust of germanium is close to that of tin. Thus, germanium might

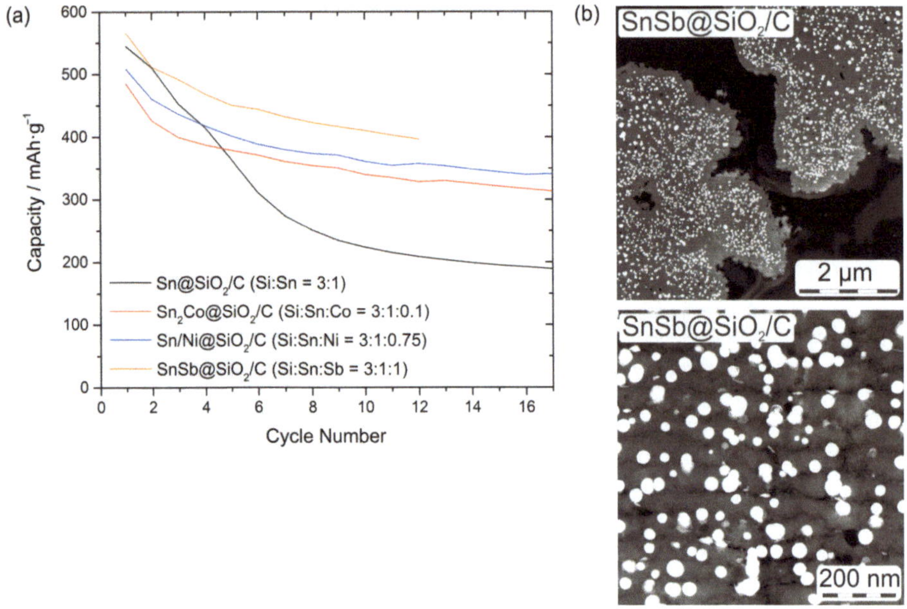

Fig. 5.11: (a) Discharge capacity of the best materials for each alloy, i.e., $Sn@SiO_2/C$ (black), $Sn_2Co@SiO_2/C$ (red), $Sn/Ni@SiO_2/C$ (blue), and $SnSb@SiO_2/C$ (orange) composite materials and (b) HAADF-TEM images of $SnSb@SiO_2/C$ composite material. The materials were received either by STP of Si-Spiro and tin(II)-2,4-dimethoxyphenylmethanolate in a 3:1 molar ratio with addition of cobalt(II) or nickel(II) carboxylate or in the case of antimony-containing composite material using additive-assisted twin polymerization by addition of $Sn(O\textit{t}Bu)_4$ and $Sb(O\textit{t}Bu)_3$. The electrochemical measurements were performed in half-cell setup at 30 mA · g^{-1}. The figure was created according to the generated data from Ref. [27]. The image in (b) was adapted from Ref. [27].

not be used in battery application for vehicles, but its use for small scale applications is feasible. Its high gravimetric capacity of 1600 mAh · g^{-1} ($Li_{17}Ge_4$) makes it an interesting target material. Nevertheless, similar to tin an expansion of 246 % upon alloy formation takes place, which has to be addressed. To attenuate the effect of volume expansion, germanium is usually embedded in a carbon matrix (Ge@C) and several approaches were reported. For example, Ge@C materials are accessible by reduction of different germanium(IV) precursors, e.g., germanium tetrachloride, germanium(IV) oxide, or germanium(IV) alkoxides in an argon/H_2 or argon/acetylene flow or by annealing of elemental germanium in an argon/acetylene atmosphere [19, 41–47]. Alternatively, thermally induced twin polymerization using either polymerizable germanates or germylenes based on salicylic alcoholates might be used. Recently, both germanates as well as germylenes, were polymerized at 200 °C to obtain composites consisting of a phenolic resin and germanium oxide. Conversion of the materials under reducing conditions (Ar/H_2: 95/5) provided microporous Ge@C [11, 26]. An example for the access to Ge@C materials using molecular germanates or germylenes (twin monomers) is depicted in Scheme 5.6. For the as-prepared nanocomposites,

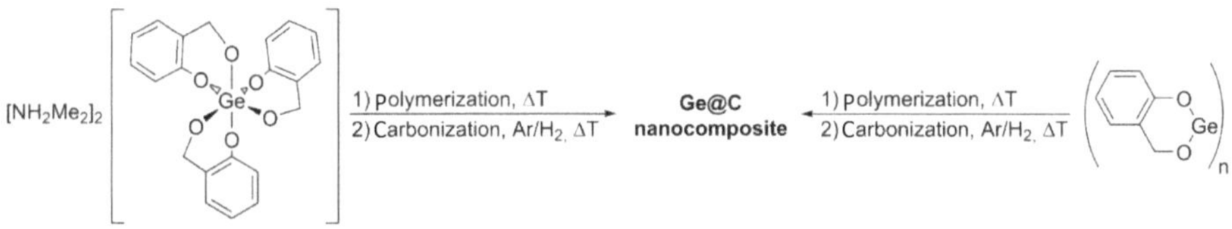

Scheme 5.6: Reaction pathways for the synthesis of Ge@C materials following the concept of twin polymerization and reduction of the as-obtained hybrid materials under argon/hydrogen atmosphere (95:5, at 600°C or 800°C) using either a germanate or a germylene precursor [11, 26].

electrochemical measurements were carried out in a half-cell setup consisting of a lithium cathode, an anode comprising 80 wt% of Ge@C material, 10 wt% carbon black, 10 wt% carboxymethyl cellulose, and a 1 M solution of $LiPF_6$ in ethylene carbonate/dimethyl carbonate (1:1 by weight and addition of 3 % of fluoroethylene carbonate) as electrolyte.

Starting from the molecular precursor bis(dimethylammonium)tris [2-(oxidomethyl)phenolate(2-)]germanate, a microporous Ge@C material was obtained after reduction at 800 °C that exhibits a BET surface area of 470 $m^2 \cdot g^{-1}$ with a micropore content of 61 %. PXRD analysis revealed the formation of crystalline germanium with an average crystallite size of 27 nm (using the Scherrer equation). Carbon and germanium contents of approximately 60 % and 20 % were determined, respectively. Electrochemical measurements of the anode material were carried out at current densities of 346 $mA \cdot g^{-1}{}_{Ge@C}$ and 1384 $mA \cdot g^{-1}{}_{Ge@C}$ (corresponding to 1C and 4C). High cycling stabilities with stable cycling for 100 and 500 cycles, depending on the current densities, were obtained with a reversible capacity of 370 $mAh \cdot g^{-1}{}_{Ge@C}$, respectively. Notably, no fading during cycling and high coulombic efficiencies of ≥99 % were determined [26]. The electrochemical tests are depicted in Figure 5.12. Given the low germanium content, these results are very promising and future research should focus on reduction of the carbon content.

A first step to reach this goal would be the use of monomers with a higher Ge:C ratio. This was realized by twin polymerization of germylenes based on salicylic alcoholates, which gave a phenolic resin/germanium oxide-containing hybrid material. Carbonization yielded a Ge@C composite with the best Ge:C ratio by weight (approximately 1:1 by EDX analysis) for the material synthesized at 800 °C. In order to study the influence of the carbonization temperature on the final properties, several temperature profiles were chosen. For example, the material carbonized at 600 °C possessed a higher micropore content (34.5 %) but a similar BET surface area (238 $m^2 \cdot g^{-1}$) compared to the material carbonized at 800 °C (25.1 % micropore content; BET surface area 268 $m^2 \cdot g^{-1}$). In general, the germanium content was increased by increasing the carbonization temperature, showing germanium contents of 39.8 % for the material carbonized at 600 °C and 46.5 % at 800 °C. As expected, smaller crystallites were obtained in the Ge@C materials carbonized at

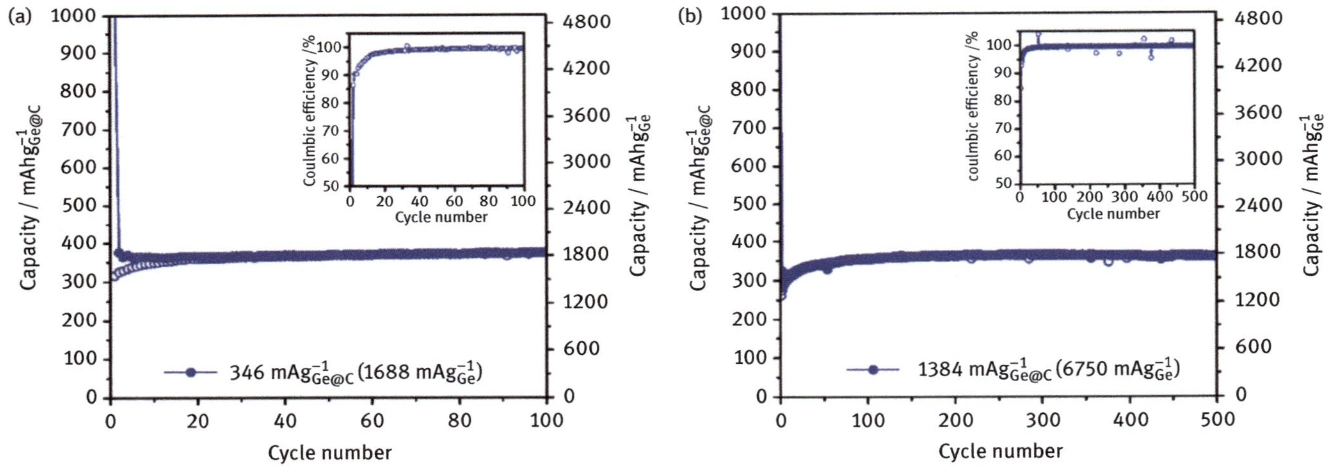

Fig. 5.12: Discharge capacity and coulombic efficiency of anode material based on Ge@C received following the concept of twin polymerization using bis(dimethylammonium)tris[2-(oxidomethyl)phenolate(2-)]germanate as monomer and carbonization in Ar/H_2 atmosphere (95:5) of the as-obtained hybrid material. The electrochemical measurements were carried out at (a) 346 $mA \cdot g^{-1}{}_{Ge@C}$ or (b) 1384 $mA \cdot g^{-1}{}_{Ge@C}$. This figure is reproduced from Ref. [26] with permission of The Royal Society of Chemistry.

600 °C (13 nm), while a crystallite size of 51 nm was determined (Scherrer equation) in the case of a material that was carbonized at 800 °C. Subsequent electrochemical tests of the as-obtained Ge@C materials as anode material indicated an exceptionally good cycling stability. Thus, no fading was observed for at least 100 (cycling at 346 $mA \cdot g^{-1}_{Ge@C}$) and 500 cycles (cycling at 1384 $mA \cdot g^{-1}_{Ge@C}$). Notably, coulombic efficiencies increased in each case to ≈ 99 % during cycling. Significantly higher capacities for the composite material synthesized at 600 °C were reported for both current values with stable capacities of 540 $mAh \cdot g^{-1}_{Ge@C}$ and 390 $mAh \cdot g^{-1}_{Ge@C}$, respectively; whereas values of 400 $mAh \cdot g^{-1}_{Ge@C}$ and 220 $mAh \cdot g^{-1}_{Ge@C}$, respectively, were measured for composite materials synthesized at 800 °C. The higher capacities of the material synthesized at 600 °C were related to smaller crystallites of germanium in the material. It is noteworthy that the capacity related to the germanium content in the composite was given by 980 $mAh \cdot g^{-1}_{Ge}$ (after 500 cycles) for the material synthesized at 600 °C, which is lower than the maximum value of 1600 $mAh \cdot g^{-1}$ ($Li_{17}Ge_4$). This was attributed to impurities such as germanium oxide (e.g., capacity of GeO_2 961 $mAh \cdot g^{-1}$), the presence of which reduces the capacity. Thus, there seems to be further potential to optimize the materials' performance. The electrochemical measurements are given in Figure 5.13 [11].

In conclusion, Ge@C materials synthesized by the twin polymerization process are highly promising as materials in high-performance anodes for rechargeable lithium-ion batteries. It was shown that cycling stabilities without fading for 500 cycles are feasible, independent from the nature of the twin monomer, which was either a germanate or a germylene. The most promising results were obtained with the Ge@C composite as synthesized from the material that was carbonized at 600 °C, starting from the germylene germanium(II) 2-(oxidomethyl)phenolate. The highest capacity and cycling stability are most presumably based on the formation of small crystallites within the carbon matrix, which act as a buffer medium for volume expansion [11, 26].

5.1.3 Summary

To overcome the cell failure in lithium-ion batteries caused by volume expansion during lithiation/delithiation, the usage of porous and nanostructured materials is an appropriate strategy. With regard to this aspect, twin polymerization provides a novel access to highly porous carbon materials with BET surface areas up to 1450 $m^2 \cdot g^{-1}$. The homogeneous embedding of active materials, e.g., metal nanoparticles (Ge, Sn), nanoparticular tin alloys (Sn_2Co, Sn_3Ni_4/Sn_2Ni_4, SnSb), or sulfur into the carbon matrix was demonstrated. It was shown that the best as-obtained anode materials possess a cycling stability for at least 500 cycles (Ge@C-based anode materials) without any fading and the best cathode materials show a cycling stability for more than 1300 cycles (sulfur-containing hollow carbon spheres) exhibiting only minor

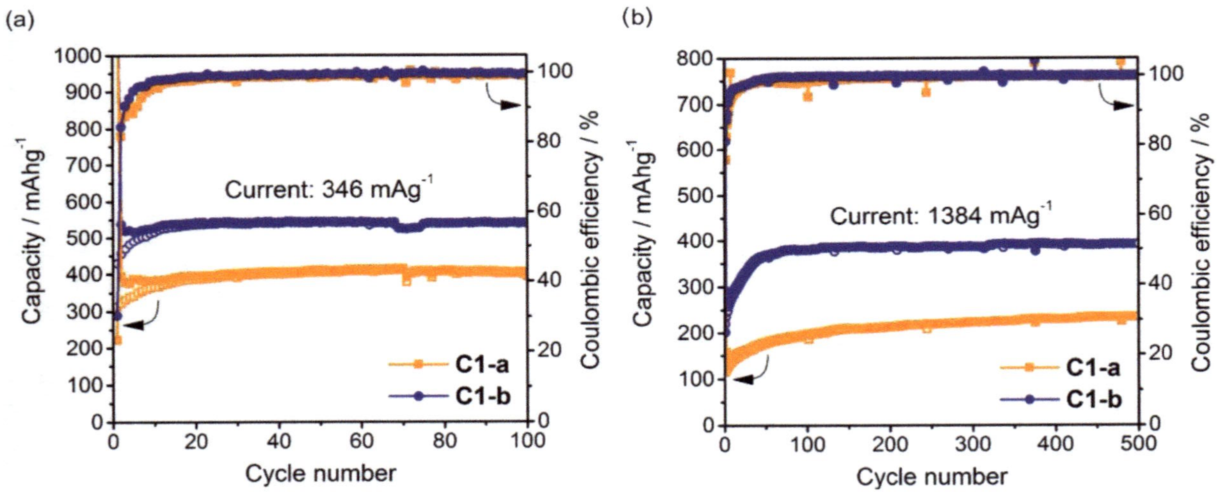

Fig. 5.13: Discharge capacity and coulombic efficiency of an anode material based on Ge@C as prepared following the concept of twin polymerization with germanium(II) 2-(oxidomethyl)phenolate as monomer and reduction of the as-obtained hybrid material at 800 °C in the case of C1-a and at 600 °C for C1-b. The electrochemical measurements were carried out at (a) 346 mA · $g^{-1}{}_{Ge@C}$ or (b) 1384 mA · $g^{-1}{}_{Ge@C}$. This figure is reproduced from Ref. [11] with permission of The Royal Society of Chemistry.

fading of 0.08 % per cycle. Thus, twin polymerization offers a versatile approach to high performance cathode and anode materials for lithium-ion batteries on the basis of carbon-based composites. Somewhat surprisingly, a poor cycling stability was obtained in the case of tin-containing SiO_2/C composites, the reasons for which are not fully understood yet. It might be assumed that a better accessibility and a higher amount of active material will increase the capacity and the cycling stability of these materials significantly. Therefore, a partial removal of silica by post-synthesis treatment seems to be a feasible approach to increase the amount of active tin within the material and to improve its accessibility.

Katja Schreiter, Matthias Birkner, Mandy Göring, Susanne Höhne, Daniel Uhlig

5.2 Functionalized twin monomers and their application in materials synthesis

Polymers can, in principle, be divided into those with structural applications and those with specific functions. Classical polymers such as polystyrene, polypropylene, or polyethylene are far more well-known than functional polymers. One reason for this is the fact that classical polymer materials are used in our daily lives in a large number of applications because of their mechanical properties and high durability. In comparison, functional polymer materials are often less visible in the environment. They do not stand out as a result of their thermomechanical properties, but much more because of their actions or effects based on highly specific chemical, physicochemical, or biophysical properties. The areas of application for functional polymers range from the life sciences to self-healing polymers and conductive materials [48–50]. A. J. Heeger, A. G. MacDiarmid, and H. Shirakawa were awarded the Nobel Prize in Chemistry in the year 2000 for the discovery and development of conductive polymers [51, 52].

There are two key strategies for the synthesis of functional polymers: (i) a polymer analogous reaction: a reaction subsequent to polymerization where functional groups are introduced into the pre-formed polymer matrix (e.g., *via* click reactions) [53, 54]; or (ii) the direct (co-)polymerization of the desired functional monomers or their precursors with corresponding selected structural and crosslinking monomers (e.g., *via* controlled radical (co-)polymerization or ring-opening (co-)polymerization). In many cases, one of these two methods can be used for the synthesis of a particular functional polymer, the choice of synthetic method depending on the state of knowledge, the availability of starting materials, handling, and economics. Polymer analogous reactions, for example, represent an attractive method for the synthesis of functional polymers which overcomes the limited functional group tolerance of some controlled polymerizations. The aim of the strategies is to obtain new polymers with a hitherto unobtainable degree of functionality with precise control of the sequence in the polymer and thus over the physicochemical properties [55].

5.2.1 Characteristics of functionalized twin monomers

The design of new functionalized building blocks for the construction of functional materials with new and/or improved properties is a continuing theme in modern materials science. In recent years, the concept of twin polymerization (TP) was established and a broad range of twin monomers (TM) for the synthesis of functional materials based on organic/inorganic hybrid materials has become available.

TP consists of complex sub-processes where two different polymers are formed from just one single monomer, the TM (A–C or B–C), in a mechanistically coupled way (see Chapter 2) [56–64]. The original reactivity of the TM can, on the one hand, be electronically and sterically modified using an additional chemical functionalization Y in a twin monomer Y–B–C. On the other hand, the coupling of the TP with further simultaneous polymerization processes is possible by means of such functionalization (see Chapter 5.2.4.1). The functionality can influence the mechanistic process starting from the actual TP and thus influences the structure and the resulting properties of the hybrid material formed [65]. The TP of functionalized TMs requires precise adjustment of the reaction components and parameters to obtain nanostructured functional hybrid materials. By introducing a functionalized TM into the TP, several scenarios are possible. In one scenario (Scheme 5.7 (a)), two homopolymers are formed from one functionalized TM Y–B–C, while the functionality Y does not necessarily influences the TP. A functionalized TM Y–B–C can polymerize simultaneously with either one or several TMs. Simultaneous twin polymerization (STP) can be controlled by the TM combination Y–B–C and A–C, where the functionality Y either prevents the copolymerization of A and B, for example, (Scheme 5.7 (a)) or promotes it (Scheme 5.7 (b)).

During TP, interaction of the functionality Y with the organic component is also theoretically possible (Scheme 5.8). For this scenario, aspects such as the spacer length and the molecular structure, and thus associated parameters such as reactivity and polarity of the TM can be important.

It is also possible to bind the unit A–B *via* the functional group Y_1 to surfaces or other polymers (Y_2-polymer), whereupon surface functionalization or crosslinking can be achieved (Scheme 5.9). The functionality Y_1 in this case serves as an anchor point and must be adjusted in accordance with molecular reactivity (Y_2) of the desired polymer.

The functionality Y can be integrated into both the organic (C, e.g., salicyl alcohol unit) or the inorganic fragment (B, e.g., silicon atom) of the TM or linked to the organic or inorganic fragment by introducing a spacer to decouple the influence on the reactivity (Fig. 5.14).

The scenarios in Schemes 5.7–5.9 are, however, only shown for TMs which have a functionalization Y in the inorganic fragment B (TM Y–B–C). The focus of the Sections 5.2.1–5.2.3 is on TMs where a functionality such as an amino group is built in *via* a spacer at the silicon atom.

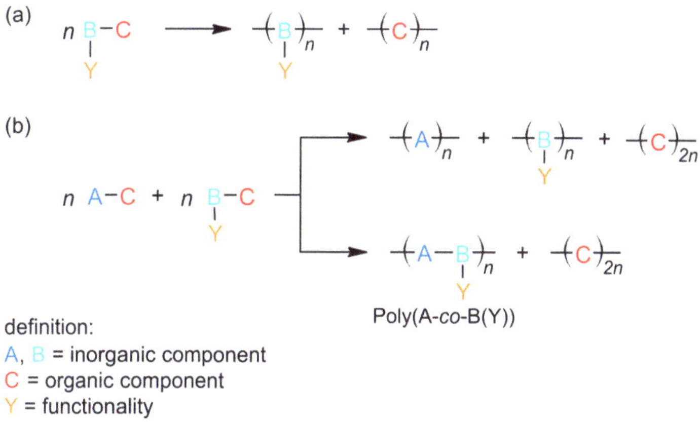

Scheme 5.7: Two scenarios for the reaction of a functionalized TM in TP (a) or in STP (b).

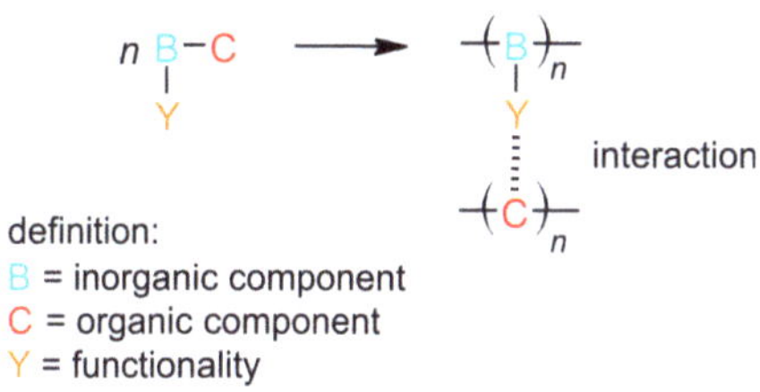

Scheme 5.8: Covalent and noncovalent interaction of the functionality Y with the organic component during TP.

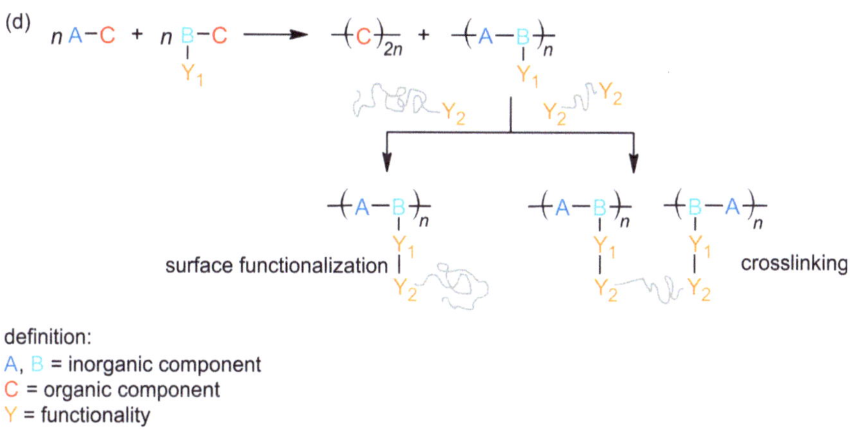

Scheme 5.9: Twin monomer (TM) with functionality Y_1 as an anchor point for grafting onto different materials.

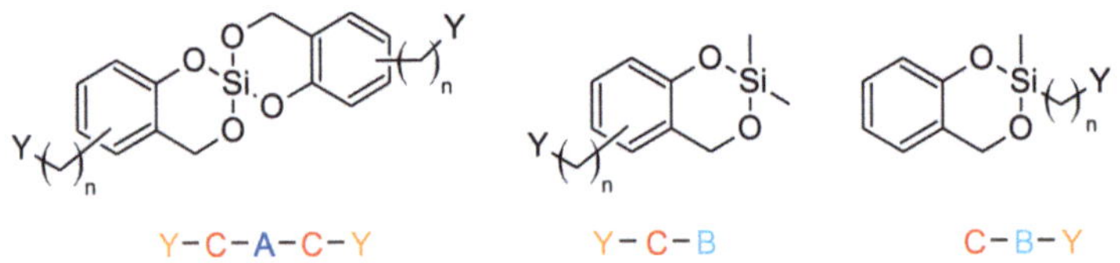

Fig. 5.14: Possibilities of connecting functionalities Y to the TM.

5.2.2 Amino-functionalized twin monomers

Amino-functionalized twin monomers can be synthesized starting from commercially available amino-*n*-propylmethyldimethoxysilanes by means of fluoride-catalyzed transesterification with salicyl alcohol. TMs with primary, secondary, or tertiary amino groups, as well as those with bifunctional character in the periphery, can be obtained [66].

The amino-functionalized TMs shown in Scheme 5.10 can be polymerized in twin polymerization and they can participate in simultaneous twin polymerization (STP) in combination with, e.g., the TM 2,2'-spirobi[4*H*-1,3,2-benzodioxasiline] (Si-Spiro).

Advantageously, the amino-functionalized TMs (see Scheme 5.10) used for the STP can serve both as components and as base catalysts for the production of ternary hybrid materials consisting of phenolic resin/silica-poly(aminopropylsiloxane). This is an elegant way of synthesis, because impurities resulting from the use of additional base catalysts can be avoided. The "ideal twin polymerization" of Si-Spiro to phenolic resin and silica starts at temperatures above 200 °C. Instead, the monomer Si-Spiro was simultaneously polymerized with different amino-functionalized twin monomers (see Scheme 5.10, (a)–(f)) at a much lower temperature (120 °C) using a stoichiometric ratio of 1:1.

The STP of Si-Spiro with amino-functionalized TMs can be triggered by heat in the temperature range from 99 °C (e.g., TM (d), Scheme 5.10) to 205 °C (e.g., TM (e), Scheme 5.10). The trigger temperature for STP depends on monomer composition and molecular structure of the amino-functionalized TM in the monomer mixture. Details on DSC investigations with regard to the effect of molecular structure and catalysts on the onset temperature of the polymerization of TMs are to be found in Chapter 4.1.

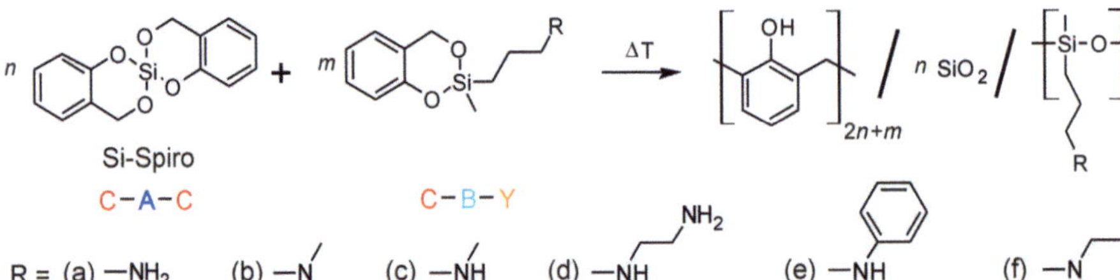

Scheme 5.10: Thermally induced STP of TM Si-Spiro with the amino-functionalized TMs (a)–(f).

According to the concept of STP [58, 63], organic/inorganic hybrid materials consisting of phenolic resin, SiO_2, and oligo-(3-amino-*n*-propyl)-methylsiloxane (OAMS) can be obtained (Scheme 5.10). The hybrid materials obtained are investigated in detail using the example of (S)TP of 2-(3-amino-*n*-propyl)-2-methyl-4*H*-1,3,2-benzodioxasiline (APSI; Scheme 5.10, R = NH_2) and Si-Spiro at different molar ratios.

APSI reacts with Si-Spiro in different molar ratios (Si-Spiro:APSI = 95:05, 85:15, 50:50, 15:85, 0:100) *via* a thermally induced (S)TP. The formation of phenolic resin and OAMS can be detected with certainty using solid state $^{13}C\{^1H\}$-CP-MAS NMR spectroscopy. The different monomer ratios are reflected in the solid state $^{13}C\{^1H\}$-CP-MAS NMR spectra in the signal intensities of the phenolic resin/OAMS. The assignment of signals to the structure is shown in Figure 5.15. The formation of *o,o'*- and *o,p'*-bonded phenolic resin structures could be proven by signals in the aromatic region (signals 3 and 4) and from the broad signal 6 of the bridging methylene group (Fig. 5.15 (left)). The signal for the methyl group of OAMS is located at 0 ppm.

The solid state $^{29}Si\{^1H\}$-CP-MAS NMR spectra of these hybrid materials show the expected Q and/or D signals arising from SiO_2 and OAMS, which originate from Si-Spiro and APSI. The different monomer ratios are reflected in the solid state $^{29}Si\{^1H\}$-CP-MAS NMR spectra and the signal intensities of SiO_2-OAMS change accordingly. In each case, Q_4 (–110 ppm) is the most intense signal in the silica region of the spectrum, although the cross polarization technique used overrated Q_2 (–90 ppm) and Q_3 (–100 ppm) signals due to polarization transfer from ^1H to ^{29}Si. This result indicates that the silica network is highly condensed. Use of a low ratio of APSI results in D(Q) species. This D(Q) structure indicates formation of Si–O–Si bonds between the polydialkylsiloxane units and the silicon dioxide network and thus to the presence of a copolymer. The corresponding ^{29}Si NMR signal arises at –17 ppm. Increasing the content of the monomer APSI in the STP leads to more intense D signals with typical chemical shifts of –20 ppm in the solid state $^{29}Si\{^1H\}$-CP-MAS NMR spectra, which indicate a D_2 structure for the hybrid material. Longer chains (or rings) of OAMS are formed upon increasing content of APSI. In TP of APSI, the resulting hybrid material only shows D signals in $^{29}Si\{^1H\}$-CP-MAS NMR spectrum.

The transparent hybrid materials do not show any macroscopically visible agglomeration. HAADF-STEM image of the hybrid material from the STP of Si-Spiro:APSI (50:50) reveals phase domains with sizes of approximately 2–4 nm (Fig. 5.16). This dimension is significantly smaller than the comparable phenolic resin/silicon dioxide composite materials synthesized using the sol-gel process [67, 68].

5.2.3 Interface design – functionalized TM as adhesion promoter

For the combination of different groups of materials in functionalized hybrids, the suitability of the reaction mixture as an adhesive system based on twin polymerization has been investigated. In particular, the potential of the functionalized twin

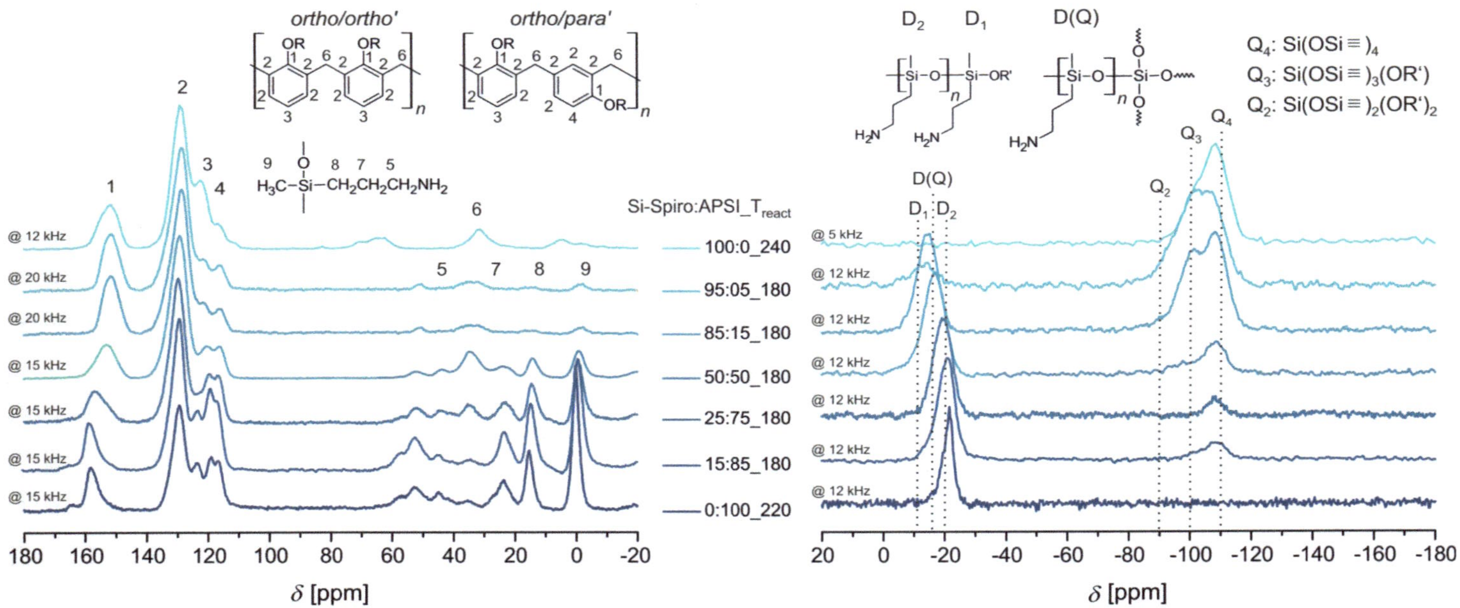

R = H, SiO_2, polydialkylsiloxane

R' = H, alkyl, aryl

Fig. 5.15: Solid state $^{13}C\{^1H\}$-CP-MAS (left) and $^{29}Si\{^1H\}$-CP-MAS NMR spectra (right) of hybrid materials obtained by (S)TP of Si-Spiro and APSI using different reactant ratios at different reaction temperatures.

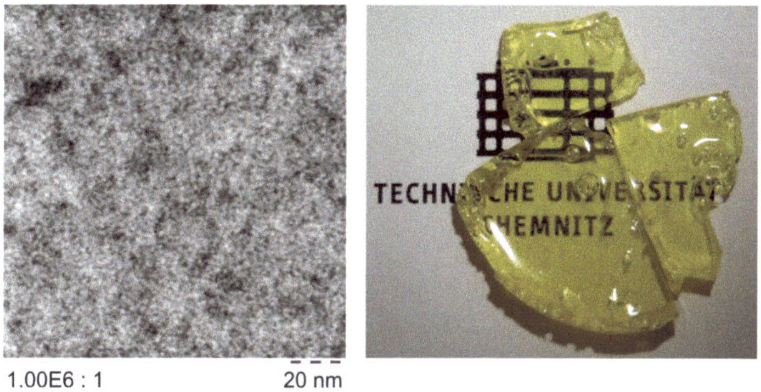

Fig. 5.16: (left) HAADF-STEM image of the hybrid material from the STP of Si-Spiro:APSI (50:50) showing phase domains of 2–4 nm in size. (right) Photographic image of the transparent, monolithic composite material [66]. Published by The Royal Society of Chemistry.

monomers has been studied as new type of adhesive systems for aluminum/PA 6 hybrid components [69]. APSI, Si-Spiro, and mixtures of the two TMs were used. Mechanical parameters of the resulting hybrid materials, determined by nanoindentation, show that an increasing proportion of the amino-functionalized TM lowers both the indentation hardness and the indentation modulus.

The properties of the hybrid materials depend on the ratio of SiO_2 to polydialkylsiloxane. If the mechanical properties of the phenolic resin content are assumed to be constant in all hybrid materials, the hardness and indentation modulus of the hybrid material increases with increasing SiO_2 and decreasing polydialkylsiloxane content. The influence of the SiO_2 component on the mechanical properties of the hybrid material is particularly visible in the creep behavior (Fig. 5.17) [69]. For the hybrid material starting from APSI (Si-Spiro:APSI = 0:100), an almost 10 times deeper penetration into the sample is observed than for the sample Si-Spiro:APSI = 15:85. In the TP of APSI, a nanostructured silicon dioxide network cannot be formed, which leads to a decrease in the mechanical load bearing of the hybrid material. Samples with higher SiO_2 contents show a similar creep retention as the sample Si-Spiro:APSI = 15:85 (Fig. 5.17).

The ability of the hybrid material to bind to an aluminum surface was determined by measuring the strength of adhesion in a pull-off test. Pure TMs and various ratios of the mixture were applied between two aluminum plates and then thermally polymerized. The binding strength of the joined samples achieved in the tension test is shown in Figure 5.18. For the aluminum/aluminum bonding, a maximum adhesive strength of 12 MPa was achieved using a Si-Spiro:APSI mixture in a molar ratio of 15:85 (see Fig. 5.18 (a)). The hybrid material from this mixture (Si-Spiro:APSI, molar ratio of 15:85) showed a slightly higher creep in the nanoindentation measurements. Accordingly, the adhesion promoter can compensate residual stress of the polymer component in the composite. The adhesive strength of polyamide 6 was investigated in the model system aluminum/polyamide 6-GF (PA 6-GF: glass fiber reinforced polyamide 6) with

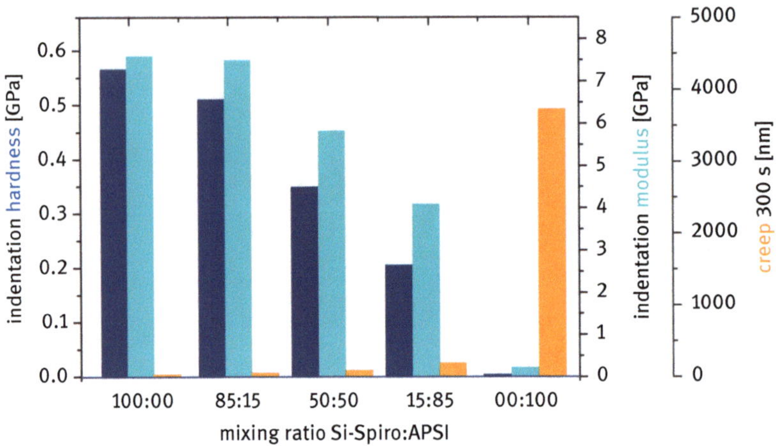

Fig. 5.17: Indentation hardness and modulus of Si-Spiro/APSI hybrid material, determined from curves with holding period of 60 s. Creep of the bulk hybrid materials with holding period of 300 s [69]. Copyright John Wiley & Sons, Inc. Reproduced with permission.

the previously established monomer mixture of Si-Spiro:APSI in a molar ratio of 15:85 at different joining temperatures and surface roughness (corundum abrasion) (see Fig. 5.18 (b)). For the preparation of the aluminum/PA 6-GF composite a minimum joining temperature of 240 °C is necessary. At 220 °C, a composite was not obtained in the experiments. The joining temperature must therefore be above the melting temperature of the PA component (T = 220 °C). In the pull-off test, adhesive strengths of the aluminum/PA 6-GF composite are achieved which hardly differ depending on the joining temperatures (240 °C, 260 °C) and in addition are smaller for blasted aluminum surface (Figure 5.18 (b)). The adhesion strengths show the affinity of the system Si-Spiro:APSI to these particular functional groups. The amino groups present in the Si-Spiro:APSI adhesive system can interact with the aluminum/aluminum oxide substrate and also with polyamides. Hydrogen bonds can be formed between the polymerized system Si-Spiro:APSI and the PA component (e.g., –NH···O = C–). Furthermore, formation of Si–O–Al bonds during the polymerization of the TM may also play a role in the adhesion mechanism. Roughening the aluminum surface by mechanical abrasion using corundum does not increase the adhesion strength, which can be attributed to the preferred binding of the adhesive to aluminum oxide/hydroxide. Mechanical abrasion with corundum leads to a change in the surface composition and thus to a different wettability between adhesion promoter and metal surface (Fig. 5.18 (b)).

In further experiments, catechol was added to the monomer mixtures as reactive additive. Catechol serves on the one hand as a bidentate ligand able to form a coordinative bond with the metal surface [70], and on the other hand it can be incorporated into the network of the organic/inorganic hybrid material *via* an electrophilic substitution reaction. While the hybrid material Si-Spiro:APSI (15:85) has an extractable

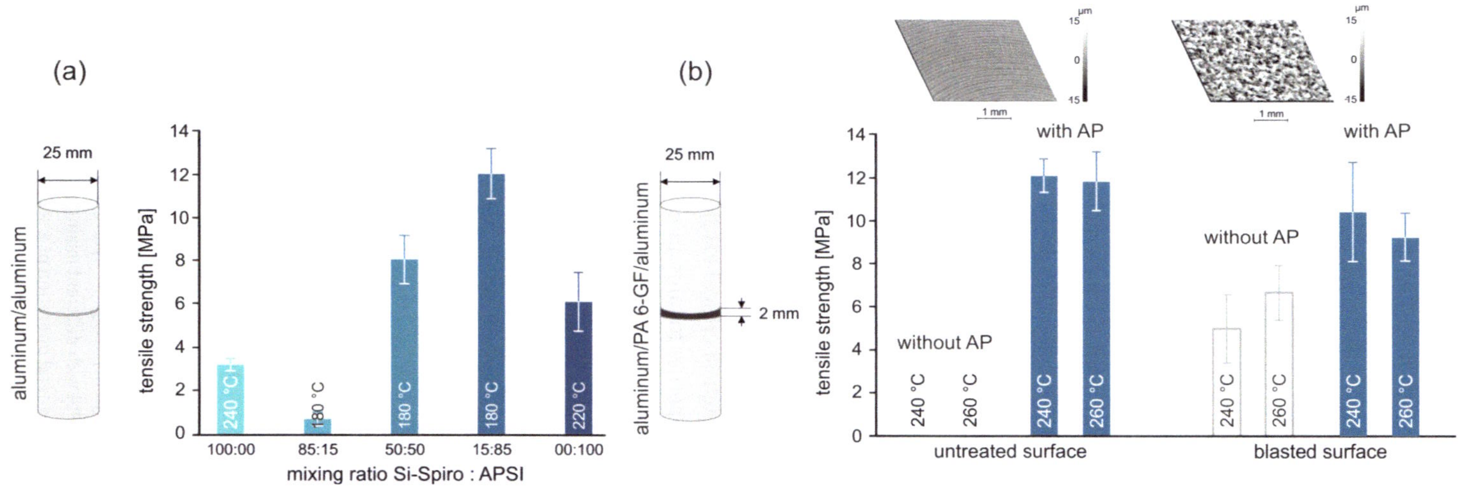

Fig. 5.18: (a) System aluminum/TP layer/aluminum: adhesive strengths from the pull-off test (axial loading tensile test: DIN EN 582; aluminum alloy AW-6082) with different monomer compositions in molar ratio; (b) System aluminum/PA 6-GF with TP layers: adhesive strengths from the pull-off test (axial loading tensile test: DIN EN 582) with monomer ratio of Si-Spiro:APSI (15:85) at different temperatures and surface roughness (AP = adhesion promoter).

fraction (extraction agent: dichloromethane) of more than 30 wt%, the extractable fraction of the hybrid material decreases to less than 1 wt% by adding 2 wt% catechol. Thus, catechol has been covalently integrated into the hybrid material, which is therefore better crosslinked. By adding 2 wt% catechol to the monomer mixture of Si-Spiro and APSI, an improved adhesion is achieved in the system aluminum/aluminum compared to the application of Si-Spiro:APSI (15:85). In aluminum/PA composites, however, these bonding agent systems are not as effective as the system Si-Spiro:APSI (15:85) without catechol. This can be partly due to the higher hardness of the system by adding catechol.

The results of scratch tests confirm that the use of a mixture of Si-Spiro and APSI in a ratio of 15:85 results in the formation of a hybrid material film with the best adhesion to the metal substrate compared to all other films produced with different monomer compositions. SEM images in Figure 5.19 show the whole scratch track as well as a detailed image of the initial damage to the film. The damage to the hybrid material film of the monomer mixture with a ratio of 85:15 (Si-Spiro:APSI) appears very rapidly after the beginning of the test. This indicates that the critical loading, which correlates with the adhesion to the metal substrate, is very small. Delamination of this film appears at the onset of flaking without microcracking directly after plastic deformation of the film. Other films tested show microcracking that develops rapidly to flaking on increased loading. When using a monomer solution of Si-Spiro and APSI with a ratio of 15:85 (Si-Spiro:APSI), the first damage occurs furthest from the beginning of the scratching, corresponding to the highest critical loading.
Hybrid materials with very good adhesion properties for aluminum/PA 6 hybrid materials are accessible *via* TP. The success of these compounds is principally attributed to the molecular composition of the resulting hybrid material. The formed interpenetrating network, consisting of a phenolic resin, a silica network, and a polydialkylsiloxane

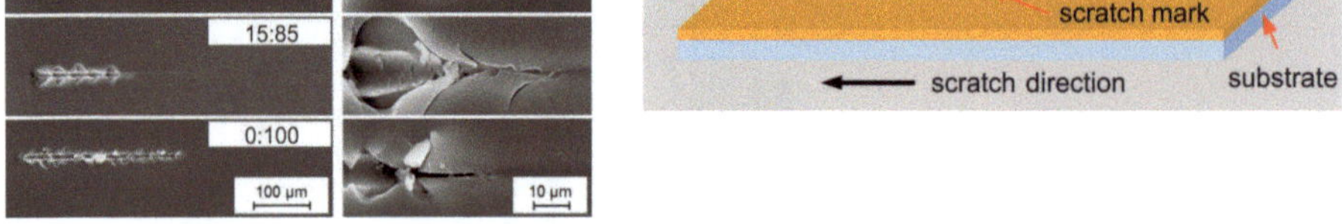

Fig. 5.19: Left: SEM micrographs of scratch tracks of organic/inorganic hybrid material films produced from chloroform solutions of Si-Spiro and APSI with different mixing ratios on polished aluminum surfaces. Right: schematic diagram of the general operation of a scratch tester taking a scratch measurement [69]. Copyright John Wiley & Sons, Inc. Reproduced with permission.

bearing primary amino groups, proves to be very effective and enables chemical bond formation to both inorganic substrates and organic matrices. In addition, the hybrid materials are almost completely free of defects, because no low molecular weight by-products were formed. Advantageously, the mechanical properties of the resulting hybrid material or adhesion promoter can be tailored by means of the chemical composition of the monomer mixtures. A high content of inorganic particles in these materials provides the necessary toughness and low shrinkage, while the organic components provide the curing properties in combination with the paste-like behavior.

5.2.4 Post-functionalization – construction of layered systems

Twin monomers or organic/inorganic hybrid materials which contain amino groups are key precursors to post-functionalization reactions, which allow access to a broad range of novel polymer structures and surface modifications of nanostructured materials. Primary, secondary, and tertiary amines can undergo diverse reactions. Scheme 5.11 gives an overview of some possible reactions and thus post-functionalizations of primary amino functions in TMs or in organic/inorganic hybrid materials. In the following Sections 5.2.4.1–5.2.4.3, the reactions of TMs or organic/inorganic hybrid materials with (bis-)epoxides (Scheme 5.11 (a)) and (bis-)isocyanates (Scheme 5.11 (g)) are described in particular.

5.2.4.1 Polymerization of APSI with bis-epoxides

The synergistic combination of twin polymerization of APSI with the step-growth polymerization of bisphenol A diglycidyl ether (BADGE) is suitable to produce homogeneous hybrid materials by a thermally induced process [71]. The monomer BADGE can react with both primary and secondary amines in the reaction mixture and thus functions as an additional bridge between the oligosiloxane chains formed during the TP. The expected reaction scenario is shown in Scheme 5.12. Generally, the two different rates of polymerization are independent of each other, but they must be in the same order of magnitude to achieve the formation of interpenetrating networks. The degree of crosslinking can be controlled by varying the ratio of APSI to BADGE [71].

The ternary formation of three different polymer structures, phenolic resin, epoxide/amine composite material, and crosslinked polydialkylsiloxane, within one procedure in the melt occurs smoothly as evidenced by solid state $^{13}C\{^{1}H\}$-CP-MAS and $^{29}Si\{^{1}H\}$-CP-MAS NMR spectroscopies (Fig. 5.20). Formation of phenolic resin *via* TP can be detected from the signals in regions I and II in solid state $^{13}C\{^{1}H\}$-CP-MAS NMR spectra, which are representative for phenolic resin structures [72, 73]. The polydialkylsiloxane formed is detected *via* an intense ^{13}C NMR signal at 0 ppm ($-CH_3$) and signals in the range from 5 to 44 ppm (propyl chain, signals 6–8). It can

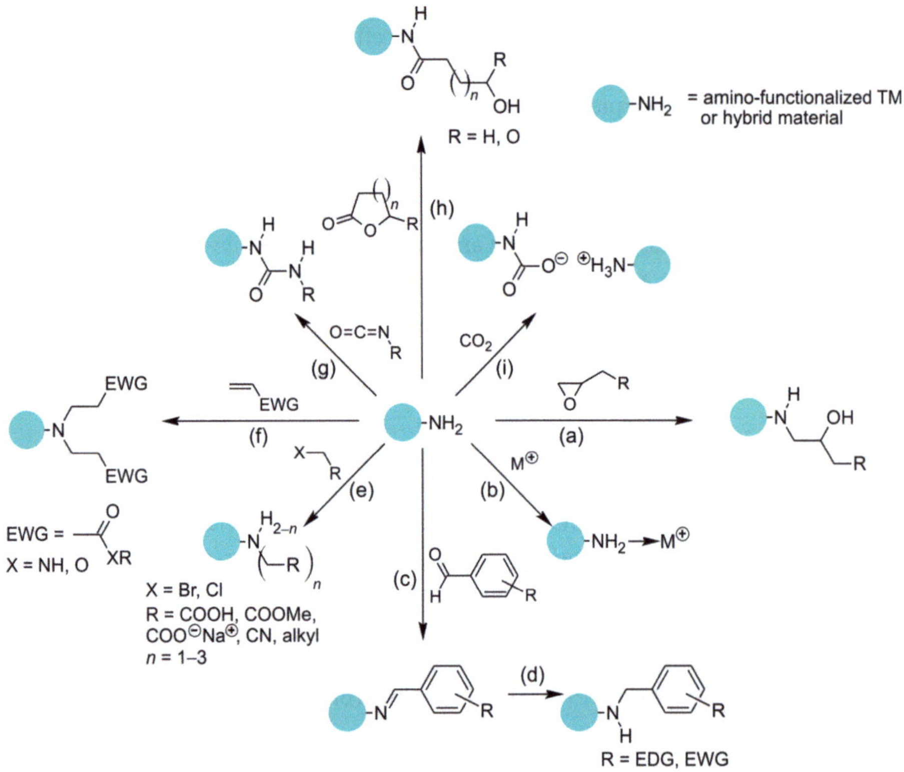

Scheme 5.11: Possible derivatization of TM/hybrid materials originating from primary amino groups; EDG = electron donor group, EWG = electron withdrawing group.

be verified that an addition reaction has taken place from the signals at 144 ppm (IV) and at 69 ppm (III) and the absence of an epoxy ring (see Fig. 5.20, signals 1 and 2). In the solid state $^{29}Si\{^{1}H\}$-CP-MAS NMR spectrum, a D_2 signal for polydialkylsiloxane is detected at –19 ppm. In comparison, the signal for polydialkylsiloxane from the polymerization of APSI occurs at –22 ppm. The down-field-shifted shoulder is associated with D_1 signals or with small ring structures.

Besides the desired reaction between the amino functionality and the epoxide group, several further reactions of the newly formed OH group, derived from the epoxide fragment, are possible which can react with functional groups of both reactants [74–76].

The polymerizations were carried out with various molar ratios of the reactants (T = 140 °C, t = 1 h), at different reaction temperatures (BADGE:APSI = 1:2, t = 1 h), and reaction times (BADGE:APSI = 1:2, T = 140 °C). The final hybrid materials were extracted with dichloromethane in a Soxhlet extractor. NMR spectroscopic studies of the extracts showed predominantly salicyl alcohol generated by transesterification of the APSI fragment. The hybrid material obtained with the monomer ratio BADGE:APSI = 1:2

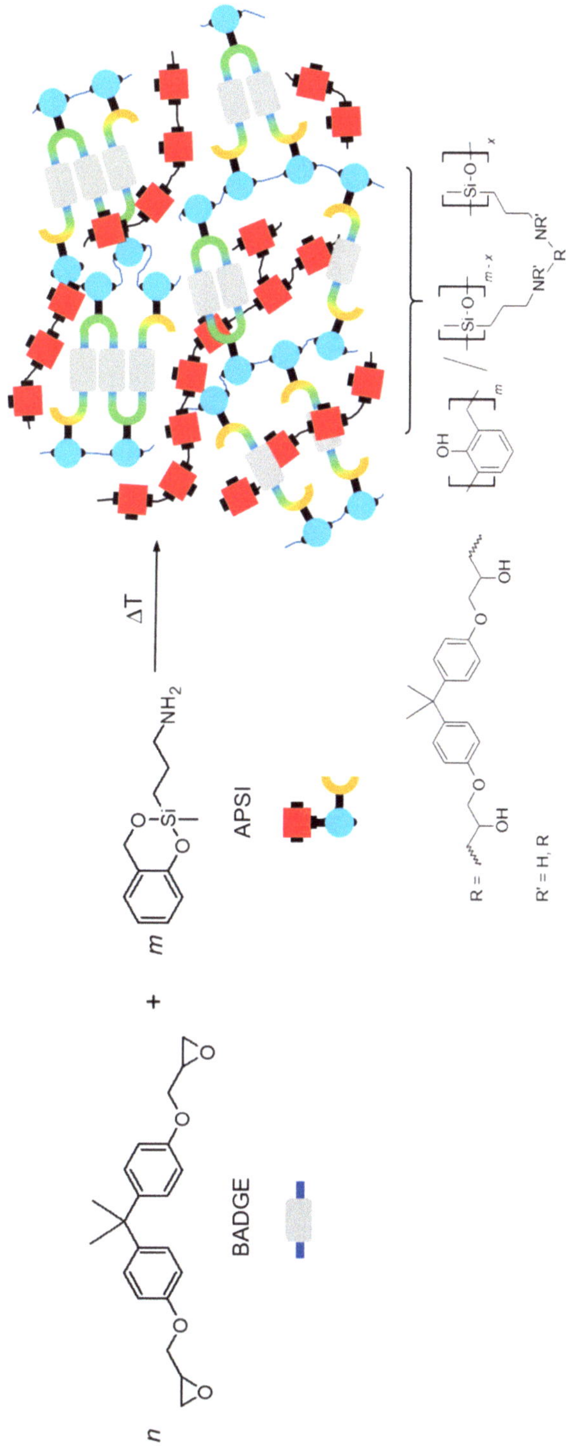

Scheme 5.12: Scenario for the thermally induced reaction of BADGE with APSI.

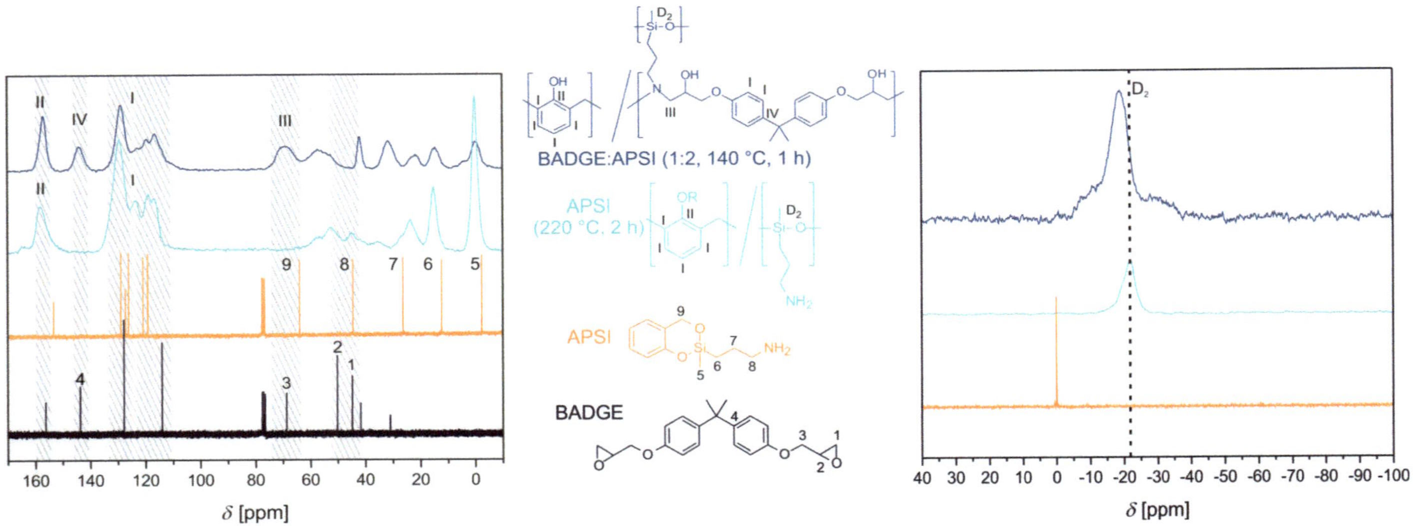

Fig. 5.20: Left: solid state $^{13}C\{^{1}H\}$-CP-MAS NMR spectrum (@15kHz) of a non-extracted sample (BADGE:APSI 1:2, 140 °C, 1 h) and polymerized APSI, liquid ^{13}C NMR spectrum of BADGE and APSI in $CDCl_3$; Right: solid state $^{29}Si\{^{1}H\}$-CP-MAS NMR spectrum (@12kHz) of a non-extracted sample (BADGE:APSI 1:2, 140 °C, 1 h) and polymerized APSI, liquid ^{29}Si NMR spectrum of APSI in $CDCl_3$. Reprinted and adapted from Ref. [71], Copyright 2017, with permission from Elsevier.

shows the lowest extractable content (3.9 wt%) of the ratios investigated. Furthermore, hybrid materials which are synthesized at low temperatures (see Fig. 5.21) or using an excess of APSI have a higher extractable content [71]. The quantity of the extractable salicyl alcohol content is significantly reduced with increasing the reaction temperature above 140 °C. This indicates an increase of crosslinking density in the hybrid material due to the polymerization of salicyl alcohol (Fig. 5.21).

Mechanical parameters of the resulting hybrid materials were determined by nanoindentation. Figure 5.22 shows the measured values of the indentation hardness and the indentation modulus of the hybrid materials made from APSI with the additive BADGE at different monomer ratios (Fig. 5.22, above) and at different temperatures (Fig. 5.22, below). The indentation hardness and modulus also increase with an increasing proportion of BADGE. The indentation hardness of the hybrid material obtained with the monomer ratio APSI:BADGE = 1:1 is about 8 times higher than that of the hybrid material with the composition APSI:BADGE = 6:1. A further increase in the proportion of bis-epoxide has very little effect on the mechanical properties investigated.

With rising reaction temperature (Fig. 5.22, below) and reaction time, the hardness and modulus of the ternary hybrid material increase. Increasing the reaction time of the system APSI:BADGE 1:1 (140 °C) from 1 h to 6 h causes an increase in the indentation hardness from 0.192 to 0.247 GPa. In comparison, hybrid materials made from pure APSI (see Section 5.2.2) have an indentation hardness of 0.003 GPa and an indentation modulus of 0.2 GPa. Due to the increasing crosslinking of the polydialkylsiloxane, the hardness of the hybrid material increases, based on the assumptions that the mechanical properties of the phenolic resin remain constant in all hybrid materials and that side reactions of the epoxide are negligible.

The thermograms of the DSC analysis of final hybrid materials (heating rate of 10 K·min^{-1} from 20–300 °C) show neither endothermic nor exothermic processes for possible post-curing reactions. According to DSC, a reaction temperature of 140 °C is thus sufficient to completely polymerize APSI in the presence of BADGE. The results obtained by thermogravimetric analysis (TGA) show that the temperature for a 5 wt%

reaction temp.	220 °C	180 °C	160 °C	140 °C	120 °C	100 °C	80 °C
amount of extractables	1.9 wt%	4.1 wt%	5.1 wt%	3.9 wt%	33.0 wt%	40.1 wt%	59.5 wt%
$T_{5\%}$	229 °C	–	–	164 °C	–	–	135 °C

Fig. 5.21: Images of monolithic samples synthesized from BADGE and APSI with a molar ratio of 1:2 at different temperatures with their extractable content. $T_{5\%}$ = temperature at which 5% weight loss is recorded by thermogravimetric analysis at a heating rate of 10 K·min^{-1} under air. Reprinted and adapted from Ref. [71], Copyright 2017, with permission from Elsevier.

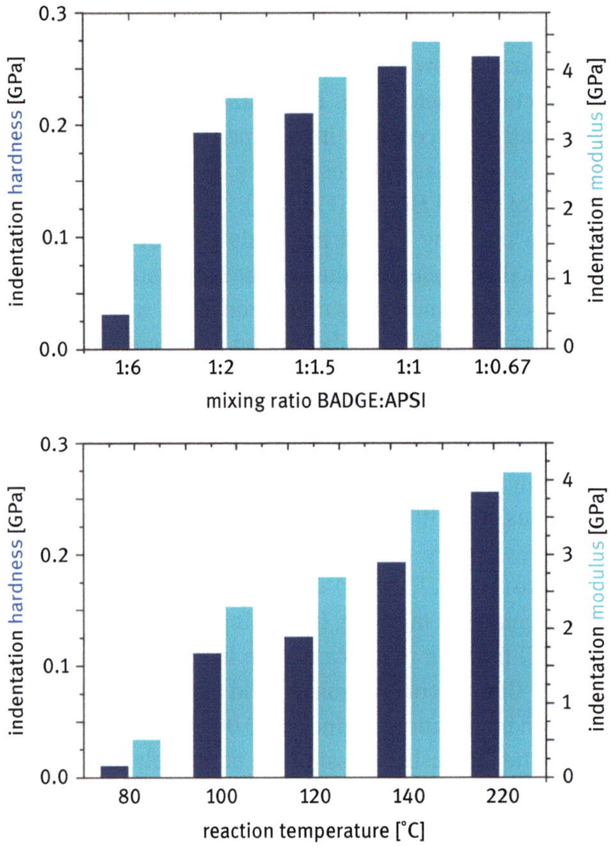

Fig. 5.22: Indentation hardness and modulus of hybrid materials of different BADGE:APSI ratios (reaction temperature: 140 °C) (above) and synthesized at different reaction temperatures (composition: BADGE:APSI = 1:2) (below), determined from curves with holding periods of 30 s.

weight loss depends on the monomer composition and the reaction temperature. With increasing BADGE content in the hybrid material and increasing polymerization temperature, the thermal stability of the hybrid material increases up to 300 °C due to the reduced amount of phenolic resin and a higher crosslinked network [71].

5.2.4.2 Reaction of hybrid layers with glycidyl reagents for functionalized surface coatings

Surfaces and surface coatings with new functional or even multi-functional properties play an important role in modern physical, chemical, and biological applications. Research in this area is particularly concentrated on coatings which are (i) easy to clean or are self-cleaning, (ii) have sensor/actuator properties, and (iii) those with defined tribological and mechanical properties as well as long-term stability [77–87].

The prerequisite for the construction of layer systems is adequate adhesion coupled with abrasion resistance on technical surfaces. In order to attach polymeric functional layers, the technical surfaces are previously equipped with reactive groups, whose density depends on the method of preparation. A common method is, for example, the silylation with functional alkoxysilanes [88]. Primer or macromolecular anchoring layers are based on the grafting of functional polymers onto surfaces [89, 90], and therefore contain a large number of functional groups. They are often used for the covalent bonding of polymer brushes. Hitherto, the combination of abrasion resistance and chemical functionality in one homogeneous layer requires a number of reaction steps or remains unresolved.

Currently, the trend in surface coatings is heading in the direction of control of the coating composition at the molecular level, and of the morphology on a micro- and nanometer scale. This should result in a more homogeneous distribution of functionalities combined with a controllable micro-hardness and durability. Twin polymerizations using amino-functionalized twin monomers are suitable for these types of application because the potential of the TP to generate nanostructured abrasion resistant hybrid materials (see Section 5.2.3) can be smartly combined with a versatile reactive organic functionality. Further derivatizations are possible with both the monomer and the hybrid material. In the following, this concept is realized using the amino-functionalized TM APSI.

The accessibility of the amino groups in the hybrid materials was first tested on bulk material resulting from the simultaneous twin polymerization (STP) of Si-Spiro and APSI by addition of *tert*-butyl glycidyl ether (*t*BGE). In a similar manner to that described in Section 5.2.4.1, the amino functionalities react with the epoxide to give the addition product (amino alcohol) (Scheme 5.13).

A comparison of solid state $^{13}C\{^{1}H\}$-CP-MAS NMR spectra as well as FTIR spectra of the starting bulk materials with the functionalized bulk materials proves the successful conversion (see Fig. 5.23). FTIR spectrum of the sample Si-Spiro_APSI_25:75 with *t*BGE shows absorption bands at 2975 cm^{-1} ($\upsilon_{asym,sym}$ (CH_3)), at 1389 cm^{-1} and 1365 cm^{-1} for the characteristic double bands of the $C(CH_3)_3$ unit, and at 1200 cm^{-1} for the ν(CO) bands which are assigned to the *tert*-butyl units and verify the conversion of the amino groups with *t*BGE.

The STP of functional TMs can be used for coating of smooth aluminum surfaces. For this purpose, Al sputtered Si-wafers were used. By applying a spin coating procedure, uniformly TP-based coatings were obtained as shown in the atomic force microscopy (AFM) height images in Figure 5.24 (b). Micro scratch tests of hybrid material layers (n:m = 1:1) on the aluminum surface show good adhesion and resistance to abrasion. In the hybrid material, with n:m = 1:1, trailing crack formation can be observed. Increasing the proportion of the monomer APSI (n:m = 1:3) results in crack formation and bursting (Fig. 5.24 (a)).

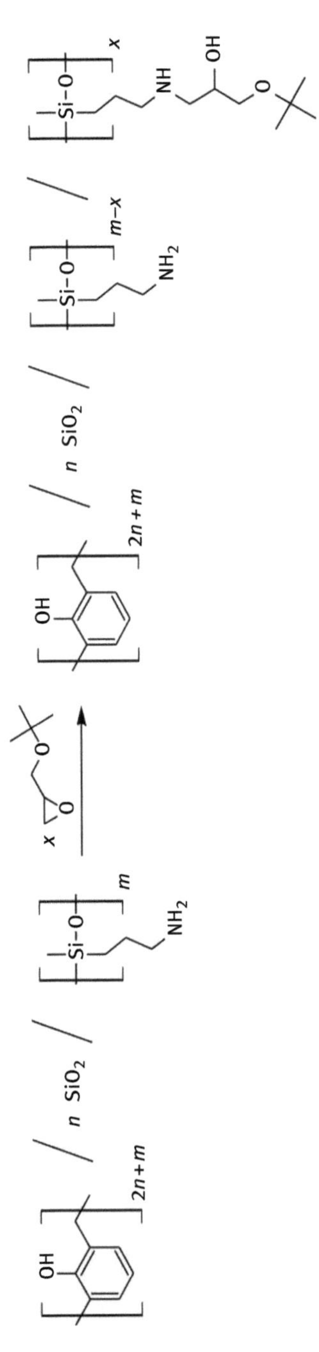

Scheme 5.13: Surface functionalization by reaction of the accessible amino groups with *t*BGE.

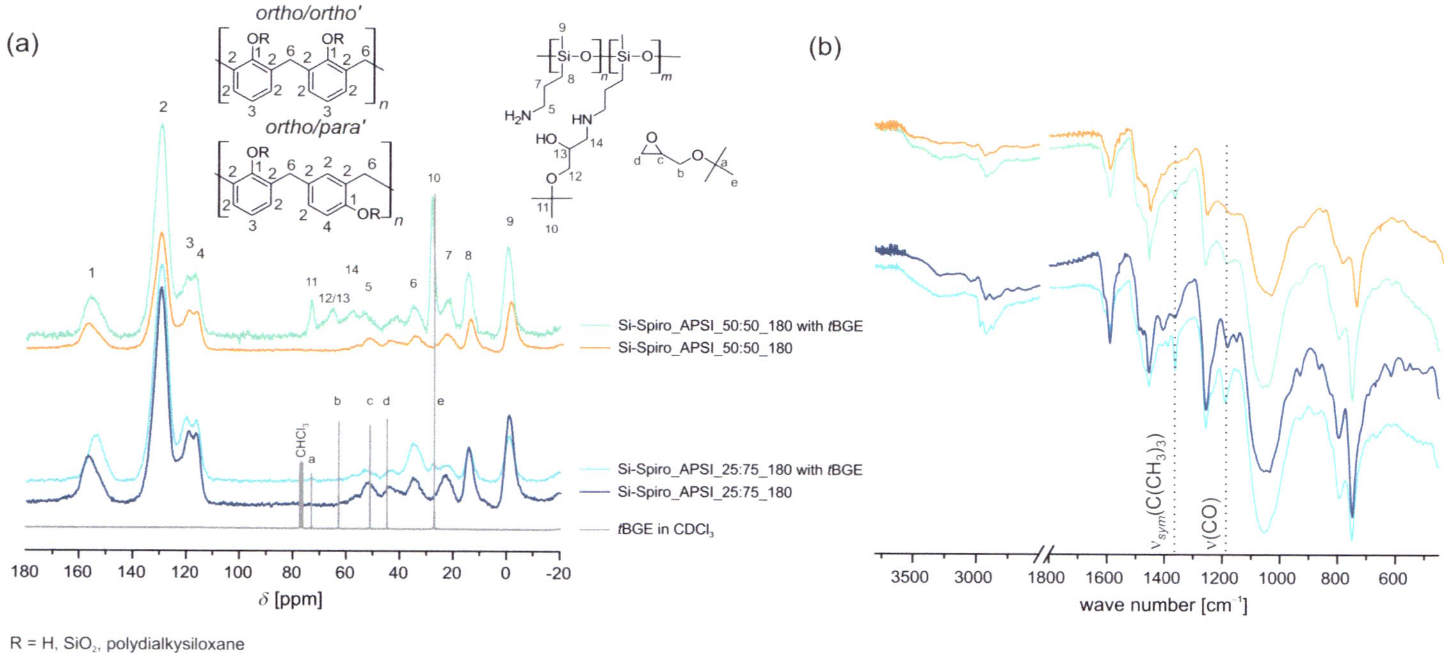

Fig. 5.23: (a) Solid state $^{13}C\{^{1}H\}$-CP-MAS NMR spectra (@15 kHz) of (functionalized) bulk material, starting bulk material and liquid ^{13}C NMR of *t*BGE in $CDCl_3$. (b) FTIR spectra of the bulk material before and after the functionalization with *t*BGE.

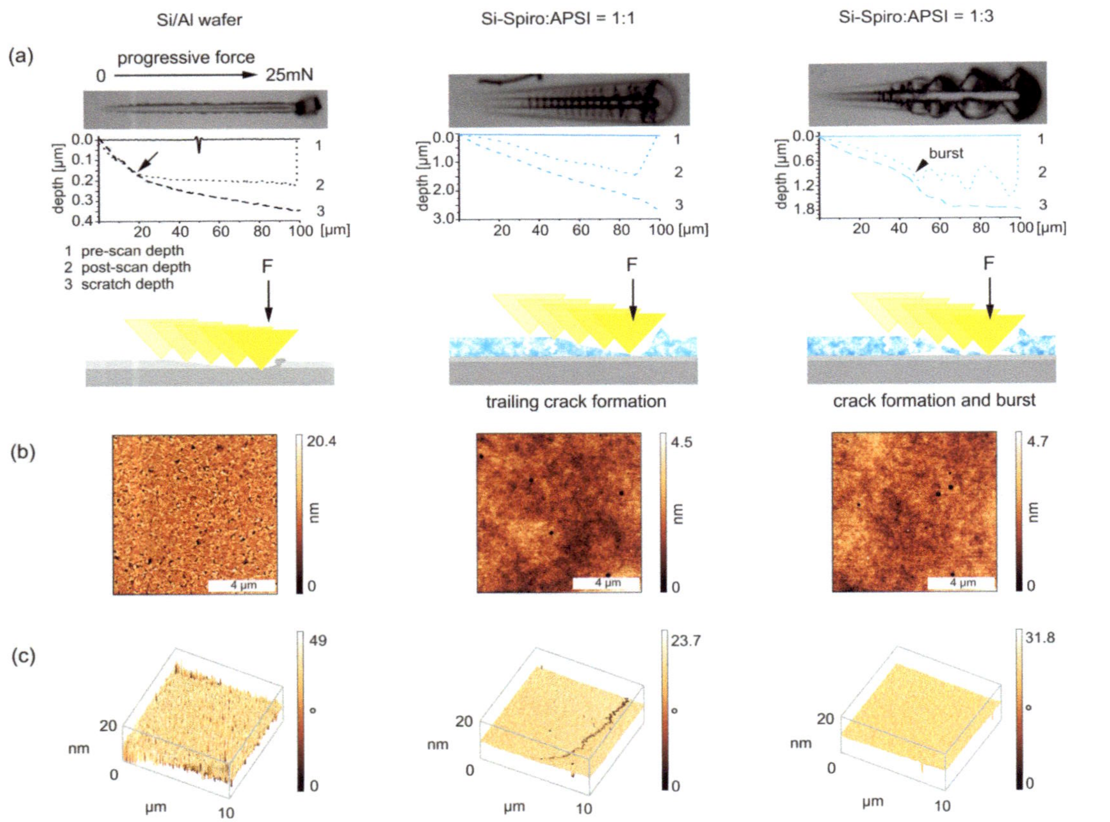

Fig. 5.24: (a) Micro scratch tests, (b) AFM height, and (c) AFM 3D phase images of TP-hybrid material layers with the compositions $n{:}m$ = 1:1 and 1:3 and of the reference Si/Al-wafer.

The accessible amino groups of the organic/inorganic hybrid material layer on the aluminum surface are suitable for attaching functional polymers to the surface (second coating process). For this purpose, glycidyl methacrylate block copolymers were used. The glycidyl groups react with the amino groups of the hybrid material producing a thin stable polymer layer. Functional polymers for the grafting process are synthesized using atom transfer radical polymerization (ATRP) similar to known procedures [91–93]. The used copolymers contain functional groups as polyethylene glycol (PEG) and *tert*-butyl moieties. The *tert*-butyl moieties, which can easily be acidolyzed after the coating step to give carboxy groups, allow to modify the character of the polymer layer further. Using X-ray photoelectron spectroscopy (XPS), the successful coating with the methacrylate polymer is first of all detected by the lack of the amino and silicon atom signals of the TP layer in the survey spectrum. This is due to the covering of the TP layer with the polymer layer. Therefore, it is concluded that the polymer layer is thicker than 5 to 8 nm, which is the information depth of XPS. The molecular composition of the formed coating is verified by the deconvolution of the C 1s spectrum. Five different binding states of the carbon atoms were proven designated as A to E in Figure 5.25 (c). The component peaks A and B represent the carbon atoms of the polymer backbone. The peaks C and D are attributed to C–O groups in ether and unreacted glycidyl groups of the functional copolymer and peak E represents the methacrylic carboxyl groups (Fig. 5.25 (c)).

In summary, two-step coating processes were carried out on aluminum surfaces: (I) an initial coating step – STP of Si-Spiro and APSI for irreversibly fixed amino-functionalized coatings; (II) a second coating step – functionalization of the organic/inorganic hybrid layer by covalent binding of accessible amino groups with functional polymers. Various polymeric structures with reactive substituents, such as block or comb copolymers, are suitable for the second coating step. With the applied polymers hydrophilic nanostructured hybrid layers are obtained, which could find an application in fields like antifouling or low ice adhesive layers.

5.2.4.3 Reaction with (bis-)isocyanates

Amino-functionalized twin monomers should allow the coupling of twin polymerization reactions of mono- or polyisocyanates. Formation of urea structures (see Ref. [94]) during the reaction of isocyanates with primary or secondary amines in the side chain can occur with both, the TMs and the final hybrid material, if accessibility is ensured. TMs and twin prepolymers (TPP) synthesized are shown in Scheme 5.14.

A twin prepolymer is a prepolymer with reactive end groups which can react with each other during the process of TP (Scheme 5.15) [63].

The TPPs are obtained by a two-step one-pot synthesis. In the first step, poly (tetramethylene ether) glycol (PTMEG) or polypropylene glycol (PPG) reacts with

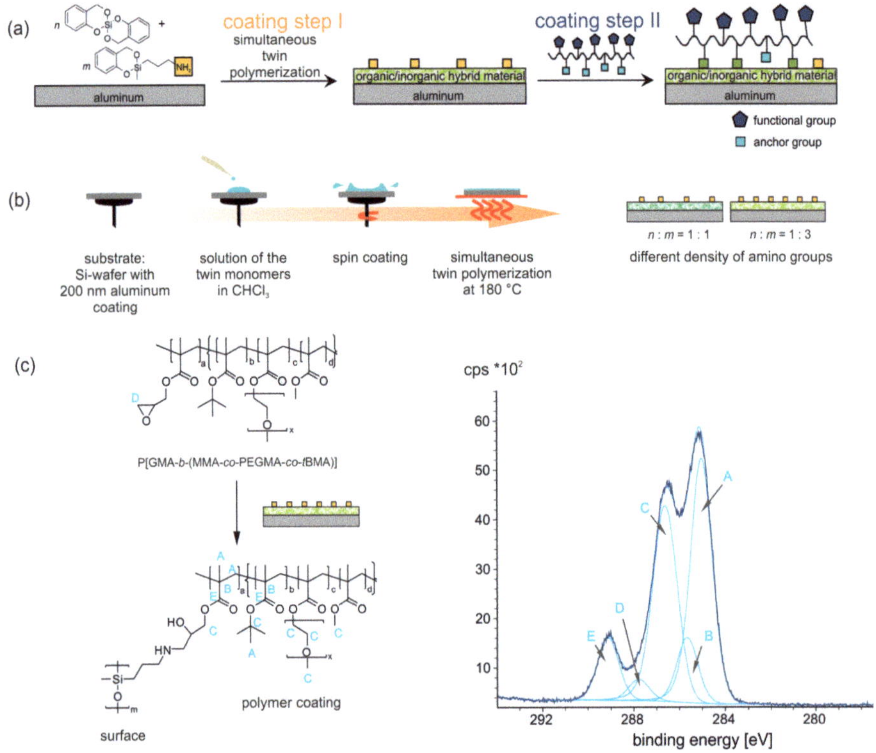

Fig. 5.25: (a) Two-step coating process on Si/Al-wafers by (b) spin coating; (c) high-resolution C 1s XPS spectrum after the second coating with glycidyl methacrylate block copolymer P[GMA-*b*-(MMA-*co*-PEGMA-*co*-*t*BMA)]. The assignment of the component peaks is shown in the formula and explained in the text. (GMA = glycidyl methacrylate, MMA = methyl methacrylate, PEGMA = poly(ethylene glycol) methacrylate, *t*BMA = *tert*-butyl methacrylate).

methylene diphenyl diisocyanate (MDI) to give a prepolymer (PP). The second step is the end capping by adding 2-(3-amino-*n*-propyl)-2-methyl-4*H*-1,3,2-benzodioxasiline (amino-functionalized TM, APSI) to give the TPP (see Scheme 5.14 (c)).

A particular feature of this is that by synthesis of so-called twin prepolymers, oligomeric or polymeric urethane structures can be integrated into the twin polymerization process and thus into nanostructured organic/inorganic hybrid materials.

The specific structural features of polyurethane (PU) and in particular PU elastomers are determined by their two-phase microstructures, resulting from the thermodynamic incompatibility between glassy or "hard" urethane segments and rubbery, or "soft", polyester or polyether segments [95–97]. Microphase separation affects the physical and mechanical properties of PU, e.g., hardness and modulus of elasticity, abrasion resistance, and scratch resistance [98]. The addition of nanosilica can improve the thermal, rheological,

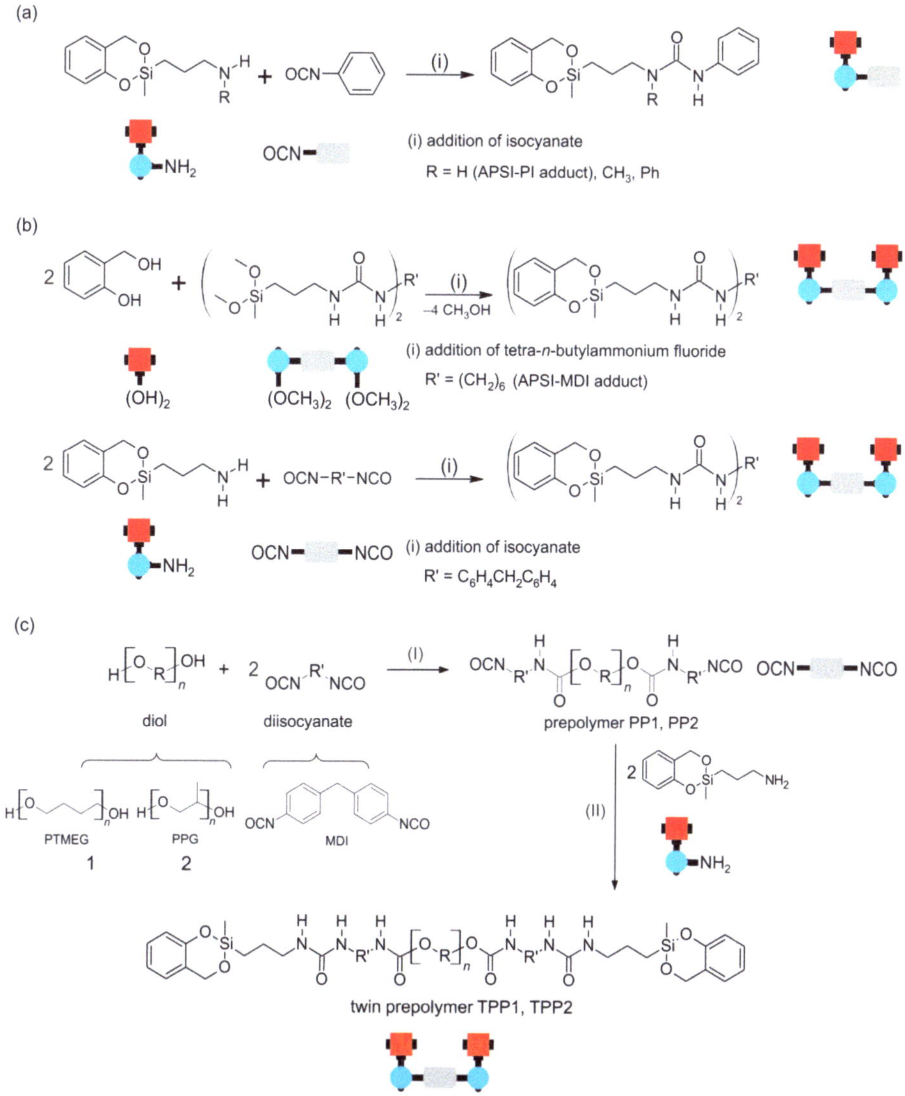

Scheme 5.14: Synthesis of TMs by (a) reaction of phenyl isocyanate with amino-functionalized TMs and by (b) reaction of salicyl alcohol with urea-functionalized dimethoxysilane and of APSI with isocyanate-containing prepolymers. (c) Synthesis of prepolymers functionalized with isocyanate end groups and subsequent synthesis of a twin prepolymer. (PTMEG = poly(tetramethylene ether) glycol, PPG = polypropylene glycol, MDI = methylene diphenyl diisocyanate).

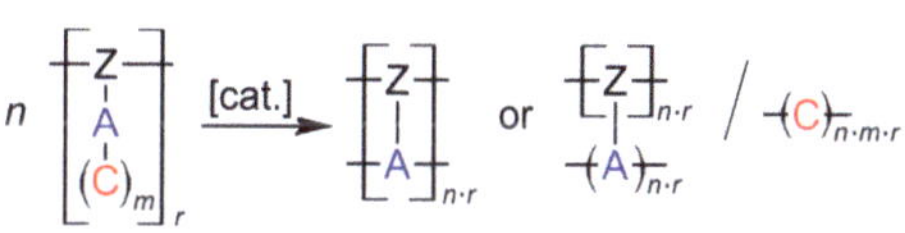

Scheme 5.15: TP of a twin prepolymer [63].

and mechanical adhesion properties of PU. TP is a completely novel synthesis route for urethane-based organic/inorganic hybrid materials with nanoscale silica.

The polymerization behavior of the TMs and TPPs was studied using differential scanning calorimetry (DSC). Various reaction scenarios are observed, depending on whether the functionalized TM or the twin prepolymer is used alone, in the presence of a second TM, or a potential catalyst affects the temperature profile (Fig. 5.26, Tab. 5.1).

On heating the twin prepolymer TPP1 to 350 °C, an exothermic peak can be observed at 279 °C. If Si-Spiro is added, the temperature of the exothermic peak is lowered to 245 °C (Fig. 5.26, (c) and (d); Tab. 5.1). Si-Spiro alone shows an exothermic peak at 234 °C (see Chapter 3.1.2).

If 1,4-diazabicyclo[2.2.2]octane (DABCO) is added as catalyst to the monomer mixture, the trigger temperature for polymerization is lowered to 142 °C of TPP1+Si-Spiro and to 148 °C of TPP2+Si-Spiro (see Tab. 5.1). The addition of lactic acid ($pK_a = 3.86$) leads to a trigger temperature of about 80 °C for the polymerization of the monomer mixture TPP1:Si-Spiro. However, no uniform product is obtained due to the high viscosity and the inherent inhomogeneity of the reaction mixture at this temperature. Therefore, acid catalysis is not the method of choice here. TPP2 shows an exothermic peak at 293 °C. The addition of Si-Spiro and DABCO results in a similar trend to lower polymerization temperatures (Tab. 5.1).

For the synthesis of organic/inorganic hybrid materials with urethane functionalities, twin prepolymers were chosen and subjected to a simultaneous twin polymerization in combination with the TM Si-Spiro. The prepolymers TPP1 or TPP2 were converted with Si-Spiro (1:1) to organic/inorganic hybrid materials at different reaction temperatures. In a second series of experiments, various equimolar quantities of TPP1 and Si-Spiro were used (Scheme 5.16). The corresponding coding for the hybrid material obtained is given as follows: HMx(n)-T-C; x = 1 or 2, hybrid material (HM) from twin prepolymer TPP1 or TPP2, respectively; n = molar ratio of twin prepolymer to Si-Spiro; T = reaction temperature; C = use of a catalyst.

According to the concept of STP, the resulting organic/inorganic hybrid material consists of phenolic resin, SiO_2, and oligodialkylsiloxane (ODAS) with integrated urethane structural units which also have a crosslinking function. The SiO_2 and ODAS groups can form Si–O–Si bonds and thus form a class II hybrid material [99] among the phenolic resin/SiO_2-ODAS hybrid compounds (Scheme 5.16).

As is shown by the DSC results and subsequent extraction tests with the hybrid material, a reaction temperature of 170 °C is sufficient for STP of TPP1 with Si-Spiro, whereas STP of TPP2 with Si-Spiro requires a reaction temperature of 190 °C (see Fig. 5.27).

The effects of polymerization temperature as well as those of the reactant ratio on the polymerization behavior are shown in Figure 5.27. Extraction of the hybrid materials with dichloromethane yielded extractable fractions that could be identified as

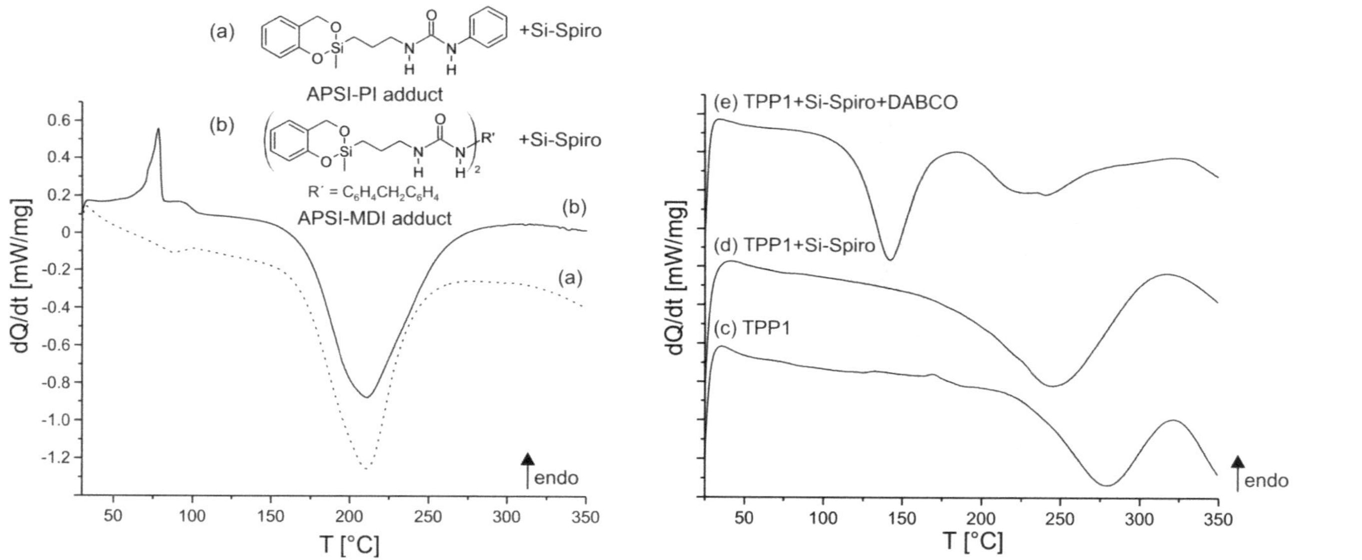

Fig. 5.26: Differential scanning calorimetry measurements of (left) the TMs together with Si-Spiro (1:1) ((a), (b)) and (right) (c) twin prepolymer TPP1, (d) twin prepolymer together with Si-Spiro (1:1), and (e) twin prepolymer TPP1 together with Si-Spiro and DABCO (1 wt%).

Tab. 5.1: Characteristic trigger temperatures for thermally induced and DABCO-catalyzed STP of different monomer mixtures of TM or TPP with Si-Spiro as determined by DSC measurements. PI = phenylisocyanate, MDI = methylene diphenyl diisocyanate.

	$T_{onset, exo}$ ($T_{peak, exo}$) [°C]			
	APSI-PI adduct	**APSI-MDI adduct**	**TPP1**	**TPP2**
pure	258 (274)	240 (270)	234 (279)	245 (293)
+Si-Spiro	169 (210)	169 (211)	150 (245)	191 (237)
+Si-Spiro+DABCO	–	–	117 (142) 195 (229)	134 (148)

prepolymer and low molecular weight products of side reactions or decomposition products such as salicyl alcohol. Depending on the polymer backbone selected, the extractable amounts can be very different. For instance, when using TPP1, an extractable fraction of 47 wt% was obtained whereas with TPP2, an extractable fraction of 84 wt% was achieved. The results of the DSC measurements and extraction experiments show that for the system TPP2:Si-Spiro, a polymerization temperature of 170 °C is too low.

The monomers Si-Spiro and TTP1 or TPP2 were thus polymerized at higher temperatures (190 °C/210 °C/230 °C) and for TPP1 at various monomer ratios (TPP1:Si-Spiro = 100:0/85:15/50:50/15:85/5:95). An increasing polymerization temperature reduces the extractable proportion of the resulting hybrid material to ≈ 35 wt% for TPP1 and ≈ 33 wt% for TPP2 (T = 210 °C, m_{TPP_1/TPP_2}:$n_{Si\text{-}Spiro}$ = 1:1). The large amount of extractable portions can be attributed to several reasons. On the one hand, the ratio of the reactive groups to the molar weight of the prepolymer is decisive. On the other hand, the polymerization process is a bulk polymerization and the increasing viscosity during the polymerization is a limiting factor for complete conversion. At a temperature of 230 °C, highly foamed partly dark-colored hybrid materials are obtained. TG-MS measurements show liberation of CO_2 and traces of water, which indicates a thermal degradation of the polymers. The urethane and urea increments are, after the biuret and allophanate substructures, the most thermally labile compounds in a PU formulation.

The solid state NMR spectra of the hybrid materials produced from TPP1 and Si-Spiro verify the formation of phenolic resin, SiO_2, and ODAS (Fig. 5.28). Overall, the solid state $^{29}Si\{^1H\}$-CP-MAS NMR spectra of the hybrid materials from Si-Spiro and the corresponding twin prepolymer show the expected Q and D signals (Fig. 5.28 (b)). The D_1, D(Q), and D_2 signals in the region around –10 to –23 ppm correspond to difunctional (D) silicon structures. The D(Q) signal at approximately –16 to –17 ppm is characteristic for products of STP, indicating reduced flexibility of the oligodialkylsiloxane (ODAS) content, and thus suggests that ODAS is bound to the SiO_2 network. According to the classification, this is thus a simultaneous twin copolymerization [58].

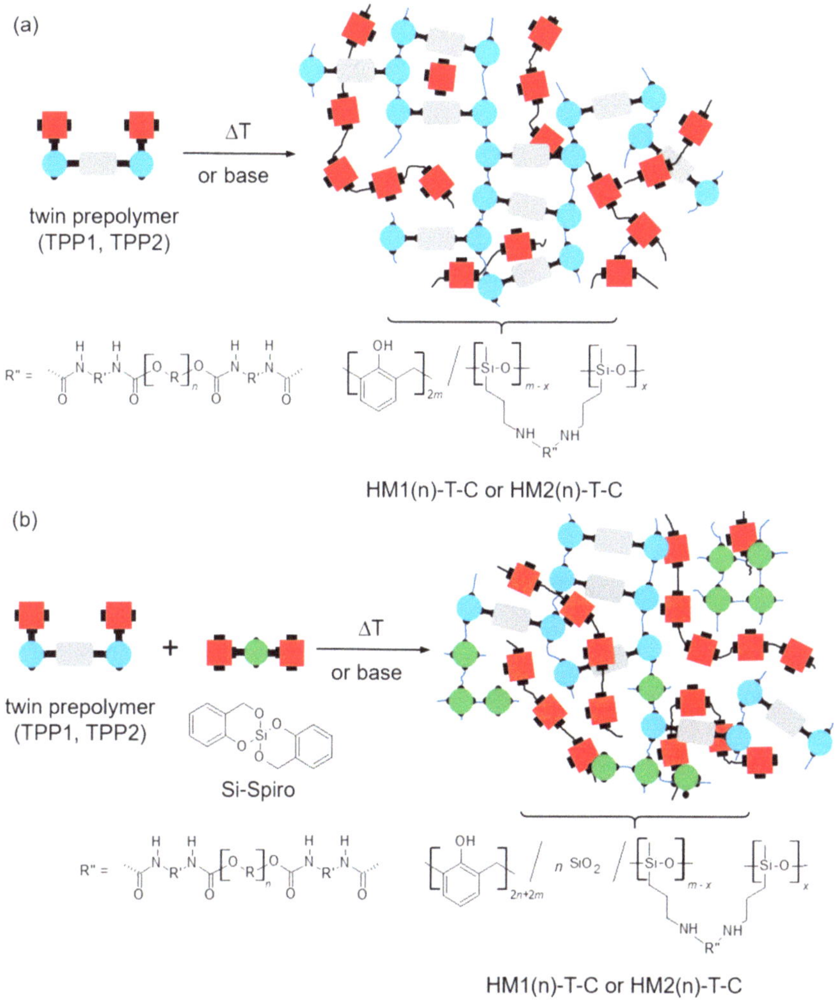

Scheme 5.16: Reaction scenario for thermally induced or base-catalyzed TP of the twin prepolymers TPP1 or TPP2 (a) or STP of the twin prepolymer TPP1 or TPP2 with Si-Spiro (b). (R = poly(tetramethylene ether) glycol (PTMEG) or polypropylene glycol (PPG), R' = methylene diphenyl diisocyanate (MDI)).

The different monomer ratios of TPP1:Si-Spiro are reflected in the NMR spectra and the signal intensities of phenolic resin/ODAS vary correspondingly. An increasing content of monomer TPP1 leads to formation of longer or generally more ODAS chains and thus also to more intense D signals, as is seen in the ^{29}Si NMR spectra (Fig. 5.28 (b)). A larger quantity of ODAS could also be extracted, as fewer covalent bonds between the SiO_2 network and ODAS had been formed. A lower ratio of monomer TPP1 gives a strong signal at approximately –17 ppm, representing D(Q) species. The hybrid material HM1(100)-170 only shows D and no Q signals because of the absence of Si-Spiro in the process.

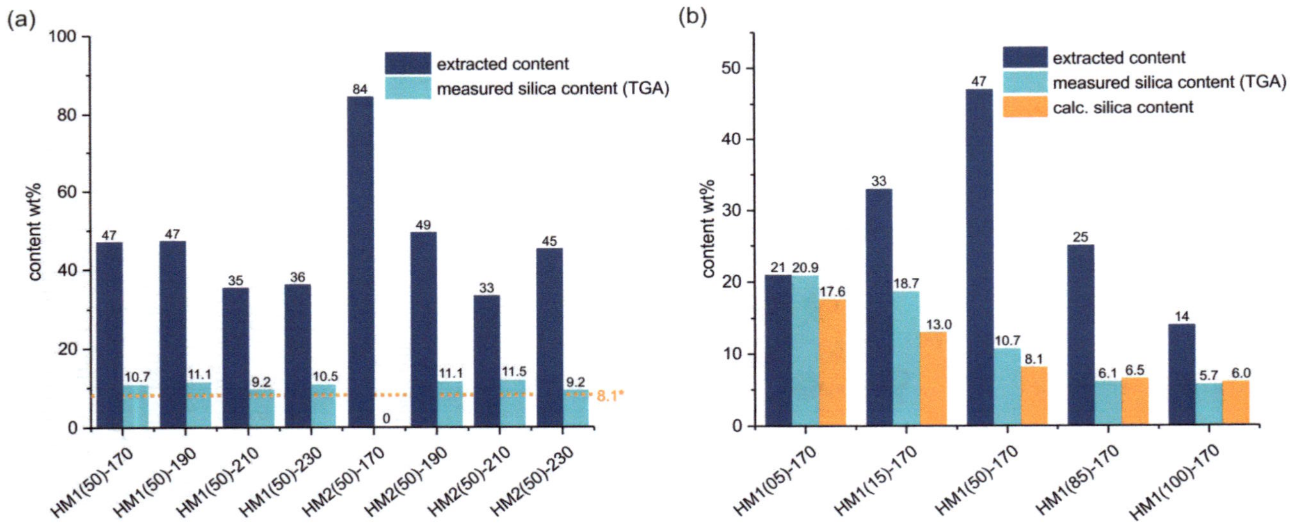

Fig. 5.27: Results of the extraction and subsequent TGA of hybrid materials, which are obtained (a) from the monomers TPP1 or TPP2 and Si-Spiro (molar ratio of 50:50) at different reaction temperatures (*calc. silica content of 8.1 wt%) and (b) from different monomer ratios of TPP1 and Si-Spiro at a reaction temperature of 170 °C.

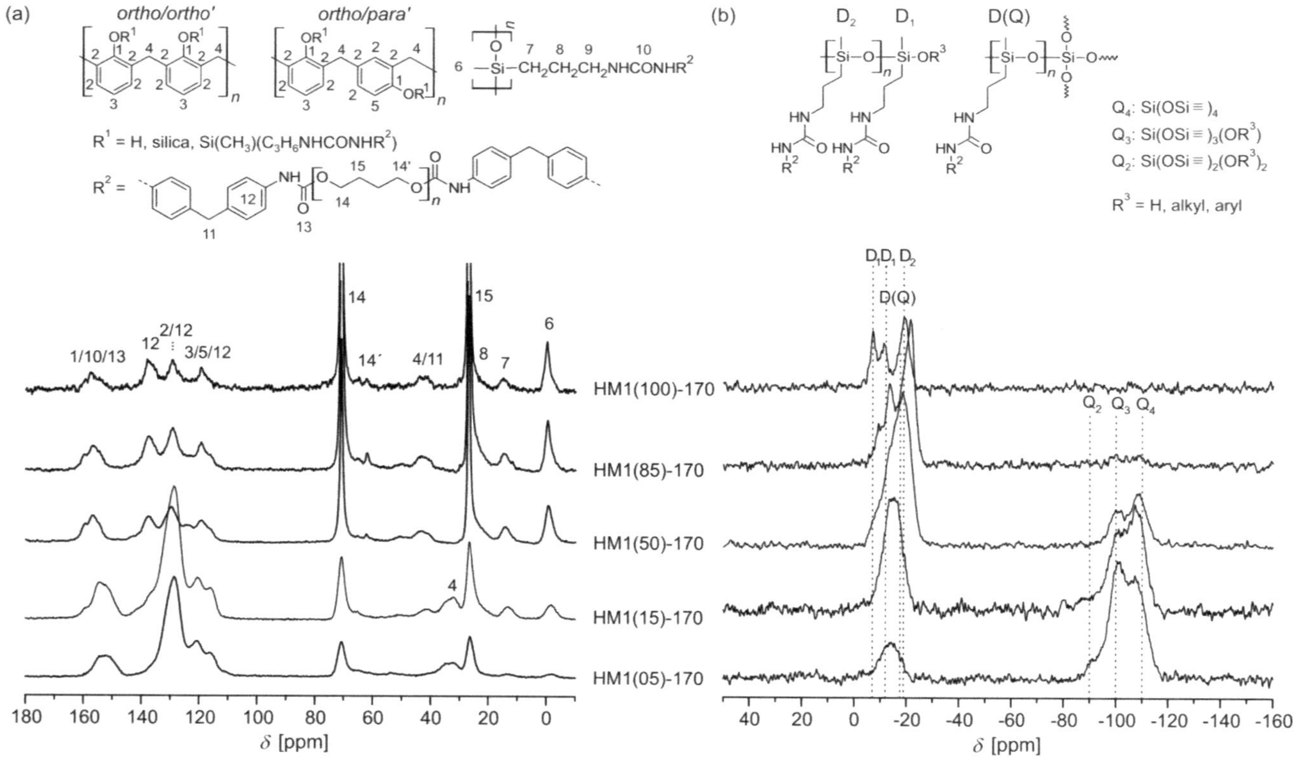

Fig. 5.28: Solid state NMR spectra of selected samples with different molar ratios of the monomers TPP1 and Si-Spiro. (a) Solid state $^{13}C\{^{1}H\}$-CP-MAS NMR spectra @15kHz and (b) solid state $^{29}Si\{^{1}H\}$-CP-MAS NMR spectra @12kHz.

The formation of phenolic resin, ODAS, and urethane structures can be detected using solid state $^{13}C\{^1H\}$-CP-MAS NMR spectroscopy. The formation of *o,o´*- and *o,p´*-bonded phenolic resin structures were proven from the signals in the aromatic region (signals 3 and 5, Fig. 5.28 (a)). The signal for the bridging methylene group (signal 4, Fig. 5.28 (a)) is overlapped by an intense signal of the CH_2 groups of the ether fragment (signal 15, Fig. 5.28 (a)).

The HAADF-STEM images of the organic/inorganic hybrid materials show bright SiO_2 clusters with a size of approximately 2 nm. The monomer ratios chosen have no significant effect on the size of the phase domains (Fig. 5.29). This SiO_2-rich domains are significantly smaller than that in comparable phenolic resin/silicon dioxide composite materials which were synthesized *via* a sol-gel process [100].

Fig. 5.29: HAADF-STEM images of hybrid materials obtained from mixtures of TPP1 and Si-Spiro in molar ratio of (a) 50:50 (HM1(50)-170), (b) 85:15 (HM1(85)-170), and (c) 100:0 (HM1(100)-170) at a reaction temperature of 170 °C.

Inorganic/organic hybrid materials are accessible by synergistic combination of twin prepolymers TPP1 or TPP2 with Si-Spiro using STP. The interpenetrating networks resulting from three different polymers deliver nanostructured phase domains of approximately 2 nm in size. The formation of the networks takes place simultaneously and in the presence of urethane structural increments, and thus represents a completely new procedure for incorporating SiO_2 into nanoscale urethane formulations. Based on these results and further work, a correlation of the molecular structure of these urethane-based organic/inorganic hybrid materials with their macroscopic properties has to be established, which is important for the development of nanocomposites for industrial applications, e.g., the construction of lightweight structures.

5.2.5 Summary

This chapter covers the further development of the concept of (simultaneous) twin polymerization to the synthesis of functionalized twin monomers and functional organic/inorganic hybrid materials. The generation of nanostructured organic/inorganic hybrid materials with good adhesion properties coupled with reactive, accessible functionalities by twin polymerization shows a completely novel approach to the development of functional materials in the emerging areas of materials science.

The focus is given to the synthesis of twin monomers, which carry a functional group covalently bonded to the silicon atom *via* an alkyl spacer. Starting from the amino-functionalized twin monomer APSI, a large number of derivatizations for generating further functionalities at the periphery of the twin monomers is possible. Urethane structures can be successfully integrated into twin prepolymers. This represents a novel concept for the integration of nanoscale SiO_2 into urethane formulations. The addition of additives like catechol or reaction components like bisphenol A diglycidyl ether can increase the crosslinking density and thus the hardness of the hybrid materials *via* twin polymerization. During the polymerization of functionalized twin monomers, a nanostructured network of phenolic resin, polydialkylsiloxane, and silicon dioxide is formed and the functionalization remains in the side chain of the polydialkysiloxane. Hence, a reaction at the amino group can be performed not only on the twin monomer, but also on the organic/inorganic hybrid material and was shown for the reaction with glycidyl methacrylate block copolymer on amino-functionalized hybrid materials, whereupon the construction of surface coatings is possible.

References

[1] Bruce PG, Scrosati B, Tarascon J-M. Nanomaterials for Rechargeable Lithium Batteries. Angew Chem Int Ed 2008;47:2930–46. doi:10.1002/anie.200702505.

[2] Ren J, Zhang Y, Bai W, Chen X, Zhang Z, Fang X, et al. Elastic and Wearable Wire-Shaped Lithium-Ion Battery with High Electrochemical Performance. Angew Chem Int Ed 2014;53:7864–9. doi:10.1002/anie.201402388.

[3] Casimir A, Zhang H, Ogoke O, Amine JC, Lu J, Wu G. Silicon-based anodes for lithium-ion batteries: Effectiveness of materials synthesis and electrode preparation. Nano Energy 2016;27:359–76. doi:10.1016/j.nanoen.2016.07.023.

[4] Nitta N, Wu F, Lee JT, Yushin G. Li-ion battery materials: present and future. Mater Today 2015;18:252–64. doi:10.1016/j.mattod.2014.10.040.

[5] Park C-M, Kim J-H, Kim H, Sohn H-J. Li-alloy based anode materials for Li secondary batteries. Chem Soc Rev 2010;39:3115–41. doi:10.1039/B919877F.

[6] Goward GR, Taylor NJ, Souza DCS, Nazar LF. The true crystal structure of $LizM_4$ (M = Ge, Sn, Pb)–revised from $Li_{22}M_5$. J Alloys Compd 2001;329:82–91. doi:10.1016/S0925-8388(01)01567-5.

[7] Reichmann TL, Gebert C, Cupid DM. Investigation of the Li solubility in the intermediate phase $Li_{17}Sn_4$ relevant to understanding lithiation mechanisms in Sn-based anode materials. J Alloys Compd 2017;714:593–602. doi:10.1016/j.jallcom.2017.04.232.

[8] Larcher D, Beaulieu LY, Mao O, George AE, Dahn JR. Study of the reaction of lithium with isostructural $A_{(2)}B$ and various Al_xB alloys. J Electrochem Soc 2000;147:1703–8. doi:10.1149/1.1393421.

[9] Choi N-S, Yao Y, Cui Y, Cho J. One dimensional Si/Sn-based nanowires and nanotubes for lithium-ion energy storage materials. J Mater Chem 2011;21:9825–40. doi:10.1039/C0JM03842C.

[10] Tian H, Xin F, Wang X, He W, Han W. High capacity group-IV elements (Si, Ge, Sn) based anodes for lithium-ion batteries. J Materiomics 2015;1:153–69. doi:10.1016/j.jmat.2015.06.002.
[11] Kitschke P, Walter M, Rüffer T, Seifert A, Speck F, Seyller T, et al. Porous Ge@C materials via twin polymerization of germanium(II) salicyl alcoholates for Li-ion batteries. J Mater Chem A 2016;4:2705–19. doi:10.1039/C5TA09891B.
[12] Winter M, Besenhard JO, Spahr ME, Novák P. Insertion Electrode Materials for Rechargeable Lithium Batteries. Adv Mater 1998;10:725–63. doi:10.1002/(SICI)1521-4095(199807)10:10<725::AID-ADMA725>3.0.CO;2-Z.
[13] Ji X, Nazar LF. Advances in Li-S batteries. J Mater Chem 2010;20:9821–6. doi:10.1039/B925751A.
[14] Choi N-S, Chen Z, Freunberger SA, Ji X, Sun Y-K, Amine K, et al. Challenges Facing Lithium Batteries and Electrical Double-Layer Capacitors. Angew Chem Int Ed 2012;51:9994–10024. doi:10.1002/anie.201201429.
[15] Bresser D, Passerini S, Scrosati B. Recent progress and remaining challenges in sulfur-based lithium secondary batteries – a review. Chem Commun 2013;49:10545–62. doi:10.1039/C3CC46131A.
[16] Evers S, Nazar LF. New Approaches for High Energy Density Lithium–Sulfur Battery Cathodes. Acc Chem Res 2013;46:1135–43. doi:10.1021/ar3001348.
[17] Rosenman A, Markevich E, Salitra G, Aurbach D, Garsuch A, Chesneau FF. Review on Li-Sulfur Battery Systems: an Integral Perspective. Adv Energy Mater 2015;5:1500212. doi:10.1002/aenm.201500212.
[18] Hudak NS. Nanostructured Electrode Materials for Lithium-Ion Batteries. Lithium-Ion Batter. Adv. Appl., Amsterdam: Elsevier; 2014, p. 57–82. doi:10.1016/B978-0-444-59513-3.00004-2.
[19] Yue GH, Zhang XQ, Zhao YC, Xie QS, Zhang XX, Peng DL. High performance of Ge@C nanocables as the anode for lithium ion batteries. RSC Adv 2014;4:21450–5. doi:10.1039/C4RA01512F.
[20] Noh M, Kim Y, Kim MG, Lee H, Kim H, Kwon Y, et al. Monomer-Capped Tin Metal Nanoparticles for Anode Materials in Lithium Secondary Batteries. Chem Mater 2005;17:3320–4. doi:10.1021/cm0504337.
[21] Yu Y, Gu L, Wang C, Dhanabalan A, van Aken PA, Maier J. Encapsulation of Sn@carbon Nanoparticles in Bamboo-like Hollow Carbon Nanofibers as an Anode Material in Lithium-Based Batteries. Angew Chem Int Ed 2009;48:6485–9. doi:10.1002/anie.200901723.
[22] Wang H, Huang H, Chen L, Wang C, Yan B, Yu Y, et al. Preparation of Si/Sn-Based Nanoparticles Composited with Carbon Fibers and Improved Electrochemical Performance as Anode Materials. ACS Sustain Chem Eng 2014;2:2310–7. doi:10.1021/sc500290x.
[23] Liu B, Abouimrane A, Ren Y, Balasubramanian M, Wang D, Fang ZZ, et al. New Anode Material Based on SiO–$Sn_xCo_yC_z$ for Lithium Batteries. Chem Mater 2012;24:4653–61. doi:10.1021/cm3017853.
[24] Böttger-Hiller F, Kempe P, Cox G, Panchenko A, Janssen N, Petzold A, et al. Twin Polymerization at Spherical Hard Templates: An Approach to Size-Adjustable Carbon Hollow Spheres with Micro- or Mesoporous Shells. Angew Chem Int Ed 2013;52:6088–91. doi:10.1002/anie.201209849.
[25] Brückner J, Thieme S, Böttger-Hiller F, Bauer I, Grossmann HT, Strubel P, et al. Carbon-Based Anodes for Lithium Sulfur Full Cells with High Cycle Stability. Adv Funct Mater 2014;24:1284–9. doi:10.1002/adfm.201302169.
[26] Kitschke P, Walter M, Rüffer T, Lang H, Kovalenko MV, Mehring M. From molecular germanates to microporous Ge@C via twin polymerization. Dalton Trans 2016;45:5741–51. doi:10.1039/C6DT00049E.

[27] Leonhardt C. Zinnalkoxide als Präkursoren für zinnhaltige Nanokomposite. Dissertation. Technische Universität Chemnitz, 2016.
[28] Choudhury S, Ebert T, Windberg T, Seifert A, Göbel M, Simon F, et al. Hierarchical porous carbon cathodes for lithium-sulfur batteries using carbon derived from hybrid materials synthesized by twin polymerization. Part Part Syst Char 2018; accepted.
[29] Bramnik K, Hanefeld P, Koenig HM, Lange A. Electroactive material, and use thereof in anodes for lithium-ion cells. WO2010112580A1, 2010.
[30] Hanefeld P, Hildebrandt N, Lange A, Leitner K, Staudt C. Porous film material, comprising at least one carbonaceous semimetal oxide phase, and the use thereof as separator material for electrochemical cells. WO2011000858A1, 2011.
[31] Cox G, Dyllick-Brenzinger R, Gronwald O, Lange A, Smit T. Process for producing a composite material. WO2012098149A2, 2012.
[32] Gronwald O, Janssen N, Lange A, Moehwald H. Composite, its production and its use in separators for electrochemical cells. WO2013110985A1, 2013.
[33] Cox G, Lange A, Leitner K, Leonhardt C, Mehring M, Wolf H. Tin Oxide-Containing Polymer Composite Materials. WO2013042034A1, 2013.
[34] Cox G, Csihony S, Lange A, Wolf H. Composite materials and process for production thereof. US20130341559A1, 2013.
[35] Cox G, Csihony S, Lange A, Wolf H. Process for producing a composite material. WO2014001273A1, 2014.
[36] Cox G, Csihony S, Lange A, Wolf H. Composite materials and methods for the production thereof. WO2014001949A1, 2014.
[37] Böttger-Hiller F, Mehner A, Anders S, Kroll L, Cox G, Simon F, et al. Sulphur-doped porous carbon from a thiophene-based twin monomer. Chem Commun 2012;48:10568–70. doi:10.1039/C2CC35112A.
[38] Schuster J, He G, Mandlmeier B, Yim T, Lee KT, Bein T, et al. Spherical Ordered Mesoporous Carbon Nanoparticles with High Porosity for Lithium–Sulfur Batteries. Angew Chem Int Ed 2012;51:3591–5. doi:10.1002/anie.201107817.
[39] Leonhardt C, Seifert A, Csihony S, Sommer H, Mehring M. Nanocomposites by the use of simultaneous twin polymerization: tin alloys in a carbon/silica matrix. RSC Adv 2016;6:3091–8. doi:10.1039/C5RA18574B.
[40] Leonhardt C, Brumm S, Seifert A, Lange A, Csihony S, Mehring M. Tin Nanoparticles in Carbon/Silica Hybrid Materials by the Use of Twin Polymerization. ChemPlusChem 2014;79:1440–7. doi:10.1002/cplu.201402137.
[41] Li L, Seng KH, Feng C, Liu HK, Guo Z. Synthesis of hollow GeO_2 nanostructures, transformation into Ge@C, and lithium storage properties. J Mater Chem A 2013;1:7666–72. doi:10.1039/c3ta11381g.
[42] Seng KH, Park M, Guo ZP, Liu HK, Cho J. Catalytic Role of Ge in Highly Reversible GeO_2/Ge/C Nanocomposite Anode Material for Lithium Batteries. Nano Lett 2013;13:1230–6. doi:10.1021/nl304716e.
[43] Liu J, Song K, Zhu C, Chen C-C, van Aken PA, Maier J, et al. Ge/C Nanowires as High-Capacity and Long-Life Anode Materials for Li-Ion Batteries. ACS Nano 2014;8:7051–9. doi:10.1021/nn501945f.
[44] Xiao Y, Cao M. High-Performance Lithium Storage Achieved by Chemically Binding Germanium Nanoparticles with N-Doped Carbon. ACS Appl Mater Interfaces 2014;6:12922–30. doi:10.1021/am502867f.
[45] Henderson EJ, Hessel CM, Cavell RG, Veinot JGC. How Processing Atmosphere Influences the Evolution of GeO_2-Embedded Germanium Nanocrystals Obtained from the Thermolysis

of Phenyl Trichlorogermane-Derived Polymers. Chem Mater 2010;22:2653–61. doi:10.1021/cm100107p.
[46] Hoffman M, Veinot JGC. Understanding the Formation of Elemental Germanium by Thermolysis of Sol-Gel Derived Organogermanium Oxide Polymers. Chem Mater 2012;24:1283–91. doi:10.1021/cm2035129.
[47] Li X, Liang J, Hou Z, Zhu Y, Wang Y, Qian Y. A synchronous approach for facile production of Ge-carbon hybrid nanoparticles for high-performance lithium batteries. Chem Commun 2015;51:3882–5. doi:10.1039/C5CC00080G.
[48] Shunmugam R, Chowdhury SR, editors. Functional polymers: design, synthesis, and applications. Oakville, ON, Canada; Waretown, NJ, USA: Apple Academic Press; 2017.
[49] Leinfelder KF. New developments in resin restorative systems. J Am Dent Assoc 1939 1997;128:573–81.
[50] Ahner J, Micheel M, Geitner R, Schmitt M, Popp J, Dietzek B, et al. Self-healing Functional Polymers: Optical Property Recovery of Conjugated Polymer Films by Uncatalyzed Imine Metathesis. Macromolecules 2017;50:3789–95. doi:10.1021/acs.macromol.6b02766.
[51] Shirakawa H, Louis EJ, MacDiarmid AG, Chiang CK, Heeger AJ. Synthesis of electrically conducting organic polymers: halogen derivatives of polyacetylene, $(CH)_x$. J Chem Soc Chem Commun 1977:578–80. doi:10.1039/c39770000578.
[52] Brédas JL, Marder SR, Salaneck WR, Alan J. Heeger, Alan G. MacDiarmid, and Hideki Shirakawa. Macromolecules 2002;35:1137–9. doi:10.1021/ma0118973.
[53] Theato P, Klok H-A, editors. Functional Polymers by Post-Polymerization Modification: Concepts, Guidelines, and Applications. Weinheim, Germany: Wiley-VCH Verlag GmbH & Co. KGaA; 2012. doi:10.1002/9783527655427.
[54] Akelah A, Sherrington DC. Recent developments in the application of functionalized polymers in organic synthesis. Polymer 1983;24:1369–86. doi:10.1016/0032-3861(83)90218-5.
[55] Elias H-G. Makromoleküle: Struktur, Eigenschaften, Synthesen, Stoffe. Basel, Heidelberg: Hüthig & Wepf; 1971.
[56] Auer AA, Richter A, Berezkin AV, Guseva DV, Spange S. Theoretical Study of Twin Polymerization – From Chemical Reactivity to Structure Formation. Macromol Theory Simul 2012;21:615–28. doi:10.1002/mats.201200036.
[57] Jenkins AD, Kratochvíl P, Stepto RFT, Suter UW. Glossary of basic terms in polymer science (IUPAC Recommendations 1996). Pure Appl Chem 1996;68. doi:10.1351/pac199668122287.
[58] Löschner T, Mehner A, Grund S, Seifert A, Pohlers A, Lange A, et al. A Modular Approach for the Synthesis of Nanostructured Hybrid Materials with Tailored Properties: The Simultaneous Twin Polymerization. Angew Chem Int Ed 2012;51:3258–61. doi:10.1002/anie.201108011.
[59] Spange S, Kempe P, Seifert A, Auer AA, Ecorchard P, Lang H, et al. Nanocomposites with Structure Domains of 0.5 to 3 nm by Polymerization of Silicon Spiro Compounds. Angew Chem Int Ed 2009;48:8254–8. doi:10.1002/anie.200901113.
[60] Kempe P, Löschner T, Adner D, Spange S. Selective ring opening of 4*H*-1,3,2-benzodioxasiline twin monomers. New J Chem 2011;35:2735–9. doi:10.1039/C1NJ20654K.
[61] Spange S, Grund S. Nanostructured Organic-Inorganic Composite Materials by Twin Polymerization of Hybrid Monomers. Adv Mater 2009;21:2111–6. doi:10.1002/adma.200802797.
[62] Ebert T, Cox G, Sheremet E, Gordan O, Zahn DRT, Simon F, et al. Carbon/carbon nanocomposites fabricated by base catalyzed twin polymerization of a Si-spiro compound on graphite sheets. Chem Commun 2012;48:9867. doi:10.1039/c2cc34775j.
[63] Ebert T, Seifert A, Spange S. Twin Polymerization – a New Principle for Hybrid Material Synthesis. Macromol Rapid Commun 2015;36:1623–39. doi:10.1002/marc.201500182.

[64] Ebert T, Wollbrink A, Seifert A, John R, Spange S. Multiple polymerization – formation of hybrid materials consisting of two or more polymers from one monomer. Polym Chem 2016;7:6826–33. doi:10.1039/C6PY01619G.

[65] Kempe P, Löschner T, Auer AA, Seifert A, Cox G, Spange S. Thermally Induced Twin Polymerization of 4*H*-1,3,2-Benzodioxasilines. Chem – Eur J 2014;20:8040–53. doi:10.1002/chem.201400038.

[66] Göring M, Seifert A, Schreiter K, Müller P, Spange S. A non-aqueous procedure to synthesize amino group bearing nanostructured organic–inorganic hybrid materials. Chem Commun 2014;50:9753. doi:10.1039/C4CC03640A.

[67] Chiang C-L, Ma C-CM, Wu D-L, Kuan H-C. Preparation, characterization, and properties of novolac-type phenolic/SiO_2 hybrid organic-inorganic nanocomposite materials by sol-gel method. J Polym Sci Part Polym Chem 2003;41:905–13. doi:10.1002/pola.10624.

[68] Haraguchi K, Usami Y, Yamamura K, Matsumoto S. Morphological investigation of hybrid materials composed of phenolic resin and silica prepared by in situ polymerization. Polymer 1998;39:6243–50. doi:10.1016/S0032-3861(98)00107-4.

[69] Göring M, Schreiter K, Schuberth A, Windberg T, Jung H, Anders S, et al. Amino Group Bearing Organic-Inorganic Hybrid Materials for Joining Aluminum Alloys and Thermoplastic Fiber-Reinforced Parts. Adv Mater Interfaces 2017;4:1601115. doi:10.1002/admi.201601115.

[70] McBride MB, Wesselink LG. Chemisorption of catechol on gibbsite, boehmite, and noncrystalline alumina surfaces. Environ Sci Technol 1988;22:703–8. doi:10.1021/es00171a014.

[71] Birkner M, Schreiter K, Trommler K, Seifert A, Spange S. Ternary hybrid material formation by twin polymerization coupled with the bis-epoxide/amine step growth polymerization. Polymer 2017;121:38–45. doi:10.1016/j.polymer.2017.06.015.

[72] Chuang IS, Maciel GE. Carbon-13 NMR investigation of the stability of a resol-type phenol-formaldehyde resin toward formalin, toward base, and toward nonoxidizing or oxidizing acid. Macromolecules 1991;24:1025–32. doi:10.1021/ma00005a009.

[73] Bryson RL, Hatfield GR, Early TA, Palmer AR, Maciel GE. Carbon-13 NMR studies of solid phenolic resins using cross polarization and magic-angle spinning. Macromolecules 1983;16:1669–72. doi:10.1021/ma00244a021.

[74] Pascault J-P, Williams RJ. Epoxy polymers new materials and innovations. Weinheim: Wiley-VCH; 2010.

[75] Al-Oweini R, El-Rassy H. Surface characterization by nitrogen adsorption of silica aerogels synthesized from various $Si(OR)_4$ and $R''Si(OR')_3$ precursors. Appl Surf Sci 2010;257:276–81. doi:10.1016/j.apsusc.2010.06.086.

[76] Agrawal G, Wang J, Brüster B, Zhu X, Möller M, Pich A. Degradable microgels synthesized using reactive polyvinylalkoxysiloxanes as crosslinkers. Soft Matter 2013;9:5380–90. doi:10.1039/c3sm50248a.

[77] Uhlmann P, Frenzel R, Voit B, Mock U, Szyszka B, Schmidt B, et al. Research agenda surface technology: Future demands for research in the field of coatings materials. Prog Org Coat 2007;58:122–6. doi:10.1016/j.porgcoat.2006.08.020.

[78] Grinthal A, Aizenberg J. Mobile Interfaces: Liquids as a Perfect Structural Material for Multifunctional, Antifouling Surfaces. Chem Mater 2014;26:698–708. doi:10.1021/cm402364d.

[79] Wong T-S, Kang SH, Tang SKY, Smythe EJ, Hatton BD, Grinthal A, et al. Bioinspired self-repairing slippery surfaces with pressure-stable omniphobicity. Nature 2011;477:443–7. doi:10.1038/nature10447.

[80] Anand S, Paxson AT, Dhiman R, Smith JD, Varanasi KK. Enhanced Condensation on Lubricant-Impregnated Nanotextured Surfaces. ACS Nano 2012;6:10122–9. doi:10.1021/nn303867y.

[81] Epstein AK, Wong T-S, Belisle RA, Boggs EM, Aizenberg J. Liquid-infused structured surfaces with exceptional anti-biofouling performance. Proc Natl Acad Sci 2012;109:13182–7. doi:10.1073/pnas.1201973109.
[82] Kim P, Wong T-S, Alvarenga J, Kreder MJ, Adorno-Martinez WE, Aizenberg J. Liquid-Infused Nanostructured Surfaces with Extreme Anti-Ice and Anti-Frost Performance. ACS Nano 2012;6:6569–77. doi:10.1021/nn302310q.
[83] Wang S, Liu K, Yao X, Jiang L. Bioinspired Surfaces with Superwettability: New Insight on Theory, Design, and Applications. Chem Rev 2015;115:8230–93. doi:10.1021/cr400083y.
[84] Stuart MAC, Huck WTS, Genzer J, Müller M, Ober C, Stamm M, et al. Emerging applications of stimuli-responsive polymer materials. Nat Mater 2010;9:101–13. doi:10.1038/nmat2614.
[85] Lv J, Song Y, Jiang L, Wang J. Bio-Inspired Strategies for Anti-Icing. ACS Nano 2014;8:3152–69. doi:10.1021/nn406522n.
[86] Sojoudi H, Wang M, Boscher ND, McKinley GH, Gleason KK. Durable and scalable icephobic surfaces: similarities and distinctions from superhydrophobic surfaces. Soft Matter 2016;12:1938–63. doi:10.1039/C5SM02295A.
[87] Kreder MJ, Alvarenga J, Kim P, Aizenberg J. Design of anti-icing surfaces: smooth, textured or slippery? Nat Rev Mater 2016;1:15003. doi:10.1038/natrevmats.2015.3.
[88] Sakeye M, Smått J-H. Comparison of Different Amino-Functionalization Procedures on a Selection of Metal Oxide Microparticles: Degree of Modification and Hydrolytic Stability. Langmuir 2012;28:16941–50. doi:10.1021/la303925x.
[89] Iyer KS, Zdyrko B, Malz H, Pionteck J, Luzinov I. Polystyrene Layers Grafted to Macromolecular Anchoring Layer. Macromolecules 2003;36:6519–26. doi:10.1021/ma034460z.
[90] Voigt I, Simon F, Estel K, Spange S. Structure and Surface Polarity of Poly(vinylformamide-*co*-vinylamine) (PVFA-*co*-PVAm)/Silica Hybrid Materials. Langmuir 2001;17:3080–6. doi:10.1021/la001366s.
[91] Matyjaszewski K, Tsarevsky NV. Nanostructured functional materials prepared by atom transfer radical polymerization. Nat Chem 2009;1:276–88. doi:10.1038/nchem.257.
[92] Höhne S, Uhlmann P. Synthesis of functional block copolymers and terpolymers containing polyglycidyl methacrylate blocks. J Polym Sci Part Polym Chem 2015;53:675–84. doi:10.1002/pola.27491.
[93] Frenzel R, Höhne S, Hanzelmann C, Schmidt T, Winkler R, Drechsler A, et al. Tunable Hydrophilic or Amphiphilic Coatings: A "Reactive Layer Stack" Approach. ACS Appl Mater Interfaces 2015;7:12355–66. doi:10.1021/am507403t.
[94] Zhang G, Dass A, Rawashdeh A-MM, Thomas J, Counsil JA, Sotiriou-Leventis C, et al. Isocyanate-crosslinked silica aerogel monoliths: preparation and characterization. J Non-Cryst Solids 2004;350:152–64. doi:10.1016/j.jnoncrysol.2004.06.041.
[95] Szycher M. Szycher's handbook of polyurethanes. 2nd ed. Boca Raton, FL: Taylor & Francis; 2013.
[96] Van Bogart JWC, Gibson PE, Cooper SL. Structure-property relationships in polycaprolactone-polyurethanes. J Polym Sci Polym Phys Ed 1983;21:65–95. doi:10.1002/pol.1983.180210106.
[97] Prisacariu C. Polyurethane Elastomers. Vienna: Springer Vienna; 2011. doi:10.1007/978-3-7091-0514-6.
[98] Krol P. Synthesis methods, chemical structures and phase structures of linear polyurethanes. Properties and applications of linear polyurethanes in polyurethane elastomers, copolymers and ionomers. Prog Mater Sci 2007;52:915–1015. doi:10.1016/j.pmatsci.2006.11.001.
[99] Blanchard J, In M, Schaudel B, Sanchez C. Hydrolysis and Condensation Reactions of Transition Metal Alkoxides: Calorimetric Study and Evaluation of the Extent of Reaction.

Eur J Inorg Chem 1998; 1115–27. doi:10.1002/(SICI)1099-0682(199808)1998:8<1115::AID-EJIC1115>3.0.CO;2-N.
[100] Chiang C-L, Ma C-CM. Synthesis, characterization, thermal properties and flame retardance of novel phenolic resin/silica nanocomposites. Polym Degrad Stab 2004;83: 207–14. doi:10.1016/S0141-3910(03)00262-3.

Michael Mehring and Stefan Spange

6 Concluding Remarks

This monograph presents the current state of knowledge of twin polymerization. The twin polymerization process has been developed in order to provide a novel strategy for the production of nanostructured organic/inorganic hybrid materials. This aspect is covered by major parts of the book. Great attention is therein given to the synthesis of nanostructured polymer resin/silica hybrid materials, the class of materials that was invented first, offering a plethora of potential applications. This class of materials has been treated in detail in theoretical, synthetic, and application-related chapters. Strong emphasis is placed on the interplay between structure and reactivity from both experimental and theoretical points of view. The construction of hierarchically structured porous carbon materials by application of twin polymerization of silicon-containing twin monomers is discussed separately. Here, it is demonstrated that pore geometry and morphology can be designed on demand, and the construction of sulfur-loaded cathode materials for application in Li-ion batteries is described as a seminal example.

The new polymerization principle is based on the development of a new monomer class, the so-called twin monomers. The structural characteristics of this new monomer class in silicon chemistry also inspired the research of new classes of inorganic compounds, which, however, are only partly usable for twin polymerization. These studies demonstrate the scope and limitations of the twin polymerization approach. Thus, we were able to identify and define the structural peculiarities necessary for the development of readily polymerizable twin monomers. In addition to siliceous hybrid materials, porous inorganic oxidic hybrid materials based on B_2O_3, TiO_2, SnO_2, GeO_2, HfO_2, or WO_3 have been made accessible by twin polymerization and twin polymerization related processes. The field of twin polymerization for the synthesis of inorganic oxides, mixed oxides, and their hybrid materials is still in its infancy, and it is expected that this field of research will advance in future.

The further development of twin polymerization processes not only relies on the availability of new twin monomers, but is also inspired by the fact that a combination with other polymerization processes or inorganic precursor chemistry is applicable. This offers the possibility to couple the twin polymerization with conventional step-growth polymerization.

The smart combination of metal particle precursor chemistry with twin polymerization has become an elegant method for the production of noble metal nanoparticles that are embedded in porous nanostructured carbon or silica matrices. This aspect holds enormous potential to bridge twin polymerization and heterogeneous catalysis.

The twin polymerization has been proven to be a versatile new synthetic concept. The restriction of twin polymerization to step-growth polymerization processes can be overcome when functionalized twin monomers are designed that contain chain

https://doi.org/10.1515/9783110499360-006

polymerizable groups. New developments are to be expected by such combined approaches.

The most important message of this book is: generation of nanostructured polymeric organic/inorganic hybrid materials succeeds through the kinetic and thermodynamic coupling of two (or more) simultaneous growth processes. This book is meant to be an inspiration for chemists to develop new materials that are otherwise inaccessible by combining conventional simultaneous polymerization processes.

Chemnitz, Fall 2018

Index

1,1',1'',1'''-silanetetrayltetrakis(azepan-2-one) 42, 196
1,1',1''-methylsilanetriyltris(azepan-2-one) 203
1,1',4,4'-tetrahydro-2,2'-spirobi[benzo[d][1,3,2]oxazasiline] 191
1,1'-(dimethylsilanediyl)bis(azepan-2-one) 203
1,3,5-trioxane 42

2,2-dialkyl-4H-1,3,2-benzoxasiline 172, 185
2,2-dimethyl-4H-1,3,2-benzodioxasiline 99, 172, 182, 208
2,2'-spirobi[3,4-dihydro-1H-1,3,2-benzodiazasiline] 191
2,2'-spirobi[4H-1,3,2-benzodioxasiline] (Si-Spiro) 20, 39, 52, 117, 136, 168, 208, 236, 298, 311, 315
– computed IR frequency 119
– computed NMR chemical shift 119
2-(3-amino-n-propyl)-2-methyl-4H-1,3,2-benzodioxasiline (APSI) 190, 299, 305, 311, 315
2-(3-morpholino-n-propyl)-2-methyl-4H-1,3,2-benzodioxasiline 190
2-aminobenzyl alcohol 93, 191
2-aminobenzylamine 93, 191
2-(aminomethyl)aniline *See* 2-aminobenzylamine
2-(aminomethyl)phenol 93, 191
2-(aminophenyl)methanol. *See* 2-aminobenzyl alcohol
2-chloro-2-methyl-4H-1,3,2-benzoxasiline 169
2-furfuryloxy-2-methyl-4H-1,3,2-benzoxasiline 178
2-hydroxybenzyl alcohol *See* Salicyl alcohol
2-(mercaptomethyl)phenolate 114
2-(N-(2-aminoethyl)-3-amino-n-propyl)-2-methyl-4H-1,3,2-benzodioxasiline 190
2-(N-methyl-3-amino-n-propyl)-2-methyl-4H-1,3,2-benzodioxasiline 190
2-(N,N-dimethyl-3-amino-n-propyl)-2- methyl-4H-1,3,2-benzodioxasiline 190
2-(N-phenyl-3-amino-n-propyl)-2-methyl-4H-1,3,2-benzodioxasiline 190
2-thiophenemethanol 33, 38, 62

3,3'-diphenyl-3,3',4,4'-tetrahydro-2,2'-spirobi[benzo[e][1,3,2]oxazasiline] 93
3-aminopropyltriethoxysilane 204

4,4',8,8'-tetra-tert-butyl-4H,4'H-2,2'-spirobi[benzo[d][1,3,2]dioxagermine] 86
4,4'-di-tert-butyl-4H,4'H-2,2'-spirobi[benzo[d][1,3,2]dioxagermine] 86
4H,4'H-2,2'-spirobi[benzo[d][1,3,2]dioxagermine] 54
4H,4'H-2,2'-spirobi[benzo[d][1,3,2]dioxastannine] 97
4H,4'H-2,2'-spirobi[benzo[d][1,3,2]dithiagermine] 54, 89
4H,4'H-2,2'-spirobi[benzo[d][1,3,2]oxathiagermine] 89
4H,4'H-2,2'-spirobi[benzo[e][1,3,2]oxathiagermine] 89
4-nitrophenol 255

6,6'-dibromo-4H,4'H-2,2'-spirobi[benzo[d][1,3,2]dioxasiline] 73
6,6'-dimethyl-4H,4'H-2,2'-spirobi[benzo[d][1,3,2]dioxasiline] 73
6,6'-di-tert-butyl-4H,4'H-2,2'-spirobi[benzo[d][1,3,2]dioxasiline] 80
6-bromo-2,2-di-tert-butyl-4H-benzo[d][1,3,2]dioxagermine 101

Activation barrier 153, 154
Acyclic diene metathesis (ADMET) 11
Additive 222
Adhesion promoter 299, 302, 305
Adhesive strength 301
ADMET. *See* Acyclic diene metathesis (ADMET)
AFM. *See* atomic force microscopy
Agglomeration 201
Allophanate substructure 320
Aluminum 301, 311
Aluminum oxide/hydroxide 302
Aluminum trifurfurylate 43
Amine
– primary 305, 315
– secondary 305, 315
– tertiary 305
Anatase 36
Anchoring layer 311
Anode
– carbon-based 273
– germanium-containing 289

https://doi.org/10.1515/9783110499360-007

Application, potential 230
APSI. *See* 2-(3-amino-n-propyl)-2-methyl-4H-1,3,2-benzodioxasiline
AROP. *See* Polymerization, addition-ringopening
Atomic force microscopy (AFM) 311
Atom transfer radical polymerization (ATRP) 315
ATRP. *See* Atom transfer radical polymerization

B3LYP-D3 120
BADGE. *See* Bisphenol A diglycidyl ether
Battery
- lithium-ion 273
- lithium-sulfur 278
Bead 139, 140
Benzyl alcohol 36, 66
Benzylic cation 65
β-hydride elimination 94
BET surface area 177, 209, 212, 225, 236, 238, 242, 245, 253, 257
BFM. *See* Bond fluctuation model
bis(dimethylammonium) tris[2-(oxidomethyl) phenolate(2-)]germanate 69, 110
bis(dimethylammonium) tris[4-bromo-2-(oxidomethyl)phenolate(2-)]germanate 110
bis(dimethylammonium) tris[4-methyl-2-(oxidomethyl)phenolate(2-)]germanate 110
(bis-)epoxide 305, 306, 311
(bis-)isocyanate 305, 315
Bisphenol A diglycidyl ether (BADGE) 305, 309
Biuret 320
Bond fluctuation model (BFM) 139
Bond fraction 141
Bond vector 139
- reaction center 140
- reactive (rBFM) 139

Calixarene 82
Capacity 273
Carbon 206
Carbon black 212, 216
Carbon dioxide release 187
Carbon hull 253
Carbonization 45, 242
Carbon matrix 274
Carbon particle 208
Carothers equation 6, 185
Catalysis 298, 318
- 1,4-diazabicyclo[2.2.2]octane (DABCO) 82, 148, 211, 318
- 1,8-diazabicyclo[5.4.0]undec-7-ene (DBU) 82, 183, 148, 154
- 4-methylbenzene-1-sulfonic acid 96
- amino-functionalized twin monomer 189
- C,C cross-coupling 238
- hydrogenation 238
- lactic acid 183
- methanesulfonic acid 214, 237, 257, 276
- p-toluenesulfonic acid 187
- pyridine 84
- reduction 238
- tetra-n-butylammonium fluoride (TBAF) 82
- trichloroacetic acid 100
- trifluoroacetic acid (TFA) 148, 183
- trifluoromethanesulfonic acid 100
Catechol 302
Cathode material 214
CCSD(T) 120
Cell-setup
- full-cell 279
- half-cell 278
Characterization 237, 242
- microporosity 277
- silver-enriched porous 236
Charge 153
Cluster 56
CMK-1 207
CMM. *See* Monomer, cleavable multifunctional
Coarse-grained approach 125
Cobalt(II,III) oxide-supported porous material 238
Composite
- carbon 208, 210, 213
- carbon/carbon 212, 216
- carbon/carbon/silica 212
- carbon/graphite 211
- carbon/silica 177, 208
- carbon/Si_3N_4 196
- metal oxide nanoparticle@carbon and silica material 236, 240, 241
- polyamide 6/SiO_2 196
- polyamide 6/SiO_2-poly(3-aminopropyl) siloxane 204
- polyamide 6/SiO_2-polymethylsiloxane 203
- Sn/C/SiO_2 245
Connectivity 144
Coordination polymer 56, 96
Copolymerization 11, 27
- alternating 11

– block 11
– statistical 11
Core shell particle 180
Corundum abrasion 302
Coulombic efficiency 278
Counter anion 149
Creep behavior 301
Creep retention 301
Critical loading 304
Crosslinking 16, 210, 295, 296, 305, 309, 318
Cyclic oligoarene 65
Cycling stability 273

DABCO. *See* Catalysis, 1,4-diazabicyclo[2.2.2] octane
Dabco® DC 193 187
DBU. *See* Catalysis, 1,8-diazabicyclo[5.4.0] undec-7-ene
Delamination 304
Delithiation 273
Density Functional Theory (DFT) 116, 148
– BP86-D3/SVP 148
– calculation 74
– PW6B95-D3/def2-QZVPP 149
– PW6B95-D3/def2-TZVPP 148
– RI-DFT 148
Depolymerization 215
Derivatization 306, 311
DFC. *See* Difurfurylcarbonate
DFOS. *See* Difurfuryloxydimethylsilane
Dialkyl group 14 salicyl alcoholate 100
Diazagermine 94
Diazasiline 94
Diethoxydimethylsilane 28
Differential scanning calorimetry (DSC) 91, 104, 113, 192, 202, 298, 309, 318, 320
Difurfurylcarbonate (DFC) 187
Difurfuryloxydimethylsilane (DFOS) 28, 51, 124
Di-iso-propylnaphtalene (DIPN) 175
Dimethylammonium cation 111
Dimethyl siloxane 102
Dioxastannine 95
DIPN. *See* Di-iso-propylnaphthalene
Dissipative particle dynamics (DPD) 125
Distribution 175
DLPNO-CCSD(T) 120
DP_n. *See* Number average degree of polymerization
DSC. *See* Differential scanning calorimetry

EDX 174, 241
EFTEM 175
Electrochemical test 275
Electron density 154
Electrophilic aromatic substitution 149
Electrostatic potential 122
E-modulus 185
Energetics 74
Energy source, renewable 273
Energy storage 273
Equilibrium 103
ε-aminocaproic acid 196
ε-caprolactam 196
Etching 210, 45
Ethoxytrimethylsilane 28
Exothermic process 154

Flaking 304
Floor temperature of S_8 277
Foam 187
– PFA 187
– PFA/silica 188
– silica 188
Formation of quinone methide 79
Fragmentation 101
FTIR spectroscopy 201
Functionality 2, 16, 295, 306, 311
Furan-2-ylmethanolate 57, 114
Furfuryl acetate 38
Furfuryl alcohol (FA) 13, 28
Furfuryloxysilane monomer 29
Furfuryloxytrimethylsilane 28

γ-aminopropylmethyldialkoxysilane 189
Ge@C 93
Ge–CH_2 connectivity 108
Gel formation 16
Germanate 54, 290
Germanium 54
– hexacoordinated 85
– pentacoordinated 85
Germanium(II) salicyl alcoholate 103
Germanium oxide 105
Germocane 100
– cyclic 108
Germylene 103, 290
Glycidyl methacrylate 315
Grafting process 315
Graphene 206

Graphite 206
Group 14 salicyl alcoholate 55

HAADF-STEM 185, 212, 240, 243, 247, 299, 324
Hammett constant 75, 115
Hardness 185
Hard template 213, 45, 216
- Aerosil 214
- Aerosil® AS90 251
- Aerosil® OX50 251
- carbon 180
- characterization 253
- external template 217
- glass fibre 182
- graphene 212
- graphite 211
- LiChrospher 214
- metal nanoparticle 182
- polystyrene particle 215
- silica 180, 182, 206, 214, 217, 276
- Stöber particle 215, 251
- TiO_2 182
- ZnO 182
HCS. *See* Hollow carbon sphere
Heck reaction 5
Heterobifunctional monomer 27, 34
Hexamethyldisiloxane (HMDS) 28
Hexamethylenetetramine (HMTA) 191
HMDS. *See* Hexamethyldisiloxane
HMTA. *See* Hexamethylenetetramine
Hollow carbon sphere (HCS) 208, 214, 251, 275
- characterization 254, 257
- gold-containing 251
- iron-oxide-filled 251, 257
- metal-loaded porous 251
- metal-nanoparticle-filled 252
- silver-containing 251
- sulfur-loaded 276
Hybrid material 1, 275
- classification 3
- organic/inorganic 1, 12, 27, 51, 99, 167, 207, 222, 233, 286, 302
- PFA-PEG/silica 180
- PFA/silica 177
- phenolic resin-PEG/silica 186
- phenolic resin-PFA/polymethylsiloxane 179
- phenolic resin/polysiloxane-silica 184
- phenolic resin/silica 20, 177, 208
- phenolic resin/silica-polysiloxaneaminopropylsilane 189
- polyaniline-formaldehyde resin/polysilazane 196
- polyaniline-formaldehyde resin/silica 193
- polyfurfuryl alcohol/silica 210
- ternary 182, 189
Hybrid material with subentry organic/inorganic 1, 12
Hydrofluoric acid 210
Hydrophobicity 203
Hysteresis 209

Indentation hardness 301, 309
Indentation modulus 301, 309
Indium(III) (2,4-dimethoxyphenyl) methanolate 225
Initiation 9
Interface design 299
Intermediate 78
Interpenetrating network 51

Lactic acid 318
Lewis acid base adduct 96
Lithiation 273
Lithium 281
Lithium metal oxide 273
Lithium-sulfur battery 214

Mass spectrometry 101
Material, highly porous 228
Mechanical parameter 301, 309
Mechanical property 295, 301, 305, 309, 310, 316
Metal additive 247
Metal alkoxide 114
Metal alloy particle 46
Metal carboxylate 46, 232, 247
- characterization 234
- decarboxylation 234
- decomposition 234
- silver(I) carboxylates 236
Metalloid alkoxide 114
- metal nanoparticle@carbon and silica material 236
Metal nanoparticle (M-NP) 231
- characterization 234
- gold 253
- silver 236, 253
- supported carbon and silica material 236
Metal-organic complex 234
- characterization 234

– decomposition 234
Metal oxide nanoparticle 232, 234, 238, 251, 257
Methylene blue 255
Methyltrichlorosilane 169
Microcracking 304
M_n. *See* Number average molecular weight
M-NP. *See* Metal nanoparticle
Molecular dynamics 135
– reactive (rMD) 144
Molecular germanate based on salicyl alcoholate 110
Molecular sieve 207
Monolith 208
Monomer
– activated 10
– cleavable multifunctional (CMM) 18
– heterobifunctional 15
– homoNfunctional 14
– multifunctional 14
Monte Carlo method
– lattice-based 135
Monte Carlo step (MCS) 139
– reactive (rMCS) 140
Morphology 51, 144, 207, 213
Mössbauer spectroscopy 242, 243
Multi-walled carbon nanotube (MWCNT) 212
MWCNT. *See* Multi-walled carbon nanotubes

Nanoindentation 301, 309
Nanostructure 274, 296, 305, 315, 316, 324
Nanostructure formation 148, 175
Natural bond orbital analysis (NBO) 89
NBO. *See* Natural bond orbital analysis
Nitrogen doping 210
Nitrogen sorption isotherm 209
NMR spectroscopy
– ^{13}C 195, 200, 106, 114, 241, 242, 299, 305, 311, 324
– ^{29}Si 174, 183, 195, 200, 242, 299, 305, 320, 321
– ^{119}Sn 95, 245, 108
Novak 34
Number average degree of polymerization (DP_n) 6
Number average molecular weight (M_n) 6

OAMS. *See* Oligo-(3-amino-n-propyl)-methylsiloxane
Octamethoxy calix[4]resorcinarene 64
ODAS. *See* oligodialkylsiloxane
Oligo-(3-amino-n-propyl)-methylsiloxane (OAMS) 299
Oligodialkylsiloxane (ODAS) 318, 320
Oligodimethylsiloxane 51
One-pot procedure 238
Organic polymer network 61
Ortho-cresol 91
Oxazagermine 94
Oxidation 219
Oxide
– inorganic 217
– metal 218
– metalloid 218
– mixed inorganic 218

PA6. *See* Polyamide 6
PDMS. *See* Polydimethylsiloxane
PEG. *See* Soft template, polyethylene glycol
Pentacoordinated silicon 77
Periphery 43
PFA. *See* Polyfurfuryl alcohol
Phase domain size 226
Phenolate 154
Phenolic resin 40, 207, 298, 299, 304, 318
– ortho/para-selectivity 153, 158
Phenylmethanolate 63, 114
pK_a value 133
Plasmon resonance 251
Polyamide 6 (PA6) 42, 196, 301, 304
Polyaniline-formaldehyde resin 193
Polycondensation 31, 39
Polydimethylsiloxane (PDMS) 28
Polyfurfuryl alcohol (PFA) 51, 28, 207
Polymer
– functional 295, 311, 315
Polymerization
– addition-ring-opening (AROP) 10
– chain-addition 9
– chain-reaction 7
– degree of 6
– free radical (FRP) 9, 34
– of heterobifunctional monomer 36
– living 9
– mechanism 4
– of monomer combinations 11
– of olefins 8
– precipitation 208, 214

Polymerization (continued)
- ring-opening metathesis (ROMP) 10, 34
- ring-opening (ROP) 10, 39
- selective 11
- simultaneous (SP) 11, 12, 18, 36
- slurry 211, 214
- step-growth 4, 31
- step-reaction 4
Poly(methoxybenzyl alcohol) 38
Polymethoxymethylbenzene 64
Polysilazane 194
Polystyrene 281
Poly(2-thiophenemethanol 33, 63, 210
Pore size distribution 144, 209
Porosity 206, 236, 242
- mesoporosity 207, 216
- microporosity 207, 216
Post-functionalization 305
Potassium fluorido silicate 82
PP. *See* Prepolymer
Precursor 207
- bifunctional 61
Precursor based on dialkyl group 14 salicyl alcoholate 99
Precursor class 56
Prepolymer (PP) 316, 317
Presence of water 67
Primer 311
Production are summarized 27
Propagation 9
Propagation step
- inorganic 156
- organic 156
$PSSO_3H$ microparticle 283
Pull-off test 301, 303
PXRD 237, 245, 251
Pyrolysis 207
Pyrrole 210

Quadruple polymerization 178
Quantum chemical study 117
Quinone methide 76, 91, 153, 154

Radial distribution function (RDF) 144
Reaction
- aromatic substitution 75
- barrier 75, 105
- bond cleavage 90
- chain propagation 60
- condensation 52, 60
- downstream 61, 82
- electrophilic aromatic substitution 65, 70
- heteropolar bond cleavage 90
- initial step 52, 76, 105
- intermolecular condensation 72
- intramolecular C–O insertion 108
- intramolecular condensation 72
- kinetics 54, 61
- propagation 60, 83
- ring-opening 149
Reaction mechanism
- 1M1B model 154
- 1M1T model 149
- 2M1B model 154
- 2M1T model 153
- 2M2T model 153
- acid catalysis 137, 146
Reaction probability 140
Reactivity 114
- position of methoxy group 109
- prediction 76
- specific 57
- study 80
Reactivity scale 115
Reactivity scale concept 83
ReaxFF atom type 146
- force field 145
- parametrization 145
- SDAT-ReaxFF 146
- TP-ReaxFF 145
ROMP. *See* Polymerization, ring-opening metathesis
ROP. *See* Polymerization, ring-opening

Salicyl alcohol 39, 296, 306, 317
Salicyl alcoholate 67, 114
SBA-15 207
Scherrer equation 253
Scratch test 304, 311
Scratch track 304
See salicyl alcohol 39
Segregation process 62
SEM 174, 188
Shrinkage 207
Shrinking 34
Silanolate 155
Silanol group 72
Silica 28, 31, 40, 51, 194, 196, 298, 320, 321

– characterization 237
– mesoporosity 236, 242
– silver-decorated porous 236
Silica molecular sieve 207
Silica network 61
Silicon alkoxide 51
Silicone 100
Silicon nitride 191
Silicon oil 28
Silicon tetrachloride 191
Silylation 311
Single crystal X-ray diffraction analysis (SCXRD) 103
Single-source methodology 233
Single-source precursor 233
Sintering 226
Si-Spiro. *See* 2,2'-spirobi[4H-1,3,2-benzodioxasiline]
Smart additive 305
Sn/M ratio 248
SnO_2 64
Sodium hydroxide 211
Soft template 45, 281
– myristic acid ester 180
– polyethylene glycol (PEG) 180
Sol-gel process 1, 13, 31, 34, 179
– coating 181
– non-aqueous 35, 37
Solid electrolyte interface 280
SP. *See* Polymerization, simultaneous
Specific surface area 44
Spin coating 311
Spirocyclic diazagermine 93
Spirocyclic diazasiline 93
Spirocyclic germanium compound 88
Spirocyclic germanium salicyl alcoholate 85
Spirocyclic germanium salicyl thiolate 85
Spirocyclic germanium thiolate 88
Spirocyclic oxazagermine 93
Spirocyclic oxazasiline 93
Spirocyclic silicon salicyl alcoholate 69, 226
Spirocyclic tin salicyl alcoholate 95
SQUID. *See* Superconducting Quantum Interference Device
Stannylene 103
STEM 175, 248
Sterically demanding 81
STP. *See* Twin polymerization, simultaneous
Structured porous carbon materials 208
Structure formation process 123, 141, 225
Structure-reactivity relationship 51
Substituent 81
– electron-density-donating 73
– electron-density-withdrawing 73
Substitution 86, 100
Sucrose 207
Sulfur 210, 214, 275
Sulfur content 210
Sulfuric acid 215
Super Conducting Quantum Interference Device (SQUID) 243, 258
Surface Coating 310
Surface functionalization 296, 312
Surface roughness 302
Surfactant 187

TBAF. *See* Catalysis, tetra-n-butylammonium fluoride
TEM 186, 203, 241, 251, 258
Template 206, 213
Tension test 301
Termination 9
Tert-butyl group at the methylene group 86
Tetraethoxysilane 28
Tetrafurfuryloxysilane (TFOS) 28, 31, 51, 188, 227
Tetraisopropyl orthotitanate 33
Tetrakis(furan-2-ylmethoxy)stannane 57
Tetrakis(phenylamino)silane 191
Tetrakis(thiophen-2-ylmethoxy)stannane 62
Tetrakis(thiophen-2-ylmethyl)orthosilicate 62
Tetramethoxysilane 39
Tetra[(N-methylpyrrol)methoxy]silane 210
Tetraphenoxysilane 42
Tetra(pyrrolmethoxy)silane 210
Tetra(thiophene-2-ylmethoxy)silane 210
TFA. *See* Catalysis, trifluoroacetic acid
TFOS. *See* Tetrafurfuryloxysilane
TGA. *See* Thermogravimetric analysis
Thermodynamical stability 158
Thermogravimetric analysis (TGA) 91, 109,202, 234, 241, 309, 322
Thiolate 67, 85
Thiolate based salicyl alcoholate 67
Thiophen-2-ylmethanolate 57, 114
Thiophene 210, 276
Tin 54
Tin-2,4-dimethoxyphenylmethanolate 247
Tin alkoxide 245

Tin alloy 247, 274
- characterization 248
- embedded in carbon and silica material 247
Tin(II) 4-methoxy-2-(oxidomethyl)phenolate 108
Tin(II) 4-methyl-2-(oxidomethyl)phenolate 108
Tin(II) 5-methoxy-2-(oxidomethyl)phenolate 108
Tin(II) salicyl alcoholate 108
Titanium dioxide 33
Titanium tetrachloride 36
TM. *See* Twin monomer
TP. *See* Twin polymerization
TPP. *See* Twin prepolymer
Transesterification 167, 306
Transesterification reaction 298
Transfer 9
Triethoxymethylsilane 28
Triethyl aluminum 43
Trifurfuryloxymethylsilane 28
Triggering
- acid anhydride-catalyzed 57
- acid-catalyzed 70, 135, 149
- base-catalyzed 52, 136, 154
- fluoride-initiated 52
- Lewis-acid-catalyzed 52, 54, 57
- Lewis-base-catalyzed 82, 102
- proton-catalyzed 57, 74, 102, 117, 118
- thermally induced 76, 118, 128, 210
- thermally induced, hypothesis for the mechanism 134
Trigger temperature 298, 320
Triple polymerization 178
Tris(furan-2-ylmethyl) borate 229
Tungsten hexachloride 38
Tungsten oxide 38
Tungsten oxychloride 39
TURBOMOLE V6.5 148
Twin monomer 20, 27, 137, 167
- amino-functionalized 189, 298, 305, 311, 315
- deficient 42, 191, 196
- functional 44, 189
- functionalized 295, 296, 301, 318
- hypothetical 39
- ideal 117
- implicit 42
- ionic 68
- neutral 68
Twin polymerization 13, 27, 36, 274
- additive 45
- additive-assisted 288
- apparent 37, 39, 42, 191
- coating 181
- condensative 39
- homo- 189
- ideal 39, 40, 52, 298
- simultaneous co- 183, 320
- simultaneous (STP) 40, 55, 99, 109, 182, 187, 189, 208, 210, 218, 236, 241, 245, 247, 284, 296, 298, 318
- in situ 38
- surface 180, 211, 275
- synergistic combination 305
Twin prepolymer (TPP) 315, 316

Urea
- increment 320
- structure 315
Urethane
- functionality 318
- increment 320
UV-Vis spectroscopy 252, 255

Volatile by-products 112
Volume expansion 274, 293

Wafer 311

XPS. *See* X-ray photoelectron spectroscopy
X-ray photoelectron spectroscopy (XPS) 315

Bei Fragen zur Produktsicherheit wenden Sie sich bitte an:
If you have any questions regarding product safety,
please contact:

Walter de Gruyter GmbH
Genthiner Straße 13
10785 Berlin
productsafety@degruyterbrill.com